D0164087

Molecular Engineering Thermodynamics

Building up gradually from first principles, this unique introduction to modern thermodynamics integrates classical, statistical, and molecular approaches, and is especially designed to support students studying chemical, biochemical, and materials engineering. In addition to covering traditional problems in engineering thermodynamics in the context of biology and materials chemistry, students are introduced to the thermodynamics of DNA, proteins, polymers, and surfaces.

It includes:

- Over 80 detailed worked examples, covering a broad range of scenarios such as fuel cell efficiency, DNA/protein binding, semiconductor manufacturing, and polymer foaming, emphasizing the practical real-world applications of thermodynamic principles.
- More than 300 carefully tailored homework problems, designed to stretch and extend students' understanding of key topics, accompanied by an online solution manual for instructors.
- All the necessary mathematical background, plus resources summarizing commonly used symbols, useful equations of state, microscopic balances for open systems, and links to useful online tools and datasets.

Juan J. de Pablo is the Liew Family Professor at the Institute for Molecular Engineering, University of Chicago, and a former Director of the Materials Science and Engineering Center on Structured Interfaces, University of Wisconsin, Madison. He has won several teaching awards, has been awarded a Presidential Early Career Award in Science and Engineering from the NSF, and is a Fellow of the APS, the American Academy of Arts and Sciences, and the Mexican Academy of Sciences.

Jay D. Schieber is Professor of Chemical Engineering in the Department of Chemical and Biological Engineering and the Department of Physics, and Director of the Center for Molecular Study of Condensed Soft Matter, at the Illinois Institute of Technology. He has been a visiting professor at universities in Europe and Asia, holds numerous teaching awards, and was the 2004 Hougen Scholar at the University of Wisconsin, Madison.

Cambridge Series in Chemical Engineering

Books in Series

Baldea and Daoutidis, *Dynamics and Nonlinear Control of Integrated Process Systems*
Chau, *Process Control: A First Course with MATLAB*
Cussler, *Diffusion: Mass Transfer in Fluid Systems, Third Edition*
Cussler and Moggridge, *Chemical Product Design, Second Edition*
De Pablo and Schieber, *Molecular Engineering Thermodynamics*
Denn, *Chemical Engineering: An Introduction*
Denn, *Polymer Melt Processing: Foundations in Fluid Mechanics and Heat Transfer*
Duncan and Reimer, *Chemical Engineering Design and Analysis: An Introduction*
Fan and Zhu, *Principles of Gas–Solid Flows*
Fox, *Computational Models for Turbulent Reacting Flows*
Leal, *Advanced Transport Phenomena: Fluid Mechanics and Convective Transport*
Lim and Shin, *Fed-Batch Cultures: Fundamentals, Modeling, Optimization, and Control of Semi-Batch Bioreactors*
Marchisio and Fox, *Computational Models for Polydisperse Particulate and Multiphase Systems*
Mewis and Wagner, *Colloidal Suspension Rheology*
Morbidelli, Gavriilidis, and Varma, *Catalyst Design: Optimal Distribution of Catalyst in Pellets, Reactors, and Membranes*
Noble and Terry, *Principles of Chemical Separations with Environmental Applications*
Orbey and Sandler, *Modeling Vapor–Liquid Equilibria: Cubic Equations of State and their Mixing Rules*
Petyluk, *Distillation Theory and its Applications to Optimal Design of Separation Units*
Rao and Nott, *An Introduction to Granular Flow*
Russell, Robinson, and Wagner, *Mass and Heat Transfer: Analysis of Mass Contactors and Heat Exchangers*
Schobert, *Chemistry of Fossil Fuels and Biofuels*
Sirkar, *Separation of Molecules, Macromolecules and Particles, Principles, Phenomena and Processes*
Slattery, *Advanced Transport Phenomena*
Varma, Morbidelli, and Wu, *Parametric Sensitivity in Chemical Systems*

"This is a book to use many times. First as a textbook that explains the principles of thermodynamics and statistical mechanics with rigour and clarity. The importance and the contemporary relevance of the subject matter is illustrated in many examples from physics, chemical engineering and biology – and it is to these examples that future readers are likely to return time and again. They illustrate how thermodynamics can be used as a framework to organize and quantify our understanding of an amazing variety of physical phenomena. A textbook to hold on to."

Daan Frenkel, University of Cambridge

"The book by Professors J. J. de Pablo and J. D. Schieber, titled *Molecular Engineering Thermodynamics*, is both sensible and innovative. They use a postulational approach to present the basic ideas of the subject, which, I believe, is the best way to teach equilibrium thermodynamics, since it is clear and concise. Their book is also important because they show how thermodynamics can be used to attack problems involving chemical reaction equilibria, properties of polymer solutions and blends, and surfaces and interfaces. They also make it clear how thermodynamics may be applied to engineering flow systems (which are not at equilibrium). A chapter on statistical mechanics shows how molecular ideas fit into the subject of thermodynamics."

R. Byron Bird, Univerity of Wisconsin-Madison

Molecular Engineering Thermodynamics

JUAN J. DE PABLO
University of Chicago

JAY D. SCHIEBER
Illinois Institute of Technology

CAMBRIDGE
UNIVERSITY PRESS

University Printing House, Cambridge CB2 8BS, United Kingdom

Cambridge University Press is part of the University of Cambridge.

It furthers the University's mission by disseminating knowledge in the pursuit of education, learning, and research at the highest international levels of excellence.

www.cambridge.org
Information on this title: www.cambridge.org/9780521765626

First published 2014

Printed and bound in the United Kingdom by TJ International Ltd. Padstow Cornwall

A catalog record for this publication is available from the British Library

Library of Congress Cataloging in Publication data
De Pablo, Juan J.
Molecular engineering thermodynamics / Juan J. de Pablo, University of Chicago, Jay D. Schieber, Illinois Institute of Technology.
pages cm
ISBN 978-0-521-76562-6 (hardback)
1. Thermodynamics. 2. Chemical engineering. 3. Molecular dynamics. I. Schieber, Jay D. II. Title.
QD504.D423 2013
621.402′1–dc23

2013001919

ISBN 978-0-521-76562-6 Hardback

Additional resources for this publication at www.cambridge.org/depablo

To Marina, Clara, and Luis, and to Jennifer and Nathaniel

CONTENTS

PREFACE

Students of engineering and the physical sciences are often introduced to thermodynamics in a way that has evolved little over the last several decades.

This book is an outgrowth of the sense that thermodynamics courses should reflect changes in the problems that engineers and scientists encounter in practice. Important industrial sectors, including the oil, chemical, semiconductor, and pharmaceutical, continue to recruit large numbers of scientists and engineers, but new demands require them to be more versatile, and able to apply fundamental thermodynamic concepts in a wider range of situations. At the same time, start-ups in emerging fields must be nimble and rely on a work force that is equally comfortable applying thermodynamic principles to rationalize observations in biology or advanced materials design. Traditional boundaries between science and engineering are becoming blurred, and versatility in engineering is necessarily built on a broader understanding of far-reaching scientific principles. Engineering students must place greater emphasis on broadly applicable scientific concepts and learn how to use these in different contexts, and students in the sciences must gain a better appreciation for how such concepts are applied in engineering practice.

One clear and common trend in most engineering and scientific disciplines is that of control over smaller length scales. Experimental methods are able to probe and manipulate the behavior of individual molecules, and large-scale processes can be used to mass produce ultra-small electronic devices. We have entered the age of "molecular engineering," and it is important to emphasize molecular-level concepts in thermodynamics texts.

This manuscript grew out of a set of lecture notes that, with the help of our students and colleagues, we assembled over the last decade. It was originally inspired by the texts and papers that we learned from and, over the course of time, the original content was re-organized in a manner that became easier to deliver in a classroom setting. As we taught and re-taught our thermodynamics courses, we had an opportunity to collect a wide variety of examples of applications of thermodynamics from different disciplines. While we have tried to acknowledge all of the original sources from which ideas and content were taken, we have been influenced by more researchers than we could know. We offer our deepest apologies to any authors who may have inadvertently been omitted from our list of references, and welcome any suggested corrections. Thermodynamics is a mature subject, and we do not make any claims to present in this text a new or original interpretation.

In teaching a two-semester undergraduate sequence, as well as our own graduate courses, we also had ample occasion to realize which concepts students could fully appreciate and remember and which ones they could not. At times, we experienced a fear of reprisal at the mere mention of entropy. Our lecture notes gradually evolved into a postulatory presentation of thermodynamics that allowed us to transition between different aspects and applications of thermodynamics without having to re-introduce some of the underlying principles. Outlining the geometrical framework of thermodynamics from the beginning made it easier to later exploit its consequences in a wide range of applications. Thermodynamic stability, colligative properties, phase equilibria, or adsorption at surfaces and interfaces could all be introduced in a seamless manner, always starting from a common approach. Also examples ranging from the entropy loss associated with polymer stretching to the extraction of heavy molecules using super-critical fluids could be more easily discussed within the same sequence of lectures.

At our own institutions, we have used the text in a two-semester sequence for undergraduate students, or for a one-semester course for graduate students. In the past, the text has been used by chemical and biological engineers, biomedical engineers, materials scientists, and chemists. Particular attention has been devoted to illustrating new concepts with examples from various disciplines, ranging from chemistry to physics and biology.

Our presentation of thermodynamics has been heavily influenced by Callen's classic text. Our manuscript begins with the statement of three postulates in Chapter 2. These postulates are then used to introduce the concept of thermodynamic surfaces, and the minimization (or maximization) of thermodynamic potentials in Chapter 3. Examples of our non-traditional approach to engineering thermodynamics are illustrated with the manipulation of individual DNA molecules using optical tweezers and the adsorption of protein molecules onto surfaces. Chapter 4 is devoted to thermodynamic stability and its consequences. These ideas are subsequently built upon to analyze thermodynamic processes, such as cycles, turbines, and fuel cells. Our discussion of such processes is necessarily more concise than that encountered in traditional engineering texts. If needed, that material can be supplemented with examples already available in the literature. This five-chapter introduction to classical thermodynamics is followed by a chapter on statistical mechanics, in which a connection is made between the thermodynamic potentials introduced earlier and the behavior of collections of molecules. The presentation of statistical mechanics can be facilitated considerably by relying on software for molecular simulations that is widely available online. This is also the chapter in which the idea of fluctuations is introduced, a subject of importance in small, especially biological, systems.

The introduction to statistical mechanics is followed by a chapter on the nature and origin of intermolecular forces that serves to establish a link between chemistry, structure, and the ensuing thermodynamic behaviour. The subsequent two chapters are devoted to a discussion of phase equilibria in a wide variety of contexts, including systems comprising gas, liquid, and solid phases. Our presentation is based on the texts by Prausnitz et al. and by Sandler. Concepts from phase equilibria are illustrated with examples ranging from super-critical extraction to the modeling of the blood–brain barrier through water–octanol partitioning. Chapter 10 is devoted to chemical equilibria and the study of reacting systems. The examples that are discussed range from reactions encountered in semiconductor manufacturing to biotechnology processes (e.g., polymerase chain reaction). Simple stochastic concepts are introduced for the study of fluctuations in small reaction volumes. Although polymers are introduced throughout the text in earlier examples, Chapter 11 is devoted to polymer solubility, including copolymers and the compressibility that arises with super-critical solvents, as used in foaming. Chapter 12 is devoted to the thermodynamics of surfaces, and builds on the original introduction by

Guggenheim with examples related to the characteristics of super-hydrophobic surfaces and the collapse of nanoscopic structures encountered in nano-lithographic processes.

There are four appendices. The first appendix reviews all the mathematical background necessary for the manipulations in the text above algebra. The text makes extensive use of partial differentiation, Taylor-series expansion, the chain rule, Jacobian transformations, and Leibniz's rule; all students are expected to have proficiency in these tools. For completeness, the Gauss divergence theorem is included, but it is essential only for a later appendix. The cubic-equation solutions are useful for numerical work in single-component vapour–liquid stability predictions, whereas combinatorics is an essential tool in statistical-mechanical calculations. Appendix B summarizes all key information about several equation-of-state models for pure systems. Appendix C derives the microscopic balances from first principles. These derivations make a strong connection with the subject of transport phenomena and provide the basis for all flowing-system analyses considered in Chapter 5. The final appendix provides a small amount of physical data necessary for performing calculations using the sorts of models presented in Appendix B. Much more complete physical data are obtainable online, or from software.

Thermodynamics is a mature, large discipline, yet its application is found in an ever-growing number of problems. A single textbook cannot be expected to cover all that important ground. This book is intended to give students a broader flavor of the subject than they might otherwise experience. We expect that this text will require continuous adaptation to our changing discipline. However, we believe that the approach taken here is sufficiently robust to facilitate incorporation of new challenges and inclusion of new topics, and is appropriate for introductory or intermediate-level courses in the sciences and engineering disciplines.

ACKNOWLEDGMENTS

We are indebted to the many students at Wisconsin and Illinois who suffered through our courses in thermodynamics and who worked with us on this text. We have benefited considerably from the advice of many of our colleagues, who have patiently worked through earlier versions of this text and used it in their lectures. We are grateful to the Department of Chemical and Biological Engineering at the University of Wisconsin for the award of a Hougen Fellowship; that fellowship allowed Jay Schieber to spend a semester in Madison, during which some of the material in this text was developed. Juan de Pablo is grateful for the Howard Curler Distinguished Professorship; without that professorship it would have been difficult to complete this text. Juan de Pablo would like to thank the Department of Chemical and Biological Engineering at the University of Wisconsin for providing a collegial and scholarly environment in which to launch an academic career. We are also grateful to our new colleagues at the University of Chicago for creating a stimulating atmosphere that will undoubtedly nurture the development of the nascent field of molecular engineering.

Specifically, we thank Professor Hamid Arastoopour for supporting the time spent on this project during his tenure as Chairman at IIT, and Professors Samira Azarin, Rafael Chavez, Horacio Corti, Fernando Escobedo, David Gidalevitz, Juan Pablo Hernandez, Thomas Knotts, Manos Mavrikakis, Paul Nealey, Jai Prakash, Rob Riggleman, Rob Selman, Eric Shusta, Amadeu Sum, Igal Szleifer, and David Venerus, who were brave enough to test drive the text during its development. Bob Bird and Charlie Hill provided invaluable advice and were always willing to share their views of thermodynamics and of what a textbook should be like. Several students went well beyond the call of duty in helping us prepare many of the examples and problems presented in this manuscript, including Manan Chopra, Ami Desai, Lisa Giammona, Bill Mustain, Deepa Nair, Thidaporn Kitkrailard, Ekaterina Pilyugina, Andrew Spakowitz, Jian Qin, Sadanand Singh, Rebeccah Stay, and Umaraj Saberwal, who provided feedback, helped develop exercises or examples, or worked on exercise solutions.

DEFINITIONS

Symbol	Name	Where introduced or defined
A	Area	Fundamental
A_{chain}	Cross-sectional area per chain	Eqn. (11.53)
a	Parameter in several models	
a_i	Activity of species i	Eqn. (9.6), p. 283
A_s	Cross-sectional area	p. 151
$A_0, A_1, \ldots$	Constants	
a_0	Parameter in various models	
a_K	Kuhn step length	Eqn. (2.84), p. 49
b	Parameter in various models	
$C(T)$	Heat capacity	Eqns. (3.16)
c	Parameter in various models	
C_L, c_L	Constant-length heat capacity	Eqn. (3.97), p. 83
C_P, c_P	Constant-pressure heat capacity	Eqn. (3.62), p. 70
C_V, c_V	Constant-volume heat capacity	Eqn. (3.84), p. 80
c_i	Molar concentration of species i	Section C.1 of Appendix C
E	Electric field	
E_{tot}	Total energy of system	Eqn. (5.5), p. 151
$\vec{\boldsymbol{F}}$	Force	Fundamental, p. 4
F	Helmholtz potential	Eqn. (3.10), p. 55
f	Specific Helmholtz potential	$:= F/N$
f_i	Fugacity	Eqn. (8.41), p. 258
f_i^{pure}	Fugacity of pure component	Eqn. (8.41), p. 258
F_0	Helmholtz potential at reference state	
F^{R}	Residual Helmholtz potential	Eqn. (4.66), p. 136
$\vec{\boldsymbol{g}}$	Gravitational vector	Fundamental, Table D.5, p. 468
g	Specific Gibbs free energy	$:= G/N$
G	Gibbs free energy	Eqn. (3.11), p. 55
G_0	Gibbs free energy at reference state	
G^{R}	Residual Gibbs free energy	Eqn. (4.66), p. 136
H	Enthalpy	Eqn. (3.9), p. 55
$\hbar$	Reduced Planck constant	Fundamental, Table D.5, p. 468
h	Specific enthalpy	$:= H/N$
H_0	Enthalpy at reference state	
$\vec{\boldsymbol{J}}_i$	Molar flux of species i	Section C.1 of Appendix C
H^{R}	Residual enthalpy	Eqn. (4.66), p. 136

k	Thermal conductivity	
k_B	Boltzmann constant	$:= R/\tilde{N}_A$, Eqn. (2.20), p. 20
K_{tot}	Total kinetic energy of system	Eqn. (5.6), p. 151
l	Distance	
L	Length (of elastic strand)	Fundamental
L_0	Length at reference state	
l_p	Persistence length	Eqn. (2.85), p. 50
log	Natural logarithm	$:= \int_1^x (1/x)dx$, $\exp(\log x) = x$
$\log_{10}$	Logarithm base 10	$10^{(\log_{10} x)} = x$
m	mass	
$\bar{m}_i, m = v, s, g, \ldots$	Partial molar property	Eqn. (8.8), p. 251
M_s	Moles of adsorption sites	Eqn. (3.109), p. 90
M_{tot}	Total mass of system	Section 5.1
$\dot{m}$	Mass flow rate	
M_w	Molecular weight	
M_E	Excess property	Eqn. (9.2), p. 283
$\vec{n}$	Unit normal vector	Eqn. (5.7), p. 152
N_i	Mole number of species i	Fundamental, p. 15
N_K	Number of Kuhn steps	Eqn. (2.84), p. 49
N_p	Number of persistence lengths	Eqn. (2.85), p. 50
P	Pressure	Eqn. (2.24), p. 23
p_j	Probability of being in microstate j	Eqn. (6.14), p. 186
P_0	Pressure at reference state	
P_s	Spinodal pressure	Section 4.2.1
P_c	Critical pressure	Section 4.2.1
P^{sat}	Saturation pressure	Section 4.2.2
Q	Heat	Eqn. (2.1), p. 10
Q	Canonical-ensemble partition function	Eqn. (6.13), p. 186
$q_{site}(T)$	Single-site partition function	Eqn. (3.109), p. 90
q_i	Charge of species i	Chapter 7
q_{ij}	Molecule-pair partition function	p. 313
R	Ideal-gas constant	Fundamental, Table D.5, p. 468
r	Number of species	p. 17
R_i	Reaction rate of species i	Section C.1 of Appendix C
S	Entropy	p. 17, defined in Eqn. (2.20), p. 20
s	Specific entropy	$:= S/N$
S_0	Entropy at reference state	
S^R	Residual entropy	Eqn. (4.66), p. 136
S_{tot}	Total entropy of system	Eqn. (5.10), p. 152
T	Temperature	Eqn. (2.25), p. 23
T_0	Temperature at reference state	

T_s	Spinodal temperature	Section 4.2.1
T_c	Critical temperature	Section 4.2.1
$\mathcal{T}$	Tension	Eqn. (2.72), p. 37
U	Internal energy	Fundamental, p. 9
u	Specific internal energy	$:= U/N$
U_j	Energy of microstate j	p. 184
U_0	Internal energy at reference state	
U_tot	Total internal energy of system	Eqn. (5.6), p. 151
U^R	Residual internal energy	Eqn. (4.66), p. 136
V	Volume	Fundamental, p. 5
v	Specific volume	$:= V/N$
V_0	Volume at reference state	
$\vec{\boldsymbol{v}}$	Velocity	$:= d\vec{\boldsymbol{r}}/dt$, p. 4
V^R	Residual volume	Eqn. (4.66), p. 136
v_c	Critical volume	Section 4.2.1
v_v	Vapor volume	Section 4.2.2
v_l	Liquid volume	Section 4.2.2
$\bar{v}_i^\infty$	Partial molar volume at infinite dilution for species i	Eqn. (8.90)
W	Work	Eqn. (1.1), p. 4
X	Unconstrained variable	Eqn. (3.23), p. 59
z	Compressibility factor	Eqn. (4.28), p. 121
z_c	Critical compressibility factor	Section 4.27
Δg_b	Base-pair denaturation	Eqn. (3.102), p. 87
Δh_b	Base-pair denaturation	Eqn. (3.102), p. 87
Δs_b	Base-pair denaturation	Eqn. (3.102), p. 87
$đ\ldots$	Imperfect differential	p. 10
λ	Unspecified generic constant	
β	Parameter in various models	
$\delta(i,j)$ or δ_{ij}	Kronecker delta function	$:= \begin{cases} 1, & i=j \\ 0, & i \neq j \end{cases}$
γ	Parameter in various models	
γ_i	Activity coefficient	Eqn. (9.7), p. 283
ϕ	Fugacity coefficient	Eqn. (8.42), p. 258
Ω	Number of microstates	Eqn. (2.20), p. 20
μ^d	Dipole moment	Chapter 7
μ_i	Chemical potential of species i	Eqn. (2.26), p. 23
μ°	Chemical potential under ideal conditions	Eqn. (3.112), p. 92
$\mathcal{T}$	Tension	Eqn. (2.72), p. 37
ψ, ψ_1, ψ_2	Various functions	
Π	Osmotic pressure	p. 302
α	Coefficient of thermal expansion	Eqn. (3.60), p. 70
α^p	Polarizability	p. 217
κ_T	Isothermal compressibility	Eqn. (3.61), p. 70
κ_S	Isentropic compressibility	Eqn. (3.123), p. 102
ϵ_0	Permittivity of vacuum	Chapter 7

$\epsilon_a, \epsilon_\perp, \epsilon_\parallel$	Langmuir adsorption parameters	Eqn. (3.110), p. 91
χ	Flory–Huggins interaction parameter	Eqn. (3.115), p. 92
Ψ	Grand canonical potential	Eqn. (6.47), p. 195
θ	Fraction of filled sites, Langmuir	$:= N/M_s$, Eqn. (3.111), p. 92
θ_i	Angle of orientation between two dipoles	
Δh^{vap}	Heat of vaporization	Eqn. (4.47), p. 128
Φ_{tot}	Total potential energy of system	Eqn. (5.6), p. 151
$\vec{\nabla}$	Vector differential operator	Eqn. (C.6), p. 452
ρ	Mass or molar density	Appendix C
$\tilde{\rho}$	Number density of molecules	
$\vec{\vec{\tau}}$	Stress tensor	Eqn. (C.9), p. 453
η_s	Newtonian viscosity	Eqn. (C.9), p. 453
D/Dt	Substantial derivative	Eqn. (C.11), p. 453
ϵ	Non-equilibrium energy density	Eqn. (C.15), p. 454
$\vec{q}$	Heat-flux vector	Section C.3 of Appendix C
Φ_v	Newtonian dissipation function	Eqn. (C.18), p. 455
$\sigma_\alpha, \alpha =$ hf, mf, sw, r	Entropy creation rates (densities)	Eqn. (C.29), p. 458
$\sigma_M, M = U, N, V$	Fluctuation variance	Eqn. (6.63), p. 200
$\Sigma_\alpha, \alpha =$ hf, mf, sw, r	Entropy-creation rates	Eqn. (5.8), p. 152
Δ	Changes in quantity (equilibrium)	
Δ	Difference in exit (equilibrium) minus inlet (flow)	Eqn. (5.2), p. 150
Δ	Grand canonical partition function	Eqn. (6.46), p. 194
η	Efficiency	Eqn. (5.49), p. 163
ε	Coefficient of performance	Eqns. (5.53) and (5.54)
Λ	De Broglie wavelength	Eqn. (6.24)
ω	Acentric factor	Eqn. (B.7)
ω_d	Degeneracy	p. 313

Ξ	Grand canonical partition function	Eqn. (6.47), p. 195
Ψ	Generalized potential	p. 195
υ_0	Electronic frequency of molecule	p. 217
σ, ϵ	Lennard–Jones parameters	p. 220
$\langle \ldots \rangle$	Average	Eqns. (6.63) and (8.3), pp. 200, 248
$:=$	"is defined as"	

1 Introduction

For good or evil, all physical processes observed in the Universe are subject to the laws and limitations of thermodynamics. Since the fundamental laws of thermodynamics are well understood, it is unnecessary to limit your own understanding of these thermodynamic restrictions.

In this text we lay out the straightforward foundation of thermodynamics, and apply it to systems of interest to engineers and scientists. Aside from considering gases, liquids and their mixtures – traditional problems in engineering thermodynamics – we consider also the thermodynamics of DNA, proteins, polymers, and surfaces. In contrast to the approach adopted by most traditional thermodynamics texts, we begin our exposition with the fundamental postulates of thermodynamics, and rigorously derive all steps. When approximations are necessary, these are made clear. Therefore, the student will not only learn to solve some standard problems, but will also know how to approach a new problem on safe ground before making approximations.

Thermodynamics gives interrelationships between the properties of matter. Often these relationships are non-intuitive. For example, by measuring the volume and heat capacity[1] as functions of temperature and pressure, we can find all other thermodynamic properties of a pure system. Then, we can use relations between different thermodynamic properties to estimate the temperature rise of a fluid when it is expanded in an insulated container, or, we can use such data to predict the boiling point of a liquid. In Chapter 2, we introduce the necessary variables to describe a system in thermodynamic equilibrium. We also discuss several assumptions that are made early on. In that chapter, the natural thermodynamic variables are energy and volume, or, for a surface, energy and area, and for a polymer, energy and length. Subsequently, in Chapter 3, we introduce additional variables that are useful for solving problems when temperature and pressure are controlled. In Chapter 4, we consider phase diagrams for pure substances. In other words, we learn how to predict the temperature or pressure at which a pure liquid will boil, or when a pure vapor will condense.

Rigorously speaking, thermodynamics applies only to systems at equilibrium, that is, systems that are stable at rest, and that are not subject to a temperature gradient or flow. However, for many systems that are out of equilibrium, it is possible to make reasonable approximations and use thermodynamics even during flow, or when the temperature and pressure are changing with time. Such approximations are considered in Chapter 5, so that we may begin to solve problems involving flows and changes.

[1] The heat capacity will be defined in Chapter 2, but roughly means something like the amount of energy necessary to raise the temperature of a substance by one degree.

Chapter 6 introduces statistical mechanics. Statistical mechanics allows us to (1) connect thermodynamics to molecular properties; (2) study small systems, and large systems near critical points, where thermodynamics does not apply; and (3) consider fluctuations in thermodynamic variables. Most of what follows in the book does not require statistical mechanics, although the final section of most remaining chapters will invoke it.

In order to make a strong intuitive connection between molecules and thermodynamics, Chapter 7 discusses many of the important molecular interactions. Chapters 8 and 9 introduce the machinery to solve vapor–liquid equilibrium problems, such as dew-point and bubble-point calculations, in mixtures. We study reactions in Chapter 10. Polymers and surfaces are considered throughout the text, but Chapters 11 and 12 cover these topics in greater detail.

Thermodynamics is an extremely broad field, and no single text can cover all of the topics important to engineering and science. Therefore, it is usually important to revisit thermodynamics repeatedly. In this book we try to give an overview of many important topics in engineering, and a flavor of how these problems are solved. Typically, more advanced models will exist to cover a given topic, but the solution techniques are essentially the same. Let us consider some questions that it is appropriate to ask of the field. These are only representative of the very broad scope of thermodynamics. Many more questions are possible than those presented here.

1.1 RELEVANT QUESTIONS FOR THERMODYNAMICS

1. What is temperature? What does it have to do with "entropy"?
2. Pick up a butane lighter that has a transparent casing. Note that it contains both liquid and gas. These phases are both butane, at the same temperature and pressure. Yet, some of it is liquid and the rest is gas. Why? How can we predict when the substance will be just one phase, and when it will separate into two? (See an example of a region where a model fluid makes two phases – called a phase diagram – in Figure 4.5 on p. 124.)
3. Press the button to open the valve in the lighter, but without striking the flint to start a flame, and measure the temperature of the exiting fluid. It is approximately the temperature of an ice cube. Could we have predicted that?
4. A refrigerator (or air conditioner) makes heat flow from a cold space to a warm one. How does it do that? How much energy must we expect to buy from the utility company to do that?
5. Compress an ideal gas (say air in a balloon) at constant temperature. The balloon pushes back, so it can be used to do work, say lift a book off of the floor. However, it is possible to prove that the compressed and uncompressed gas has the same energy. So, how can it do work? A traveler once said that energy is "the ability to do work." Was he wrong?
6. Stretch a rubber band and move it quickly to your lips; it feels warm. Let it contract, and it feels cool. Why does a rubber band do that? It is also an experimental fact that a rubber band's tension at constant length increases with temperature, as can be shown with a hair

dryer. That is not true for a metal spring. It certainly does not seem obvious that these observations are related – could thermodynamics tell us why?

7. Some pure-component fluids, or mixtures of fluids, refract white light in beautiful, opalescent ways, showing many colors. This thermodynamic point (specific temperature, density, etc.) is called the *critical point*. How do these fluids do that?
8. Creutzfeldt–Jakob disease was initially transmitted from one surgery patient to another by surgical tools, although the tools had supposedly been sterilized by heat. Many researchers now believe that the disease is not caused by an organism, but by an errant protein called a *prion*. How can a single errant protein out of countless proteins in the brain cause such a degenerative disease?
9. We have all learned that "like likes like," e.g., olive oil dissolves in vegetable oil, but not in water. What is the explanation for this?
10. Contrary to the previous statement, polymers do not easily dissolve in solvents of similar chemical makeup, and do not mix with nearly identical polymers. In fact, Figure 11.3 on p. 369 shows that hydrogenated and deuterated polybutadiene are not miscible, although they are chemically nearly identical! Why?
11. We learn early on in physics that energy is conserved. Yet, most economic analyses revolve around the cost of energy. If it is always around, why are we so concerned with it?
12. Mechanical laws would allow the existence of a perpetual-motion device. How do the laws of thermodynamics prohibit its existence?
13. Air flows into the Ranque–Hilsch vortex tube of Figure 5.3 on p. 159 at room temperature, but exits in a hot stream and a cold stream. Although energy is conserved, it seems like we are getting a free lunch. How is this physically possible? Couldn't we build a perpetual-motion machine from it?
14. Certain molecules in a gas phase can react only when they are adsorbed on a catalytic surface. If we increase the pressure in the gas phase, how will the amount of adsorption change? How can this experiment be used to estimate surface area in porous materials?
15. If the pressure "acting" on a substance is increased at constant temperature, will its volume always decrease?
16. If your equipment tells you that the heat capacity of your new wonder compound is negative, is it time to call the manufacturer for technical support?
17. We find that the ground water for our drinking supply has been contaminated. How much energy must we expend to remove the contaminant?
18. What is the best separation we can expect from distillation? What is the minimum amount of energy necessary to achieve this separation?
19. If we burn one gallon of gasoline, what is the greatest amount of work we can expect to get out?
20. In primary school, I learned that there were three phases of matter: gas, liquid and solid. This is incorrect. What other kinds of phases are there?
21. Someone once told me that a fuel cell has a higher theoretical efficiency than an engine. Is that true?
22. If one uses optical tweezers to pull on a segment of DNA, why does the strand pull back?

23. A helical coil of DNA in a solvent will uncoil and separate into two strands if the temperature is raised. This is *denaturing*. Actually, there is a range of temperatures within which the DNA is partially coiled and attached, and partially separated. What is this temperature range, and how does it change with solvent?

The following chapters should help you answer all of these questions. For the moment, let us review a couple of basic concepts that are important to begin with.

1.2 WORK AND ENERGY

The mathematics background necessary to follow this book is reviewed in Appendix A. We assume that the reader is familiar with the typical units and dimensions of length, mass, and energy. Work and energy play key roles in thermodynamics, so let us review these physical concepts briefly.

Recall that work is defined as force times distance[2]

$$\text{Work} := \text{Force} \times \text{Distance}. \tag{1.1}$$

If I lift an object of mass m off the floor by a distance l, then the work I have done on the object is the constant force $m\vec{g}$, times the distance l. Hence, $W = m|\vec{g}|l$, where $|\vec{g}|$ is the gravitational constant, and W is work.

We also know that energy is conserved, so the potential energy $E_{\text{potential}}$ of the object must have increased, $\Delta E_{\text{potential}} := E^{\text{final}}_{\text{potential}} - E^{\text{initial}}_{\text{potential}} = m|\vec{g}|l$.

By similar arguments, we can find other forms of energy, such as kinetic energy. For example, if we exert a constant force $\vec{F}$ on a baseball of mass m, neglecting friction with air, the ball will undergo constant acceleration according to Newton's classical mechanics,

$$\vec{F} = m\frac{d\vec{v}}{dt}, \tag{1.2}$$

where $\vec{v}$ is the velocity of the ball, and t is time. During an infinitesimal time dt, the ball is displaced by $\vec{v}\,dt$. Hence, the infinitesimal change in kinetic energy of the ball E_{kinetic} is the infinitesimal amount of work

$$dE_{\text{kinetic}} = \vec{F} \cdot \vec{v}\,dt = m\frac{d\vec{v}}{dt} \cdot \vec{v}\,dt = m\vec{v} \cdot d\vec{v}. \tag{1.3}$$

If we integrate each side from zero initial speed (and zero kinetic energy) to the final speed $|\vec{v}|$, we find

$$E_{\text{kinetic}} = \frac{1}{2}m|\vec{v}|^2. \tag{1.4}$$

We see that, if we know the force and displacement of an object, we can find its change in kinetic energy. You may already be familiar with the result, but notice how it was obtained – from the definition of work and the conservation of energy. Similar ideas will be used in Chapter 2.

[2] The symbol := means "is defined as."

Note that power is the rate at which we do work. Hence, the quantity $\vec{F} \cdot \vec{v}$ is the power exerted on the ball.

EXAMPLE 1.2.1 A piston in a box changes the volume V in the box. How much work does it take to compress the piston by a distance L?

Solution. Since the pressure P inside the box might be changed by the act of compression, it is best to begin with an infinitesimal compression of the piston. The infinitesimal work (ΔW) is the force times the infinitesimal distance (ΔDistance, where positive ΔDistance means a decrease in volume) the piston moves. However, the force is the pressure times the area A of the piston,

$$\Delta W = PA\,\Delta\text{Distance} = -P\,\Delta V, \tag{1.5}$$

where ΔV is the change in volume of the box. Make careful note of the minus sign. If we integrate all of these infinitesimal changes, we obtain

$$W = -\int_{V_0}^{V_\mathrm{f}} P\,dV, \tag{1.6}$$

where V_0 and V_f are the initial and final volumes of the box.

We can test our answer by examining the dimensions. Work has units of energy. The potential energy is given by $m|\vec{g}|l$, so it has units of (mass $\times$ length^2/time^2), since $|\vec{g}|$ has dimensions of acceleration. The integral on the right side has dimensions of pressure times volume. Pressure is (force/area), or (mass $\times$ length/time^2/length^2), or (mass/time^2/length). Hence, the integral has dimensions (mass $\times$ length^2/time^2) and our analysis is dimensionally consistent. □

In the previous example, we found the work done to compress a gas in a box. But what form of energy changed? Is there an additional way to change this form of energy? The answers to these questions are given in Chapter 2.

Exercises

1.1.A By searching other texts, find examples of problems where thermodynamics plays an important role in understanding. Cite your reference(s). (Actually, it would be much more difficult to find a problem where thermodynamics does not apply.)

1.1.B In later chapters (see Section 3.6) we discover that all thermodynamic properties can be reduced to just a few measurable quantities. Two of these quantities are called "the coefficient of thermal expansion," which is defined by Eqn. (3.60), and "the isothermal compressibility," Eqn. (3.61). Design simple experiments to measure these two quantities for a substance. Although it has not yet been introduced, assume that temperature is easily measured by a thermocouple, whose probe is a thin wire.

1.1.C Aside from gas, liquid and solid, what other kinds of phases are there? There are actually several kinds of solid phases, so you may choose to name and describe one of these. Please give a reference from which you learned of this phase of matter. The more obscure and bizarre the phase is, the better.

1.2.A If we exert a torque $\mathcal{T}_\omega$ on a stirring paddle such that it rotates with angular velocity ω, how much power (work per time) are we exerting on the paddle?

1.2.B What other kinds of energy can you think of? Cite some examples where such forms might be relevant, and specify how they might inter-convert with some other form of energy.

1.2.C How much work does it take to stretch a spring from a length L_0 to L if you know how the tension $\mathcal{T}$ varies with L? In particular, what if the tension is zero at L_0, and varies linearly with displacement?

1.2.D A lever arm allows one to exert large forces. In other words, one applies a small force at one end of the arm, and a large force is given by the end nearer the fulcrum. Is work "conserved" in this case? In other words, does the lever arm exert the same work at the short end as it receives at the long end? Show your answer quantitatively.

2 The postulates of thermodynamics

Thermodynamics is the combination of a structure plus an underlying governing equation. Before designing plays in basketball or volleyball, we first need to lay down the rules to the game – or the *structure*. Once the structure is in place, we can design an infinite variety of plays and ways that the game can run. Some of these plays will be more successful than others, but all of them should fit the rules. Of course, in sports you can sometimes get away with breaking the rules, but Mother Nature is not so lax. You might be able to convince your boss to fund construction of a perpetual-motion machine, but the machine will never work.

In this chapter we lay the foundation for the entire structure of thermodynamics. Remarkably, the structure is simple, yet powerfully predictive. The cost for such elegance and power, however, is that we must begin somewhat abstractly. We need to begin with two concepts: energy and entropy. While most of us feel comfortable and are familiar with energy, entropy might be new. However, entropy is no more abstract than energy – perhaps less so – and the approach we take allows us to become as skilled at manipulating the concept of entropy as we are at thinking about energy. Therefore, in order to gain these skills, we consider many examples where an underlying governing equation is specified. For example, we consider the fundamental relations that lead to the ideal-gas law, the van der Waals equation of state, and more sophisticated equations of state that interrelate pressure, volume, and temperature.

Although the postulates are fairly simple, their meaning might be difficult to grasp at first. In fact, most students will need to revisit the postulates many times, perhaps over several years. We recommend that you think about the postulates the way you might vote in Chicago – early and often (or even after you die). Throughout most of the book, however, we will use the postulates only indirectly; in other words, we will derive some important results in this chapter, and then use these results throughout the book. It is therefore important to remember the results derived here and summarized at the end of the chapter.

2.1 THE POSTULATIONAL APPROACH

Where does Newton's law of motion $\vec{F} = (d/dt)(m\vec{v})$ come from? How did Einstein discover $E = mc^2$? These equations were not derived from something – they were guessed in a flash of brilliant insight. We have come to accept them for several reasons. First, because they describe a great many experiments, and were used to predict previously unknown phenomena. Secondly, they are simple and straightforward to comprehend, although perhaps sometimes difficult to implement. Thirdly, and more importantly, we accept these assertions, or *postulates*, of Newton and Einstein because we have never seen them violated. It is for these reasons that we believe that energy is always conserved.

In this chapter we describe the postulates that make up the theory of thermodynamics. Just like Newton and Einstein, we must be willing to abandon our theory if experiments ever contradict the postulates. The postulates given here are not the most general possible, but are designed to be easily understood, and applicable to most systems of interest to engineers and scientists. In a few sections we briefly consider generalizations. The approach in this chapter is essentially that of Callen [21]. Hence, the same restrictions apply – namely, the system must be isotropic, homogeneous, and large enough for us to neglect surface effects if we are talking about the bulk, or that we may neglect edge effects if we are talking about surfaces, and there must be no external forces.

In what follows we make five fundamental postulates. The first postulate (first law) posits the existence of an additional form of energy called "internal energy," which, along with kinetic, potential, electromagnetic, and other energies, obeys a conservation principle. The second postulate assumes that every system has equilibrium states that are determined by a few macroscopic variables. The third postulate introduces a quantity called "entropy" on which internal energy depends. The fourth postulate (the second law) and the fifth (Nernst) postulate prescribe properties of entropy. The rest of thermodynamics follows from these straightforward ideas which have far-reaching consequences.

2.2 THE FIRST LAW: ENERGY CONSERVATION

What is the definition of energy? Despite using the word and the concept nearly every day, most engineers and scientists stop short when asked this question. Nonetheless, we still find the concept very useful. Consider a few simple thought experiments about energy. (1) How much energy does this book have if you hold it above your head before letting it drop to the floor? You might answer that it has potential energy $m|\vec{g}|l$, where $|\vec{g}|$ is the gravitational constant, and l is the height of the book above the floor. (2) If the book is flying with speed $|\vec{v}|$ while it is at height l above the floor, is its energy $m|\vec{g}|l + \frac{1}{2}m|\vec{v}|^2$? (3) What if I place an ice cube on the book? We note that the ice melts, and we assume (correctly) that heat was transferred from the book to the ice.

What do these experiments tell us? First, we notice that the energy we ascribed to the book changed depending on the situation. That is because the situations made us think about several different *degrees of freedom* or variables that we used to define the book's energy. When we held the book over our head, we instinctively thought of position, and then calculated the potential energy of the book. When we pictured the book flying, we then thought of position and velocity, and added the kinetic energy. When the ice melted on the book, we thought about concepts like temperature, heat, or maybe internal energy.[1]

Note that, before we can talk about energy accurately, it is important to *specify the system*, and to *specify what variables we are using*. In our first example the system is the book,

[1] Our postulates will allow us to distinguish clearly among internal energy, heat, and temperature.

and the position of the book is the variable. The other important point is the first law, or first postulate.

Postulate I (the first law). Macroscopic systems possess an *internal energy U* that is subject to a conservation principle and is extensive.

An **extensive variable** is one that is linearly dependent on system size, and, conversely, an **intensive variable** is one that is independent of system size.[2] From a statistical-mechanical, or atomistic, point of view, we do not need this postulate, since we know that atoms store energy. We also know that all forms of energy obey a conservation principle. However, we choose to make this starting point clear in our framework by expressing it as a postulate.

This internal energy is somehow transferred from the book to the ice in our thought experiment. Internal energy has properties just like other forms of energy in that it can be exchanged between different systems, converted to and from other forms of energy, or used to extract work. By "conservation principle," we mean that, if we add up all of the forms of energy in an isolated system, the sum total of those energies is constant, although one form might have increased while another decreased, say some kinetic energy became internal energy. It is worthwhile to quote extensively from Callen [21, p. 11].

> The development of the principle of conservation of energy has been one of the most significant achievements in the evolution of physics. The present form of the principle was not discovered in one magnificent stroke of insight but was slowly and laboriously developed over two and a half centuries. The first recognition of a conservation principle repeatedly failed, but in each case it was found possible to revive it by the addition of a new mathematical term – a "new kind of energy." Thus consideration of charged systems necessitated the addition of the Coulomb *interaction energy* (Q_1Q_2/r) and eventually of the energy of the electromagnetic field. In 1905 Einstein extended the principle to the relativistic region, adding such terms as the relativistic rest-mass energy. In the 1930s Enrico Fermi postulated the existence of a new particle called the *neutrino* solely for the purpose of retaining the energy conservation principle in nuclear reactions ...

Where is internal energy stored? Although it is not necessary to introduce atoms and molecules into the classical theory of thermodynamics, we might be bothered by this question. The answer might be in the vibrations of the atoms (kinetic energy) and the spring-like forces between atoms (potential energy), or in the energies in the subatomic particles. So why do we call it "internal energy" instead of kinetic plus potential energy, for example? Recall our thought experiments above, where we found that the variables used to describe our system are essential to define energy. If we knew the precise positions and velocities of all the atoms in the book, we could, in principle, calculate all $\sim 10^{23}$ kinetic and potential energies of the book, and we would not need to think about internal energy. However, that approach is not just impractical, but, it turns out, not even necessary. We can carry out meaningful calculations of internal energy by using just a few variables that are introduced in Postulate III.

[2] A precise mathematical definition will be given shortly, in Postulate III.

2.3 DEFINITION OF HEAT

If we do work on a system, then its energy must be increased. However, we often observe that, after we have done work on a system, its internal energy returns to its original state, although the system has done no work on its environment. For example, push the book across the table, then wait a few minutes. Although you performed work on the book, it has the same initial and final potential, kinetic, and internal energies. The reason why the internal energy of a system can change without work is that energy can also be transferred in the form of **heat**.

Since the internal energy of a system may be changed either by work or by heat transfer, and we have postulated that energy is conserved, we can define the heat transfer to, or from, a system as

$$đQ := dU - đW. \tag{2.1}$$

It is worthwhile to commit this equation to memory, or, better yet: $dU = đQ + đW$. In this notation, work is positive when done on the system, and heat is positive when transferred to the system. We write the differentials for heat and work using $đ$ instead of d because they are **imperfect differentials**.

Perfect differentials are not dependent upon the path taken from the "initial" to the "final" states – they depend only on the initial and final values of the independent variables. **Imperfect differentials** $đ$ have two important properties that distinguish them from perfect differentials d. First, imperfect differentials do depend upon the path. For example, if we set this book on the floor and push it a short distance $d\vec{x}$ away using force $\vec{F}$ and then the same distance $-d\vec{x}$ back to its original position using force $-\vec{F}$, then the sum (perfect) differential for its position is zero – exactly the same as if we had left the book sitting there: $d\vec{x}_{\text{tot}} = d\vec{x}_1 + d\vec{x}_2 = d\vec{x} + (-d\vec{x}) = 0$. However, the work done on the book was positive in both moves, so the sum (imperfect) differential of the work is positive, even though the book ended up where it began: $đW_{\text{tot}} = \vec{F}_1 \cdot d\vec{x}_1 + \vec{F}_2 \cdot d\vec{x}_2 = \vec{F} \cdot d\vec{x} + (-\vec{F}) \cdot (-d\vec{x}) = 2\vec{F} \cdot d\vec{x}$.

Secondly, if we integrate a perfect differential, we obtain a difference between final and initial states. For example, if we integrate $d\vec{x}$ from $\vec{x}_0$ to $\vec{x}_1$, we obtain the difference $\Delta\vec{x} = \vec{x}_1 - \vec{x}_0$. When we integrate $đW$, for example, we simply obtain the total work in the path W, not a difference.

In order to control, and therefore measure, the internal energy of a system, we need to control both the heat flow and the work done on the system. This manipulation is accomplished primarily through control of the *walls* of a system. The following definitions establish ideal conditions that we may approach approximately in real situations.

An **adiabatic wall** is one that does not allow heat to flow through it. In real situations this is approximated by heavily insulated walls. On the other hand, a **diathermal wall** is one that permits heat to flow through it freely. In addition, we often assume that the wall has a sufficiently small mass that its thermodynamic effects are negligible. Such a wall might be approximated by one that is thin and made of metal.

When no matter is exchanged with the environment, we say the system is **closed**. When no mass, heat, or work is exchanged, we say the system is **isolated**. When both energy and mass can be exchanged, we say that the system is **open**.

EXAMPLE 2.3.1 A gas is placed in an insulated container that contains a frictionless piston and stir paddles attached to a falling weight (Figure 2.1). The piston is attached to a scale, so that we can calculate the pressure of the gas inside by measuring the force on the piston and dividing by its surface area. We can also measure the rate at which the weight falls. With this setup, we perform two experiments.

Experiment #1. When we move the piston slowly, we find a relationship between the pressure and volume of the form

$$P^3V^5 = \text{constant}. \tag{2.2}$$

Experiment #2. A stirrer inside the container is attached to a falling weight, which spins the stirrer; at constant volume, the pressure changes with time according to the following relation:

$$\frac{dP}{dt} = -\frac{2}{3}\frac{m|\vec{\mathbf{g}}|}{V}\frac{dl}{dt}, \tag{2.3}$$

where m is the mass of the weight and l is the height of the weight. Find the internal energy as a function of volume and pressure $U(V, P)$, relative to its value at some reference volume and pressure $U_0 := U[V_0, P_0]$, assuming that the internal energy of the gas is a function of V and P only.

Solution. Our system is the gas in the container, and our state variables are volume and pressure. Note that the first experiment allows us to do work *on* the system by decreasing its volume,

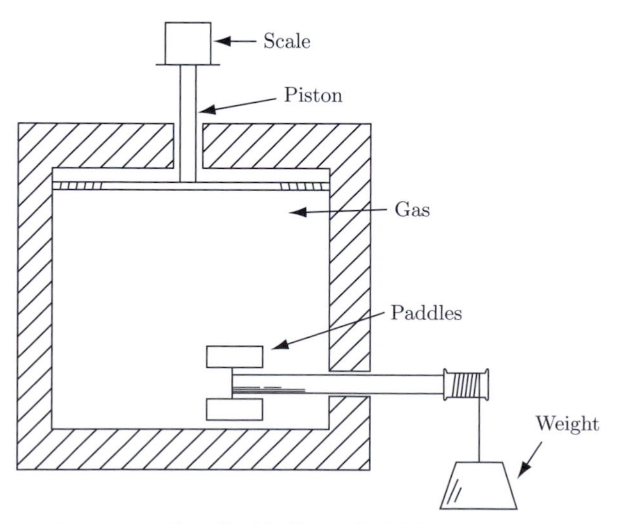

Figure 2.1 The experimental apparatus described in Example 2.3.1.

or extract work *from* the system by increasing its volume. The second experiment can only do work *on* the system. The conservation of energy (Eqn. (2.1)) allows us to determine the changes in the internal energy by integrating the work along an appropriate path, since the system is insulated, making $đQ = 0$, or by calculating the loss in potential energy of the falling weight.

If we consider the state of the system on a plane with volume as the x axis and pressure as the y axis, then we see that each experiment allows us to move in the plane only in a specific manner. If the system begins at a specific pressure and volume, then the second experiment allows us to increase the pressure, but not change the volume. The first experiment allows us to change both the pressure and the volume, but only such that P^3V^5 stays constant. However, these two experiments are sufficient to allow us to move between any two points on the state plane.

For example, the points A and B in Figure 2.2 can be connected by two different paths. First, we can draw a vertical line through point A, indicating the second experiment, and a line of constant P^3V^5 through point B, indicating the first experiment. We can then proceed along the path using the two line segments that connect these two points. Alternatively, we could draw a vertical line through point B and a line of constant P^3V^5 through point A and connect the points. Once we have a path connecting the points, we integrate the work required to move between them. Consider the first path.

Since the point C is connected to point A by Experiment #2 and to point B by Experiment #1, its pressure and volume must satisfy the two equations

$$\begin{aligned} V_A &= V_C, \\ P_C^3 V_C^5 &= P_B^3 V_B^5, \end{aligned} \tag{2.4}$$

from which we may readily solve for P_C:

$$P_C = P_B \left(\frac{V_B}{V_A} \right)^{5/3}. \tag{2.5}$$

Similarly, for any point on the line connecting B and C, the pressure can be found as a function of volume:

$$P(V) = P_B \left(\frac{V_B}{V} \right)^{5/3}, \quad \text{curve connecting B and C.} \tag{2.6}$$

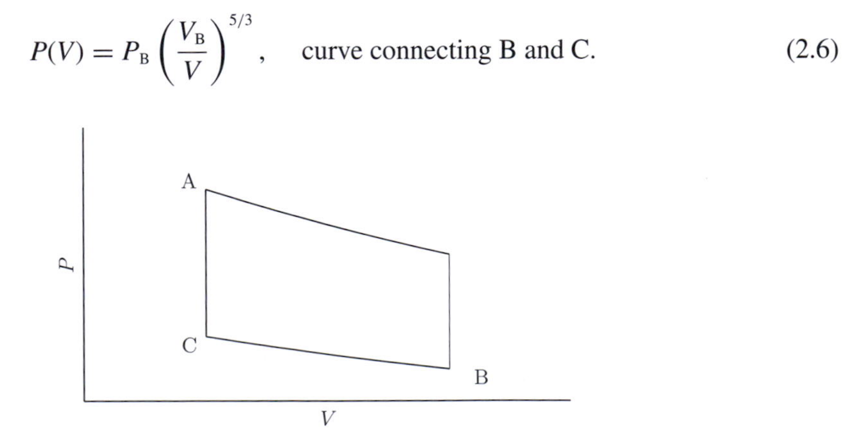

Figure 2.2 The pressure–volume state plane of Example 2.3.1.

Since Experiment #1 is performed quasi-statically, we can assume that the pressure in the container is that measured on the piston. Therefore, Eqn. (2.1) can be written as

$$\begin{aligned} dU &= đW_{qs} \\ &= -P\,dV. \end{aligned} \tag{2.7}$$

Using Eqn. (2.6), we obtain

$$dU = -P_B \left(\frac{V_B}{V}\right)^{5/3} dV. \tag{2.8}$$

We can integrate Eqn. (2.8) from C to B to obtain

$$U[V_B, P_B] - U[V_C, P_C] = -\frac{3}{2}P_B V_B \left[\left(\frac{V_B}{V_A}\right)^{2/3} - 1\right]. \tag{2.9}$$

Note that we use the square brackets [...] for U to indicate where the quantity is being evaluated, as opposed to parentheses (...) to indicate a functional dependence. For example, we would write $U(V, P)$ to indicate that U depends upon V and P, and $U[P_A]$ to indicate that we are evaluating U at P_A, but arbitrary V.

To integrate U along the path CA, we need to find the work done during the experiment. Recalling that the potential energy of the weight is $m|\vec{\mathbf{g}}|l$ and neglecting any kinetic energy of the falling weight, an energy balance on the container plus weight yields $d(U + m|\vec{\mathbf{g}}|l) = 0$. We can therefore write

$$\begin{aligned} dU &= -m|\vec{\mathbf{g}}|\dot{l}\,dt \\ &= \frac{3}{2}V\,dP, \end{aligned} \tag{2.10}$$

where we have used Eqn. (2.3) to obtain the second line. Since the volume is constant along CA, we can integrate Eqn. (2.10) from C to A to give

$$\begin{aligned} U[V_A, P_A] - U[V_C, P_C] &= \frac{3}{2}V_A(P_A - P_C) \\ &= \frac{3}{2}V_A\left[P_A - P_B\left(\frac{V_B}{V_A}\right)^{5/3}\right], \end{aligned} \tag{2.11}$$

where we have used Eqn. (2.5). Let us call A our reference state: $(P_A, V_A) \to (P_0, V_0)$; and let B be any arbitrary state: $(P_B, V_B) \to (P, V)$. If we subtract Eqn. (2.11) from Eqn. (2.9), we obtain our desired expression for the internal energy of the system:

$$U(V, P) - U_0 = \frac{3}{2}(PV - P_0V_0). \tag{2.12}$$

Equation (2.12) is an example of an **equation of state** – an equation that relates several thermodynamic properties with one another. Another example is the well-known ideal-gas equation of state ($PV = NRT$) that relates pressure, volume, mole number, and temperature to one another. □

The equation of state derived in Example 2.3.1 is used in the following example to calculate heat flows in a system.

EXAMPLE 2.3.2 The gas from Example 2.3.1 is placed in a container with a diathermal wall and a frictionless piston. Find the amount of heat and work necessary for two different steps. The first step decreases the pressure from P_1 to P_2 at constant volume V_1. The second process increases the volume from V_1 to V_2 at constant pressure P_2. Assume each step is quasi-static.

Solution. In the first step, the volume is held constant, so $dV = 0$, and the amount of work done on the system is zero: $W^{\mathrm{I}} = 0$. Therefore, the definition for the heat flux, Eqn. (2.1), becomes

$$đQ = dU, \tag{2.13}$$

which we can integrate along any path from the initial (P_1, V_1) to the final (P_2, V_1) conditions to give

$$\begin{aligned} Q^{\mathrm{I}} &= \int_{(P_1,V_1)}^{(P_2,V_1)} dU(P,V) \\ &= U[P_2, V_1] - U[P_1, V_1] \\ &= \frac{3}{2} V_1(P_2 - P_1). \end{aligned} \tag{2.14}$$

The third line follows from Eqn. (2.12). Since $P_1 > P_2$, we have $Q^{\mathrm{I}} < 0$. To perform this step, we must extract heat from the system.

In the second step, the volume changes, so work is done. If we integrate Eqn. (2.1) over the second step, $(P_2, V_1) \to (P_2, V_2)$, we obtain

$$\begin{aligned} Q^{\mathrm{II}} &= \int_{V_1}^{V_2} P_2\, dV + \int_{V_1}^{V_2} dU[V, P_2] \\ &= P_2 \int_{V_1}^{V_2} dV + U[V_2, P_2] - U[V_1, P_2] \\ &= P_2(V_2 - V_1) + \frac{3}{2} P_2(V_2 - V_1) \\ &= \frac{5}{2} P_2(V_2 - V_1). \end{aligned} \tag{2.15}$$

Note that the first term on the right-hand side of the third line of Eqn. (2.15) represents the work done in the second step: $W^{\mathrm{II}} = P_2(V_2 - V_1)$. □

Here we have used straightforward experiments to find heat exchanges for a system, and how the internal energy depends on pressure and volume. Then we used these results to predict how to manipulate the system in a desired way. In Example 2.6.2 we give the complete characterization of this fluid, which is called a simple, ideal gas. At low densities, all gases behave ideally;

later we will see that simple gases are monatomic. At higher densities, more complicated equations of state are necessary in order to accurately describe the behavior of liquids and gases. Several of these are given in the appendix, and we will call on them throughout the book.

2.4 EQUILIBRIUM STATES

The density of pure water at 1 atm and 25 °C is always 1 g/cm^3, no matter where the water comes from. This observation holds if I start with ice from Lake Mendota in the winter, or with steam from a teapot in Shanghai. We call this stable state of water an **equilibrium state**. Thermodynamics deals only with these equilibrium states, not with the dynamics of the system between such states. (However, it does tell us which equilibrium states are available to a system that is not at equilibrium and which states are not.) Thermodynamics just does not say how long it will take to reach equilibrium, or by what path the system will attain equilibrium. These observations lead us to the second, rather sensible, postulate.

Postulate II. There exist **equilibrium states** of a macroscopic system that are characterized completely by the internal energy U, the volume V, and the mole numbers of the m species $N_1, N_2, \ldots, N_m$ in the system.

If we know U, the volume and the mole numbers, then the equilibrium state is fixed.

When we do work on a system, we often wish to consider processes that occur slowly. In such a case, we assume that the system is nearly always in an equilibrium state. We call such a slow change a **quasi-static process**. For example, we might change the pressure in a container by changing the force on a piston slowly, perhaps by removing grains of sand that are resting on the piston one at a time. When we changed the pressure slowly in Example 2.3.1, we were changing it quasi-statically.

2.5 ENTROPY, THE SECOND LAW, AND THE FUNDAMENTAL RELATION

Since we do not have information on the positions and velocities of all of the atoms, we need to determine what set of variables is necessary to describe the internal energy of our system. A few observations can guide us.

- When touching a hot stove, we notice that it transfers some internal energy to us (don't try this experiment at home without the supervision of an adult). Intuitively we know that this mechanism of energy storage has something to do with temperature and heat flow.
- In order to make an air balloon smaller, we have to do work on the balloon by squeezing it. This mechanism has something to do with applied force and volume.
- To place more air molecules in the balloon, we do work on the balloon by blowing, or pumping. This mechanism has something to do with the number of moles of gas in the balloon.

Our observations have suggested a number of possible independent variables for the internal energy: one involving heat or temperature, a second involving volume or pressure, and a third involving mole number. We do not wish to use heat, however, since it is handled using an imperfect differential. When a system is changed quasi-statically, then the work described in the second bullet above can be written as $đW = -P\,dV$. We need to introduce a second quantity called *entropy* that will allow us to use only perfect differentials for changes to U from heat flow.

We save time and effort by jumping directly to the correct answer. The justification for these choices will have to come from the ability of the theory to explain experimentally observed phenomena.

At this point it is instructive to digress momentarily and elaborate on some of the underlying mathematical ideas behind our next postulate. Generally speaking, when facing a complex problem involving multiple variables, a common strategy is to define an objective function that one tries to maximize or minimize subject to several constraints until a solution is found. As an example, a city might want to alleviate congestion by controlling the size of the streets, the location of traffic lights, the duration and sequence of green lights, etc. To do so, city planners might construct an objective function that they might choose to maximize or minimize. Various choices are possible; a reasonable possibility could be to create a function that describes the average idle time per driver. Mathematically speaking, idle time would be a function of the variables listed above. For a given number of cars or "flow rate"and a given street layout, one would then attempt to minimize that function by controlling the arrangement of traffic lights. Thermodynamics does the same thing; a function of several "natural variables," such as the volume and the size of the system, is created and maximized or minimized (depending on the nature of that function). One difference is that Nature seems to have already chosen the function that is to be maximized. That function must satisfy several criteria, which is really what the next postulate is about.

Postulate III. Complete thermodynamic information is contained at equilibrium in the internal energy as a function of the quantity called **entropy** S, the volume V, and the mole numbers of its r constituents $N_1, N_2, \ldots, N_r$, which are all *extensive quantities*. The functional form of $U(S, V, N_1, N_2, \ldots, N_r)$ satisfies the following properties.

- It is additive over its constituents (it is a first-order, homogeneous function of its arguments, or extensive):

$$U(\lambda S, \lambda V, \lambda N_1, \lambda N_2, \ldots, \lambda N_r) = \lambda U(S, V, N_1, N_2, \ldots, N_r).$$

- It is continuous and differentiable:

$$\left(\frac{\partial U}{\partial S}\right)_{V,N_1,N_2,\ldots,N_r}$$

  is well defined everywhere.

- It is a monotonically increasing function of S:

$$\left(\frac{\partial U}{\partial S}\right)_{V,N_1,N_2,\ldots,N_r} \geq 0.$$

This postulate is true only for large systems – the so-called **thermodynamic limit**. Small systems, such as proteins, might not satisfy the extensivity condition (the first bullet).

We shall see later that entropy is intimately related to heat transfer and temperature, and that pressure is associated with volume changes.

The properties of Postulate III can be used to make several key observations.

- If we pick $\lambda = 1/N$, then we obtain for pure-component systems

$$Nu(s, v) = U(S, V, N), \tag{2.16}$$

  where using a small letter indicates a **specific** or **molar** property, i.e., $u := U/N$ is the internal energy per unit mole. Thus, for pure-component systems, all thermodynamic information is contained in the specific properties, and we need not know N.
- The second property allows us to use the usual calculus manipulations (which are summarized in Appendix A).
- The second and third properties allow us to invert the relation $U = U(S, V, N_1, \ldots, N_r)$ to find a unique function for S,

$$S = S(U, V, N_1, \ldots, N_r), \tag{2.17}$$

  which enjoys properties similar to those enjoyed by U. Namely, we can write that

$$\left(\frac{\partial S}{\partial U}\right)_{V,N_1,\ldots,N_r} > 0, \tag{2.18}$$

  and is well defined. Property (2.18) follows from Eqn. (A.18) in Appendix A.
- Since we can invert U to find an expression for S, we can also show that S is a homogeneous, first-order function of its arguments:

$$S(\lambda U, \lambda V, \lambda N_1, \lambda N_2, \ldots, \lambda N_r) = \lambda S(U, V, N_1, N_2, \ldots, N_r). \tag{2.19}$$

  Therefore, we may deal either with S or with U as the dependent variable. The appropriate choice is dictated merely by convenience; there is no more information in $U(S, V, N_1, \ldots, N_r)$ than there is in $S(U, V, N_1, \ldots, N_r)$. The first form is called the **fundamental energy relation**, and the second is called **the fundamental entropy relation**. These are sometimes called **constitutive relations**, since there often exist different fundamental relations to describe a single substance. For example, to describe oxygen we might use an ideal gas (at low densities), a virial relation (at moderate densities), or a Peng–Robinson relation (at high densities), depending on the necessary accuracy or the simplicity of the calculation.

Most importantly, as we will soon see, if we know either the fundamental entropy or the fundamental energy relation for a system, then we have *complete thermodynamic information about that system*. From either $S(U, V, N_1, \ldots, N_r)$ or $U(S, V, N_1, \ldots, N_r)$ we can find the system's *PVT* relation, its constant-volume heat capacity, its phase behavior, everything. This fact has far-reaching ramifications. For example, in Section 3.6 we show that, if we know the material properties of a substance (the heat capacity, coefficient of thermal expansion, and isothermal compressibility) for a range of temperatures and pressures, we can calculate all possible thermodynamic quantities at any temperature and pressure. We can also show that the change in temperature with respect to volume at constant internal energy must equal the change in pressure over temperature with respect to internal energy at constant volume for a substance. Although these relations are rigorously derived from thermodynamics, they are far from being intuitively obvious.

Let's return to the idea of "equilibrium states" for a moment, and ask *How does the system find its equilibrium state?* Experimental observation suggests an essential property of all systems: each system finds one equilibrium state for given values of internal energy, volume, and mole number. This observation means that, if you have two non-equilibrium systems that have different phases and different conditions but the same energy, volume, and mole numbers, these two systems will reach the same equilibrium point. In other words, many non-equilibrium systems will all converge on the same final equilibrium condition. Clearly, the equilibrium point must be something special. As for other natural systems, we find that the following postulate explains our observations.

Postulate IV (the second law). The unconstrained variables of an isolated system arrange themselves so as to maximize entropy within the constrained equilibrium states.

Consider two examples of systems with unconstrained variables.

- A flask contains a solution of species A and B, and the covering of the flask is a membrane that is permeable to species A, but not permeable to B (Figure 2.3). The flask is placed in a large vat that also contains some mixture of A and B. The flask and membrane are rigid, but allow heat to pass through. For this system, V and N_{B} are constrained for each of the two subsystems (the flask and the vat). Therefore, the amount of A and the internal energy in the flask will change until S for the composite system is maximized, since $N_{\mathrm{A}}^{\mathrm{flask}}$ and U^{flask} are unconstrained.

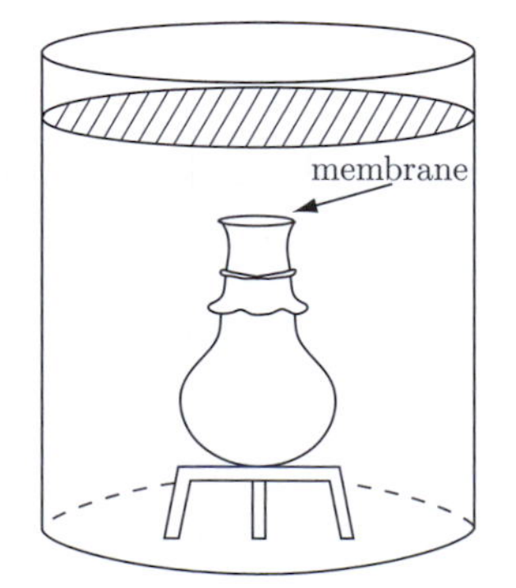

Figure 2.3 A flask whose opening is covered by a semi-permeable membrane sits in a large vat. The flask and the vat contain different concentrations of species A and B. The membrane allows A to pass through, but not B. Maybe A is water, and B is a large protein, for example.

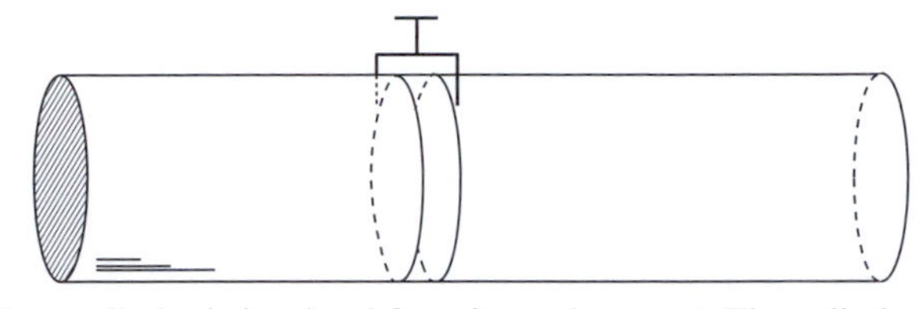

Figure 2.4 A rigid, hollow cylinder is insulated from its environment. The cylinder is divided by a partition, and each side contains some gas. When the pin is removed the position of the inner partition is no longer constrained, and it moves.

- A rigid cylinder is divided into two compartments by a movable partition whose position is momentarily fixed by a constraining pin. See Figure 2.4. The cylinder is insulated, and is impermeable to either mass or energy flux. We remove the pin and allow the movable wall to reposition itself. The movable wall will position itself so as to make the entropy of the total system maximum, given that the total volume, internal energy, and mole numbers of the system are fixed. In this example, the system consists of both compartments of the cylinder, and each compartment would be a subsystem. Note that the change in the total entropy of the system would be either positive or zero, but never negative.

The last statement above is actually general: the total entropy change of an isolated system (such as the Universe) is always non-negative. Entropy is always increasing, although energy is always constant in the Universe. Therefore, there is always plenty of energy around, but the entropy might not be low enough for you to be able to get the energy to do what you want.

There is considerable ambiguity in identifying precisely what the second law of thermodynamics is in the literature – whether it is embodied in some or all of the properties in Postulate III, or just in Postulate IV, for example. Rudolf Clausius, who apparently first coined the term *entropie*, meaning "transformation" in Greek, wrote the second law as *Die Entropie der Welt strebt einem Maximum zu.*[3] Following his lead, we prefer to call Postulate IV the second law of thermodynamics.

For completeness, we now state the final postulate of thermodynamics [92].

Postulate V (the Nernst postulate). The entropy is zero when and only when the substance is a crystal with $(\partial U/\partial S)_{V,N_1,\ldots,N_m} = 0$.

In the following section we will see that the fifth postulate states that the entropy is zero when the absolute temperature is zero. Although it is important for fundamental questions, and in statistical mechanics, we will not use the fifth postulate much in this book. In fact, despite the postulate's importance, we will often use fundamental relations that violate the Nernst law. Why? Because a fundamental relation is usually valid only over some range of thermodynamic conditions. So, if we use a fundamental relation for, say, nitrogen, in the regions where it is liquid, gas, or super-critical, but never where it is crystalline, it need not satisfy the Nernst postulate.

[3] The entropy of the Universe strives towards a maximum [28].

An aside about entropy and statistical mechanics

Entropy always seems strange at first sight – it usually involves symbols such as "<" or ">" rather than "=," and it is not conserved. Also, the term *statistical mechanics* sounds rather intimidating. However, the *ideas* in statistical mechanics are rather straightforward, even though their implementation is often difficult. Although a quantitative understanding of statistical mechanics is not necessary in order to use most of this book, it is sometimes still enlightening to be aware of its ideas. A fundamental idea of statistical mechanics – so important that it is on the gravestone of Ludwig von Boltzmann, is the *definition* of entropy. That is correct – in statistical mechanics, entropy is not a fundamental quantity, but is actually *defined*. There is no definition for energy, so entropy is actually less abstract than energy! Boltzmann's definition for entropy is stunningly simple:[4]

$$S(U, V, N) := k_B \log \Omega(U, V, N). \tag{2.20}$$

The first term is called, appropriately enough, the **Boltzmann constant**, and is equal to the ideal-gas constant divided by Avogadro's number, $k_B = R/\tilde{N}_A$. The next term, Ω, is just a number. It is the number of ways in which the molecules in a system can arrange themselves while still keeping the energy fixed at U, the volume fixed at V, and the mole numbers at N.

Imagine a solid crystal of atoms regularly arrayed on a lattice. The positions of the atoms are more or less fixed. If we swap the atoms about, we still have the same microscopic state. So, the only possible microscopic arrangements arise from how the energy is distributed. Each atom can vibrate in its position, or several atoms can vibrate together. The greater the vibration, the more energy an atom has. Maybe one atom is vibrating with all the energy of the crystal, or maybe each atom vibrates in the same way as all the others. If we could count up all the ways in which the energy could distribute itself, we could find the entropy. Statistical mechanics deals with calculating such large numbers. Even without tackling these sorts of problems, though, we can still think qualitatively about entropy.

An ideal gas consists of molecules flying about in a container. The gas is very dilute, so the molecules rarely see each other. They are independent. The energy is stored either in the kinetic energy of the molecules or in the vibrations in the interatomic bonds, but only a negligible amount is stored in intermolecular forces. If the molecules are monatomic, they have only kinetic energy. The possible arrangement of the molecules, then, is not just how energy is distributed, but where the molecules are in the container. If we keep the energy constant, how can we lower the entropy? Well, if we decrease the volume, the molecule has fewer allowed locations. Hence, Ω goes down. Therefore, the entropy of an ideal gas decreases with decreasing volume, at constant energy.

As a third example, rubber is made up of cross-linked polymer chains. Between the chemical cross-links, the chain can take many different conformations. If we stretch the rubber, the cross-links become separated in space. This stretching deformation decreases

[4] Here and throughout the text, "log" refers to the natural logarithm, and "$\log_{10}$" is logarithm to the base 10.

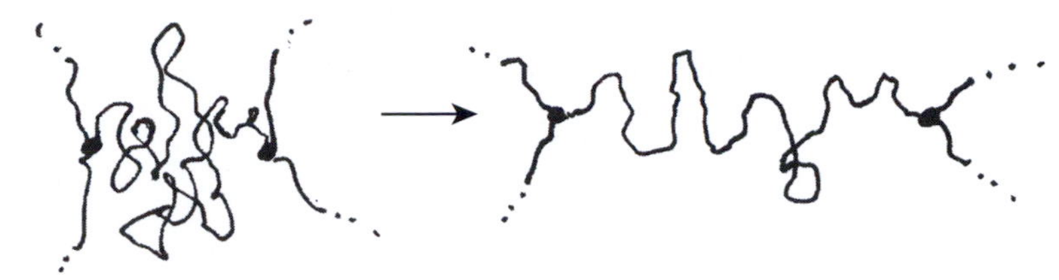

Figure 2.5 A polymer strand between two chemical cross-links. The sketch on the left characterizes a strand in an unstretched rubber band, and the sketch on the right shows the same strand when the rubber is stretched. For a fixed separation of cross-links, the strand on the left can sample more configurations and therefore has larger entropy.

the number of ways in which the chains can arrange themselves (see Figure 2.5). Hence, stretching rubber decreases its entropy. For a rubber band, the length is more important than the volume, so we substitute L for V in our list of independent variables. Considering Postulate IV, can you explain why the stretched rubber snaps back when its length is no longer constrained, or why a gas under pressure expands?

The reader might find it useful to keep these qualitative ideas in mind when trying to understand the thermodynamic behavior of materials.

EXAMPLE 2.5.1 Van der Waals postulated the following fundamental entropy relation for a system:

$$S = Ns_0 + NR \log\left[\left(\frac{U/N + aN/V}{u_0 + a/v_0}\right)^c \left(\frac{V/N - b}{v_0 - b}\right)\right], \tag{2.21}$$

where a, b, c, R, u_0, v_0, and s_0 are constants. Does it satisfy the postulates?

Solution. The internal energy U does play a role. If we also stipulate that U is conserved, then the first postulate is satisfied. Postulate II is satisfied by examination. Clearly, there is only one mole number, or a single species. In order to check Postulate III, we could invert Eqn. (2.21) to find an explicit expression for U. Alternatively, we can check the equivalent criteria for S. We choose the latter. We first check that S is a homogeneous, first-order function of its arguments (that it is extensive):

$$\begin{aligned} S(\lambda U, \lambda V, \lambda N) &= \lambda N s_0 + \lambda NR \log\left[\left(\frac{\lambda U}{\lambda N} + \frac{a\lambda N}{\lambda V}\right)^c \left(\frac{\lambda V}{\lambda N} - b\right)\right] \\ &\quad - \lambda NR \log\left[(u_0 + a/v_0)^c (v_0 - b)\right] \\ &= \lambda\left\{Ns_0 + NR \log\left[\left(\frac{U/N + aN/V}{u_0 + a/v_0}\right)^c \left(\frac{V/N - b}{v_0 - b}\right)\right]\right\} \\ &= \lambda S(U, V, N), \end{aligned}$$

which proves that the first property of Postulate III is satisfied. We check the second and third properties of Postulate III by differentiating Eqn. (2.21) with respect to U:

$$\left(\frac{\partial S}{\partial U}\right)_{V,N} = \frac{cR}{U/N + aN/V}. \tag{2.22}$$

The second and third properties of Postulate III are satisfied only if $cR > 0$ and $a > 0$. The proposed fundamental relation says nothing about the fourth postulate, but any problems we solve using Eqn. (2.21) must also satisfy Postulate IV.

The fifth postulate applied to this example requires that, as $U/N + aN/V \to 0$, then $S = 0$. However, Eqn. (2.21) shows that the entropy goes to negative infinity in the limit. Hence, the proposed relation cannot be valid for very small values of S. In particular, for sufficiently small values of U and large values of V, the entropy is predicted to become negative, which is unphysical. Therefore, this fundamental relation is valid in a limited range of thermodynamic states. However, as long as we stay in the region within which the entropy is physical, we may use this equation. In other words, at moderate values for the entropy, our system might obey this model. However, as the entropy is lowered, our system can no longer obey this model, since it violates the Nernst postulate. □

2.6 DEFINITIONS OF TEMPERATURE, PRESSURE, AND CHEMICAL POTENTIAL

The postulates have not yet defined the important thermodynamic quantities of temperature or pressure, which might seem surprising. However, these quantities follow as definitions that are based on the fundamental quantities introduced in the postulates. It is important that these definitions fit our observations about temperature and pressure. In the following two sections we show that the definitions given in this section lead to two important intuitive predictions: if two systems with different temperatures are placed in thermal contact, thermodynamic equilibrium will be attained when the high-temperature body has given enough heat to the low-temperature body that they are at the same temperature; two subsystems at different pressures in mechanical contact reach equilibrium when the high-pressure subsystem has expanded at the expense of the low-pressure body until the two subsystems are at equal pressures.

We come to the definitions of T and P by considering changes in U. For example, we have already seen in Chapter 1 that work can be done on a system to change its internal energy by changing its volume. Similarly, we can change U by adding heat or moles. Starting with our fundamental energy relation $U(S, V, N_1, \ldots, N_r)$, and using the definition for the differential, Eqn. (A.3), we can make the mathematical observation

$$dU = \left(\frac{\partial U}{\partial S}\right)_{V,\{N_i\}} dS + \left(\frac{\partial U}{\partial V}\right)_{S,\{N_i\}} dV + \sum_i^r \left(\frac{\partial U}{\partial N_i}\right)_{S,V,\{N_{j\neq i}\}} dN_i, \tag{2.23}$$

where, for convenience, we use $\{N_i\}$ to mean $N_1, \ldots, N_r$, and $\{N_{j\neq i}\}$ means $N_1, \ldots, N_{i-1}, N_{i+1}, \ldots, N_r$. We already know that the (reversible) work done on a system is $-P\,dV$. We also know from experience that volume changes are driven by pressure differences between subsystems. Similarly, from experience we know that heat transfers are driven by temperature differences. Finally, the third term on the right-hand side of Eqn. (2.23) suggests that there is some other property that should drive changes in mole numbers, or mass fluxes. It turns out that these observations can be predicted by the postulates when we make the following definitions for the **pressure** P, **temperature** T, and **chemical potential of species** i, μ_i.

$$P := -\left(\frac{\partial U}{\partial V}\right)_{S,N_1,\ldots,N_r}, \tag{2.24}$$

$$T := \left(\frac{\partial U}{\partial S}\right)_{V,N_1,\ldots,N_r}, \tag{2.25}$$

$$\mu_i := \left(\frac{\partial U}{\partial N_i}\right)_{S,V,N_{j\neq i}}. \tag{2.26}$$

Note that the third property of Postulate III and the definition for temperature Eqn. (2.25) require that $T \geq 0$. Also, we now see that Postulate V requires that the entropy be zero when the temperature is absolute zero. The following two sections show how these definitions fit our intuitive understanding of these quantities.

Putting Eqns. (2.24)–(2.26) into the original differential equation (2.23) leads to the oft-used differential expression for U

$$dU = T\,dS - P\,dV + \sum_i \mu_i\,dN_i. \tag{2.27}$$

This equation should be committed to memory, since it is straightforward to write down the definitions of pressure, temperature, and chemical potential from it. (Try it!) It is difficult to overemphasize the importance of this equation.

The first term on the right-hand side of Eqn. (2.27) is associated with heat fluxes or irreversible parts of work, the second term is the reversible work on the system, and the third term is associated with "chemical" work. Note that Eqn. (2.27), unlike the definition for heat flow Eqn. (2.1), contains only perfect differentials.

If we are given a fundamental energy relation, then we can find the temperature, pressure, and chemical potentials as functions of entropy, volume, and mole numbers. Such expressions are called the **equations of state** of a system. Unlike the fundamental entropy or energy relations, these equations do not individually contain complete thermodynamic information, but rather are derivatives (this point is further illustrated in Section 3.1). A relation between U and T is sometimes called a **thermal equation of state**, and a relation between P and V is sometimes called a **mechanical equation of state**, although these distinctions are not clear-cut.

All three quantities (T, P, and μ_i) are **intensive properties**, meaning that they are independent of system size. If we, for example, double each of the extensive properties $(S, V, N_1, \ldots, N_r)$, then T, P, and μ_i remain unchanged.

Alternative but equivalent definitions for temperature, pressure, and chemical potential can be derived for the fundamental entropy relation. Beginning with the fundamental entropy relation $S(U, V, N_1, \ldots, N_r)$, we can write

$$dS = \left(\frac{\partial S}{\partial U}\right)_{V,\{N_i\}} dU + \left(\frac{\partial S}{\partial V}\right)_{U,\{N_i\}} dV + \sum_i^r \left(\frac{\partial S}{\partial N_i}\right)_{U,V,\{N_{j\neq i}\}} dN_i. \tag{2.28}$$

If we solve the differential of U, Eqn. (2.27), for dS, we find

$$dS = \frac{1}{T}\,dU + \frac{P}{T}\,dV - \sum_i^r \frac{\mu_i}{T}\,dN_i. \tag{2.29}$$

When we compare Eqn. (2.29) with Eqn. (2.28) we find the relations in the entropy formulation that are equivalent to Eqns. (2.24)–(2.26):

$$\frac{P}{T} = \left(\frac{\partial S}{\partial V}\right)_{U,\{N_i\}}, \tag{2.30}$$

$$\frac{1}{T} = \left(\frac{\partial S}{\partial U}\right)_{V,\{N_i\}}, \tag{2.31}$$

$$\frac{\mu_i}{T} = -\left(\frac{\partial S}{\partial N_i}\right)_{U,V,\{N_{j\neq i}\}}. \tag{2.32}$$

These last relations can also be found by straightforward algebra from the differential for U, Eqn. (2.27). Given a fundamental energy relation $U(S, V, \{N_i\})$, one can use the definitions Eqns. (2.24)–(2.26) to find equations of state for pressure (the mechanical equation of state), temperature (the thermal equation of state), or chemical potential (the chemical equation of state) as functions of S, V, and $\{N_i\}$. Alternatively, given the fundamental entropy relation $S(U, V, \{N_i\})$, we can use the equivalent definitions Eqns. (2.30)–(2.32) to find equations of state for P, T, and μ_i as functions of U, V, and $\{N_i\}$.

EXAMPLE 2.6.1 Find the thermal and mechanical equations of state for the fundamental entropy relation of Example 2.5.1. Also find the chemical potential as a function of temperature and volume.

Solution. Since we are given an entropy relation, it is more convenient to use the alternative definitions Eqns. (2.30)–(2.32) rather than those based on internal energy. From Eqn. (2.22) of Example 2.5.1 and Eqn. (2.31) we have already found the temperature:

$$\frac{1}{T} = \left(\frac{\partial S}{\partial U}\right)_{V,N} = \frac{cR}{U/N + aN/V}. \tag{2.33}$$

If we solve for U explicitly, we obtain

$$U = cNRT - \frac{aN^2}{V}, \tag{2.34}$$

which is a thermal equation of state. If we use Eqn. (2.30), we find

$$\begin{aligned} P &= T\left(\frac{\partial S}{\partial V}\right)_{U,N} \\ &= \frac{RT}{V/N - b} - \frac{acN^2R}{V^2}\frac{T}{U/N + aN/V} \\ &= \frac{RT}{V/N - b} - \frac{aN^2}{V^2}, \quad \text{van der Waals equation of state.} \end{aligned} \tag{2.35}$$

In going from the first to the second line, we have used the proposed fundamental entropy relation Eqn. (2.21). In going to the third line, we used the thermal equation of state, Eqn. (2.33). The resulting mechanical equation of state Eqn. (2.35) is called the **van der Waals equation of state** for a fluid.

To find the chemical potential, we use the alternative definition which is based on entropy, Eqn. (2.32),

$$\begin{aligned} \mu &= -T\left(\frac{\partial S}{\partial N}\right)_{U,V} \\ &= -RT\log\left[\left(\frac{U/N + aN/V}{u_0 + a/v_0}\right)^c\left(\frac{V/N - b}{v_0 - b}\right)\right] \\ &\quad - \frac{cNRT}{U/N + aN/V}\left(-\frac{U}{N^2} + \frac{a}{V}\right) - \frac{NRT}{V/N - b}\left(-\frac{V}{N^2}\right) - Ts_0 \\ &= -RT\log\left[\left(\frac{T}{T_0}\right)^c\left(\frac{v - b}{v_0 - b}\right)\right] + cRT - \frac{2a}{v} + \frac{RTv}{v - b} - Ts_0, \end{aligned} \tag{2.36}$$

where we have used the thermal equation of state to eliminate the internal energy in the second line. We have also used the notation that a lower-case variable is a specific or molar quantity: $v := V/N$.

These results will be useful in the next example. □

Some important things to note from this example are as follows.

- Simple ideal-gas behavior can be recovered from these results by setting $a = b = 0$.
- For an ideal gas, the internal energy depends only on temperature, and not on volume. However, for the van der Waals fluid, the internal energy decreases with decreasing volume at fixed temperature (and mole number). Hence, the parameter a represents attractive forces between molecules, which are neglected in the ideal gas.
- The specific volume v is required to be greater than b. Hence, b represents the repulsive forces between molecules that keep the fluid from becoming infinitely dense.
- We cannot simultaneously use the van der Waals mechanical equation of state and assume that the internal energy is independent of volume. The thermal equation of state might have a different temperature dependence, but the mechanical and thermal equations of state come from a single fundamental relation, and must always be used together.

- We call these equations of state because each of them is derived from a fundamental relation, but the fundamental relation cannot be derived from a single equation of state. We can say that the fundamental relation has more information than does an equation of state.[5]

EXAMPLE 2.6.2 If a single-component, simple ideal gas is expanded **isentropically** (at constant entropy) until the pressure is halved, what happens to the temperature?

Solution. An ideal gas is one that obeys the mechanical equation of state

$$PV = NRT, \quad \text{ideal-gas equation of state,} \tag{2.37}$$

and the internal energy can be written as a function of temperature only, $U = U(T)$. If the internal energy is also a linear function of temperature, then we call it a **simple ideal gas**,

$$U = cNRT, \quad \text{simple ideal gas, thermal equation of state,} \tag{2.38}$$

where c is a constant. We note that the van der Waals fluid of Examples 2.5.1 and 2.6.1 reduces to a simple ideal gas when we set $a = b = 0$. Hence, the fundamental entropy relation for a simple ideal gas can be found from Eqn. (2.21) to be

$$S = Ns_0 + NR \log\left[\left(\frac{U}{Nu_0}\right)^c \frac{V}{Nv_0}\right], \quad \text{simple ideal gas.} \tag{2.39}$$

In this problem, we are changing T and P while keeping S constant. Therefore, it is useful to find $S = S(T, P)$. We can obtain this relation by eliminating U and V from Eqn. (2.39) using Eqns. (2.37) and (2.38):

$$S = Ns_0 + NR \log\left[\left(\frac{T}{T_0}\right)^{c+1} \frac{P_0}{P}\right], \quad \text{simple ideal gas,} \tag{2.40}$$

where $T_0 := u_0/(cR)$ and $P_0 := RT_0/v_0$. From this expression, we see that, in order to keep the process isentropic, we must keep T^{c+1}/P constant. Hence, if the pressure is halved, then the final temperature $T_{\rm f}$ is

$$T_{\rm f} = \frac{T_{\rm i}}{2^{1/(c+1)}}, \tag{2.41}$$

where $T_{\rm i}$ is the initial temperature. □

EXAMPLE 2.6.3 Find the work necessary to complete this process, if it is done quasi-statically. How much heat is transferred?

Solution. If it is done quasi-statically, then the work is

$$đW_{\rm qs} = -P\,dV. \tag{2.42}$$

[5] Later we will see (in Chapter 4) that a fundamental relation for a pure species can be derived from two equations of state.

In order to integrate this equation, we need to determine how the pressure is changing with the volume. Since the gas is ideal, we can write

$$P = \frac{NRT}{V}.$$

However, the temperature is also changing with volume. From Example 2.6.2, we determined that an isentropic change in pressure requires that T^{c+1}/P be held constant. Hence,

$$T = T_{\rm i}\left(\frac{P}{P_{\rm i}}\right)^{1/(c+1)}.$$

Inserting this equation into the ideal-gas law yields

$$P = \left(\frac{NRT_{\rm i}}{V}\right)^{(c+1)/c}\frac{1}{P_{\rm i}^{1/c}}.$$

If we insert this expression into our differential equation for quasi-static work, Eqn. (2.42), we obtain

$$đW_{\rm qs} = -\left(\frac{NRT_{\rm i}}{V}\right)^{(c+1)/c}\frac{1}{P_{\rm i}^{1/c}}\,dV.$$

We may now integrate this equation from the initial volume $V_{\rm i}$ to the final volume; we obtain, after some simplification,

$$\begin{aligned} W_{\rm qs} &= -\frac{(NRT_{\rm i})^{(c+1)/c}}{P_{\rm i}^{1/c}}\int_{V_{\rm i}}^{V_{\rm f}}\frac{dV}{V^{(c+1)/c}} \\ &= cNRT_{\rm i}\left[\left(\frac{V_{\rm i}}{V_{\rm f}}\right)^{1/c} - 1\right]. \end{aligned} \tag{2.43}$$

But

$$\frac{V_{\rm i}}{V_{\rm f}} = \frac{NRT_{\rm i}}{P_{\rm i}}\frac{P_{\rm f}}{NRT_{\rm f}} = \frac{1}{2^{c/(c+1)}}.$$

Hence,

$$W_{\rm qs} = cNRT_{\rm i}\left[\frac{1}{2^{1/(c+1)}} - 1\right]. \tag{2.44}$$

Because $W_{\rm qs}$ is negative, we find that the gas does work on its surroundings during the process. The process might be accomplished by gradually decreasing the pressure in the environment around the gas, allowing it to expand very slowly. Or, we might remove the weight on a piston very gradually.

We can find the energy transferred during the process from the conservation of energy, or from the definition of Q, Eqn. (2.1);

$$\begin{aligned} Q_{\rm qs} &= U_{\rm f} - U_{\rm i} - W_{\rm qs} \\ &= cNR\,(T_{\rm f} - T_{\rm i}) - cNRT_{\rm i}\left[\frac{1}{2^{1/(c+1)}} - 1\right] \\ &= 0. \end{aligned} \tag{2.45}$$

Therefore, an adiabatic, quasi-static process for an ideal gas is isentropic. Had the process been carried out rapidly, the required work would have been higher. However, the change in internal energy would be the same, so it would have been necessary to remove heat from the system in order to keep it isentropic. □

An isentropic process might seem strange at first sight; how can one control entropy? However, we can prove that the final result of Example 2.6.3 is actually very general: *an adiabatic, isomolar, quasi-static process is isentropic*. For a constant-molar process, the differential for the internal energy, Eqn. (2.27), can be written

$$dU = T\,dS - P\,dV, \qquad \text{closed system (isomolar).}$$

If the process is done quasi-statically, then energy conservation, Eqn. (2.1), allows us to write this equation as

$$đQ_{\text{qs}} + đW_{\text{qs}} = T\,dS - P\,dV, \qquad \text{closed system.}$$

Since $đW_{\text{qs}} = -P\,dV$, we find that

$$đQ_{\text{qs}} = T\,dS, \qquad \text{isomolar,} \tag{2.46}$$

which proves that an adiabatic, quasi-static, isomolar process is isentropic for any material. In fact, some equivalent approaches to thermodynamics treat temperature as a fundamental quantity and define entropy by this relation. We prefer Callen's way, because it allows a much easier connection later to statistical mechanics.

EXAMPLE 2.6.4 What is the chemical potential of a simple ideal gas as a function of temperature and pressure?

Solution. We can find the chemical potential from the fundamental relation, Eqn. (2.39), using the alternative definition for μ,

$$\begin{aligned}
\frac{\mu}{T} &= -\left(\frac{\partial S}{\partial N}\right)_{U,V} \\
&= -s_0 - R\log\left[\left(\frac{U}{u_0 N}\right)^c \frac{V}{v_0 N}\right] + (c+1)R \\
&= -s_0 - R\log\left[\left(\frac{T}{T_0}\right)^c \frac{v}{v_0}\right] + (c+1)R.
\end{aligned} \tag{2.47}$$

To complete the solution, we multiply each side by T. However, we still need to eliminate v in favor of T and P, which we can do using the ideal-gas mechanical equation of state $v = RT/P$:

$$\begin{aligned} \mu &= -s_0 T - RT \log\left[\left(\frac{T}{T_0}\right)^c \frac{RT}{Pv_0}\right] + (c+1)RT \\ &= \mu^\circ(T) + RT \log P. \end{aligned} \tag{2.48}$$

We have split the result in this way to emphasize the dependence of μ on pressure. Note that the chemical potential of an ideal gas goes to negative infinity as the pressure goes to zero. □

EXAMPLE 2.6.5 Find the change in temperature for the adiabatic compression of a simple van der Waals gas. Use this result to find the work required for such a compression.

Solution. If we assume that the process is quasi-static, then the work is related to the equilibrium pressure by

$$đW_{\text{qs}} = -P\,dV. \tag{2.49}$$

If we use the van der Waals equation of state, Eqn. (2.35), to replace pressure, we see that we cannot integrate this equation, because the temperature is not constant during the integration. However, we can solve this problem using the thermal equation of state for a simple van der Waals fluid. If we take the differential of each side of Eqn. (2.34), we obtain

$$dU = cNR\,dT + \frac{aN^2}{V^2}\,dV. \tag{2.50}$$

When we insert Eqn. (2.50) into Eqn. (2.49) (using $dU = đW$ for an adiabatic process), and use the mechanical equation of state, Eqn. (2.35), we obtain

$$cNR\,dT = -\frac{RT}{V/N - b}\,dV, \quad \text{simple van der Waals, adiabatic.} \tag{2.51}$$

Note the cancelation that occurred in order to obtain this result. If we divide each side by NRT, we can then integrate each side from the initial thermodynamic state (T_0, V_0),

$$\begin{aligned} c\int_{T_0}^{T} \frac{dT}{T} &= -\int_{V_0}^{V} \frac{dV}{V - bN}, \\ c\log\left(\frac{T}{T_0}\right) &= -\log\left(\frac{V - bN}{V_0 - bN}\right), \\ T &= T_0\left(\frac{V_0 - bN}{V - bN}\right)^{1/c}, \quad \text{simple van der Waals, adiabatic.} \end{aligned} \tag{2.52}$$

The second line is obtained by performing the integrations, and the third line by taking the exponential of each side. Now that we have the temperature as a function of volume for the adiabatic compression of a simple van der Waals fluid, we can find the work necessary, just as we did in Example 2.6.3. The details are left as an exercise. □

EXAMPLE 2.6.6 We use a container of ideal gas as a crude refrigerator in the following way. A simple ideal gas held in a container with a piston (to manipulate volume), and a diathermal wall can be used as a refrigerator. We also have insulation available to make the diathermal wall adiabatic at will. The idea is straightforward: when the gas is compressed isothermally, it expels heat; when it is allowed to expand, it takes in heat from its surroundings. By expanding the gas in one environment (the refrigerator), and compressing it in another (outside), heat can be transferred from a cold space to a warm one. The gas is cycled through compression and expansion and is called the refrigerant. The work done to compress the gas is the effective cost of refrigerating. Find the amount of heat that can be removed from an environment at 100 °C and expelled to an environment at 50 °C per mole of gas per cycle.

Solution. The path used in the refrigeration cycle is sketched in Figure 2.6. Step a is an adiabatic compression, step b is **isobaric** (i.e. carried out at constant pressure) cooling, and step c is isothermal expansion. We assume all steps are performed reversibly, so our calculation will be of the maximum possible cooling. We also assume that the gas is a simple one, with $c = 3/2$. To help us do the calculation, it is useful to make two tables, one for the thermodynamic states at each point on the diagram, 1–3, and another for the work and energies:

Point	T (°C)	P (atm)	V (l)
1	50	1	
2	100		
3	50		
Step	**ΔU (J)**	**Q (J)**	**W (J)**
a		0	
b			
c	0		

Note that we have filled in the states which are already known. Similarly, we have created on the right-hand side a table for the changes in each of the steps. Note that we have assumed an

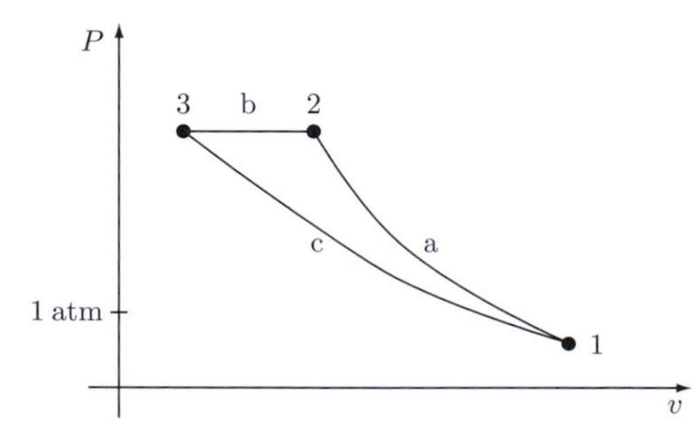

Figure 2.6 The thermodynamic path for the ideal-gas refrigeration cycle studied in Example 2.6.6.

adiabatic step a. The change in step c for the internal energy is zero because the internal energy for an ideal gas depends only on temperature.

We now proceed to fill in the table entries one by one. The reader might wish to try to complete these tables on his or her own first (probably using a pencil).

Because we know the temperature and pressure at point 1, we can use our PvT equation of state to find the volume. For an ideal gas, the equation of state is

$$\begin{aligned} V_1 &= \frac{NRT_1}{P_1} \\ &= \frac{(1\ \text{mol})(82.06\ \text{cm}^3 \cdot \text{atm}/(\text{mol}\cdot\text{K}))(323.15\ \text{K})}{1\ \text{atm}} \\ &= 26.5\ \text{l}. \end{aligned} \tag{2.53}$$

Note that we assumed one mole of gas, and a value for the ideal-gas constant available in Table D.5 in Appendix D.

From here there are several ways to proceed. For the simple ideal gas, a straightforward way is to find the change in internal energy in step a:

$$\begin{aligned} \Delta U_{\text{a}} &= cNRT_2 - cNRT_1 \\ &= cNR(T_2 - T_1) \\ &= \frac{3}{2}(1\ \text{mol})(8.314\ \text{J}/(\text{mol}\cdot\text{K}))(50\ \text{K}) \\ &= 623.6\ \text{J}. \end{aligned} \tag{2.54}$$

By conservation of energy, or the definition of heat, Eqn. (2.1), we know that $Q_{\text{a}} = \Delta U_{\text{a}} = 623.6\,\text{J}$. Since step b has the same magnitude in temperature change, but opposite sign, we also know that $\Delta U_{\text{b}} = -623.6\,\text{J}$. Alternatively, we could note that the system is run in a cycle so that $\Delta U_{\text{a}} + \Delta U_{\text{b}} + \Delta U_{\text{c}} = 0$, to find ΔU_{b}.

However, we still do not know the final thermodynamic state at point 2. This state can be found either by noting that a reversible adiabatic expansion is isentropic, as we proved earlier, or by using the following derivation:

$$\begin{aligned} dU &= đQ + đW, \\ cNR\,dT &= -P\,dV, \quad \text{quasi-static, adiabatic, simple ideal gas,} \\ \frac{c\,dT}{T} &= -\frac{dV}{V}, \\ c\log\left(\frac{T_2}{T_1}\right) &= -\log\left(\frac{V_2}{V_1}\right), \\ V_2 &= V_1\left(\frac{T_1}{T_2}\right)^c. \end{aligned} \tag{2.55}$$

To obtain the second line, we used the fact that the step is adiabatic ($đQ = 0$), and that for a simple ideal gas $U = cNRT$. For the second line we used the ideal-gas equation of state, and integrated from state 1 to state 2 in the third line. Taking the exponential of the fourth line gives us the needed expression:

$$\begin{aligned} V_2 &= (26.5\ \text{l})\left(\frac{323.15\ \text{K}}{373.15\ \text{K}}\right)^{3/2} \\ &= 21.4\ \text{l}. \end{aligned} \tag{2.56}$$

Then, using the ideal-gas equation of state, we can find the pressure at point 2 to be 1.43 atm, which is also the pressure at point 3, from our assumption of isobaric cooling in step b.

Now that we know the temperature and pressure at point 3, we can find the volume, $V_3 = NRT_3/P_3 = 18.5$ l. It is straightforward to find the work done in step b, since it is isobaric, and we assume reversibility:

$$\begin{aligned} đW_{qs} &= -P\,dV, \\ W_b &= P_2(V_2 - V_3) \\ &= (1.43\ \text{atm})(21.4 - 18.5\ \text{l})\frac{8.314\ \text{J/(mol}\cdot\text{K)}}{0.082\,06\ \text{l}\cdot\text{atm/(mol}\cdot\text{K)}} \\ &= 420.2\ \text{J}. \end{aligned} \tag{2.57}$$

Note our trick in the third line to effect the unit change: the fraction is just the ratio of the ideal-gas constant R in one set of units divided by R in another set (hence the ratio is unity). By virtue of energy conservation, $Q_b = \Delta U_b - W_b = -1044$ J. Finally, to find the work done in step c, we assume reversible work only, and use our PvT equation of state

$$\begin{aligned} đW &= -P\,dV \\ &= -NRT\frac{dV}{V}, \\ W_c &= NRT\log(V_3/V_1) \\ &= (1\ \text{mol})(8.314\ \text{J/(mol}\cdot\text{K)})(323.15\ \text{K})\log\left(\frac{18.5\ \text{l}}{26.5\ \text{l}}\right) \\ &= -966\ \text{J}. \end{aligned} \tag{2.58}$$

By virtue of energy conservation we know that $Q_c = -W_c = 966$ J. Hence, we have now completed both tables, which look like

Point	***T* (°C)**	***P* (atm)**	***V* (l)**
1	50	1	26.5
2	100	1.43	21.4
3	50	1.43	18.5
Step	**Δ*U* (J)**	***Q* (J)**	***W* (J)**
a	623.6	0	623.6
b	–623.6	–1044	420.2
c	0	966	–966

One can define the coefficient of performance as the heat extracted divided by the work needed (assuming that no work is recovered during the expansion). Heat is extracted from the environment when Q is positive, which is in step b. It is necessary to perform work on the gas during steps a and b. Hence, our coefficient of performance is $\varepsilon = Q_c/(W_a + W_b) = 0.925$. Of course, this assumes complete reversibility. In reality, there are several reasons why this would not be attained, such as friction in the piston. We also know, from derivations in Appendix C, that any heat flux, such as that necessarily through the diathermal wall, will lead to entropy generation, and hence irreversibility. □

2.7 TEMPERATURE DIFFERENCES AND HEAT FLOW

It may seem odd at first to use something abstract like entropy as a fundamental quantity, and then define an everyday quantity like temperature on the basis of entropy. Therefore, it is important to show that the definition for temperature and the postulates do indeed lead to predictions about temperature that fit our experience. From experience, we know the following.

- Temperature is intensive.
- Two systems in thermal contact reach the same temperature at equilibrium.
- If two objects have different temperatures and are placed in thermal contact, then heat flows from the object of higher temperature to the colder object.
- If we raise the temperature of something either at constant pressure or at constant volume, we expect its energy to go up. Or equivalently, adding heat raises the temperature.

The first observation has already been shown in the definition. We now show that the postulates predict the second and third observations. The fourth observation is not shown until Section 4.1.

Consider two objects that are in thermal contact with each other, but are isolated from the rest of the Universe. Initially the two objects are at equilibrium while isolated from one another, and have different extensive properties S, U, V, and N (see Figure 2.7). Then, we suddenly place them in thermal contact. Since the system is closed, the total internal energy of the system must be constant: $dU = dU^{(1)} + dU^{(2)} = 0$, where the superscripts indicate object 1 and object 2. Hence,

$$dU^{(1)} = -dU^{(2)}. \tag{2.59}$$

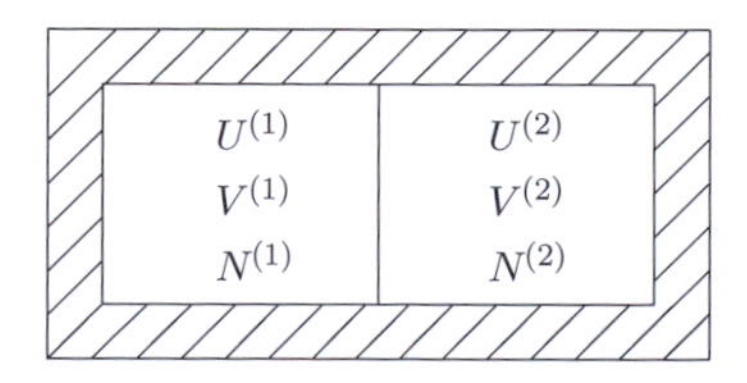

Figure 2.7 We consider two initially isolated systems with different energy, volume, and mole number (U, V, N). The partition allows heat to flow between the two systems, but is rigid and impermeable.

From Postulate IV, we know that the equilibrium state is attained when the entropy reaches a maximum. If each subsystem is **isochoric** (i.e. maintained at constant volume, $dV^{(1)} = dV^{(2)} = 0$) and isomolar ($dN^{(1)} = dN^{(2)} = 0$), then the entropy differential equation, Eqn. (2.29), becomes

$$\begin{aligned} dS &= dS^{(1)} + dS^{(2)} \\ &= \frac{1}{T^{(1)}} dU^{(1)} + \frac{1}{T^{(2)}} dU^{(2)} \\ &= \left(\frac{1}{T^{(2)}} - \frac{1}{T^{(1)}} \right) dU^{(2)}, \quad \text{isomolar, isochoric,} \end{aligned} \tag{2.60}$$

where we have used Eqn. (2.59) to obtain the third line. The postulates claim that the unconstrained variables – in this case $U^{(2)}$ – will arrange themselves in such a way as to maximize the entropy. In order for it to be a maximum, the total entropy must satisfy two conditions.

First, at equilibrium $(\partial S/\partial U^{(2)})_{V^{(1)},V^{(2)},N^{(1)},N^{(2)}}$ must be zero. Hence, Eqn. (2.60) leads to $T^{(2)} = T^{(1)}$ at equilibrium, which is the second observation we wished to prove.

Secondly, in order for the entropy to be a maximum, entropy must change from a lower to a higher value on going from the initial to the final configuration:

$$\left(\frac{\partial S}{\partial U^{(2)}} \right)_{V^{(1)},V^{(2)},N^{(1)},N^{(2)}} = \left(\frac{1}{T^{(2)}} - \frac{1}{T^{(1)}} \right). \tag{2.61}$$

The inequality follows from the postulate that the entropy of an isolated system must increase on going from one equilibrium state to another, $dS > 0$. Equation (2.61) says that, if $T^{(2)} > T^{(1)}$, then $\left(\partial S/\partial U^{(2)}\right)_{V^{(1)},V^{(2)},N^{(1)},N^{(2)}} < 0$; hence $dU^{(2)} < 0$ and heat flows from object 2 to object 1. On the other hand, if $T^{(2)} < T^{(1)}$, then $\left(\partial S/\partial U^{(2)}\right)_{V^{(1)},V^{(2)},N^{(1)},N^{(2)}} > 0$, and heat flows from object 1 to object 2. This result is shown graphically in Figure 2.8. The equilibrium state is the maximum of the curve. If the initial state lies to the left of the maximum, then $U^{(2)}$ will increase. From Eqn. (2.61), we see that this is the case when $T^{(1)} > T^{(2)}$. This prediction agrees with our third observation about the intuitive properties of temperature: heat flows from hot to cold objects.

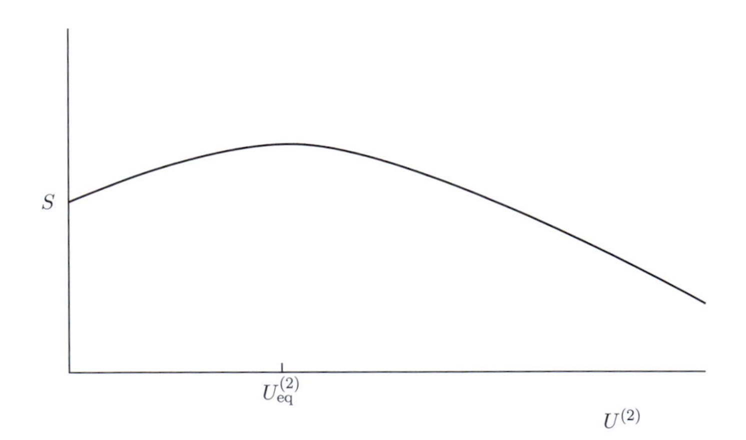

Figure 2.8 The general form of entropy as a function of the internal energy of one subsystem for the setup in Figure 2.7.

EXAMPLE 2.7.1 Two tanks containing a gas insulated and isolated from the environment are connected by a pipe with a valve. The valve is initially shut, so the two tanks are not equilibrated with one another. Tank 1 has volume $V_1 = 5\,\mathrm{l}$, initial pressure $P_1 = 3\,\mathrm{atm}$, and initial temperature $T_1 = 250\,\mathrm{K}$. The second tank has volume $V_2 = 2\,\mathrm{l}$, initial pressure $P_2 = 5\,\mathrm{atm}$, and initial temperature $T_2 = 323\,\mathrm{K}$. Assuming that the gas is a simple ideal gas with $c = 3/2$, find the final temperature of the tanks.

Solution. The walls of the tanks are rigid, so the volume of each tank remains fixed. The valve allows the gas to pass between the tanks, so the number of moles in each tank is *not* conserved. However, the total number of moles $N = N_1 + N_2$ is fixed. The valve also allows energy to pass between the two tanks, so the energy of each tank changes, but the total energy $U = U_1 + U_2$ is fixed. The constants of the process are then V_1, V_2, N, and U. Everything else $(N_1, N_2, T_1, T_2, S_1, S_2, P_1, P_2)$ can change in reaching the final state.

We can find the initial mole numbers from the ideal-gas equation of state, Eqn. (2.37),

$$P_i V_i = N_i R T_i, \quad i = 1, 2, \quad \text{simple ideal gas.} \tag{2.62}$$

Hence, $N_1 = 0.7312\,\mathrm{mol}$, $N_2 = 0.3773\,\mathrm{mol}$, and $N = 1.108\,\mathrm{mol}$. We can find the initial internal energies from the thermal equation of state for a simple ideal gas, Eqn. (2.38),

$$U_i = c N_i R T_i, \quad i = 1, 2, \quad \text{simple ideal gas.} \tag{2.63}$$

Therefore, $U_1 = 2.280\,\mathrm{kJ}$, $U_2 = 1.520\,\mathrm{kJ}$, and $U = 3.800\,\mathrm{kJ}$. The final state occurs when the temperatures in the two tanks are the same. Since the resulting gas still satisfies the thermal equation of state for a simple ideal gas, we can find the final temperature T straightaway:

$$T = \frac{U}{cNR} = \frac{(3.80\,\mathrm{kJ})(1000\,\mathrm{J/kJ})}{\frac{3}{2}(1.108\,\mathrm{mol})(8.3144\,\mathrm{J/mol\cdot K})} = 275\,\mathrm{K}. \tag{2.64}$$

□

Note the usefulness of this strategy to solve thermodynamics problems: *identify which variables are constant.*

2.8 PRESSURE DIFFERENCES AND VOLUME CHANGES

In Section 2.7 we found that our intuition about temperature is predicted by the postulates: temperature differences lead to heat flow from hot to cold objects. Similarly, we show here that pressure differences lead to changes in volume. In other words, if two objects are in mechanical contact, the one at higher pressure will expand at the expense of the volume of the other object, until the pressures are equal.

Consider two containers of gas that have different temperatures and different pressures. Also, we relax the constraint on volume, and allow the volumes of the two subsystems to change. (See Figure 2.4.) The two containers are separated by a movable, diathermal partition. Initially, the partition is insulating, and fixed. Then, we allow the partition to move and transfer heat between the two compartments. Throughout the process, the composite system is closed to the Universe.

Since the system is closed, again the internal energy is conserved, and Eqn. (2.59) holds. Since the composite system is closed to the Universe, the total volume must also be conserved, and

$$dV^{(1)} = -dV^{(2)}. \tag{2.65}$$

Again we hold the mole numbers in each subsystem constant, so the differential for S, Eqn. (2.29), becomes

$$\begin{aligned} dS &= dS^{(1)} + dS^{(2)} \\ &= \frac{1}{T^{(1)}} dU^{(1)} + \frac{P^{(1)}}{T^{(1)}} dV^{(1)} + \frac{1}{T^{(2)}} dU^{(2)} + \frac{P^{(2)}}{T^{(2)}} dV^{(2)} \\ &= \left(\frac{1}{T^{(1)}} - \frac{1}{T^{(2)}}\right) dU^{(1)} + \left(\frac{P^{(1)}}{T^{(1)}} - \frac{P^{(2)}}{T^{(2)}}\right) dV^{(1)}, \end{aligned} \tag{2.66}$$

where we have used Eqns. (2.59) and (2.65) to obtain the third line. Since the variables $U^{(1)}$ and $V^{(1)}$ are unconstrained, each of the terms on the right-hand side of the last line of Eqn. (2.66) must be zero at equilibrium. Hence, we again find that $T^{(1)} = T^{(2)}$. However, we also find that $P^{(1)} = P^{(2)}$ at equilibrium. Consider the special case when the two compartments are at equal temperature T. Then, the change in volume for the compartments is given by the stipulation that $\Delta S > 0$, according to the postulates. Thus,

$$T\left(\frac{\partial S}{\partial V^{(1)}}\right)_{U,V,N^{(1)},N^{(2)}} = P^{(1)} - P^{(2)}. \tag{2.67}$$

If $P^{(1)} > P^{(2)}$, then the volume of $V^{(1)}$ must increase to satisfy the postulates. Likewise, $P^{(1)} < P^{(2)}$ requires that $V^{(1)}$ decreases. These results agree with our observation about pressure.

EXAMPLE 2.8.1 Find the final pressure of the gas in Example 2.7.1.

Solution. We have now proven that the final state occurs when the temperatures and pressures in the two tanks are the same. Since the resulting gas still satisfies the mechanical equation of state for a simple ideal gas, we can find the final pressure P:

$$P = \frac{NRT}{V_1 + V_2} = \frac{(1.108 \text{ mol})(0.082\,06 \text{ l} \cdot \text{atm/mol} \cdot \text{K})(275 \text{ K})}{7 \text{ l}} = 3.572 \text{ atm}. \tag{2.68}$$

□

In Exercise 2.8.A, one finds that there is a role played by chemical potential for the number of moles that is analogous to that played by temperature for energy, and to that played by pressure for volume. Hence, we find that, for two systems in thermal, mechanical, and chemical contact, equilibrium implies

$$T^{(1)} = T^{(2)}, \quad \text{thermal equilibrium,} \tag{2.69}$$

$$P^{(1)} = P^{(2)}, \quad \text{mechanical equilibrium,} \tag{2.70}$$

$$\mu_1^{(1)} = \mu_1^{(2)}, \quad \text{chemical equilibrium.} \tag{2.71}$$

These results are completely general,[6] and should be remembered.

We also find that, if the chemical potential of a species is non-uniform, that species will move from the region of high chemical potential to the region of low chemical potential. Roughly speaking, the chemical potential depends strongly on concentration, so we sometimes say that moles diffuse to offset differences in concentration. However, as stability analysis shows in Chapter 4, that approximation is not always correct.

2.9 THERMODYNAMICS IN ONE DIMENSION

So far we have considered systems where the independent variables are (S, V, N), or their conjugates (T, P, μ). However, some important systems are only two-dimensional, such as adsorption on a surface, or a thin film, and, hence, volume plays no role. It is also possible to apply thermodynamics to such systems with a slight modification to the development we have seen so far. In this section, we consider a one-dimensional system – the length of a rubber band. In later sections, we will also consider two-dimensional systems, such as a catalytic surface and a thin film.

To consider the rubber band, we must use an appropriate replacement for volume, and then the quantity analogous to pressure. For the rubber band, we replace volume with the length of the rubber band, $V \rightarrow L$, and pressure with tension, $P \rightarrow -\mathcal{T}$. The sign is different, because the work done on a rubber band is $+\mathcal{T}\, dL$, as opposed to $-P\, dV$ for a fluid. Hence, we can write the thermodynamic definition for tension that is consistent with the mechanical work as

$$\mathcal{T} := \left(\frac{\partial U}{\partial L}\right)_S. \tag{2.72}$$

Rubber is comprised of cross-linked polymeric chains; an ideal rubber is made up of one large molecule, so $N = 1/\tilde{N}_A$, where $\tilde{N}_A \cong 6.023 \times 10^{23}$ is **Avogadro's number**. Therefore, one cannot easily change the size of the system, and the number of moles is fixed. Hence, a fundamental entropy relation for a rubber band must be of the form $S = S(U, L)$, and extensivity is a little different.

A few simple experiments one can easily perform at home show two important trends: (1) the tension in a rubber band increases with length; and (2) the tension *increases* with increasing temperature.[7] Hence, any reasonable fundamental relation should predict these two

[6] These are completely general for a three-dimensional system in the thermodynamic limit, that is. For two-dimensional systems, there may also exist an analog to equal pressure, which is covered in greater detail in Chapter 12. Small systems are more subtle [60].

[7] Try suspending a weight with a rubber band, and heating it with a hair dryer. However, be careful to make sure that the rubber band is well equilibrated. Stretch and heat the rubber band a bit first. Then, keeping the weight attached, let the rubber band cool to room temperature. Mark the height of the weight, and then begin warming the rubber band with the dryer. As the temperature increases, the tension should increase, raising the height of the weight.

observations. The latter observation is curious, since it is the opposite of that seen for metal springs, where the tension decreases with temperature.

In the following example we consider a fundamental entropy relation that was derived elsewhere from straightforward statistical-mechanical arguments.

EXAMPLE 2.9.1 Treloar [134] has suggested that the fundamental entropy relation for a rubber band is

$$S = s_0 L_0 - A_c L_0 R \left[\frac{1}{2} \left(\frac{L}{L_0} \right)^2 + \left(\frac{L_0}{L} \right) \right] + c_L L_0 \log \left[1 + \frac{U - u_0 L_0}{c_L L_0 T_0} \right], \tag{2.73}$$

where s_0 and u_0 are the specific entropy and internal energy at our reference state, but these are per rubber-band rest length, not per mole. Our reference state is the reference temperature T_0 and rest length L_0. We have two more constants, A_c and c_L. The former has something to do with the modulus or rigidity of the rubber band, and the latter is called the specific constant-length heat capacity. What does this model predict for tension as a function of length and temperature? Does it agree with experiment?

Solution. To find the tension and temperature, we derive two equations of state from the fundamental relation. We can find temperature in the usual way from its (alternative) definition

$$\begin{aligned} T &:= 1 \Big/ \left(\frac{\partial S}{\partial U} \right)_L . \\ &= T_0 \left[1 + \frac{U - u_0 L_0}{c_L L_0 T_0} \right], \end{aligned} \tag{2.74}$$

where we used the fundamental relation Eqn. (2.73) to obtain the second line. From this result we can find the thermal equation of state

$$U = U_0 + c_L L_0 (T - T_0). \tag{2.75}$$

Curiously, we find that the internal energy does *not* depend upon length at constant temperature.

To find the tension in the rubber band, we could use the thermodynamic definition Eqn. (2.72). However, it is more convenient to use the alternative expression

$$\begin{aligned} \mathcal{T} &= -T \left(\frac{\partial S}{\partial L} \right)_U \\ &= A_c R T \left[\left(\frac{L}{L_0} \right) - \left(\frac{L_0}{L} \right)^2 \right], \end{aligned} \tag{2.76}$$

where we have again used the fundamental relation Eqn. (2.73). We find several important features of the model from this result.

- The length dependence of the tension given by Eqn. (2.76) has a reasonable shape, but is not exactly what is found from experiment.
- Our thermal equation of state shows that lengthening the rubber band at constant temperature does not change the internal energy.

- The tension increases linearly with temperature, again in accord with experiment.[8]
- The tension in the rubber band is zero only when its length is L_0, its natural rest length.
- The second term in the square brackets of Eqn. (2.76) is related to incompressibility of the rubber band, and shows that a compressive force is necessary if one wants to decrease the length of the rubber band from its rest length. □

Hence, an isothermal stretch of the rubber band stores no energy in the rubber band. Nonetheless, the rubber band can do work on its environment by pulling up a weight, say. It does this by *increasing its entropy*. Actually, there is a strong analogy with compression of an ideal gas. The student should be able to prove that *isothermal compression of an ideal gas does not change its internal energy, but rather decreases its entropy.*

The fundamental relation considered in the previous example captures many of the key experimental features. However, a real rubber band shows richer behavior.

- The internal energy does not grow linearly with temperature.
- At sufficiently low temperatures, the rubber can crystallize, and shows no strong elastic behavior.
- The tension deviates from the expression given because of strand entanglements and the finite extensibility of the strand. Also, at large extensions, the rubber can eventually crystallize, or break at large stretches. The physical meaning of the constants is now clear. T_0 is a reference temperature. The rest length of the rubber band (the length for which the tension is zero) is L_0.[9]

Given the limitations and successes of this fundamental relation, one can draw many analogies with an ideal gas, which also fails at low temperatures and low volumes. In other words, the rubber band behaves like an ideal gas, except that increasing the length of the rubber band is like decreasing an ideal gas's volume. For example, compressing an ideal gas isothermally changes the gas's entropy, but not its internal energy.

Therefore, we choose to call Eqn. (2.73) a **fundamental entropy relation for a simple, ideal rubber band.**[10] We have shown that this fundamental relation yields

$$\begin{aligned} \mathcal{T} &= T\psi(L), \\ \left(\frac{\partial U}{\partial L}\right)_T &= 0, \\ \left(\frac{\partial S}{\partial L}\right)_T &< 0. \end{aligned} \tag{2.77}$$

In words, the last two relations say that isothermal stretching of the rubber band decreases the entropy, but leaves the internal energy unchanged. In Section 3.7 we show that experimental

[8] We will see in Section 3.7.1 the interesting result that the linear dependence on temperature is tied directly to the second note above. There we also show how isothermal stretching affects the entropy, but not the energy, of a real rubber band.

[9] From Eqn. (2.75) we see that C_L is the constant-length heat capacity. Heat capacities are rigorously defined in Section 3.5.

[10] In the language of mechanics, it is the equation that arises for an incompressible, neo-Hookean solid in extensional deformation.

observation of the first relation leads inescapably to the second and third relations. The results are not too surprising in light of the definition of entropy on p. 20.

Summary

In this chapter we introduced or used several fundamental quantities:

- volume V, area A, or length L;
- mole number N (or numbers, $\{N_i\}$);
- internal energy U;
- entropy S; and
- equilibrium states.

The properties of U, S, and equilibrium states are described in the four fundamental postulates that form the basis of all thermodynamics. We know that a function of the form $U = U(S, V, N)$ or one of the form $S = S(U, V, N)$ completely determines the thermodynamic properties of a substance.

From these fundamental quantities, we derived

- temperature, $T := (\partial U/\partial S)_{V,N}$;
- pressure, $P := -(\partial U/\partial V)_{S,N}$;
- chemical potential, $\mu := (\partial U/\partial N)_{S,V}$; and
- heat flow, Q: $đQ := dU - đW$.

The ideas in the first three definitions are neatly summarized in the differential for internal energy

$$dU = T\,dS - P\,dV + \sum_{i=1}^{r} \mu_i\,dN_i. \tag{2.27}$$

We proved the following statements.

- In an isomolar, quasi-static process $đQ_{\text{rev}} = T\,dS$.
- The fundamental postulates say that heat flows from hot to cold objects until the temperatures are equal.
- Compressible objects in mechanical contact change their volumes until the pressures are equal, with the higher-pressure object expanding at the expense of the other.
- Compounds diffuse across permeable boundaries from regions of high chemical potential to regions of low chemical potential.

The last three results were written mathematically for two subsystems (1) and (2) in thermal, mechanical, and chemical contact:

$$T^{(1)} = T^{(2)}, \quad \text{thermal equilibrium,} \tag{2.69}$$

$$P^{(1)} = P^{(2)}, \quad \text{mechanical equilibrium,} \tag{2.70}$$

$$\mu_1^{(1)} = \mu_1^{(2)}, \quad \text{chemical equilibrium.} \tag{2.71}$$

All of the above results in the summary should be memorized. In fact, most problems of engineering thermodynamics can be solved using only these results plus some of the results in the following chapters.

We also proved that experimental results imply that

- compressing an ideal gas isothermally decreases its entropy, but leaves its internal energy constant; and
- stretching a rubber band isothermally decreases its entropy, but leaves its internal energy constant.

For examples, we used fundamental relations for a simple ideal gas, a simple van der Waals fluid, and an incompressible, neo-Hookean rubber band. We also performed calculations for work and heat flow of expansion (or contraction), and work required for adiabatic volume (or length) changes. We showed how an ideal gas can be used as a refrigerator, and estimated the minimum amount of work required to do so.

Exercises

2.3.A The experiments from Example 2.3.1 are repeated for another gas, which exhibits different results. The first experiment reveals that

$$\left(P + \frac{\tilde{a}}{V^2}\right)^c \left(V - \tilde{b}\right)^{c+1} = \text{constant.}$$

The second experiment (using the falling weight) reveals that

$$\frac{dP}{dt} = -\frac{mg}{c(V - \tilde{b})}\frac{dl}{dt},$$

where $\tilde{a}, \tilde{b}$, and c are constants. Find the equation of state $U = U(P, V)$.

2.3.B Two points on the pressure–volume plane of Example 2.3.1 are connected by a line $PV =$ constant. Find the work and heat flow necessary to change the volume and pressure of the container in this example to follow this line from $[P_0, V_0]$ to $[P_f, V_f]$. Find the heat and work necessary to make the system follow a straight line between the points. How do your answers compare?

2.3.C A gas is placed in a container identical to the one in Example 2.3.1. It is reported that the equation of state for the gas is given by

$$U(P, V) = U_0 + A(PV^2 - P_0V_0^2).$$

If the volume is changed slowly (quasi-statically) and adiabatically, how will the pressure change?

2.3.D The gas in Example 2.3.1 begins at a pressure of 1 atm, and a volume of 1 l. If it is compressed adiabatically to half its original volume, how much work is done? If a rifle bullet of mass 4 g has the same kinetic energy, how fast is it moving in miles/h? How high must a golf ball (of mass 42 g) be raised to gain the same amount of potential energy?

2.3.E How much work does it take to compress $N = 3$ mol of an ideal gas adiabatically from $V = 1$ l at $T = 24\,°\mathrm{C}$ to a volume of 10 ml? The pressure of an ideal gas can be found from the temperature and volume by application of the ideal-gas equation of state $PV = NRT$.

2.3.F The gas from Example 2.3.1 is now to be used as part of a refrigeration cycle. To move the fluid in the cycle (look at the tubes in the back of your refrigerator to see how this fluid is carried), a pump raises its pressure from approximately 20 psig at 25 °C to 32 psig. Assuming that this pump operates adiabatically, and that the gas is pumped at 23 ft^3/min (derived from the upstream volume), estimate the minimum power (work/time) necessary to pump the fluid. Note that a flowing system necessarily has irreversibilities that make the necessary power greater than you would predict from a quasi-static process.

2.3.G We repeat the experiments of Example 2.3.1 with a different gas, and observe different behavior. The second experiment reveals that the pressure changes with time according to

$$\frac{dP}{dt} = -f_0(V)m|\vec{\mathbf{g}}|\frac{dl}{dt},$$

where $f_0(V)$ is some measured function of volume with dimensions of inverse volume. The first experiment reveals

$$[P - P_1(V)]\,\frac{\exp\left[-\int_{V_0}^{V} f_0(V')dV'\right]}{cf_0(V)} \sim \text{constant},$$

where c is a constant, and $P_1(V)$ is another measured function of volume, with dimensions of pressure. V_0 is some constant reference volume. Find the equation of state $U = U(P, V)$. Note that Exercise 2.3.A is a special case, which might provide a check for your answer.

2.4.A A substance called a *glassy polymer* is placed into a container that allows us to control the work done on and the heat added to the polymer. In state 1, the polymer has a certain energy U, volume V and mole number N. We measure the pressure under these conditions to be P_1, which appears to be stationary with time.

Then, we perform various steps of work on the system, and heat additions and extractions such that the volume and energy are the same as before. We again measure the pressure, and find $P_2 \neq P_1$, but P_2 also appears to be stationary.

Are states 1 and 2 at equilibrium? Explain.

2.5.A Determine which of the following proposed fundamental relations satisfy the postulates of thermodynamics.

1. $S = s_0 N + AU^{1/4}V^{1/2}N^{1/4}$,
2. $S = \sum_{i=1}^{r} N_i s_{i,0} + \left(\sum_{i=1}^{r} N_i c_i\right) \log\left(\frac{U}{N u_0}\right) + \sum_{i=1}^{r} N_i R \log\left(\frac{V}{v_0 N_i}\right)$,
3. $S = AUN^{\beta}\left(\frac{V^2 B}{N^2}\right)^{1/3}$,
4. $U = \frac{AS^3}{NV} + B$,
5. $U = \frac{AV^3}{NS}$,
6. $U = \frac{AN^2}{V}\exp\left(\frac{S}{NR}\right)$,
7. $S = Ns_0 + NR\log[U^A V^B N^C]$,

where $A, B, C, R, c_i, s_0, s_{i,0}, u_0$, and v_0 are constants.

2.5.B A simple van der Waals fluid is compressed isentropically to half its volume. If the initial state is (T_0, v_0), find the final temperature, the work required, and the heat flow.

2.5.C One mole of oxygen at 25 °C and 15 atm is compressed **iso-energetically** (i.e. at constant energy) to 35 atm. Calculate the change in entropy using both a simple ideal gas ($c = 5/2$) and a simple van der Waals fluid. The other parameters for the van der Waals fluid can be found from the critical properties of oxygen, as shown in Section B.3 of Appendix B. Compare the two answers. Why are they different? How does your answer fit with the definition of entropy, Eqn. (2.20)?

2.5.D Do Exercise 2.5.C above, but this time use the simplified Redlich–Kwong model, which is given in Eqn. (2.78) below, instead of the van der Waals model. You can estimate the parameter by using the expression $A \cong -6.83RT_c^{3/2}$.

2.6.A Find $U(P, V, N)$ for the van der Waals fluid whose fundamental entropy relation is given by Eqn. (2.21). Compare your answer with the result from Exercise 2.3.A.

2.6.B Check the postulates, and find $P(T, V)$ for the following fundamental entropy relation:

$$S = NR\log\left(\frac{v-b}{v_0}\right) + \frac{4b^2u^3N}{3[3a\log(v/(v+b)) - 4Ab]^2}, \tag{2.78}$$

where R, b, v_0, and a are constants. The resulting relation is called the *Redlich–Kwong equation of state*. Note that the thermal equation of state resulting from Eqn. (2.78) is not particularly accurate. In Chapter 3, we derive a more realistic thermal equation of state that is compatible with the Redlich–Kwong mechanical equation of state.

2.6.C Check the postulates, and find $P(T, V)$ for the following fundamental energy relation:

$$U = Na_0\psi(v) - \frac{N[R\log[(v-b)/v_0] - a_1\psi(v) - s]^2}{4a_2\psi(v)}, \tag{2.79}$$

where a_0, v_0, R, b, a_1, a_2, and A are constants, and

$$\psi(v) := A + \frac{1}{2\sqrt{2}b} \log\left[\frac{v + b + \sqrt{2}b}{v + b - \sqrt{2}b}\right].$$

The resulting relation is similar to the *Peng–Robinson equation of state*. Note that the thermal equation of state resulting from Eqn. (2.79) is not particularly accurate. In Chapter 3, we derive a more realistic thermal equation of state that is compatible with the Peng–Robinson mechanical equation of state.

2.6.D Check the postulates, and find $P(T, V)$ for the following fundamental energy relation:

$$U = Na_0\psi(v) - \frac{N[R\log(v - b) - a_1\psi(v) - S/N]^2}{4a_2\psi(v)}, \tag{2.80}$$

where a_0, a_1, a_2, R, b, and A are constants, and

$$\psi(v) := A + \frac{1}{\gamma - \beta} \log\left(\frac{v + \beta - b}{v + \gamma - b}\right).$$

The resulting relation is called *Martin's generalized cubic equation of state*. Note that the thermal equation of state resulting from Eqn. (2.80) is not particularly accurate. In Chapter 3, we derive a more realistic thermal equation of state that is compatible with Martin's generalized mechanical equation of state.

2.6.E Check the postulates, and find $P(T, V)$ for the following fundamental entropy relation:

$$S = Ns_0 + NR\left\{\log\left[\left(\frac{U + aN^2/V}{u_0 + a/v_0}\right)^c \left(\frac{V/N - b}{v_0}\right)\frac{1}{N^c}\right] + \frac{b(V - Nv_0)}{(V - bN)v_0}\left[4 + \frac{b(V + Nv_0 - 2Nb)}{(V - bN)v_0}\right]\right\}, \tag{2.81}$$

where a, v_0, R, b, and c are constants. The resulting relation is called the *Carnahan–Starling generalization* to the van der Waals equation of state. Note that the thermal equation of state resulting from Eqn. (2.81) has a constant heat capacity. In Chapter 3, we derive a thermal equation of state that yields a better thermal equation of state.

2.6.F Check the postulates, and find $P(T, V)$ for the following fundamental entropy relation:

$$S = Ns_0 + NR\log\left(\frac{V}{Nv_0}\right) - \frac{N^2R}{V}\left(B_0 + \frac{bN}{2V}\right) - \frac{2\psi_2(U, V, N)^{3/2}}{3\sqrt{N\psi_1(V, N)}}, \tag{2.82}$$

where $A_0, B_0, C_0, B_1, R, a, b, c, s_0, v_0, \alpha$, and γ are constants,

$$\psi_1(V, N) := \frac{3NC_0}{V} + 3c\left(\frac{1}{\gamma} + \frac{N^2}{2V^2}\right)\exp\left(-\frac{\gamma N^2}{V^2}\right) + \frac{B_1}{2},$$

and

$$\psi_2(U, V, N) := Nu_0 - U - \frac{N^2A_0}{V} + \frac{aN^3}{V^2}\left(\frac{\alpha N^3}{5V^3} - \frac{1}{2}\right).$$

The resulting relation, which is called the *Benedict–Webb–Rubin equation of state*, is commonly used for alkanes. Note that the thermal equation of state resulting from Eqn. (2.82) is not very accurate. In Chapter 3, we derive a more realistic thermal equation of state that is compatible with the PVT equation of state which arises from Eqn. (2.82).

2.6.G Find the work and heat flow necessary to compress a van der Waals fluid isothermally.

2.6.H One mole of a simple ideal gas ($c = 3/2$) is placed into a container with a nearly frictionless piston to be used as a refrigerator. The cycle of the refrigerator has three steps. In step (a), the gas is compressed adiabatically from 50 °C, 1atm to 150 °C. In step (b), the gas is cooled isobarically back to 50 °C. In step (c), the gas is expanded back to its original thermodynamic conditions. During step (c), the gas can be used to extract heat from an object at 50 °C. Find the work necessary to operate the refrigerator, and the amount of heat that can ideally be extracted.

2.6.I Find the equations of state for those fundamental relations in **Exercise** 2.5.A that satisfy the first four thermodynamic postulates.

2.6.J If you compress a van der Waals fluid isentropically until its volume is halved, what happens to the temperature? Find the quasi-static work, and show that the process is also adiabatic.

2.6.K From Eqn. (2.35), we see that the van der Waals equation of state offers a correction to the ideal gas through the constants a and b. These constants may be found from the so-called critical properties of a fluid P_c and T_c using Eqn. (4.22). From its critical properties, find a and b for nitrogen. If you have 1 mol of nitrogen at room temperature and pressure, what does the ideal-gas model predict for the volume? What does modeling the system as a van der Waals fluid predict? What is the percentage difference? Now do the same calculations for a pressure of 5 atm. For the van der Waals fluid, you must solve a cubic equation in volume. You may find the equations in Section A.7 of Appendix A useful.

2.6.L For toluene, we estimate the parameters of the van der Waals equation of state to be $a = 2.487\,643\,2\,\mathrm{Pa \cdot m^6/mol^2}$ and $b = 0.000\,149\,797\,\mathrm{m^3/mol}$. Construct a plot of specific volume versus pressure for $T = 580\,\mathrm{K}$, and the range $P < 3.78 \times 10^6\,\mathrm{Pa}$ and $6.223 \times 10^{-4}\,\mathrm{m^3/mol} < v < 0.007\,\mathrm{m^3/mol}$. Note that the van der Waals equation of state is cubic in volume. Hence, you might find the equations in Section A.7 of Appendix A useful. What difficulties arise as the pressure approaches $3.78 \times 10^6\,\mathrm{Pa}$? (*Hint: the range of volumes given corresponds to the physically acceptable roots to the cubic equation for a vapor, as we discuss in detail in Section 4.2.2.*)

2.6.M For oxygen, we estimate the parameters of the Peng–Robinson equation of state to be $a = 0.164\,995\,\mathrm{Pa \cdot m^6/mol^2}$ and $b = 1.984\,14 \times 10^{-5}\,\mathrm{m^3/mol}$ at $T = 120\,\mathrm{K}$. If you wish to store 1000 moles of oxygen in a container of volume 1 $\mathrm{m^3}$, what will the pressure be? If you triple the pressure, how many moles will be stored in the same container? Note that the Peng–Robinson equation of state is cubic in volume. Hence, you might find the equations in Section A.7 of Appendix A useful.

2.6.N From statistical mechanics it is known that the ideal-gas law holds only when the individual molecules rarely interact with one another – i.e. at low densities. However, we can also see that this is true from the equations of state. For example, in the van der

Waals equation of state,

$$P = \frac{RT}{v - b} - \frac{a}{v^2},$$

we see that the pressure is well approximated by RT/v only for large values of v. To make this criterion more quantitative, we could say that a/v^2 should be small relative to RT/v. Suppose that we arbitrarily choose an accuracy of 10% as a "good" approximation by an equation of state, then we could require that $(a/v^2)/(RT/v) < 0.10$ in order for the ideal-gas law to hold.

Construct plots of v_{min}, the minimum value of the specific volume allowing an ideal gas to be a good assumption, as functions of temperature (50–400 K). Use the van der Waals, Redlich–Kwong, and Peng–Robinson equations of state for estimates of non-ideal behavior, and assume that the gas is nitrogen. Two of these equations of state are known to be more accurate. Which is the outlier (i.e. the bad equation of state)?

Note that the parameters for the equation of state can be estimated using the critical properties of the fluid, as shown in Appendix B.

2.6.O Launching potatoes with an air gun appears to be a popular pastime in rural parts of the USA. From a thermodynamics point of view, one uses the work available from expanding air to give kinetic energy to a potato. Here we wish to estimate the minimum compression of air necessary to launch a 500-g potato a distance of 100 m. For the sake of analysis, we consider a simple setup as shown in Figure 2.9. A real air gun does *not* work in the way illustrated, but the analysis is similar.

To calculate the minimum work, we make the following simplifying assumptions. First, the air is assumed to be described by a simple ideal gas, with $c = 3/2$. Hence, we can use the analysis in Examples 2.6.2 and 2.6.3. Secondly, we neglect all irreversible losses, such as the sliding of the piston, non-quasi-static expansion, and the drag on the potato as it flies through the air.

Assuming that the air begins at room temperature, and that there is 0.1 mol of gas, find the initial volume and pressure, and the final temperature and volume of the air, assuming that the process is adiabatic. What do you expect to happen to any humidity that might be in the air after the expansion?

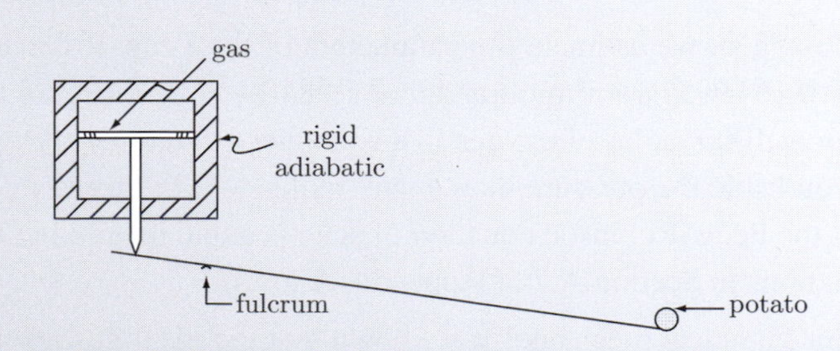

Figure 2.9 A schematic diagram of a device that uses compressed air to launch a potato. The air is held in a rigid container with adiabatic walls. The friction and inertia of all devices but the potato are neglected. The analysis of this device is described in Exercise 2.6.O.

2.6.P Finish Example 2.6.5. In other words, find the work necessary to compress a simple van der Waals gas adiabatically.

2.6.Q Using the van der Waals equation of state, calculate the amount of work necessary to compress one mole of nitrogen from 25 ml to 3 ml at 25 °C. How much does the internal energy change in the process?

2.6.R What does the van der Waals equation of state predict for the pressure of 3 mol of oxygen at room temperature when it has a volume of 10 ml? Is the ideal-gas law a good approximation for this value?

2.6.S Using the Peng–Robinson model from Exercise 2.6.C, predict how the pressure would change for the adiabatic, reversible compression of the fluid.

2.6.T In Exercise 2.3.F you calculated the minimum power necessary in order to compress an ideal gas. Now we consider the adiabatic compression of a gas that is too dense to be ideal, and use the van der Waals model to estimate the work. Find the minimum power (work/time) necessary in order to pump hydrogen from 13 atm to 15 atm assuming that the pump operates adiabatically, and is pumped at 23 ft^3/min (derived from the upstream volume). For a diatomic gas (like hydrogen) at sufficiently high temperatures, we can assume that the gas is simple, and $c = 5/2$. The other parameters are found from the expressions in Section B.3 of Appendix B and the critical properties of the fluid. The critical properties for some fluids are to be found in Table D.3 in Appendix D, and many more can be found in the NIST Web Book.

You might use the following steps.

- Begin by working in specific quantities only. Find the initial pressure, specific volume, and temperature. From the thermal equation of state, find the specific internal energy.
- Find the final state of the system assuming reversible, adiabatic compression. Example 2.6.5 might be useful here.
- How is the work (per mole) related to the change in specific internal energy?
- Now find the number of moles per minute pumped through the system, in order to find the minimum work per unit time required of the pump.

2.6.U Estimate the minimum amount of work necessary in order to use nitrogen as a refrigerant in the manner shown in Example 2.6.6. However, in this case, the initial pressure is 25 atm, so we cannot assume that the gas is ideal. Instead, use the simple van der Waals model, with $c = 5/2$. The other parameters in the model can be found from the critical properties, as given in Section B.3 of Appendix B.

2.6.V What is the change in chemical potential predicted by modeling the system as a simple van der Waals fluid on compressing one mole of carbon dioxide isothermally from 15 to 25 bar at 25 °C?

2.6.W Using the Redlich–Kwong model of Exercise 2.6.B, derive an expression to predict the work necessary for an adiabatic, reversible, isomolar compression of a fluid.

2.6.X A very realistic PvT equation of state for simple fluids is called the Peng–Robinson model. This is given by Eqn. (B.45) in Appendix B. Write this equation of state in terms of the compressibility factor $z := Pv/(RT)$, as a function of v and T. At low densities $1/v$, all real gases behave ideally, and for moderate densities they can be described by a virial expansion, Eqn. (B.3) in Appendix B. Expand the Peng–Robinson model in a Taylor series of $1/v$ around zero. From this expansion, find expressions for $B(T)$ and $C(T)$ predicted by the Peng–Robinson model.

2.8.A Consider an isolated system that is separated into two compartments by a rigid, diathermal wall that is permeable to species 1, but not to other species in the system. Show that the postulates indicate that at equilibrium the chemical potential of species 1 must be the same in each compartment. Also show that moles of species 1 move from the compartment of large chemical potential to the compartment of low chemical potential.

2.8.B Find the final temperature of the gas in Example 2.7.1 if it is described by a simple van der Waals fluid.

2.8.C A tank of volume 1 l is connected by a pipe fitted with a valve to another tank of volume 2 l. Both tanks are insulated and contain argon, but the first tank is at a pressure of 5 bar and temperature of 220 K, whereas the second tank is at a pressure of 10 bar and a temperature of 250 K. What are the final temperature and pressure when the valve is opened? Assume that the gas is described by a Peng–Robinson mechanical equation of state, as given in Section B.6 of Appendix B. Also assume that the thermal equation of state is

$$u = u_0 + \frac{a_0}{2\sqrt{2}b}\log\left[\frac{v + (1+\sqrt{2})b}{v + (1-\sqrt{2})b}\right] + \frac{3}{2}R(T - T_0).$$

2.8.D Assuming that a simple ideal gas with $c = 5/2$ is used as the refrigerant, design a refrigerator to maintain a temperature of 10 °C with an outside temperature of 22 °C. The cycle of the refrigerant has four steps. In the first step, the piston is used to expand the gas isothermally at the lower temperature. Secondly, the gas is compressed adiabatically until it reaches the higher temperature. Thirdly, it is compressed further at the higher temperature, isothermally. Finally, in the fourth step, the gas is expanded adiabatically back to its original thermodynamic conditions. Calculate the amount of work required at each step, and how much heat is transferred. How much force would you require in your design and how much distance in order to obtain the necessary work? You might find it useful to sketch the process on a pressure–volume diagram.

If the refrigerator leaks 1 J/h to the surroundings, how many times per hour must you cycle the refrigerant for your design?

A real refrigerator is only slightly more complicated than this conceptually. First, the refrigerant is cycled by pumping it around a tube, so the process is continuous. Secondly, the refrigerant undergoes phase changes between liquid and gas, which give greater heat transfers. The details for the design of a real system are given in Chapter 5.

2.8.E. Two tanks are isolated from the environment and connected by a valve. Initially, the two tanks contain oxygen at volumes 10 l and 25 l, and pressures 40 atm and 27 atm, respectively. Both are at temperature 298 K. At these pressures, we cannot assume that the gases are ideal. Instead, we use the simple van der Waals fluid (with $c = 5/2$), and determine the a and b parameters from the critical properties of O_2 (see Section B.3 of Appendix B and Table D.3 in Appendix D, or use the NIST Web Book). Determine the final temperature and pressure of each of the tanks after the valve has been opened, and equilibrium has been reached.

2.9.A Does the fundamental relation in Example 2.9.1 satisfy the postulates of thermodynamics?

2.9.B The following fundamental entropy relation is proposed for a rubber band:

$$S = s_0 L_0 + c_L L_0 \log\left[1 + \frac{U - u_0 L_0}{c_L L_0 T_0}\right] + \frac{A_c L_0 R}{2} \log\left[1 - \left(\frac{L - L_0}{L_1 - L_0}\right)^2\right]. \tag{2.83}$$

What does this fundamental relation predict for $\mathcal{T}(T, L)$? What physical interpretation does L_1 have? Is this fundamental relation in closer agreement with experiment?

2.9.C What is the tension as a function of length for a reversible, adiabatic stretch of the rubber band from Example 2.9.1, if the initial length is L_0 and the initial temperature is T_0?

2.9.D A rubber band can also be used as a refrigerator, as in Exercise 2.6.H. By following the cycle drawn in Figure 2.10 in a clockwise fashion, heat can be extracted at the lower temperature on going from A to B, and expelled at a higher temperature on going from C to D. Calculate the amount of work necessary if the refrigerator is operated at a high temperature T_H and a low temperature T_C. Assume that $L_A = 3L_0$ and $L_B = 2L_0$. Also find the amount of heat extracted during AB and CD for each cycle.

2.9.E Recent experiments have measured the tension in a single strand of DNA in water by attaching one end of the strand to a glass slide, and the other end to a magnetic bead of diameter 3 μm. By applying a known magnetic field and measuring the extension of the strand under an optical microscope, a plot of end-to-end length as a function of tension can be generated [18]. Using straightforward statistical-mechanical ideas, fundamental relations can be found for a single DNA molecule. The *Gaussian chain model* in simplified form is

$$U(S, L, N_K) = u_0 L_0 + c_L L_0 T_0 \left\{\exp\left[\frac{S - s_0}{c_L} - \frac{k_B L^2}{2 c_L L_0 N_K a_K^2}\right] - 1\right\}, \tag{2.84}$$

where u_0, T_0, and s_0 are constants with straightforward interpretation, k_B is Boltzmann's constant (equal to the ideal-gas constant divided by Avogadro's number), c_L is the specific (per rest length) constant-length heat capacity of the strand, L is the end-to-end length of the DNA strand, N_K is the number of "segments" in the strand, and a_K is the length of a segment.

A somewhat more sophisticated model is called the worm-like chain [81], with the simplified fundamental relation

$$U(S, L, N_p) = u_0 L_0 + c_L L_0 T_0 \left\{ \exp\left[\frac{S - s_0 L_0}{c_L L_0} - \frac{k_B N_p}{2 c_L L_0} \left[\left(\frac{L}{N_p l_p} \right)^2 - \frac{L}{2 N_p l_p} + \frac{N_p l_p}{4(N_p l_p - L)} \right] \right] - 1 \right\}. \tag{2.85}$$

Find expressions for the tension as a function of length for these two fundamental relations. Construct plots of length versus the logarithm of the force, and compare your results with the data in Table 2.1, assuming that (1) they were taken at room temperature, (2) the contour length ($N_p l_p$) is 32.7 μm, and (3) the persistence length l_p is 53 nm. These properties can be found from other experiments unrelated to the elasticity. Note that $a_K = 2 l_p$ and $N_K = N_p/2$.

Table 2.1 Force as a function of stretching as measured by Bustamante *et al.* [18] for stretching a strand of DNA at 298 K

L (μm)	F (fN)	L (μm)	F (fN)	L (μm)	F (fN)
8.0	30.04	24.1	301.07	29.5	1850.54
9.3	29.01	25.1	301.07	29.4	2242.39
10.0	35.15	24.3	442.08	29.7	2446.95
10.8	39.04	25.8	412.25	30.3	3920.71
16.4	82.71	26.3	535.69	30.3	3920.71
17.9	84.16	27.0	535.69	30.6	4587.87
17.8	103.78	27.0	733.50	31.2	5463.12
19.0	103.78	27.7	828.86	31.6	7746.39
18.3	117.27	27.4	936.62	31.8	10793.82
21.2	181.45	27.5	1195.98	31.6	15040.12
22.0	178.31	28.9	1195.98	32.4	25394.48
22.7	223.75	29.5	1449.23	33.0	56696.27
23.0	248.46	28.7	1609.29	33.0	67512.45

2.9.F Isothermally compress an ideal gas to half its initial volume, and calculate the changes in energy and entropy. Can the resulting system do work? How does this relate to the stretching of a rubber band? What is the analogy with Eqns. (2.77)?

2.9.G A recent paper [141] recommends a new fundamental relation for the extension of a single polymer,

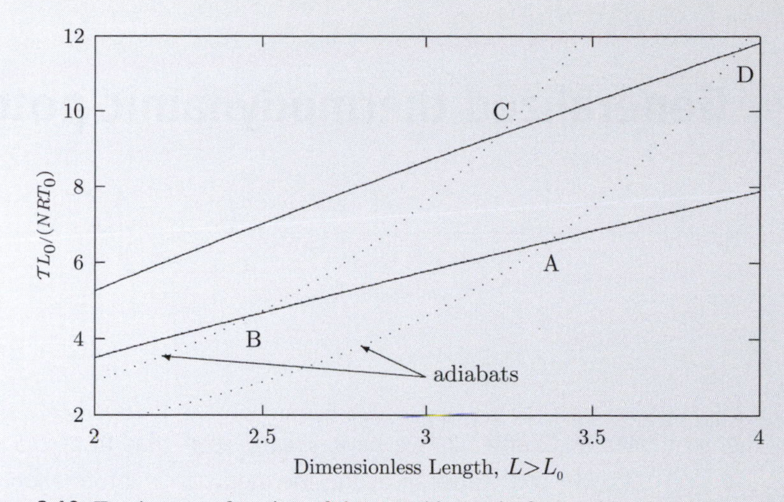

Figure 2.10 Tension as a function of the stretching ratio for a rubber band as modeled by Treloar. The solid lines are isothermal, and the dashed lines are isentropic. The rubber band may be used as a refrigerator by going clockwise around the cycle shown: A → B → C → D → A.

$$S = s_0L_0 - \frac{N_K k_B}{2}\left\{\left(1 + \frac{2}{3N_K} + \frac{10}{27N_K^2}\right)\left(\frac{L}{N_K a_K}\right)^2 - \left(2 - \frac{4}{N_K}\right)\log\left[1 - \left(\frac{L}{N_K a_K}\right)^2\right]\right\} + c_L L_0 \log\left[1 + \frac{U - u_0 L_0}{c_L L_0 T_0}\right], \tag{2.86}$$

where s_0, T_0, k_B, N_K, a_K, and c_L are constants with the same meaning as in Exercise 2.9.E. How does this expression compare with the data in Table 2.1?

2.9.H What extensibility condition does the rubber band of Example 2.9.1 satisfy? Does it satisfy the "usual" one, or is there some issue?

3 Generalized thermodynamic potentials

From the postulates in Chapter 2, we have already seen that from $U(S, V, N_1, \ldots, N_m)$ or $S(U, V, N_1, \ldots, N_m)$ we can derive *all* of the thermodynamic information about a system. For example, we can find mechanical or thermal equations of state. However, we also know that it is sometimes convenient to use other independent variables besides entropy and volume. For example, when we perform an experiment at room temperature open to the atmosphere, we are manipulating temperature and pressure, not entropy and volume. In this case, the more natural independent variables are T and P. Then, the following question arises: *Is there a function of $(T, P, N_1, \ldots, N_m)$ that contains complete thermodynamic information?* In other words, is there some function, say $G(T, P, N_1, \ldots, N_m)$, from which we could derive all the equations of state? It turns out that such functions do exist, and that they are very useful for solving practical problems.

In the first section we show how to derive such a function for any complete set of independent variables using something called Legendre transforms. In this book we introduce three widely used potentials: the enthalpy $H(S, P, N_1, \ldots, N_m)$, the Helmholtz potential $F(T, V, N_1, \ldots, N_m)$, and the Gibbs free energy $G(T, P, N_1, \ldots, N_m)$. These functions which contain complete thermodynamic information using independent variables besides S, V, and $N_1, \ldots, N_m$ are called **generalized thermodynamic potentials**.

These quantities are essential for engineering or applied thermodynamics. For example, we already know that an isolated system attains equilibrium when the entropy is maximized. However, a system in contact with a thermal reservoir attains equilibrium when the Helmholtz potential F is minimized. We show below (Examples 3.2.1 and 3.2.2) that the work necessary to compress any gas isothermally is just the change in $F(T, V, N)$. In a fuel cell, $G(T, P, N)$ is important. Or, in the open (flowing) systems considered in Chapter 5, we see that another such function (called enthalpy) is also important. In other words, we need not consider the entropy of the reservoir explicitly to find equilibrium.

In Section 3.2, we show how subsystems in contact with thermal and/or mechanical reservoirs are handled more naturally using the generalized potentials. In Section 3.6 we show how to express any derivative in terms of the measurable quantities introduced in Section 3.5.

In Section 3.3, we see that the generalized thermodynamic potentials have a set of relationships between derivatives of the independent variables called the Maxwell relations. These relations are essential for the thermodynamic manipulations discussed in the next chapter, and are used in the remainder of the text.

3.1 LEGENDRE TRANSFORMS

Recall Example 2.6.1, where we found $U(T, V, N)$ from $S(U, V, N)$. Although we could find $P(T, V, N)$ from the S expression, it is not possible to find it from $U(T, V, N)$. Hence, we lose information when we go from the fundamental relation to the equation of state. In this section we show why that is, and introduce a method to find a function $F(T, V, N)$ that has all the information that was in our original fundamental relation $S(U, V, N)$.

The naïve approach

On first reflection, we might think that finding a function with complete thermodynamic information that has T instead of S as an independent variable is straightforward. We would naïvely take the derivative of $U(S)$ with respect to S to find $T(S)$, invert, and insert $S(T)$ into $U(S(T))$ to find $\tilde{U}(T)$. However, we can show that such a naïve approach would lead to a loss of information along the way.

Consider a fundamental energy relation $U_1(S)$ as shown on the left-hand side of Figure 3.1 that we wish to manipulate in the naïve manner. We take the derivative to find the thermal equation of state:

$$T = T_1(S) := \frac{\partial U_1}{\partial S}. \tag{3.1}$$

We invert this expression (somehow) to find $S = T_1^{-1}(T)$. If we now plug this (inverted) expression into the original fundamental energy relation, we find

$$\tilde{U}_1(T) = U_1(T_1^{-1}(T)), \tag{3.2}$$

which is shown on the right-hand side of Figure 3.1.

Now suppose that we had a somewhat different fundamental energy relation $U_2(S) := U_1(S+C_0)$, also shown on the left-hand side of Figure 3.1, where C_0 is not a function of S. Note that, at a given value for U, the two fundamental relations have the same slope and, hence, the same temperature. Let us find the expression $\tilde{U}_2(T)$ and prove that it equals $\tilde{U}_1(T)$. The thermal equation of state for the new fundamental energy relation is

$$T = \frac{\partial U_2(S)}{\partial S} = \frac{\partial U_1(S+C_0)}{\partial S} = \frac{\partial (S+C_0)}{\partial S}\frac{\partial U_1(S+C_0)}{\partial (S+C_0)} = T_1(S+C_0). \tag{3.3}$$

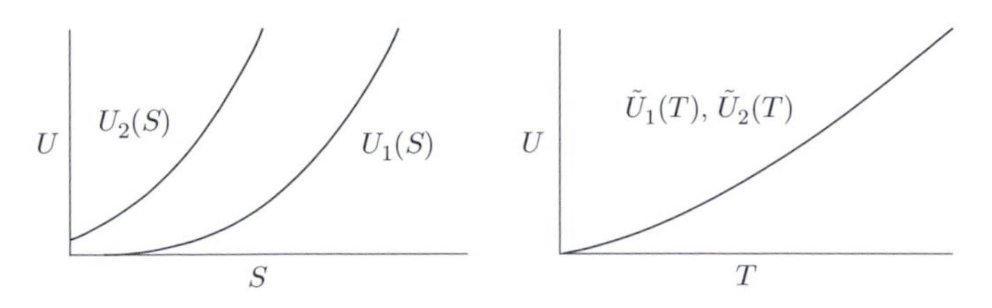

Figure 3.1 Left-hand side: two different fundamental energy relations as functions of entropy, where $U_2(S) = U_1(S + C_0)$. Right-hand side: the internal energy as a function of temperature as predicted by the two fundamental energy relations. Since these curves are indistinguishable, there is no way to recreate the original fundamental relations from these equations of state.

We invert this expression to find $S(T) = T_1^{-1}(T) - C_0$, which we can insert into the fundamental energy relation to find

$$\begin{aligned}\tilde{U}_2(T) &= U_2(S(T)) = U_1(S(T) + C_0) \\ &= U_1(T_1^{-1}(T)) = \tilde{U}_1(T).\end{aligned} \tag{3.4}$$

Note that the resulting functions $\tilde{U}_1$ and $\tilde{U}_2$ are identical, as can be seen on the right-hand side of Figure 3.1, although the fundamental relations were different. In other words, our method does not distinguish between the two different fundamental energy relations U_1 and U_2 since both lead to the same function of T. Information is lost in the transformation. That is why we call $U(T, V)$ an equation of state, rather than a fundamental relation.

Put another way, note from Example 2.6.1 that there is no way to derive the van der Waals PVT equation of state from (2.34) alone – one needs the fundamental relation (2.21).

The correct approach

Now we consider a Legendre transformation, where we wish to convert a relationship of the kind $U = U(S)$, to find some function $F(T)$ of the slope T

$$T(S) := \frac{\partial U(S)}{\partial S} \tag{3.5}$$

that contains the same information as $U(S)$. In other words, knowing $F(T)$ we want to be able to find $U(S)$ uniquely.

Instead of representing the curve as a family of paired points $U = U(S)$, we may also represent the curve as a family of tangent lines, as shown in Figure 3.2. Each curve has a tangent line of slope T at any given point on the curve. Using the equation of a line $y = mx + b$, we can find a relationship between the intercept F and the slope T,

$$F(S) := U(S) - T(S)S. \tag{3.6}$$

From Eqns. (3.5) and (3.6), we can eliminate S to find $F(T)$. To reconstruct $U(S)$ from $F(T)$, note that

$$S(T) = -\frac{\partial F}{\partial T}, \tag{3.7}$$

so that we can solve Eqn. (3.6) for U,

$$U(T) = F(T) - T\frac{\partial F}{\partial T}. \tag{3.8}$$

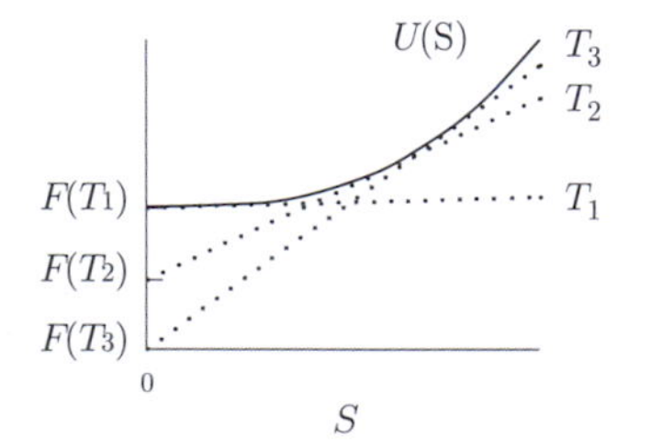

Figure 3.2 The internal energy $U(S)$ represented as both a curve and a family of tangential lines. The straight lines each have slope T_i and intercept $F(T_i)$. The slope is the first derivative of U at some point S_i. Knowing F for all T is sufficient to reconstruct the curve.

Table 3.1 How to recover the fundamental energy relation $U(S, V)$ from the generalized thermodynamic potentials F, G, and H as functions of their canonical independent variables. Given a fundamental relation of the quantity in the first column, use the equations in the second and third columns to eliminate the variables in the fourth column, leaving $U = U(S, V)$.

From	Find canonical independent variables	To find internal energy $U(S, V)$	Eliminate
$H(S,P)$	$V(S,P) = (\partial H/\partial P)_{S,N}$	$U(S,P) = H(S,P) - PV(S,P)$	P
$F(T,V)$	$S(T,V) = -(\partial F/\partial T)_{V,N}$	$U(T,V) = F(T,V) + TS(T,V)$	T
$G(T,P)$	$S(T,P) = -(\partial G/\partial T)_{P,N}$	$U(T,P) = G(T,P) + TS(T,P) - PV(T,P)$	T,P
	$V(T,P) = (\partial G/\partial P)_{T,N}$		

From Eqns. (3.7) and (3.8) we can eliminate T to find $U(S)$. Hence, knowing $F(T)$ allows us to reconstruct the entire curve.

Therefore, we define the **Helmholtz potential** F to be the Legendre transform of $U(S)$ by Eqn. (3.6). Similarly, if we perform the Legendre transform on $U(V)$, we obtain the **enthalpy** H as the Legendre transform; we can perform the Legendre transform simultaneously on both arguments of $U(S, V)$ to obtain the **Gibbs free energy** G:

$$H(S,P,N_1,\ldots,N_m) := U + PV, \tag{3.9}$$

$$F(T,V,N_1,\ldots,N_m) := U - TS, \tag{3.10}$$

$$G(T,P,N_1,\ldots,N_m) := U - TS + PV. \tag{3.11}$$

Note that the sign is different for the transformation from V to P, because the slope of $U(V)$ is $-P$. For each of the three transforms, we can reconstruct the original fundamental energy relation by Eqns. (3.7) and (3.8). The results of the inversion are summarized in Table 3.1. For a given general potential, only one set of independent variables gives complete thermodynamic information. We call the members of this set the **canonical independent variables** for that potential.

EXAMPLE 3.1.1 Find the Helmholtz potential as a function of temperature and volume for a pure-component, simple, ideal gas, which has the fundamental entropy relation

$$S = Ns_0 + NR\log\left[\left(\frac{U}{u_0}\right)^c\left(\frac{V}{v_0}\right)\left(\frac{1}{N^{1+c}}\right)\right], \tag{3.12}$$

where s_0 is the specific entropy at the reference point (u_0, v_0, N_0).

Solution. From the definition of the Helmholtz potential Eqn. (3.10), we see that we need both U and S as functions of T and V. Hence, we need the thermal equation of state. Since we have a fundamental entropy formulation, we use the alternative definition of temperature Eqn. (2.31) to find the thermal equation of state

$$\frac{1}{T} := \left(\frac{\partial S}{\partial U}\right)_{V,N} = \frac{cNR}{U}, \tag{3.13}$$

which we may use to find $U = U(T)$. When Eqn. (3.13) is inserted into the fundamental entropy relation for an ideal gas, Eqn. (3.12), we obtain an expression for $S(T, V, N)$,

$$S = Ns_0 + NR \log\left[\left(\frac{T}{T_0}\right)^c \left(\frac{V}{Nv_0}\right)\right], \tag{3.14}$$

where $T_0 := u_0/cR$. We now have the sought-after expressions for $U = U(T, V)$ and $S = S(T, V)$. When we insert the expression we found for $U(T)$, Eqn. (3.13), and Eqn. (3.14) into the definition for F, Eqn. (3.10), we find

$$F = Nf_0 + (cNR - Ns_0)(T - T_0) - NRT \log\left[\left(\frac{T}{T_0}\right)^c \left(\frac{V}{Nv_0}\right)\right], \tag{3.15}$$

where $f_0 := u_0 - s_0T_0$ is the specific Helmholtz potential at the reference state (T_0, v_0). Equation (3.15) contains all of the thermodynamic information about a simple ideal gas, as does Eqn. (3.12). □

EXAMPLE 3.1.2 The Helmholtz potential for a system is given as

$$F = Nf_0 - Ns_0(T - T_0) + N \int_{T_0}^{T} \left(\frac{T' - T}{T'}\right) c_v^{\text{ideal}}(T')dT' - NRT \log\left(\frac{V}{v_0N}\right), \tag{3.16}$$

where s_0, T_0, f_0, and v_0 are constants with straightforward interpretation, and $c_v^{\text{ideal}}(T)$ is some known function of temperature. Can you find the fundamental energy relation?

Solution. We use the prescription in Table 3.1. Using the second row, we find $S(T, V)$ from Eqn. (3.16):

$$S = -\left(\frac{\partial F}{\partial T}\right)_{V,N} = Ns_0 + N \int_{T_0}^{T} \frac{c_v^{\text{ideal}}(T')}{T'} dT' + NR \log\left[\frac{V}{v_0N}\right]. \tag{3.17}$$

We used the Leibniz rule for differentiating an integral, Eqn. (A.25) in Appendix A, to obtain the second line. We find an expression for U from the second column of Table 3.1:

$$U = F + TS$$
$$= Nf_0 + Ns_0T_0 + N\int_{T_0}^{T} c_v^{\text{ideal}}(T')dT'. \tag{3.18}$$

If we could invert Eqn. (3.17) to find $T(S, V, N)$, we could insert this expression into the upper limit of the integral in Eqn. (3.18) to obtain $U(S, V, N)$, and we would have finished. However, practically speaking, the inversion is impossible without specifying $c_v^{\text{ideal}}(T)$, and in practice is difficult to do for nearly all expressions $c_v^{\text{ideal}}(T)$. This is as close as we can come to finding a fundamental energy relation, since Eqns. (3.17) and (3.18) are sufficient to determine $U(S, V, N)$. □

This example shows that it is more natural to use F when T and V are the independent variables; the single equation (3.16) gives complete thermodynamic information, whereas we require both Eqn. (3.17) and Eqn. (3.18) in order to express the fundamental energy relation.

Note that we have internal energy as a function of temperature only, $U = U(T)$, in the thermal equation of state, Eqn. (3.18), but independent of volume. We see in Example 3.1.3 below that the mechanical equation of state is simply the ideal-gas law. Hence, Eqn. (3.16) is the fundamental Helmholtz relation for a *general* ideal gas, whose internal energy has an arbitrary dependence on temperature (specified by giving $c_v^{\text{ideal}}(T)$).

The Gibbs potential has a special relationship to the chemical potential. This relationship is made clear in the **Euler relation**, which we derive here. Recall that the first property of Postulate III requires that the internal energy be a first-order, homogeneous function of its arguments,

$$\lambda U(S, V, N_1, N_2, \ldots, N_r) = U(\lambda S, \lambda V, \lambda N_1, \lambda N_2, \ldots, \lambda N_r).$$

We take the partial derivative with respect to λ of each side of this equation (holding $S, V, N_1, \ldots, N_m$ constant), using the chain rule for partial differentiation, Eqn. (A.8) in Appendix A, on the right-hand side

$$\begin{aligned} U(S, V, N_1, N_2, \ldots, N_r) = {} & \frac{\partial(\lambda S)}{\partial \lambda}\frac{\partial U(\lambda S, \lambda V, \lambda N_1, \lambda N_2, \ldots, \lambda N_r)}{\partial(\lambda S)} \\ & + \frac{\partial(\lambda V)}{\partial \lambda}\frac{\partial U(\lambda S, \lambda V, \lambda N_1, \lambda N_2, \ldots, \lambda N_r)}{\partial(\lambda V)} \\ & + \sum_{i=1}^{r}\frac{\partial(\lambda N_i)}{\partial \lambda}\frac{\partial U(\lambda S, \lambda V, \lambda N_1, \lambda N_2, \ldots, \lambda N_r)}{\partial(\lambda N_i)}. \end{aligned} \tag{3.19}$$

The first derivative of each term on the right-hand side is straightforward to find. The second derivative is the conjugate variable of each term, the definitions of which are given by Eqns. (2.24)–(2.26). Therefore, the above equation becomes the **Euler relation**:

$$U = TS - PV + \sum_{i=1}^{r}\mu_i N_i, \quad \text{Euler relation.} \tag{3.20}$$

The Euler relation just follows from the fact that the internal energy is extensive in entropy, volume, and mole number. Recalling the definition of the Gibbs free energy Eqn. (3.11), we can write

$$G = \sum_{i=1}^{r} \mu_i N_i. \tag{3.21}$$

Hence, for a pure substance, the specific Gibbs free energy is identical to the chemical potential.

If we take the differential of each side of the Euler relation (3.20), and subtract from it the differential for internal energy, Eqn. (2.27), we obtain the well-known **Gibbs–Duhem relation**:

$$0 = S\,dT - V\,dP + \sum_{i=1}^{r} N_i\,d\mu_i, \quad \text{Gibbs–Duhem relation.} \tag{3.22}$$

Finally, we note that one may also take the Legendre transform of the entropy. The resulting functions are called **Massieu functions** [21, Section 5-4]. Exercise 3.3.D considers a transform on N, for example. A systematic way to handle any such transform may be found in the textbook by Modell and Reid and the later edition by Tester and Modell [87, 133, Sections 5-4 and 5].

EXAMPLE 3.1.3 Prove that the Helmholtz potential given by Eqn. (3.16),

$$F = Nf_0 - Ns_0(T - T_0) + N\int_{T_0}^{T}\left(\frac{T' - T}{T'}\right)c_v^{\text{ideal}}(T')dT' - NRT\log\left(\frac{V}{v_0 N}\right),$$

leads to the ideal-gas mechanical equation of state.

Solution. Using Eqn. (3.52), we can find the mechanical equation of state from the Helmholtz potential,

$$\begin{aligned} P &= -\left(\frac{\partial F}{\partial V}\right)_{T,N} \\ &= \frac{NRT}{V}, \end{aligned}$$

which is the ideal-gas law. □

3.2 EXTREMUM PRINCIPLES FOR THE POTENTIALS

Our postulates in Chapter 2 told us that equilibrium states of an isolated system are determined by the maximization of the entropy within the unconstrained, independent variables. In Section 3.1 we saw that the generalized thermodynamic potentials F, H, and G as functions of their canonical independent variables are equivalent to either the fundamental energy or the fundamental entropy relations. Therefore, it would be natural to wonder whether there are extremum principles for the internal energy and the generalized potentials – like entropy maximization for an isolated system. In this section, we find these principles, and show how they can be very useful for finding the equilibrium states of systems that are kept isothermal or isobaric.

Before considering the potentials, we first need to prove that entropy *maximization* for a closed system,

$$\left(\frac{\partial S}{\partial X}\right)_{U,V,N} = 0, \quad \left(\frac{\partial^2 S}{\partial X^2}\right)_{U,V,N} < 0, \tag{3.23}$$

is mathematically equivalent to internal-energy *minimization* at constant entropy,

$$\left(\frac{\partial U}{\partial X}\right)_{S,V,N} = 0, \quad \left(\frac{\partial^2 U}{\partial X^2}\right)_{S,V,N} > 0, \tag{3.24}$$

where X is the unconstrained variable (e.g. $U^{(1)}$ in Section 2.7). Graphically, we are saying the following: the postulates state that X_{eq} is the equilibrium state in Figure 3.3(a), and we are now going to prove that Figure 3.3(b) follows. For example, in Section 2.7, $X \equiv U^{(1)}$, and the final energy in container 1 is X_{eq}. We proved that Figure 3.3(a) means equal temperature in the two containers. We now show that this state is also an internal-energy minimum at fixed entropy.

To show that an extremum in entropy at constant internal energy is an extremum in internal energy, we begin with the first derivative of internal energy,

$$\begin{aligned}\left(\frac{\partial U}{\partial X}\right)_S &= \frac{\partial(U,S)}{\partial(X,S)} \\ &= \frac{\partial(U,S)}{\partial(U,X)} \Big/ \frac{\partial(X,S)}{\partial(U,X)} \\ &= -\left(\frac{\partial S}{\partial X}\right)_U \Big/ \left(\frac{\partial S}{\partial U}\right)_X \\ &= -T\left(\frac{\partial S}{\partial X}\right)_U,\end{aligned} \tag{3.25}$$

where it is implied that V and N are held constant in all derivatives. Since the entropy is an extremum at equilibrium and $(\partial S/\partial X)_U = 0$, Eqn. (3.25) shows that the internal energy is also

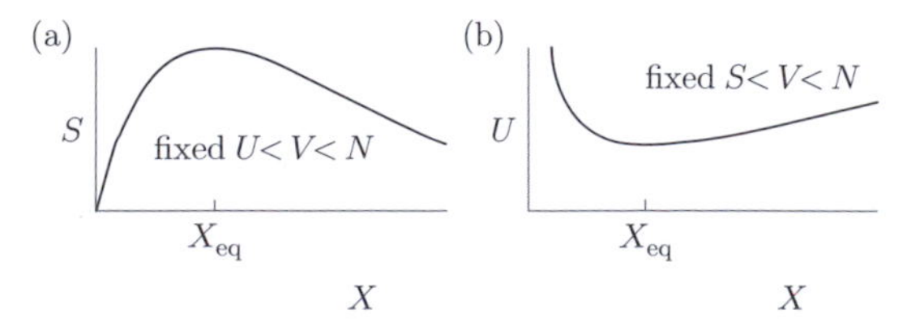

Figure 3.3 Entropy as a function of the unconstrained variable X at fixed internal energy (a), and the internal energy as a function of the unconstrained variable X at fixed entropy (b).

an extremum. To show that a *maximum* in S corresponds to a *minimum* in U, it is convenient to introduce temporarily the notation $\psi := (\partial U/\partial X)_S$. Then, we can write

$$\begin{aligned}\left(\frac{\partial^2 U}{\partial X^2}\right)_S &= \left(\frac{\partial \psi}{\partial X}\right)_S \\ &= \frac{\partial(\psi, S)}{\partial(X, S)} \\ &= \frac{\partial(\psi, S)}{\partial(X, U)}\frac{\partial(X, U)}{\partial(X, S)} \\ &= T\left[\frac{1}{T}\left(\frac{\partial \psi}{\partial X}\right)_U - \left(\frac{\partial S}{\partial X}\right)_U\left(\frac{\partial \psi}{\partial U}\right)_X\right] \\ &= \left(\frac{\partial \psi}{\partial X}\right)_U, \quad \text{when } S \text{ is maximum.}\end{aligned} \tag{3.26}$$

Note that the second Jacobian on the third line is just T. We expand the determinant of the first one to get the fourth line. The last line follows because we are considering the state for which S is a maximum, so $(\partial S/\partial X)_U = 0$. To finish the proof, we insert the definition of ψ into the last line above, and use Eqn. (3.25) to obtain

$$\begin{aligned}\left(\frac{\partial^2 U}{\partial X^2}\right)_S &= \left(\frac{\partial}{\partial X}\left(\frac{\partial U}{\partial X}\right)_S\right)_U \\ &= -T\left(\frac{\partial^2 S}{\partial X^2}\right)_U - \left(\frac{\partial S}{\partial X}\right)_U\left(\frac{\partial T}{\partial X}\right)_U \\ &= -T\left(\frac{\partial^2 S}{\partial X^2}\right)_U, \quad \text{when } S \text{ is maximum.}\end{aligned} \tag{3.27}$$

We used Eqn. (3.25) to obtain the second line, and the fact that entropy is an extremum to obtain the last line. Equation (3.27) shows that, when S is maximized for a closed system, the internal energy is minimized at constant entropy.

To characterize the potentials, we think about how to make a process isothermal in practice. We do this by splitting an isolated system into our subsystem of interest and the rest of the isolated system, which we call a heat reservoir, which is at temperature T_{res}. The subsystem of interest and the heat reservoir are not in mechanical contact (i.e. the pressure is not necessarily the same as the pressure in the reservoir, and $dV_{res} = dV_{sys} = 0$). A **heat reservoir** is a large subsystem whose temperature cannot be changed appreciably by the interactions with the subsystem of interest. For example, a reaction taking place in a single living cell in the human body might be a reasonable approximation of a system (the cell) in contact with a heat reservoir (the body). From now on we call this subsystem of interest simply the system. We show that the equilibrium state of the system is an extremum in the Helmholtz potential of the system *only*.

Since the system is at equilibrium, we know from Eqn. (2.69) that the temperature of the system is $T_{sys} = T_{res}$. From our postulates, we have proven that an equilibrium state is determined by the minimization of U at constant S, Eqn. (3.24). Since internal energy is extensive, we can write for our combined system plus reservoir

$$\begin{aligned} 0 &= \left(\frac{\partial U}{\partial X}\right)_{S,V,N}, \\ &= \left(\frac{\partial U^{sys}}{\partial X}\right)_{S,V,N} + \left(\frac{\partial U^{res}}{\partial X}\right)_{S,V,N} \\ &= \left(\frac{\partial U^{sys}}{\partial X}\right)_{S,V,N} + T^{res}\left(\frac{\partial S^{res}}{\partial X}\right)_{S,V,N} - P\left(\frac{\partial V^{res}}{\partial X}\right)_{S,V,N} + \mu\left(\frac{\partial N^{res}}{\partial X}\right)_{S,V,N}, \end{aligned}$$

where we have used the differential for U to obtain the third line. However, we now use the fact that $T^{sys} = T^{res}$ is a constant, and that the mole number and volume of the reservoir and system must also be constants. The total system is at equilibrium, so $dS_{tot} = 0$, and hence $dS^{res} = -dS^{sys}$, and we find

$$\begin{aligned} 0 &= \left(\frac{\partial U^{sys}}{\partial X}\right)_{T^{sys},V^{sys},N^{sys}} - T^{sys}\left(\frac{\partial S^{sys}}{\partial X}\right)_{T^{sys},V^{sys},N^{sys}} \\ &= \left(\frac{\partial}{\partial X}\left[U^{sys} - T^{sys}S^{sys}\right]\right)_{T^{sys},V^{sys},N^{sys}} \\ &= \left(\frac{\partial F^{sys}}{\partial X}\right)_{T^{sys},V^{sys},N^{sys}}. \end{aligned} \tag{3.28}$$

The last line follows from the definition of F. In a similar way, one can prove that this extremum is a minimum by examining the second-order derivative of F with respect to X.

It is worthwhile to reflect on the importance of the above result. It means that the equilibrium state of a system in contact with a temperature reservoir can be found simply by studying the Helmholtz potential of the system – *independently of what is happening in the reservoir*.

Similarly, one can show that a system in contact with a pressure reservoir reaches equilibrium when $dH_{sys} = 0$, and a system in contact with both a pressure and a temperature reservoir reaches equilibrium when $dG_{sys} = 0$. In Chapter 4 we show how stability requires that these functions be minimized in the intensive variables. The results are summarized in Table 3.2.

Table 3.2 When considering a system in mechanical or thermal contact with a larger system that acts as a bath, one must consider the appropriate potential for the system only.

Type of system	Equilibrium attained when
Isolated system	Entropy of system is maximized, or energy of system is minimized
System + heat bath	Helmholtz potential of system is minimized
System + pressure bath	Enthalpy of system is minimized
System + heat and pressure baths	Gibbs potential of system is minimized

EXAMPLE 3.2.1 Prove that the amount of work done on a system in contact with a heat bath is equal to the change in the system's Helmholtz potential, *if the process is done reversibly.*

Solution. We first find the differential for F, which is

$$\begin{aligned} dF &= d(U - TS) \\ &= dU - T\,dS - S\,dT \\ &= -S\,dT - P\,dV + \mu\,dN, \end{aligned} \tag{3.29}$$

where we used the definition for F in the first line, and the differential for U to obtain the last line. Since the system is isothermal, the first term on the right-hand side is zero. The third term is also zero, since the system is closed. Since $đW_{\text{rev}} = -P\,dV$, we have proven our result

$$đW_{\text{rev}} = dF, \quad \text{isothermal, closed system.} \tag{3.30}$$

Therefore, one need only find the change in free energy of an isothermal system to determine the amount of reversible work necessary. This is especially useful if a thermodynamic chart is available.

Oops! Did you notice how we called F the free energy? Many books, papers, and technical reports use that term loosely in this context. □

EXAMPLE 3.2.2 Using the thermodynamic diagram for ammonia on p. 176, find the work necessary to compress 3 l of ammonia at 4 MPa and 200 °C isothermally to 25 MPa. Find the work necessary for a compression performed adiabatically to the same pressure.

Solution. From Example 3.2.1 we know that the amount of work necessary to compress a substance isothermally is the change in the Helmholtz potential,

$$W_{\text{rev}} = \Delta F, \quad \text{isothermal, closed.} \tag{3.31}$$

We see from the thermodynamic diagram that we can obtain the enthalpy, entropy, and density for known temperature and pressure. This is sufficient information to obtain the Helmholtz potential, using its definition

$$F := U - TS = H - PV - TS = H - \frac{PM_{\text{w}}N}{\rho} - TS. \tag{3.32}$$

We also used the definition of the enthalpy, and converted from volume V to density ρ. From the diagram, we get the initial conditions

$$\begin{aligned} H[T &= 200\,^\circ\text{C}, P = 4\,\text{MPa}] = 915\,\text{kJ/kg}, \\ S[T &= 200\,^\circ\text{C}, P = 4\,\text{MPa}] = 10.55\,\text{kJ/(kg}\cdot\text{K)}, \\ \rho[T &= 200\,^\circ\text{C}, P = 4\,\text{MPa}] = 19\,\text{kg/m}^3. \end{aligned} \tag{3.33}$$

Therefore, we obtain the initial Helmholtz potential to be

$$\begin{aligned} f_{\mathrm{i}} &= (915\,\mathrm{kJ/kg}) - \frac{(4\times 10^3\,\mathrm{kPa})(1\,\mathrm{kJ/(kPa\cdot m^3)})}{19\,\mathrm{kg/m^3}} \\ &\quad - (273 + 200\,\mathrm{K})(10.55\,\mathrm{kJ/(kg\cdot K)}) \\ &= -4286\,\mathrm{kJ/kg}. \end{aligned} \tag{3.34}$$

Note that the particular choices adopted for zero enthalpy and zero entropy in creating the diagram make the Helmholtz potential negative here. The (arbitrary) choice for zeroing the internal energy cancels out when taking the difference. From the fifth postulate, we know that one cannot arbitrarily choose entropy to be zero for some state. Therefore, the values for entropy given in the diagram should all have an additive value for S_0 at the reference point. Since we take a difference here *at constant temperature*, this constant will vanish, so our answer is OK.

For the final state, $T = 200\,°\mathrm{C}$, $P = 25\,\mathrm{MPa}$, we likewise find

$$\begin{aligned} H[T &= 200\,°\mathrm{C}, P = 25\,\mathrm{MPa}] = 430\,\mathrm{kJ/kg}, \\ S[T &= 200\,°\mathrm{C}, P = 25\,\mathrm{MPa}] = 8.75\,\mathrm{kJ/(kg\cdot K)}, \\ \rho[T &= 200\,°\mathrm{C}, P = 25\,\mathrm{MPa}] = 220\,\mathrm{kg/m^3}. \end{aligned} \tag{3.35}$$

Therefore, the final specific Helmholtz potential is

$$\begin{aligned} f_{\mathrm{f}} &= (430\,\mathrm{kJ/kg}) - \frac{(25\times 10^3\,\mathrm{kPa})(1\,\mathrm{kJ/(kPa\cdot m^3)})}{220\,\mathrm{kg/m^3}} \\ &\quad - (273 + 200\,\mathrm{K})(8.75\,\mathrm{kJ/kg\cdot K}) \\ &= -3822\,\mathrm{kJ/kg}, \end{aligned} \tag{3.36}$$

and the reversible work required is

$$\begin{aligned} W_{\mathrm{rev}} &= \Delta F = \Delta f\, N = \Delta f\, V\rho \\ &= (-3822 + 4286\,\mathrm{kJ/kg})(3\,\mathrm{l})(19\,\mathrm{kg/m^3})(10^{-3}\,\mathrm{m^3/l}) \\ &= 26.4\,\mathrm{kJ}. \end{aligned} \tag{3.37}$$

Had we assumed that the gas were ideal, we would have obtained the result

$$\begin{aligned} W_{\mathrm{IG}} &= -NRT\log\left(\frac{V_{\mathrm{f}}}{V_{\mathrm{i}}}\right) = P_{\mathrm{i}}V_{\mathrm{i}}\log\left(\frac{\rho_{\mathrm{f}}}{\rho_{\mathrm{i}}}\right) \\ &= (4000\,\mathrm{kPa})(0.003\,\mathrm{m^3})(1\,\mathrm{kJ/(kPa\cdot m^3)})\log\left(\frac{220\,\mathrm{kg/m^3}}{19\,\mathrm{kg/m^3}}\right) \\ &= 29.4\,\mathrm{kJ}. \end{aligned} \tag{3.38}$$

Hence, the ideal-gas assumption introduces an error of only 11%.

To find the adiabatic compression, we recall that an adiabatic, reversible compression is also isentropic. Hence, the final state must have $P_{\mathrm{f}} = 25\,\mathrm{MPa}$, and specific entropy equal to the initial $10.55\,\mathrm{kJ/(kg\cdot K)}$. From the diagram, we see that the final state is $T_{\mathrm{f}} = 395\,\mathrm{K}$,

$\rho_f = 80\,\text{kg/m}^3$, and $h_f = 1350\,\text{kJ/kg}$. The required work is just the change in internal energy. From the definition of enthalpy, we can find u_i:

$$\begin{aligned} u_i &= h_i - P_i v_i = h_i - \frac{P_i}{\rho_i} \\ &= (915\,\text{kJ/kg}) - \frac{4000\,\text{kPa}}{19\,\text{kg/m}^3} \\ &= 704.5\,\text{kJ/kg}. \end{aligned} \tag{3.39}$$

Similarly, we find the final specific internal energy to be $u_f = 1038$ kJ/kg. Hence, the required work for the adiabatic compression is

$$\begin{aligned} W_{\text{adiab}} &= \Delta u N = \Delta u\, \rho_i V_i \\ &= (1037.5 - 704.5\,\text{kJ/kg})(19\,\text{kg/m}^3)(0.003\,\text{m}^3) \\ &= 19.0\,\text{kJ}. \end{aligned} \tag{3.40}$$

This value is smaller, because the final volume is much larger than that for the isothermal compression. □

The following example analyzes the maximum theoretical efficiency of a heat engine, which illustrates the importance of enthalpy in combustion.

EXAMPLE 3.2.3 A heat engine exploits heat flowing from a hot object to a colder one. Typically, the energy is provided by burning fuel, and the environment is the colder object. An example of such an engine is the steam cycle discussed in detail in Section 5.2.1. Find the maximum theoretical work available from one mole of hydrogen gas as a function of temperature.

Solution. A simplified schematic diagram of a heat engine is shown in Figure 3.4. The process is broken down into two subsystems as drawn. We begin with an energy balance on the furnace, assuming that the work on the environment is purely mechanical and reversible:

$$\begin{aligned} dU &= đQ_{\text{rev}} - P\,dV \\ d(U + PV) &= đQ_{\text{rev}} \\ dH &= đQ_{\text{rev}} = -đQ_f \\ \rightarrow Q_f &= -\Delta H_{\text{rxn}}(T). \end{aligned} \tag{3.41}$$

The second line follows because the system is operated isobarically. The third follows from the definition of H and noting that the sign convention for Q_f is different from the usual, as is shown in Figure 3.4. The last line comes from integrating the differential from the initial state to the final state. The difference in enthalpy between the products and the reactants, both at temperature T and pressure P, is called the **heat of reaction**, indicated by $\Delta H_{\text{rxn}}(T)$.

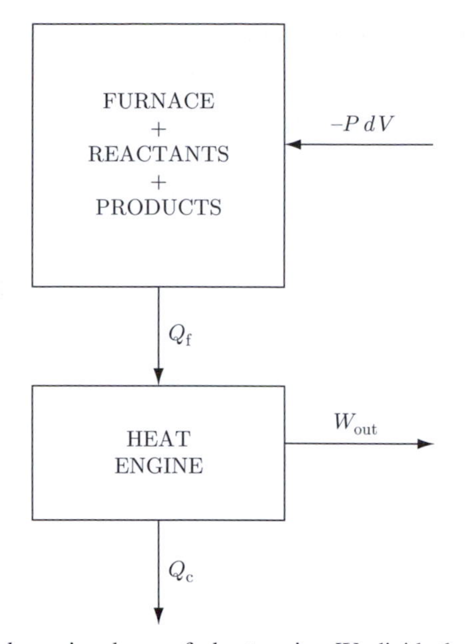

Figure 3.4 A simplified thermodynamic scheme of a heat engine. We divide the overall system into two subsystems. The top subsystem contains the furnace, reactants, and products. It operates isothermally and isobarically, so that it might exchange mechanical work with the environment W_{environ} through expansion (or contraction). It passes heat to the heat engine, which is the second subsystem. The heat engine involves cyclical behavior, so its final thermodynamic state must equal its initial state. Aside from taking heat from the furnace, it also expels heat to the environment, while producing usable mechanical work, W_{out}. To accomplish this work production, it is necessary that the reaction take place at a higher temperature than the environment: $T > T_{\text{atm}}$.

Next, performing an energy balance on the heat engine yields (being careful with the proper signs for heat and work flows, as shown in Figure 3.4)

$$dU = đQ_{\text{f}} - đQ_{\text{c}} - đW_{\text{out}}. \tag{3.42}$$

We can integrate this equation from the initial to the final state. Since the engine operates as a cycle, its final internal energy must be the same as its initial internal energy, making the left-hand side zero. Hence,

$$W_{\text{out}} = Q_{\text{f}} - Q_{\text{c}} = -\Delta H_{\text{rxn}}(T) - Q_{\text{c}}, \tag{3.43}$$

where we have used the result from the energy balance on the furnace, Eqn. (3.41).

To finish, we need to find the heat flow Q_{c}. Clearly, the smaller Q_{c}, the more work will be available. The limitation on Q_{c} comes from an overall entropy balance. Here we perform an entropy balance over just the heat engine, assuming reversible heat transfer with its environment. The entropy balance is then

$$dS_{\text{total}} = dS_{\text{environ}} + dS_{\text{engine}}, \tag{3.44}$$

where the subscripts are "total" for the Universe, "environ" for the environment, and "engine" for the engine. Still assuming reversibility, we know that the change in entropy for the environment from heat flow is $đQ_{\text{c}}/T_{\text{atm}} - đQ_{\text{f}}/T$. Note that the arrows in the sketch determine our signs here, and that we assume that the heat delivered from the furnace is at temperature T. We thus obtain

$$0 = -\frac{đQ_f}{T} + \frac{đQ_c}{T_{atm}} + dS_{engine}, \tag{3.45}$$

assuming reversibility. We integrate this equation noting that the temperatures are constant,

$$0 = \frac{Q_c}{T_{atm}} - \frac{Q_f}{T}, \tag{3.46}$$

where we have again exploited the fact that the engine operates as a cycle. We can now find the amount of work available for the heat engine:

$$W_{out} = (-\Delta H_{rxn}(T))\left(1 - \frac{T_{atm}}{T}\right). \tag{3.47}$$

While we assumed that the reaction was exothermic above, this expression actually holds for an endothermic reaction as well. In that case, each term in the parentheses is negative, and work is still available. Also note that, the greater the temperature differences, the larger the amount of work available.

These results require that the entire system operates reversibly. Hence, entropy may not be generated from heat flow across finite temperature differences, all processes must be quasi-static, there is no friction, etc. Of course, real processes always operate irreversibly. So these results allow one to estimate how close a real system is to ideal performance, or to get rough estimates when designing.

Also, many cyclical engines do not operate in this manner. An automobile engine, for example, exploits the expansion of the reaction of fuel to produce work, rather than exploiting heat flow. Hence, the analysis done here does not apply to typical combustion engines. □

3.3 THE MAXWELL RELATIONS

In the last section, we saw that the enthalpy, H, the Helmholtz potential, F, and the Gibbs free energy, G, contain complete thermodynamic information when expressed in terms of their canonical independent variables. In Chapter 2, from the definitions of temperature, pressure, and chemical potential, we have already found the total differentials for both internal energy and entropy, Eqns. (2.27) and (2.29). By simple manipulation we found for the Helmholtz potential in Example 3.2.1 the total differential

$$\begin{aligned} dF &= dU - T\,dS - S\,dT \\ &= -S\,dT - P\,dV + \sum_i^m \mu_i\,dN_i, \end{aligned}$$

where we used the differential for the internal energy Eqn. (2.27) to obtain the second line. We can perform similar manipulations on the other potentials to find their total differentials. The results are

$$dU = T\,dS - P\,dV + \sum_i \mu_i\,dN_i, \tag{3.48}$$

$$dH = T\,dS + V\,dP + \sum_i \mu_i\,dN_i, \tag{3.49}$$

$$dF = -S\,dT - P\,dV + \sum_i \mu_i\,dN_i, \tag{3.50}$$

$$dG = -S\,dT + V\,dP + \sum_i \mu_i\,dN_i. \tag{3.51}$$

From these differential expressions we can also see how to find equations of state from the generalized potential. In addition to those shown in the second column of Table 3.1, we can write

$$T = \left(\frac{\partial H}{\partial S}\right)_{P,N}, \qquad P = -\left(\frac{\partial F}{\partial V}\right)_{T,N}. \tag{3.52}$$

Therefore, all of the items of thermodynamic information can be retrieved from the generalized potentials as functions of their canonical variables, just as we have already done from the fundamental energy relation.

We can also derive some other useful, general relations, called the **Maxwell relations**. Since H, F, and G are constructed from simple manipulations of the analytic function U, they are also analytic (see Section A.1 in Appendix A). Since they are analytic functions, their second-order derivatives must be independent of the order of differentiation. Hence, we can write

$$\frac{\partial^2 H}{\partial S\,\partial P} = \frac{\partial^2 H}{\partial P\,\partial S}. \tag{3.53}$$

Recall from Appendix A that we must be careful to remember what independent variables are being held constant in second-order partial derivatives. Writing the term on the left-hand side more explicitly and using Eqn. (3.49) gives

$$\begin{aligned}\frac{\partial^2 H}{\partial S\,\partial P} &= \left(\frac{\partial}{\partial S}\left(\frac{\partial H}{\partial P}\right)_{S,N}\right)_{P,N} \\ &= \left(\frac{\partial V}{\partial S}\right)_{P,N}.\end{aligned} \tag{3.54}$$

Similarly,

$$\begin{aligned}\frac{\partial^2 H}{\partial P\,\partial S} &= \left(\frac{\partial}{\partial P}\left(\frac{\partial H}{\partial S}\right)_{P,N}\right)_{S,N} \\ &= \left(\frac{\partial T}{\partial P}\right)_{S,N}.\end{aligned} \tag{3.55}$$

Putting Eqns. (3.54) and (3.55) into Eqn. (3.53) yields our first **Maxwell relation**:

$$\left(\frac{\partial V}{\partial S}\right)_{P,N} = \left(\frac{\partial T}{\partial P}\right)_{S,N}.$$

Of course, we can use the same procedure for U, F, and G to obtain three more relations. When we do, we find the four Maxwell relations

$$\left(\frac{\partial T}{\partial V}\right)_{S,N} = -\left(\frac{\partial P}{\partial S}\right)_{V,N}, \tag{3.56}$$

$$\left(\frac{\partial V}{\partial S}\right)_{P,N} = \left(\frac{\partial T}{\partial P}\right)_{S,N}, \tag{3.57}$$

$$\left(\frac{\partial P}{\partial T}\right)_{V,N} = \left(\frac{\partial S}{\partial V}\right)_{T,N}, \tag{3.58}$$

$$\left(\frac{\partial S}{\partial P}\right)_{T,N} = -\left(\frac{\partial V}{\partial T}\right)_{P,N}. \tag{3.59}$$

It is not necessary to memorize these equations. They are straightforward to derive from the differentials, or can be generated from the thermodynamic square in the next section, when needed. Also, we use only the last relation with any regularity in this text. However, any generalized potential – including Massieu functions – can be used to generate similar relations.

It is already clear how the Maxwell relations might be useful. For example, it might be extremely difficult to measure a change in entropy with pressure for a system at constant temperature. However, the last Maxwell relation allows us to measure the change in volume with temperature at constant pressure to find the same quantity. When the Maxwell relations are used in combination with the techniques outlined in Chapter 4, we will be able to find any conceivable thermodynamic quantity in terms of such measurable quantities.

3.4 THE THERMODYNAMIC SQUARE

In this section we outline how to reconstruct all of the results of Sections 3.1–3.3 in a straightforward manner using a diagram constructed from a mnemonic device. The phrase to remember is the unpleasant sentence ***V****ery* ***F****ew* ***T****hermodynamics* ***U****nder-****G****raduate* ***S****tudents* ***H****ave* ***P****assed.*[1]

First we draw a square, and place all of the bold-faced letters at the corners and sides of the square beginning at the upper left-hand corner. Then, we draw two arrows from the corners of the square, both going upwards. The finished result is shown in Figure 3.5.

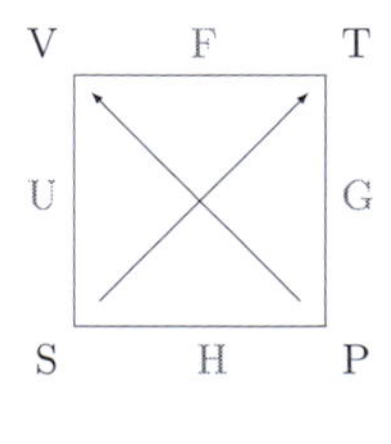

Figure 3.5 The thermodynamic square for remembering Eqns. (3.48)–(3.51) and (3.56)–(3.59).

[1] One may also choose among the phrases *Va Falloir Trimer Grave Pour Harmoniser Son Univers* and *Una Verdadera Función Thermodinamica Guarda Premisas Hondamente Simples.*

The diagram is used to construct the total differentials by the following procedure. Note that the energies are at the sides of the square, with the canonical independent variables at the adjacent corners. The independent variables make up the differentials on the right-hand side of the equation. The corresponding coefficients are connected by the arrows. Pick out the energy whose differential you would like to find. Write down the differential of that energy on the left-hand side of your equation. If an arrow points from the coefficient to the independent variable (in other words towards the side with the energy of interest), then the sign of that differential on the right-hand side is negative. Otherwise, it is positive.

EXAMPLE 3.4.1 Find the differential for internal energy using the thermodynamic square.

Solution. The square shows that the canonical independent variables for U are (S, V); hence the differentials dS and dV must show up on the right-hand side. Their coefficients according to the square are T and $-P$, respectively. Hence, we can immediately write down Eqn. (2.27):

$$dU = -P\,dV + T\,dS.$$

Note that the square does not deal with changes in mole number, so one must remember always to add $+\sum_i \mu_i\, dN_i$. □

The Maxwell relations may also be found from the square. First, note that all of the Maxwell relations involve derivatives of a variable on one corner with respect to the variable on an *adjacent* corner, while the variable on the opposite corner is held constant. The corresponding derivative in the Maxwell relation involves swapping the independent variables. The new dependent variable is opposite the new fixed variable. The sign of the relation is again given by the arrows; if the arrows both point to (or both point away from) the dependent variables, the sign is positive. If one arrow points towards one of the dependent variables, but the other arrow points away from the other dependent variable, the sign is negative.

EXAMPLE 3.4.2 Find a Maxwell relation for $(\partial S/\partial V)_T$.

Solution. We note that the fixed variable T lies opposite to the dependent variable S, and that the independent variable V lies adjacent to S, so there exists a Maxwell relation for the derivative.

The differential related to this one by a Maxwell relation must have V as the fixed variable and T as the changing variable. The intensive variable opposite V must be the dependent variable, or P. Hence, our second differential is

$$\left(\frac{\partial P}{\partial T}\right)_V.$$

Both arrows point away from the dependent variables S and P, so the sign must be the same on each. Hence, we have just found Eqn. (3.58) using the thermodynamic square. □

It is highly recommended that the reader attempt to reconstruct all of the differentials and the Maxwell relations with a blank sheet of paper and a pencil, remembering that very few thermodynamics undergraduate students have passed.

3.5 SECOND-ORDER COEFFICIENTS

Everyday experience shows that many material properties are functions of the first-order derivatives T, P, and μ_i. For example, when objects become warmer, their volume increases. Gases and liquids can shrink considerably under an increase of pressure, and ice freezes at a lower temperature when the chemical potential of salt in the ice is increased. These effects are all examples of *second-order derivatives*.[2] The three most frequently tabulated second-order derivatives for pure substances are

$$\alpha := \frac{1}{v}\left(\frac{\partial v}{\partial T}\right)_P, \quad \text{“coefficient of thermal expansion,”} \tag{3.60}$$

$$\kappa_T := -\frac{1}{v}\left(\frac{\partial v}{\partial P}\right)_T, \quad \text{“isothermal compressibility,”} \tag{3.61}$$

$$c_P := T\left(\frac{\partial s}{\partial T}\right)_P, \quad \text{“constant-pressure heat capacity.”} \tag{3.62}$$

These definitions should be memorized. Using the thermodynamic transformations introduced in Section 3.6, it is possible to relate all other second-order derivatives to these three. These quantities can be found from measurement, tables, or equations of state like those given in Appendix B and Section D.2 of Appendix D. It is straightforward to imagine experiments to measure α or κ_T. To measure C_P, we use Eqn. (2.46) to write

$$C_P := T\left(\frac{\partial S}{\partial T}\right)_P = \left(\frac{\partial Q_{\text{qs}}}{\partial T}\right)_P. \tag{3.63}$$

Therefore, if one has information on how much heat is required in order to raise the temperature of a substance by a small amount at constant pressure, the ratio of heat divided by temperature change is the heat capacity. Such measurements are routinely performed in a **differential scanning calorimeter**, or DSC. These devices raise the temperature of a substance at a controlled rate, and measure the rate of heat required for the change. If these rates are sufficiently slow that the process is quasi-static, then the ratio of $(d/dt)Q$ to $(d/dt)T$ yields the heat capacity as a function of temperature.[3]

As an aside, we point out that intuitively we expect each of these quantities to be positive. In fact, that is how the signs were chosen for the definitions. In Section 4.1 we will see that the postulates do indeed show that κ_T and c_P are required to be positive.

[2] Since pressure and temperature are first-order derivatives of the fundamental variable $U(S, V, N)$, we call quantities that are derivatives of these second-order derivatives.

[3] Glassy liquids are ones that relax extremely slowly, and can fall out of equilibrium upon cooling. Hence, most DSCs change the temperature too fast to be quasi-static for these, and spurious results can be obtained.

EXAMPLE 3.5.1 Find the isothermal compressibility for the simple van der Waals fluid.

Solution. From Examples 2.5.1 and 2.6.1, we have seen that the proposed fundamental entropy relation given in Eqn. (2.21) leads to the van der Waals mechanical equation of state,

$$P = \frac{RT}{V/N - b} - \frac{aN^2}{V^2}, \tag{2.35}$$

and the simple van der Waals thermal equation of state Eqn. (2.34). We begin with the definition of κ_T, Eqn. (3.61):

$$\begin{aligned}
\kappa_T &:= -\frac{1}{v}\left(\frac{\partial v}{\partial P}\right)_T \\
&= -\frac{1}{v(\partial P/\partial v)_T} \\
&= \frac{v^2\,(v-b)^2}{RTv^3 - 2a(v-b)^2}.
\end{aligned} \tag{3.64}$$

To obtain the second line we applied the relation Eqn. (A.18) from Appendix A. Then, in order to find the derivative in the denominator, we take the derivative of each side of the van der Waals mechanical equation of state (2.35) with respect to v, and insert the result into the second line above. After some algebraic manipulation, the third line is the result.

Note that the denominator can pass through zero for some values of v. This means that the van der Waals fluid's isothermal compressibility can sometimes be infinity! Is this result physical? The surprising answer to this question must wait until the next chapter. □

EXAMPLE 3.5.2 Find the constant-pressure and constant-volume heat capacities for the general ideal gas given in Example 3.1.2.

Solution. The definition for the constant-pressure heat capacity Eqn. (3.62) requires that we have $s(T, P)$. However, Eqn. (3.17) gives the entropy as a function of the temperature and volume. If we use the ideal-gas law found in Example 3.1.3 to eliminate volume from Eqn. (3.17), we obtain

$$S = Ns_0 + N\int_{T_0}^{T} \frac{c_v^{\text{ideal}}(T')}{T'}\,dT' + NR\log\left[\frac{RT}{Pv_0}\right], \quad \text{general ideal gas.} \tag{3.65}$$

Inserting this expression into the definition Eqn. (3.62) gives us the constant-pressure heat capacity for the general ideal gas described by Eqn. (3.16)

$$c_P(T) = R + c_v^{\text{ideal}}(T), \quad \text{general ideal gas.}$$

Note that it was necessary to use the Leibniz rule, Eqn. (A.25) from Appendix A, to obtain this result.

The constant-volume heat capacity has the logical definition

$$c_V := T\left(\frac{\partial s}{\partial T}\right)_v, \tag{3.66}$$

which, when applied to Eqn. (3.65), shows that the function "$c_v^{\text{ideal}}(T)$" is aptly named. □

$c_v^{\text{ideal}}(T)$ is actually called the "ideal constant-volume heat capacity." Polynomial expressions for several pure-component substances are given in Table D.2 in Appendix D.

The second-order derivatives are extremely useful for solving problems. In the following example, we show how to extract the work available from a hydrogen fuel cell, which operates isothermally and isobarically using the heat capacity, and a result from Section 3.2.

EXAMPLE 3.5.3 Prove that the amount of electrical work available from a fuel cell is equal to the change in the system's Gibbs potential, *if the process is carried out isothermally, isobarically, and reversibly*. Use these results to find the maximum work available from a hydrogen fuel cell utilizing the reaction

$$H_2 + \frac{1}{2}O_2 \to H_2O(g),$$

for $300\,\text{K} < T < 1000\,\text{K}$.

Solution. A fuel cell takes reactants (e.g. hydrogen and oxygen) and reacts them to make electricity. Figure 3.6 shows a picture of a fuel cell, and Figure 3.7 shows its scheme of operation. The oxidation reaction and the reduction reaction take place at different electrodes, and the transport of the electrons takes place not through the liquid phase, but rather through an electrical circuit. The circuit can then extract the work available from the reaction.

These systems are *open*, since reactants and products flow through the fuel cell. However, we will simplify the system by looking at the final state (the products + fuel cell), the initial state (the reactants + fuel cell), the work done, and the heat flow. The system is still open, however, since electrons flow across the system boundary, providing the electrical work (Figure 3.8).

For our system, we begin with the conservation-of-energy equation

$$dU = đQ + đW.$$

For our system, there are two possible kinds of work: mechanical (from compression or expansion) and electrical work. We are assuming here that everything happens reversibly. It is important to note here that we have written the electrical work as *positive* in Figure 3.8 when it provides work *to* the surroundings.

Figure 3.6 A photograph of an experimental fuel cell using methanol as a fuel. Visible in front are the fuel ports, and, in back on the left, one can see the connection to a copper electrode. Photo by William Mustain.

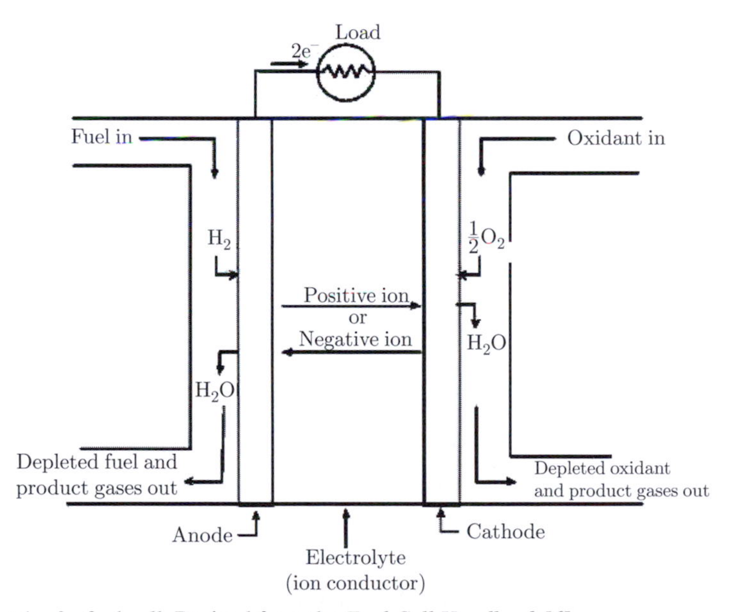

Figure 3.7 A sketch of a fuel cell. Derived from the *Fuel Cell Handbook* [6].

We begin with the differential for internal energy,

$$\begin{aligned} dU &= T\,dS - P\,dV + \sum_i \mu_i\,dN_i \\ &= T\,dS - P\,dV + \mu_e^{\text{in}}\,dN_e^{\text{in}} + \mu_e^{\text{out}}\,dN_e^{\text{out}} \\ &= T\,dS - P\,dV + (\mu_e^{\text{in}} - \mu_e^{\text{out}})dN_e^{\text{in}}, \end{aligned} \tag{3.67}$$

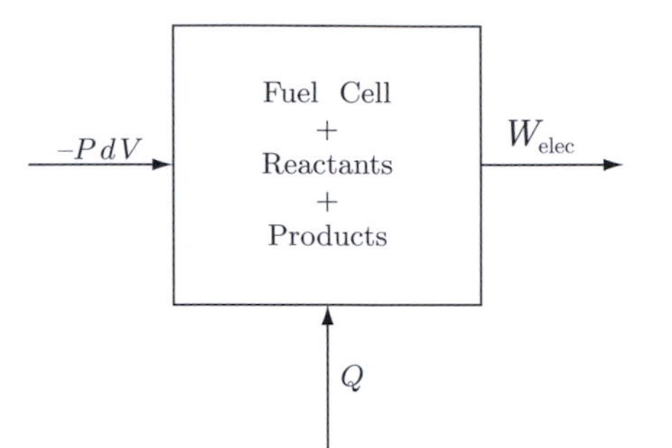

Figure 3.8 To analyze the maximum theoretical efficiency of a fuel cell, we can lump together the reactants, products, and fuel cell in our system. Here we allow no mass to pass across the boundary of our system, which operates at constant pressure and temperature. Because the system is isobaric, it may exchange work with the environment, which is assumed reversible. It also produces useful electrical work. Because it is isothermal, it may exchange heat with the environment.

since only the electrons cross the boundary, and the electrons entering one electrode must balance those exiting the other. Clearly, the first term on the right-hand side is the reversible heat exchange with the environment, the second term is the reversible work, and the third term is the electrical work:

$$\begin{aligned} đQ_{\text{rev}} &= T\,dS, \\ đW_{\text{volume}} &= -P\,dV, \\ đW_{\text{elec}} &= -\mu_e^{\text{in}}\,dN_e^{\text{in}} - \mu_e^{\text{out}}\,dN_e^{\text{out}}. \end{aligned} \tag{3.68}$$

Rearranging the energy balance yields

$$dU + P\,dV - T\,dS = -đW_{\text{elec}}. \tag{3.69}$$

We recognize that the left-hand side of this equation is just the differential for G when the system is isothermal and isobaric, since $d(U - TS + PV) = dG$:

$$dG = -đW_{\text{elec}}, \quad \text{isothermal, isobaric, reversible.} \tag{3.70}$$

Therefore, if we wish to know how much electrical work it is possible to extract from our reactants, we need only know the change in Gibbs free energy. Remember that work obtained *from* the fuel cell would be positive, so we require a reaction that has negative ΔG_{rxn}.

We may use experimental data available from JANAF [25], or from the **NIST Chemistry Web Book** to plot the specific Gibbs potential, which is the maximum allowed work per mole of hydrogen available from the fuel cell. The relevant available data are the heats of formation of each compound $\Delta H^\circ_{\text{f},i}[T_{\text{ref}}]$ at a reference temperature T_{ref} and pressure, and the ideal heat capacities $C_{P,i}^{\text{ideal}}$. We will assume here that the pressure is sufficiently low that the gases are ideal. The heat of formation of any compound at temperature T is the difference in enthalpy between the compound and its constituent elements. For example, for water

$$\Delta H_{\text{f},\text{H}_2\text{O}}[T] := H_{\text{H}_2\text{O}}[T] - H_{\text{H}_2}[T] - \frac{1}{2}H_{\text{O}_{2(\text{g})}}[T]. \tag{3.71}$$

From this definition of the heat of formation, the heat of reaction for the hydrogen reaction is the same as the heat of formation for water. However, we now need to correct for the change in temperature.

The temperature correction is achieved through the heat capacity. Beginning with the definition for H, we write

$$\begin{aligned} dH &= d(U + PV) \\ &= dU + P\,dV + V\,dP \\ &= T\,dS + V\,dP + \mu\,dN \\ &= T\,dS, \quad \text{isobaric, isomolar,} \end{aligned} \tag{3.72}$$

where we used the chain rule of differentiation to obtain the second line, the differential for U, Eqn. (2.27), to obtain the third, and the shown restrictions to obtain the last line. Dividing by dT, and holding P and N constant, yields

$$\begin{aligned} \left(\frac{\partial H}{\partial T}\right)_{P,N} &= T\left(\frac{\partial S}{\partial T}\right)_{P,N} \\ &= C_P \\ &\cong C_P^{\text{ideal}}, \end{aligned} \tag{3.73}$$

where we used the definition of heat capacity, and our assumption of ideality. We then integrate each side over T from T_{ref} to T, and use the fundamental theorem of calculus (A.24) from Appendix A to obtain

$$H[T] \cong H[T_{\text{ref}}] + \int_{T_{\text{ref}}}^{T} C_P^{\text{ideal}}[T']dT'. \tag{3.74}$$

Using Eqn. (3.74) for each of the substances in Eqn. (3.71), the heat of formation at a reference temperature, and the heat capacities, we can find the heat of formation for water at any temperature. The result is shown in Figure 3.9.

Our analysis of the fuel cell's operation is not yet complete, because some fuel cells also require an input of heat to keep them operating isothermally. This heat input also costs money, so it must either be taken from the available work or arise from irreversibility, during which some of the fuel must be burned. We can find the heat flow necessary for our example from the conservation-of-energy equation

$$đQ_{\text{rev}} = T\,dS. \tag{3.75}$$

We use this result in the next example

$$Q = T\,\Delta S_{\text{rxn}} = \Delta H_{\text{rxn}} - \Delta G_{\text{rxn}}. \tag{3.76}$$

Figure 3.9 shows that $\Delta G_{\text{rxn}} > \Delta H_{\text{rxn}}$. Therefore, $Q < 0$, and heat flows out of the hydrogen fuel cell. Since, at these temperatures, the fuel cell is operating at above room temperature, there is no additional cost of heating or refrigerating. In Figure 3.9 we also show the maximum theoretical work possible for a heat engine that burns hydrogen, assuming $T_{\text{atm}} = 298$ K. We see that the heat engine has a higher theoretical efficiency, but only if there is an extremely cold reservoir to expel the heat.

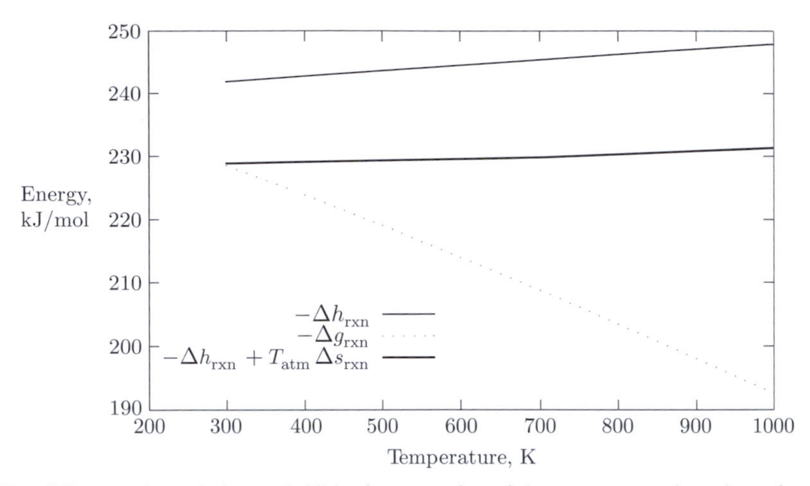

Figure 3.9 The difference in enthalpy and Gibbs free energies of the reactants and products for the hydrogen reaction considered in Example 3.5.3. In that example we show how the maximum work available for a fuel cell per mole of hydrogen is $-\Delta g_{\text{rxn}}$. In Example 3.2.3 we show that the maximum work available from a heat engine burning hydrogen is $-\Delta h_{\text{rxn}} + T_{\text{atm}}\,\Delta s_{\text{rxn}}$. Data courtesy of the NIST. □

In the previous example, we found the maximum electrical work available from a hydrogen fuel cell operating isothermally and isobarically at a range of temperatures. Fuel cells are sometimes touted as having greater theoretical efficiency than heat engines. Applying the results of the last examples to Example 3.2.3, we can then find the conditions under which work available from a fuel cell is greater than work available from a heat engine for the same amount of fuel. This comparison is made in the next example.

EXAMPLE 3.5.4 When does a fuel cell have a higher theoretical efficiency than a heat engine for a given reaction?

Solution. In Example 3.5.3, we analyzed a hydrogen fuel cell. From that analysis we discovered that the fuel cell expelled heat to its environment. Since the fuel cell was operating at temperatures well above ambient, this incurred no more cost. However, if the fuel cell *required* energy to maintain its constant temperature, or if it operated at lower temperatures, the answer would be different. Therefore, several possible outcomes are possible for the comparison.

First of all, however, the fuel cell and heat engine both require that $\Delta G_{\text{rxn}} < 0$, or the reaction will not take place (see Table 3.2). On the other hand, ΔS_{rxn} can be positive or negative. Also, knowledge of the operating temperature relative to the ambient temperature T_{atm} is important in order to determine whether we must incur the costs of heating or cooling the fuel cell. We consider the resulting four cases one at a time.

1. $\boldsymbol{\Delta S_{\text{rxn}} > 0}$ **and** $\boldsymbol{T > T_{\text{atm}}}$
   Since $\Delta S_{\text{rxn}} > 0$, the heat into the fuel cell $Q = T\,\Delta S_{\text{rxn}} > 0$ is positive. Since the temperature of the cell is above ambient, it requires heat to maintain its temperature, Eqn. (3.75).

To continue the comparison, we make a simplifying assumption that some amount of the fuel is burned outside of the fuel cell to supply the heat. Let us find the amount of fuel that would be burned. For every mole of fuel reacted in the cell, $1 - x$ is burned to keep the cell warm. The amount of heat necessary to warm the cell per mole of fuel is $Q/N = xT\,\Delta s_{\text{rxn}}$ from Eqn. (3.76). The heat available from burning the $1 - x$ moles of fuel is $Q = -\Delta h_{\text{rxn}}(1 - x)$. Setting these two expressions for Q equal tells us what fraction of fuel must be burned to keep the cell operating:

$$x = \frac{\Delta H_{\text{rxn}}}{\Delta G_{\text{rxn}}}. \tag{3.77}$$

Therefore, the amount of work available is

$$W_{\text{elec}} = -x\,\Delta G_{\text{rxn}} = -\frac{\Delta H_{\text{rxn}}}{\Delta G_{\text{rxn}}}\,\Delta G_{\text{rxn}} = -\Delta H_{\text{rxn}}. \tag{3.78}$$

Hence, the fuel cell has the same theoretical maximum of work as a heat engine, but only if there exists a heat reservoir at zero temperature.

In reality, the inefficiency of a fuel cell probably generates heat, reducing x in our analysis, but simultaneously reducing W. At any rate, the analysis here is the theoretical maximum.

2. $\boldsymbol{\Delta S_{\text{rxn}} > 0}$ **and** $\boldsymbol{T < T_{\text{atm}}}$
   Here, the fuel cell still requires heat. However, since it operates below ambient temperature, no fuel must be burned, and heat can be absorbed from the environment. Also, since $\Delta S_{\text{rxn}} > 0$,

$$W_{\text{elec}} = -\Delta G_{\text{rxn}} > -\Delta H_{\text{rxn}}, \tag{3.79}$$

   and the fuel cell has a higher maximum theoretical efficiency than a heat engine. In practice, existing fuel cells all operate at elevated temperatures, so this advantage is not exploited. The elevated temperature is required for a practical reason: faster kinetics. In theory, a fuel cell could be designed to be more efficient than a heat engine at low temperatures.
3. $\boldsymbol{\Delta S_{\text{rxn}} < 0}$ **and** $\boldsymbol{T > T_{\text{atm}}}$
   For this case, $Q < 0$, and heat is dissipated to the environment. Since the fuel cell is hotter than the environment, no additional cost is incurred for cooling. Therefore, the available work is

$$W_{\text{elec}} = -\Delta G_{\text{rxn}} < -\Delta H_{\text{rxn}}, \tag{3.80}$$

   and the fuel cell has less available work than a heat engine. A real comparison requires knowing the reaction and ambient temperatures, however.
4. $\boldsymbol{\Delta S_{\text{rxn}} < 0}$ **and** $\boldsymbol{T < T_{\text{atm}}}$
   The fuel cell still expels heat, but now it will not flow to the environment, except perhaps using a refrigerator, which costs money. This is the worst case of all. Not only is $-\Delta G_{\text{rxn}} < -\Delta H_{\text{rxn}}$, but also we do not even get all of the work from $-\Delta G_{\text{rxn}}$, because we have to pay for a refrigerator.

A summary of these results is shown in Table 3.3. Currently, all fuel cells operate at temperatures well above ambient, so only cases 1 and 3 are typical. The hydrogen fuel cell

Table 3.3 This analysis assumes isothermal, isobaric, and reversible operation of both machines. The fuel cell and furnace operate at temperature T, in an environment of temperature T_{atm}. The entropy of reaction may be either positive or negative at this temperature, as indicated.

	Fuel cell	**Heat engine**
$\Delta S_{rxn}(T) > 0, T > T_{atm}$	$W_{elec} = -\Delta H_{rxn}$	$W_{engine} = (-\Delta H_{rxn})(1 - (T_{atm}/T))$
$\Delta S_{rxn}(T) < 0, T > T_{atm}$	$W_{elec} = -\Delta G_{rxn}$ $= -\Delta H_{rxn} + T\,\Delta S_{rxn}$	$W_{engine} = (-\Delta H_{rxn})(1 - (T_{atm}/T))$
$\Delta S_{rxn}(T) > 0, T < T_{atm}$	$W_{elec} = -\Delta G_{rxn}$ $> -\Delta H_{rxn}$	$W_{engine} = (-\Delta H_{rxn})(1 - (T_{atm}/T))$
$\Delta S_{rxn}(T) < 0, T < T_{atm}$	Not possible without refrigeration	–

considered in Example 3.5.3 is a special case of the second line of Table 3.3. In Exercise 3.3.F we consider a methanol fuel cell, which is described by the first line of Table 3.3. □

In the examples here, we calculated the maximum possible work, using a closed system operating isobarically and isothermally. In Chapter 5 we consider the analogous *open* systems. There we consider systems not operating reversibly, and calculate the efficiency. Still, the analyses here form the basis for such comparisons.

3.6 THERMODYNAMIC MANIPULATIONS

The application of thermodynamics requires estimation of many sorts of derivatives. For example, in the last section we found that analysis of a fuel cell required estimates of changes in Gibbs free energy, enthalpy, and entropy. These estimates were calculated using just the constant-pressure heat capacity. Since the heat capacity is straightforward to measure experimentally, it was possible to find such data. However, sometimes we require quantities that are not so easy to measure, and are, therefore, typically not tabulated. For example, in turbine-engine design we assume that part of the cycle is done isentropically, and we need to know how the temperature changes with pressure. Hence, we need to find

$$T[P_f] = T[P_0] + \int_{P_0}^{P_f} \left(\frac{\partial T}{\partial P}\right)_{S,N} dP.$$

But it is very unlikely that you will find an equation of state for temperature as a function of pressure and entropy for many substances. On the other hand, we are aware from the structure of thermodynamics that all equations of state are derivable from a single equation (any of the fundamental relations). Hence, there must be only a finite number of equations of state that are independent, and the equation of state $T = T(S, P)$ must be expressible in terms of this independent set.

Up to this point, we have derived equations of state from fundamental relations. However, it turns out that we need integrate *only two equations of state to fix the fundamental relations for a pure substance [21, pp. 59–65]*. Therefore, if we have either two equations of state or equivalent tabulated data for our system, the derivative of interest is fixed. We can also show that, *for a single-component system, only three second-order derivatives are independent – all others can be derived from these*. This is very important, so we repeat it: *complete thermodynamic information is contained either in two equations of state or in the dependence of three second-order derivatives on temperature and pressure*. This means that sometimes we will not need to have a fundamental relation in order to make predictions – some measurement of second-order derivatives might be sufficient. This fact will be exploited in the following chapter.

In this section, we show how to reduce all derivatives (except those involving N) to the three second-order derivatives introduced in Section 3.5. The procedure suggested here might not be the most direct method for all cases, but is guaranteed to work. We use the method of Jacobians reviewed in Appendix A and the thermodynamic square introduced in Section 3.4, and one Maxwell relation.

The procedure is as follows.

1. Using the Jacobian manipulations, reduce all derivatives to those using T and P as the independent variables. This reduction may also be done by use of the three relations derived in Appendix A, but is straighforward with Jacobians.
2. All derivatives of potentials (U, H, F, G) may be reduced by use of their differentials, Eqns. (3.48)–(3.51). Recall that these are easily found from the thermodynamic square, Figure 3.5.
3. You should now be left only with derivatives of S or V with respect to T or P. The derivatives of V, and $(\partial S/\partial T)_P$ are the second-order derivatives, Eqns. (3.60)–(3.62). The other derivative, $(\partial S/\partial P)_T$, can be found from the fourth Maxwell relation, Eqn. (3.59):

$$-\left(\frac{\partial S}{\partial P}\right)_T = \left(\frac{\partial V}{\partial T}\right)_P = V\alpha.$$

Note that it is sometimes more convenient to exchange the first and second steps. The general procedure is used in the following illustrative examples.

EXAMPLE 3.6.1 Find $(\partial T/\partial P)_S$ in terms of tabulated quantities.

Solution. The first step of the procedure is to make T and P the independent variables. We use the first property of the Jacobian Eqn. (A.14) to write

$$\begin{aligned}
\left(\frac{\partial T}{\partial P}\right)_{S,N} &= \frac{\partial(T,S)}{\partial(P,S)} \\
&= \frac{\partial(T,S)}{\partial(P,T)}\frac{\partial(P,T)}{\partial(P,S)} \\
&= -\left.\frac{\partial(S,T)}{\partial(P,T)}\right/\frac{\partial(S,P)}{\partial(T,P)} \\
&= -\left.\left(\frac{\partial S}{\partial P}\right)_T\right/\left(\frac{\partial S}{\partial T}\right)_P,
\end{aligned}$$

which can also be found from Eqn. (A.20). This completes step 1. We have no derivatives of potentials, so we go on to step 3. We note that the denominator is $1/T$ times the constant-pressure heat capacity. The numerator can be eliminated using the fourth Maxwell relation,

$$-\left(\frac{\partial S}{\partial P}\right)_T = \left(\frac{\partial V}{\partial T}\right)_P = V\alpha,$$

where α is the coefficient of thermal expansion, defined by Eqn. (3.60). Hence,

$$\left(\frac{\partial T}{\partial P}\right)_S = \frac{\alpha VT}{C_P}. \tag{3.81}$$

□

EXAMPLE 3.6.2 Find $(\partial U/\partial T)_P$ in terms of tabulated quantities.

Solution. The independent variables are already T and P, so we go on to step 2. We have a derivative of the potential U, so we use its differential from the thermodynamic square. By dividing each side of the differential for U Eqn. (2.27) by dT and holding P and N constant we obtain

$$\left(\frac{\partial U}{\partial T}\right)_P = T\left(\frac{\partial S}{\partial T}\right)_P - P\left(\frac{\partial V}{\partial T}\right)_P. \tag{3.82}$$

At step 3 we recognize that we have only second-order derivatives left. Therefore, using their definitions, Eqns. (3.60) and (3.62), we find

$$\left(\frac{\partial U}{\partial T}\right)_P = C_P - PV\alpha. \tag{3.83}$$

□

EXAMPLE 3.6.3 Find the constant-volume heat capacity in terms of tabulated quantities.

Solution. The constant-volume heat capacity is defined as

$$C_V := T\left(\frac{\partial S}{\partial T}\right)_V. \tag{3.84}$$

We use the Jacobian manipulations to find

$$\begin{aligned} C_V &= T\frac{\partial(S,V)}{\partial(T,V)} \\ &= T\frac{\partial(S,V)}{\partial(T,P)} \Big/ \frac{\partial(T,V)}{\partial(T,P)} \\ &= T\begin{vmatrix} (\partial S/\partial T)_P & (\partial S/\partial P)_T \\ (\partial V/\partial T)_P & (\partial V/\partial P)_T \end{vmatrix} \Big/ \left(\frac{\partial V}{\partial P}\right)_T . \end{aligned}$$

To obtain the last line we used the definition of the Jacobian. Since there are no derivatives of potentials, we go to step 3 of the procedure. Note that all of the terms are second-order derivatives, except the upper right-hand term of the determinant; however, we can use the Maxwell relation to reduce it. Using first the Maxwell relation in expanding the determinant, and then the definitions of the second-order derivatives, we find

$$\begin{aligned} C_V &= \frac{T}{(\partial V/\partial P)_T}\left[\left(\frac{\partial S}{\partial T}\right)_P\left(\frac{\partial V}{\partial P}\right)_T + \left(\frac{\partial V}{\partial T}\right)_P^2\right] \\ &= C_P - \frac{V\alpha^2 T}{\kappa_T}. \end{aligned} \tag{3.85}$$

□

The last result simplifies to a well-known expression for an ideal gas. For the ideal gas, it is straightforward to prove that $\kappa_T = NRT/(P^2V)$ and $\alpha = NR/(PV)$. Hence, we find $C_V^{\text{ideal}} = C_P^{\text{ideal}} - R$.

EXAMPLE 3.6.4 Using the data from the steam tables, find the difference in chemical potential between steam at 2 MPa and steam at 5 MPa, both at 300 °C.

Solution. We can write the difference of the chemical potential as the integral over a derivative

$$\mu[P_1] - \mu[P_0] = \int_{P_0}^{P_1} \left(\frac{\partial \mu}{\partial P}\right)_{T,N} dP. \tag{3.86}$$

The fact that temperature and pressure are the independent variables suggests using the Gibbs potential. In fact, from the Euler relation, Eqn. (3.20), we can write

$$\mu(T,P) = g(T,P). \tag{3.87}$$

On taking the derivative with respect to P of each side

$$\begin{aligned} \left(\frac{\partial \mu}{\partial P}\right)_{T,N} &= \left(\frac{\partial g}{\partial P}\right)_T \\ &= v, \end{aligned} \tag{3.88}$$

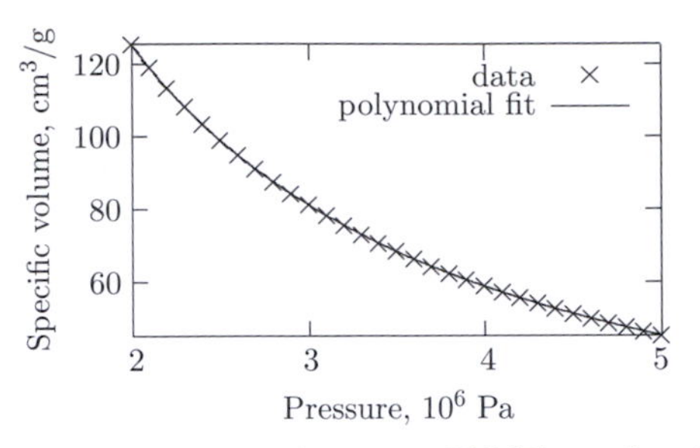

Figure 3.10 The specific volume for supersaturated steam at 300 °C as a function of pressure. Also shown is a polynomial fit to these data, Eqn. (3.90), with parameters given by Eqn. (3.92).

where we used the differential for g to obtain the second line. Equation (3.86) becomes

$$\mu[P_1] - \mu[P_0] = \int_{P_0}^{P_1} v(T, P)dP. \tag{3.89}$$

The steam tables provide numerical values for the specific volume for many values of temperature and pressure. These must be integrated numerically. There exist many simple methods for estimating the area under the curve. Here we choose to integrate a polynomial fit to the data. For other methods of integration, see [103].

Figure 3.10 shows the data taken from the steam tables in Section D.4 of Appendix D at the temperature and range of pressures of interest. Also shown is a polynomial fit to the data. We find that the expression

$$v \approx A_0 + A_1 P + A_2/P \tag{3.90}$$

describes the data extremely well. Insertion of this expression into our result Eqn. (3.89) yields the approximation

$$\mu[P_1] - \mu[P_0] \approx A_0(P_1 - P_0) + \frac{1}{2}A_1(P_1^2 - P_0^2) + A_2 \log(P_1/P_0). \tag{3.91}$$

Using a Levenberg–Marquardt method[4] for fitting a nonlinear function to data, we obtain estimates for the parameters:

$$\begin{aligned} A_0 &\approx -5.9024\,\text{cm}^3/\text{g}, \\ A_1 &\approx -3.2222 \times 10^{-7}\,\text{cm}^3/\text{g}\cdot\text{Pa}, \\ A_2 &\approx 2.609 \times 10^8\,\text{cm}^3\cdot\text{Pa/g}. \end{aligned} \tag{3.92}$$

Using the fact that $1\,\text{MPa}\cdot\text{cm}^3 = 1\,\text{J}$, we find that the chemical potential difference is

$$\mu[P_1] - \mu[P_0] \approx 217.5\,\text{J/g}. \tag{3.93}$$

□

[4] See, for example [103, Section 15.5]. Note that a power-law expression $v = A_0\,P^a$ may work even better, and is easier to fit.

3.7 ONE- AND TWO-DIMENSIONAL SYSTEMS

In Section 2.9 we considered the fundamental entropy relation of a simple, ideal rubber band. In this section, we generalize the structure of thermodynamics a little bit more to incorporate the generalized thermodynamic potentials introduced in this chapter. In particular, we revisit the rubber band to allow non-simple and non-ideal behavior. We also consider Langmuir adsorption isotherms to describe the attachment of molecules to adsorption sites on a solid surface. In the following subsection we derive our first fundamental relation from two equations of state that are found experimentally.

3.7.1 A non-ideal rubber band

Experiments show that the tension in a rubber band increases linearly with temperature at a fixed length, for stretches greater than approximately 10% [85]. We can use this experimental fact to show that stretching a rubber band isothermally decreases the entropy of the rubber band, but does not change its internal energy. We write the experimental observation as

$$\mathcal{T} = T\psi(L), \tag{3.94}$$

where $\psi(L)$ is some arbitrary, but measurable, function of length, such that the tension is zero at the rest length L_0, $\psi(L_0) = 0$, but positive for $L > L_0$. Since T and L are the independent variables (the variables easily manipulated in experiment), it is natural to use F as our fundamental relation. Using the relationship between pressure and Helmholtz potential implied by its differential, we can integrate Eqn. (3.94) from L_0 to arbitrary L at constant temperature to obtain

$$\begin{aligned} F(T,L) - F[L_0] &= \int_{L_0}^{L} \left(\frac{\partial F}{\partial L}\right)_T dL \\ &= \int_{L_0}^{L} \mathcal{T}(T,L) dL \\ &= T\int_{L_0}^{L} \psi(L')dL'. \end{aligned} \tag{3.95}$$

If we take the derivative with respect to T of each side of this equation, we can obtain an equation of state for entropy,

$$S(T,L) - S[L_0] = -\int_{L_0}^{L} \psi(L')dL'. \tag{3.96}$$

Note that the entropy decreases with stretching. We can find $S[L_0]$ from an integration over temperature,

$$\begin{aligned} S[L_0] &= S[T_0, L_0] + \int_{T_0}^{T} \left(\frac{\partial S}{\partial T}\right)_L\bigg|_{L=L_0} dT \\ &= S[T_0, L_0] + \int_{T_0}^{T} \frac{C_L[T', L_0]}{T'} dT', \end{aligned} \tag{3.97}$$

where $C_L := T(\partial S/\partial T)_L$ is the constant-length heat capacity for the rubber band.[5] Similarly, we can find $F[L_0]$ from an integration

$$F[L_0] = F[T_0, L_0] + \int_{T_0}^{T} \left(\frac{\partial F}{\partial T}\right)_L dT$$
$$= F[T_0, L_0] - \int_{T_0}^{T} S[T', L_0]dT'. \quad (3.98)$$

We can combine Eqns. (3.95), (3.97), and (3.98) to find

$$F(T, L) = T\int_{L_0}^{L} \psi(L')dL' + F_0 - (T - T_0)S_0 - \int_{T_0}^{T}\int_{T_0}^{T'} \frac{C_L[T'', L_0]}{T''} dT''\, dT'. \quad (3.99)$$

The double integral can be simplified by changing the order of integration to arrive at

$$F(T, L) = F_0 - (T - T_0)S_0 + T\int_{L_0}^{L} \psi(L')dL' - \int_{T_0}^{T} \left(\frac{T - T'}{T'}\right) C_L[T', L_0]dT'. \quad (3.100)$$

Note that we have derived a fundamental relation from the two measurable quantities $\mathcal{T}(T, L)$ and $C_L(T, L_0)$. We can use these results and the definition for F to find an equation of state for U:

$$U = F + TS$$
$$= U_0 + \int_{T_0}^{T} C_L[T', L_0]dT', \quad (3.101)$$

where $U_0 := F[L_0, T_0] + T_0S[T_0, L_0]$. Hence, we have proven that the internal energy does not depend on length at constant temperature for a rubber band whose tension changes linearly with temperature.

This result may seem surprising, given the fact that a stretched rubber band stores the ability to do work. However, the apparent paradox can be understood in the following way.[6] When a rubber band is stretched isothermally, work is done on it, and, hence, in order for it to remain isothermal, heat is dissipated to the surroundings ($đQ = -đW$). At the same time, the entropy decreases from the stretching, despite the heat flow out, as we see from Eqn. (3.96). The rubber band can do work by pulling a weight, say. However, in order for it to remain isothermal, the internal energy remains constant, and heat must be drawn in from the surroundings ($đQ = -đW$), while the weight is being pulled. Therefore, the rubber band uses heat from the surroundings to perform work. It simply uses its lowered level of entropy to drive the heat flow.

Although we have allowed an arbitrary dependence of the tension on length through the function $\psi(L)$, we have still idealized the situation somewhat by assuming that the tension is

[5] Actually, it is not straightforward to measure the constant-length heat capacity, since the rest length tends to change with temperature. A constant-tension heat capacity is easier to measure.

[6] Note that this observation appears to be a paradox only if one defines energy as "the ability to do work."

strictly linear with T. In Exercise 3.7.C we consider how to determine the amounts of energetic and entropic contributions to the elasticity from experimental data. If one accounts for thermal expansion, linearity in T is an excellent approximation.

3.7.2 Unzipping DNA

In order to replicate, DNA must first separate its two strands from the helix. This process is called **denaturation**. In most aqueous solutions DNA and proteins denature only at elevated temperature – well above body temperature. However, in the body these processes occur all the time. Therefore, there is an interest in understanding how external forces on DNA can cause the strands to denature at lower temperatures.

In a recent experiment [30], workers unzipped the two strands of a DNA double helix by pulling them apart. The process is shown schematically in Figure 3.11. One strand of the DNA is attached to a magnetic bead, and the other is attached to a solid surface (via a spacer). A desired tension can be applied to the strand by turning on a magnetic field of known strength.

At constant temperature, the tension is gradually increased. As the tension increases, more of the chain unzips, until the chain completely denatures. In this way, they were able to construct a "temperature–tension phase diagram." The data for this phase diagram are shown in Table 3.4.

These data show that the DNA can be denatured at body temperature if a force is exerted to pull the strands apart. In the following example, we use a statistical mechanical model derived in a later chapter to predict these data.

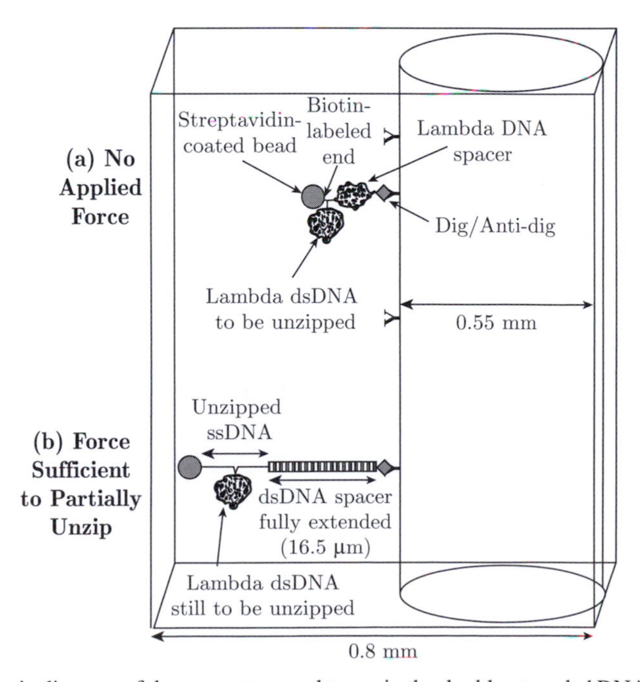

Figure 3.11 A schematic diagram of the apparatus used to unzip the double-stranded DNA segment. After [30].

Table 3.4 For a given force (in piconewtons), the temperature (in degrees Celsius) is found at which lambda phage DNA can be pulled apart in an aqueous solution, by using the apparatus sketched in Figure 3.11. At zero tension, the "melting temperature" (the temperature at which the DNA spontaneously denatures, or separates) is given. These data are taken from [30].

Tension, $\mathcal{T}$ (pN)	Melt temperature, T_m (°C)
3.44	49.9
3.62	45.0
4.47	42.0
5.41	40.1
6.64	40.4
8.89	36.0
9.93	30.0
9.93	34.4
9.93	36.3
10.4	27.3
10.4	31.4
11.1	30.0
11.6	24.0
12.0	27.0
15.5	21.9
16.5	20.0
16.9	20.2
17.7	18.1
35.9	15.1

EXAMPLE 3.7.1 A simple statistical-mechanical model may be constructed for the partially peeled DNA strand shown in Figure 3.11. If the ends of the peeled strand are pulled distance L apart with tension $\mathcal{T}$ at temperature T, the Gibbs potential is derived to be

$$G(T,\mathcal{T},\tilde{M}) = G_{\text{den}}(T,\tilde{M}) - k_B T \log\left\{\frac{[\sinh(x)/x]^{2(\tilde{M}+1)/n_b} - \lambda(T)^{\tilde{M}+1}}{[\sinh(x)/x]^{2/n_b} - \lambda(T)}\right\}, \tag{3.102}$$

where $\tilde{M}$ is the number of base pairs in the DNA, $x := \mathcal{T}a_K/(k_B T)$, a_K is the Kuhn step length of a denatured strand, $k_B = R/\tilde{N}_A$ is Boltzmann's constant, n_b is the number of base pairs per Kuhn step, and $\lambda(T) := \exp(\Delta g_b/(k_B T))$. We have also used the difference in Gibbs potential between a single attached and a single unattached base pair $\Delta g_b(T) := g_u - g_a$. Equation (3.102) assumes that the free end of the coiled chain has the two strand ends covalently bonded. Therefore, when all of the base pairs are detached, it is assumed that the DNA is one long single strand of $2\tilde{M}/n_b$ Kuhn steps. The real DNA strand has no such covalent bonding at its ends, of course. Hence, the model predicts complete peeling when all the base pairs are detached, and one long strand is made.

Using the above fundamental relation, find the distance L between the two ends of the unpeeled strands as a function of tension and temperature. For a given temperature, at a critical tension, this separation goes to infinity, which is the sought-after "temperature–tension phase diagram." Compare your result with the experimental results given in Table 3.4. Assume that $\Delta g_b(T) = \Delta h_b - T\,\Delta s_b$, where Δh_b and Δs_b are temperature-independent. Assume that $a_K = 15\,\text{Å}$, $n_b = 5$ (these estimates are from [126]). What values of Δs_b and Δh_b give the best fit? How does your estimate of Δs_b compare with the reported value (estimated using a different method, and in a different solvent) of $\Delta s_b = 85\,\text{J/K}\cdot\text{mol}$? What do you predict for the melting temperature at zero tension?

Solution. Note that the fundamental relation given here is in terms of Kuhn steps of size a_K, whereas Exercise 2.9.E used the persistence length l_p. The persistence length is equal to half the Kuhn step length, $a_K = 2l_p$. The contour length of the two models should remain the same, so $N_p = 2N_K$.

First we find the separation length L as a function of $(T,\mathcal{T},\tilde{M})$. Recall the differential for the Gibbs potential, $dG = -S\,dT + V\,dP + \mu\,dN$, which, for the one-dimensional case, becomes $dG = -S\,dT - L\,d\mathcal{T} + \mu\,d\tilde{M}$. Hence, we find the length from

$$\begin{aligned}
L &= -\left(\frac{\partial G}{\partial \mathcal{T}}\right)_{T,\tilde{M}} \\
&= -\left(\frac{\partial x}{\partial \mathcal{T}}\right)_T \left(\frac{\partial G}{\partial x}\right)_{T,\tilde{M}} \\
&= \left[\coth(x) - \frac{1}{x}\right]\frac{2a_K}{n_b}\left[\frac{\tilde{M}+1}{1-\hat{\lambda}(T,\mathcal{T})^{\tilde{M}+1}} - \frac{1}{1-\hat{\lambda}(T,\mathcal{T})}\right],
\end{aligned} \tag{3.103}$$

where we have introduced

$$\hat{\lambda}(T, \mathcal{T}) := \lambda(T)\left[\frac{x}{\sinh(x)}\right]^{2/n_{\rm b}}$$
$$= \exp\left(\frac{\Delta g_{\rm b}}{k_{\rm B}T}\right)\left[\frac{\mathcal{T} a_{\rm K}/(k_{\rm B}T)}{\sinh(\mathcal{T} a_{\rm K}/(k_{\rm B}T))}\right]^{2/n_{\rm b}}. \tag{3.104}$$

We used the chain rule to obtain the second line of Eqn. (3.103), and we skipped a step of slightly messy algebra to obtain the last line.

Equation (3.103) gives the average distance between the two free strand ends at a given tension and a given temperature. This length has two coupled contributions, however: the tension pulls the detached segment of strands straight, and the temperature and tension detach additional base pairs. To find the relative magnitude of these contributions, we can completely detach all the base pairs theoretically by letting the free energy of attachment become large and negative: $\Delta g_{\rm b} \to -\infty$. In that case, $\hat{\lambda} = 0$, and the length becomes

$$L_{\rm IL} = \left[\coth(x) - \frac{1}{x}\right]\frac{2\tilde{M}a_{\rm K}}{n_{\rm b}}. \tag{3.105}$$

This is the length the chain would have for a given tension if all the base pairs were detached, and is called the **inverse Langevin force law** (see Exercise 3.7.B). If we take the ratio of the two lengths, we obtain the expression

$$\frac{L}{L_{\rm IL}} = \left[\frac{1 + 1/\tilde{M}}{1 - \hat{\lambda}(T, \mathcal{T})^{\tilde{M}+1}} - \frac{1/\tilde{M}}{1 - \hat{\lambda}(T, \mathcal{T})}\right]. \tag{3.106}$$

This ratio now includes only the effect of detaching base pairs. When it goes to 1, all the base pairs must be detached. We can construct a plot of this ratio as a function of $\hat{\lambda}$, as shown in Figure 3.12 for $\tilde{M} = 1000$. From this figure we see that the chain "melts" when $\hat{\lambda}$ approaches 1. The value never becomes exactly 1, except for very small $\hat{\lambda}$, which means very negative values for $\Delta g_{\rm b}$, or large tensions. However, whenever $\hat{\lambda}$ is near 1, it becomes very easy for the number of attached base pairs to fluctuate to zero, at which point the two strands will be irreversibly separated. It makes sense, then, to assume that the chain becomes completely unpeeled when $\hat{\lambda} \cong 1$.

Hence, the "modified melting temperature," $T_{\rm m} = T_{\rm m}(\mathcal{T})$, is determined by setting $\hat{\lambda}(T_{\rm m}, \mathcal{T}) = 1$. Or, taking the logarithm of each side, we can write

$$-\frac{\Delta h_{\rm b}}{k_{\rm B}T_{\rm m}} + \frac{\Delta s_{\rm b}}{k_{\rm B}} = \frac{2}{n_{\rm b}}\log\left[\frac{\mathcal{T} a_{\rm K}/(k_{\rm B}T_{\rm m})}{\sinh(\mathcal{T} a_{\rm K}/(k_{\rm B}T_{\rm m}))}\right], \tag{3.107}$$

where we replace the change in Gibbs potential per base pair upon melting with the corresponding change in enthalpy and entropy, $\Delta g_{\rm b} = \Delta h_{\rm b} - T\,\Delta s_{\rm b}$. If we neglect the temperature dependences of these two quantities, then we have four parameters, $\Delta h_{\rm b}$, $\Delta s_{\rm b}$, $n_{\rm b}$, and $a_{\rm K}$, three of which have already been estimated from other experiments. To fit the remaining parameter, we plot $\log[x/\sinh(x)]$ vs. $1/T_{\rm m}$. For this plot it is useful to convert Boltzmann's constant as follows:

$$k_B = (1.380\,66 \times 10^{-23}\,\text{J/K})(1\,\text{N}\cdot\text{m/J})(10^{12}\,\text{pN/N})(10^{10}\,\text{Å/m})$$
$$= 0.138\,066\,\text{pN}\cdot\text{Å/K}. \qquad (3.108)$$

Figure 3.13 shows the fit of our equation to the data (excluding the last point). The slope and intercept of the line are, then, $-n_b\,\Delta h_b/(2k_B) \cong -11\,600 \pm 900\,\text{K}$ and $n_b\,\Delta s_b/(2k_B) \cong 36 \pm 3$, respectively. Or, $\Delta h_b \cong 6.4 \times 10^{-20}\,\text{J}$ and $\Delta s_b \cong 2.0 \times 10^{-22}\,\text{J/K}$. This latter value compares favorably with other estimates. The melting temperature at zero tension is, according to Eqn. (3.107), $T_m(\mathcal{T} = 0) = \Delta h_b/\Delta s_b \cong 320\,\text{K}$, which is too low.

The last data point in Table 3.4 does not fit well with the others. This may be because (1) fluctuations are not accounted for in thermodynamics, and a large fluctuation at high tensions could lead to irreversible peeling,[7] (2) the free energy per base pair may be temperature-dependent,

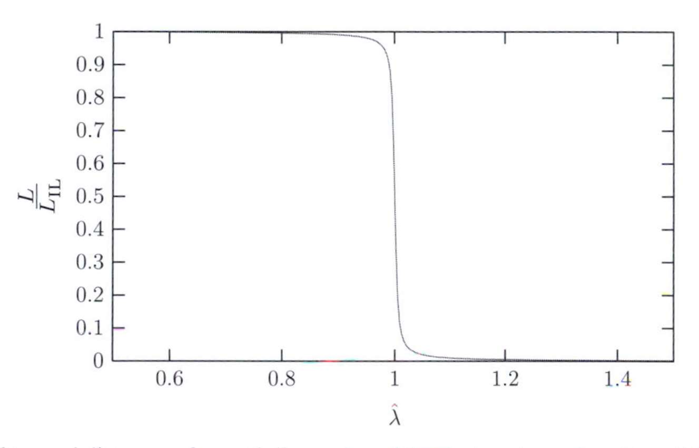

Figure 3.12 The end-to-end distance of a partially unzipped DNA strand as a function of the tension-modified base-pair partition function, $\tilde{\lambda}$. The distance is made dimensionless by dividing it by the length the chain would have if it were completely unzipped for $\tilde{M} = 1000$ base pairs. See Eqn. (3.106).

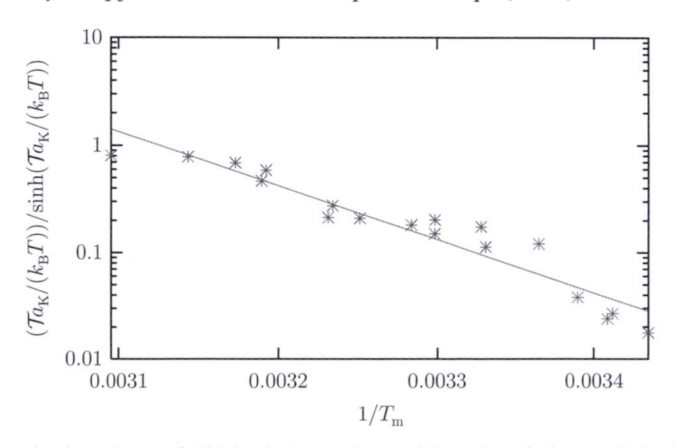

Figure 3.13 DNA-unzipping data of Table 3.4 together with a fit of the statistical-mechanical model, Eqn. (3.102). The fit to the data yields the change in enthalpy and entropy per base pair for denaturation. Note that the last data point in the table has been omitted both from the fit and from the plot here.

[7] Fluctuations are covered in Section 6.7.

(3) the model is too simple, in that it does not account for so-called stacking of the adjacent nucleotides, or (4) there may be experimental error at that high tension. □

3.7.3 Langmuir adsorption

Here we consider a solid surface that has M_s moles of sites available for binding $N \leq M_s$ moles of species A. Species A exists in an ideal-gas phase above the solid surface, but the molecules may attach themselves to these adsorption sites, detach, or hop between adjacent sites on the surface. See Figure 3.14. Such a system is important in catalysis and in some separation processes called *pressure-swing adsorption* (PSA). Adsorption measurements are also commonly used to calculate the available surface area (see Exercise 3.7.D).

Pressure-swing adsorption (PSA) is used to separate oxygen from air for asthma patients, to separate methane from waste decomposition to produce fuel, and to remove moisture from gas streams [112]. One such PSA process is shown schematically in Figure 3.15. A tank containing porous solid substrate is fed a mixture of gases from the top. One of the species, say oxygen, is preferentially adsorbed onto the solid. At the bottom end of the tank, the nitrogen-enriched gas exits. Before the solid adsorption sites are filled with oxygen, the flow is stopped, and the pressure in the tank is lowered, which allows the oxygen to desorb. An inert gas is then fed in at the bottom of the tank to flush the oxygen-enriched gas out at the top. In order to design the operation and size of such a process, it is necessary to be able to predict the amount of gas that can be adsorbed on the solid substrate.

In order to model the adsorption on the solid surface, we replace the volume with the more natural thermodynamic variable M_s. Alternatively, we could use the area and the density, or number of sites per unit area, but the development is the same. Therefore, we require a fundamental relation of the type $F = F(T, M_s, N)$:

$$F(T, M_S, N) = M_s RT \log\left(1 - \frac{N}{M_s}\right) + NRT \log\left(\frac{N}{M_s - N}\right) - NRT \log\left[q_{\text{site}}(T)\right]. \quad (3.109)$$

In the language of statistical mechanics the function $q_{\text{site}}(T)$ is called the **single-site partition function**. Its derivation is outside the scope of this textbook, since knowledge of quantum mechanics is required.[8] However, it contains information about the possible energies

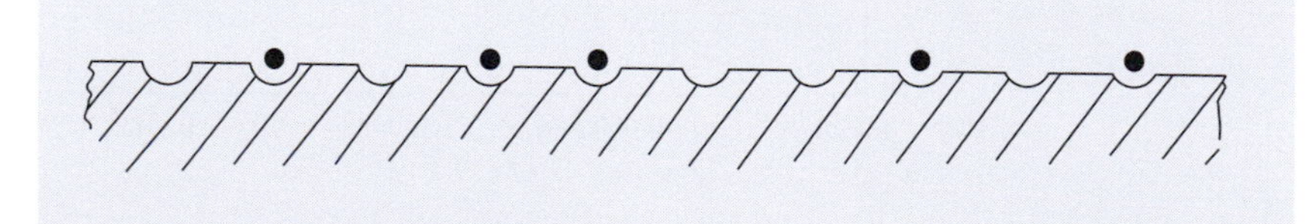

Figure 3.14 A sketch of a crystalline surface that can adsorb molecules from a gas phase. There exists a discrete number of adsorption sites on the solid surface, where the molecules (drawn here as circles) can sit.

[8] The details can be found in [59, Chapter 7].

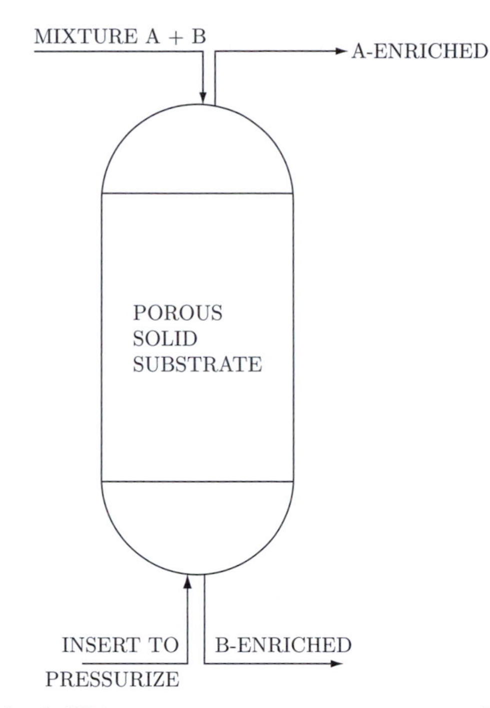

Figure 3.15 A sketch of a simple PSA process to separate two components of a gas. The tank contains solid, porous substrate, which preferentially adsorbs one of the gas species on its surface. The process has three steps. In the first step one pressurizes the tank with the mixture to be separated. The gas, now rich in the non-adsorbing species, is purged from the tank. When the pressure in the tank is lowered, the other species now desorbs, creating a gas phase rich in adsorbing species.

of interaction between one molecule and one adsorption site. The sites are assumed to have a simple potential-energy interaction with a molecule. Namely, the molecules are assumed not to interact energetically with one another.

An explicit expression for $q_{\rm site}$ can be found if a few assumptions are made, as follows. There is an energy difference per mole of ϵ_a (such that $\epsilon_a < 0$) between the minimum of the energy well for an adsorption site and a molecule infinitely far away from the surface. A molecule sitting in this well will have only two vibration frequencies from thermal motion, namely one for vibrations perpendicular to the surface, $\omega_\perp$, and one for vibrations parallel to the surface, $\omega_\parallel$. Associated with these vibrations are two characteristic vibration energies per mole, $\epsilon_\perp = \frac{1}{2}h\omega_\perp\tilde{N}_{\rm A}$ and $\epsilon_\parallel = \frac{1}{2}h\omega_\parallel\tilde{N}_{\rm A}$. Planck's constant is $h = 2\pi h$. Finally, there is an energy barrier V_0 between adjacent sites. In the Langmuir adsorption model, we also neglect the ability of the molecules to hop between adsorption sites, and characterize the system with only the three fundamental energies $\epsilon_\perp$, $\epsilon_\parallel$, and ϵ_a. This assumption will work at lower energies when $k_{\rm B}T \ll V_0$, or when T is less than approximately 250 K for most systems. With these approximations it is possible to derive

$$q_{\rm site}(T) = \frac{\exp[-\epsilon_a/(RT)]}{8\sinh^2[\epsilon_\parallel/(RT)]\sinh[\epsilon_\perp/(RT)]}. \tag{3.110}$$

From this relation, all thermodynamic properties of the lattice gas can be found. In particular, we can find the percentage of occupied adsorption sites as a function of the pressure in the gas phase above the surface. In Exercise 3.7.D, we use the results of the following example to analyze adsorption of N_2 and O_2.

EXAMPLE 3.7.2 Find the so-called **Langmuir adsorption isotherm** – the fraction of filled sites as a function of pressure at constant temperature, for an adsorption surface in equilibrium with an ideal gas.

Solution. Since the gas molecules are free to be either adsorbed or in the gas phase, their chemical potential and temperature must be the same in the two phases. Note that the surface has no volume, so there is no pressure for the molecules adsorbed.[9]

From the differential for the Helmholtz potential, we can find the chemical potential of the adsorbed phase from the fundamental Langmuir relation, Eqn. (3.109):

$$\begin{aligned}\mu &= \left(\frac{\partial F}{\partial N}\right)_{T,M_s} \\ &= RT\log\left[\left(\frac{\theta}{1-\theta}\right)\frac{1}{q_{\mathrm{site}}(T)}\right].\end{aligned} \tag{3.111}$$

where $\theta := N/M_s$ is the fraction of adsorption sites occupied.

To find the chemical potential in the ideal-gas phase, we multiply both sides of Eqn. (B.2) in Appendix B by N, and take the derivative of each side with respect to N:

$$\begin{aligned}\mu^{\mathrm{IG}} &= \left(\frac{\partial F}{\partial N}\right)_{T,V} \\ &= \mu^{\circ}(T) + RT\log P,\end{aligned} \tag{3.112}$$

where

$$\mu^{\circ}(T) := f_0 - s_0(T-T_0) + \int_{T_0}^{T}\left(\frac{T'-T}{T'}\right)c_v^{\mathrm{ideal}}(T')dT' + RT - RT\log\left(\frac{RT}{Nv_0}\right). \tag{3.113}$$

Setting the two chemical potentials equal, and solving for θ yields

$$\theta = \frac{\chi(T)P}{1+\chi(T)P}, \tag{3.114}$$

where

$$\chi(T) := \exp\left[\frac{\mu^{\circ}(T)}{RT}\right]q_{\mathrm{site}}(T). \tag{3.115}$$

Figure 3.16 shows some representative adsorption isotherms predicted by the Langmuir model.

[9] The surface does have an area, however. Hence, one can find the derivative of the internal energy with respect to area at constant entropy and mole number – analogous to pressure. The resulting quantity is called the **surface tension**, and is covered in greater detail in Chapter 12.

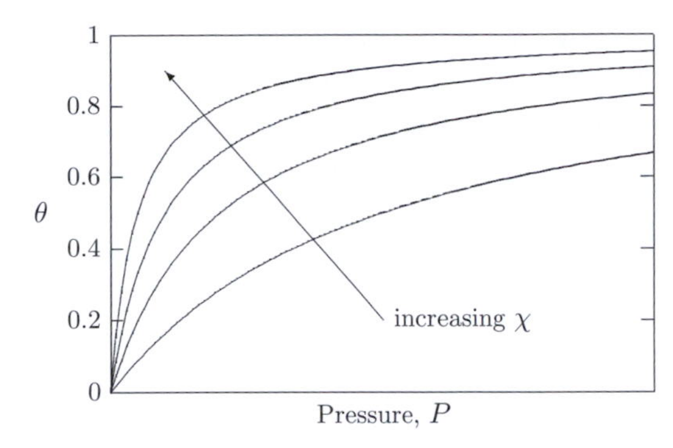

Figure 3.16 The fraction of filled sites as a function of pressure for several values of χ, as predicted by the Langmuir model, Eqn. (3.114). □

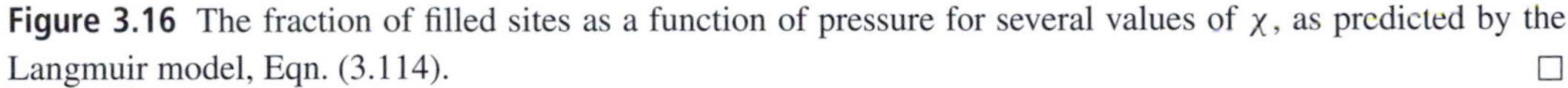

Table 3.7 in Section 3.9 shows adsorption data for nitrogen and oxygen on zeolite. Exercise 3.7.D asks the student to fit these data to the equation for the Langmuir adsorption isotherm.

The simple statistical-mechanical model used to derive the Langmuir adsorption model (given in Section 6.4) does not require that the adsorption sites be on a solid surface. Primarily, it assumes that adsorption sites do not interact energetically with one another. Hence, the result is much more general than just molecules adsorbing on solid surfaces. In fact, the equation is often used to model other adsorption phenomena, such as ligands adsorbing onto proteins, or polymer chains adsorbing onto colloidal particles. These examples are considered in the exercises.

In the following example, we use the Langmuir model to describe the binding of proteins to sites on DNA. However, an additional subtlety must be taken into account. For classical Langmuir adsorption, the chemical potential in the gas phase is not affected by the degree of adsorption θ. However, in solution, every molecule adsorbed decreases the concentration of unadsorbed molecules.

EXAMPLE 3.7.3 Proteins bind to specific sites of DNA. It is believed that the presence of other species can affect this adsorption, and we wish to examine that effect here. Heyduk and Lee [57] recently synthesized a 32-base-pair-long fragment of DNA, which they tagged with a fluorescent dye. The dye makes it possible to measure optically the fraction of DNA fragments that are bound by a specific protein. We won't be concerned here with how this measurement works, but rather just assume that it does.

Table 3.5 shows optical data for solutions of the *E. coli* cyclic AMP receptor protein (total concentration $c_{\mathrm{r}}^{\mathrm{T}}$) and a 32-base-pair fragment DNA ligand of total concentration $c_{\mathrm{D}}^{\mathrm{T}} = 0.0111\ \mu\mathrm{M}$. Also present is cyclic adenosine monophosphate (cAMP), in two different concentrations, which is believed to play a role in binding, by forming complexes with the protein. We wish to study this role by using the Langmuir model, and seeing how the binding parameter χ is affected by the concentration of cAMP.

Table 3.5 Anisotropy data A for DNA ligand binding to an *E. coli* cAMP receptor protein, as measured by fluorescence. In this table, c_r^T is the total concentration of receptor protein (bound and unbound). The first two columns are for a concentration of cAMP equal to 0.5 μM, and the right two columns are for cAMP concentration 500 μM. The data are taken from [57].

c_r^T (nM)	Anisotropy, A	c_r^T (nM)	Anisotropy, A
0	0.170	0.131	0.169
3.92	0.174	2.091	0.175
7.97	0.177	3.92	0.179
11.76	0.179	5.88	0.182
15.68	0.180	7.84	0.185
19.6	0.182	9.80	0.188
25.35	0.183	11.76	0.189
33.06	0.185	13.59	0.191
40.77	0.185	15.81	0.191
59.84	0.189	17.38	0.192
78.79	0.190	21.43	0.193
97.47	0.191	25.22	0.194
116.2	0.192	30.97	0.194
134.6	0.192	36.72	0.196

The anisotropy A is fluorescence data giving a measure of the amount of binding, and is assumed to be linearly proportional to the concentration of bound complex, c_B, and to that of unbound DNA fragment, c_D,

$$A = A_D c_D + A_B c_B, \tag{3.116}$$

where A_D and A_B are the (constant) anisotropies of the free and bound DNA, respectively. Use the data in Table 3.5 to estimate the binding parameter χ for each concentration of ligand. Does its value depend upon the cAMP concentration? Can anything be inferred about the involvement of cAMP in DNA/protein binding?

Solution. Assuming constant volume, we note that the total concentration of protein c_r^T is constant, and given by the sum of free protein c_r and bound protein c_B concentrations: $c_r^T = c_r + c_B$. Similarly, one can balance the total concentration of DNA, $c_D^T = c_D + c_B$. The generalization of Langmuir adsorption to concentration is

$$\frac{c_B}{c_r^T} = \frac{\chi c_D}{1 + \chi c_D}, \tag{3.117}$$

assuming that the DNA is in excess, and equating the chemical potentials of bound and unbound DNA. From these three equations, it is possible to relate the concentration of bound complex c_B to the total concentration of DNA ligand and protein. Inserting the balances into Eqn. (3.117) and solving for c_B yields

$$c_B = \frac{1}{2\chi}\left[1 + \chi\left(c_r^T + c_D^T\right) + \sqrt{\left[1 + \chi(c_r^T + c_D^T)\right]^2 - 4\chi^2 c_D^T c_r^T}\right]. \tag{3.118}$$

Inserting this equation into Eqn. (3.116) and replacing c_D with c_D^T gives the anisotropy A as a function of χ, A_D, A_B, and the total concentrations c_r^T and c_D^T:

$$A = A_D c_D^T + \frac{A_B - A_D}{2\chi}\left[1 + \chi\left(c_r^T + c_D^T\right) + \sqrt{\left[1 + \chi(c_r^T + c_D^T)\right]^2 - 4\chi^2 c_D^T c_r^T}\right]. \tag{3.119}$$

The concentrations are known (from Table 3.5), and we seek an estimate for χ. We can find A_D from the first entry of the table, where $c_r^T = 0$, giving us $A_D = 0.169\,54/(0.0111\,\mu\text{M}) = 15.3\,\mu\text{M}^{-1}$. Hence, we have two parameters to fit to the data: A_B and χ. Using a Levenberg–Marquardt fit[10] we are able to fit Eqn. (3.119) to the data, and find estimates for these remaining two parameters.

We fit Eqn. (3.119) to the first two columns, for which the cAMP concentration is $0.5\,\mu$M. The result is shown in Figure 3.17. Our fit gave $\chi = 0.0216 \pm 0.0014\,\mu\text{M}^{-1}$ and $A_B = 20.04 \pm 0.1\,\mu\text{M}^{-1}$. For the higher concentration of cAMP, we obtained $\chi = 0.055 \pm 0.005\,\mu\text{M}^{-1}$ and

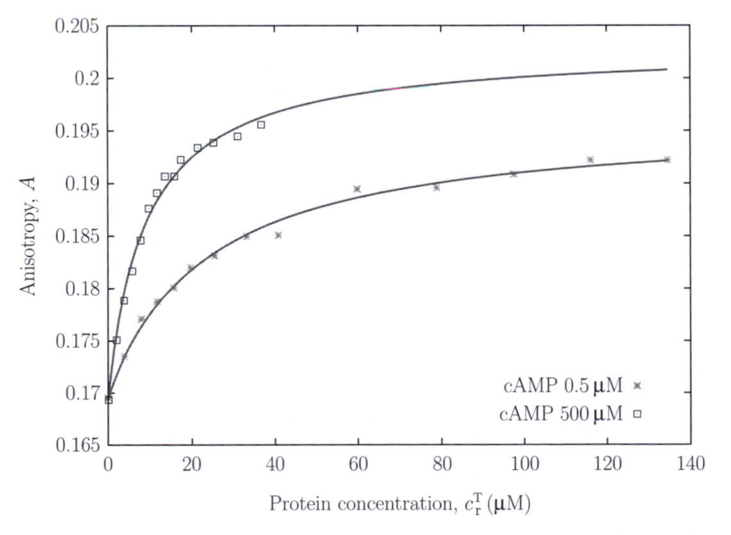

Figure 3.17 Fits of Eqn. (3.119) to the data given in Table 3.5. The lower curve is for a cAMP concentration of $0.5\,\mu$M, for which our nonlinear fit gave values of $\chi = 0.0216 \pm 0.0014\,\mu\text{M}^{-1}$ and $A_B = 20.04 \pm 0.1\,\mu\text{M}^{-1}$. The upper curve is for a cAMP concentration of $500\,\mu$M, for which the parameters were found to be $\chi = 0.055 \pm 0.005\,\mu\text{M}^{-1}$ and $A_B = 21.3 \pm 0.2\,\mu\text{M}^{-1}$.

[10] Such a fit is straightforward with a software package called gnuplot, using the command "fit." Gnuplot is freeware under the GNU license, and is available for most computer operating systems. Spreadsheets, Mathematica, MATLAB, or similar software can also accomplish this fit.

$A_B = 21.3 \pm 0.2\,\mu M^{-1}$. The value of A_B is not strongly dependent on cAMP concentration, as we expect. In fact, if we assume that $A_B = 20.6$, the average of these two values, then we also obtain reasonable fits for $\chi = 0.0162 \pm 0.0007$ and 0.076 ± 0.004, respectively. On the other hand, χ does depend strongly on the concentration of cAMP, indicating that cAMP does play a role in the binding of the protein to the DNA fragment. Apparently, cAMP enhances the ability of the protein to bind. Exercise 3.7.I uses a reaction model for the same binding process. □

Summary

In the previous chapter, we covered internal energy U, entropy S, temperature T, pressure P, chemical potential μ_i, heat flow Q, work W, and the five postulates of thermodynamics. We proved that these postulates and definitions fit our intuition about temperature and heat flow, pressure and expansion, and mass flow.

In this chapter, we used *Legendre transforms* to define new thermodynamic quantities with complete thermodynamic information:

- the enthalpy $H := U + PV$,
- the Helmholtz potential $F := U - TS$, and
- the Gibbs free energy $G := U - TS + PV$.

We proved that an isolated system at equilibrium has minimum internal energy at constant entropy, volume, and mole number. The other subsystems are summarized in Table 3.2.

Each of the generalized potentials has a canonical differential, which can be used to derive equations of state, or Maxwell relations. These differentials are

$$dU = T\,dS - P\,dV + \sum_i \mu_i\,dN_i, \tag{3.48}$$

$$dH = T\,dS + V\,dP + \sum_i \mu_i\,dN_i, \tag{3.49}$$

$$dF = -S\,dT - P\,dV + \sum_i \mu_i\,dN_i, \tag{3.50}$$

$$dG = -S\,dT + V\,dP + \sum_i \mu_i\,dN_i. \tag{3.51}$$

The Maxwell relations are derived from these differentials by using the analytic property of the potentials, Eqn. (A.5) in Appendix A. The differentials and the Maxwell relations are conveniently memorized through the thermodynamic square, Figure 3.5.

These are not the only such generalized potentials, but are the most common. For example, in Exercise 3.3.D we introduce another such potential that is common in statistical mechanics.

Along the way, we derived the Euler relation

$$U = TS - PV + \sum_{i=1}^{r} \mu_i N_i, \quad \text{Euler relation,} \tag{3.20}$$

and the Gibbs–Duhem relation

$$0 = S\,dT - V\,dP + \sum_{i=1}^{m} N_i\,d\mu_i, \quad \text{Gibbs–Duhem relation.} \tag{3.22}$$

The Euler relation shows that the specific Gibbs potential is identical to the chemical potential for pure substances. The Gibbs–Duhem relation is useful for mixtures, which will be considered later in this book.

We also proved that the generalized potentials are useful for calculating work done under isothermal and/or isobaric conditions.

We defined three (of many possible) second-order derivatives:

- the coefficient of thermal expansion $\alpha := (1/v)(\partial v/\partial T)_P$,
- the isothermal compressibility $\kappa_T := -(1/v)(\partial v/\partial P)_T$, and
- the constant-pressure heat capacity $c_P := T(\partial s/\partial T)_P$.

All other second-order derivatives can be found in relation to these, for a single-component system. In fact, *any thermodynamic quantity for a single-component system* can be found from these quantities (Section 3.6).

Complete thermodynamic information for a pure substance is also contained in two equations of state (e.g. $P = P(T, v)$ and $u = u(T, v)$), or in three relations for second-order derivatives (e.g. $\alpha = \alpha(T, P)$, $\kappa_T = \kappa_T(T, P)$, and $c_P = c_P(T, P)$).

As example models, we considered the general ideal gas, Eqn. (3.16), and a nonlinear elastic strand, Eqn. (3.100).

We also analyzed the following problems:

- adiabatic and isothermal compression of ammonia, Example 3.2.2;
- a heat engine, Example 3.2.3;
- a fuel cell, Example 3.5.4;
- deriving a fundamental relation for a non-ideal, elastic strand, Section 3.7.1;
- unzipping a single strand of DNA by using tension and heat, Example 3.7.1;
- Langmuir adsorption on a solid surface, Example 3.7.2; and
- binding of a protein to a DNA fragment, Example 3.7.3.

In Section 3.7.1, we showed how to generate a fundamental relation from two equations of state for a general elastic string. We also predicted the fractional adsorption of molecules on a solid surface as a function of pressure, the so-called "Langmuir isotherm."

Exercises

3.1.A Verify that the two fundamental energy relations $U_1(S) = AS^c$ and $U_2(S) = A(S + C_0)^c$ (A and C_0 may be functions of V but not S; c is a constant) give the same equation of state for $U(T)$, but give different fundamental Helmholtz relations.

3.1.B Show that the fundamental enthalpy relation

$$H = ANP^{\gamma} \exp\left[\frac{S\gamma}{NR}\right] \tag{3.120}$$

is that of a simple ideal gas. How are the constants here related to those in Eqn. (2.39)?

3.1.C The fundamental Helmholtz potential for a fluid is given as

$$F(T, V, N) = Nf_0 - Ns_0(T - T_0) - \frac{aN^2}{V} - NRT \log\left(\frac{V/N - b}{v_0}\right) + N\int_{T_0}^{T}\left(\frac{T' - T}{T'}\right)c_v^{\text{ideal}}(T')dT' \tag{B.16}$$

Find the thermal and mechanical equations of state for this fluid, and explain why it might be called a general, van der Waals fluid.

3.1.D Show that the fundamental Helmholtz relation

$$f = f_0 - s_0(T - T_0) - RT \log\left(\frac{v - b}{v_0}\right) + \frac{a}{b\sqrt{T}} \log\left[\left(\frac{v}{v + b}\right)\right] + \int_{T_0}^{T}\left(\frac{T' - T}{T'}\right)c_v^{\text{ideal}}(T')dT'. \tag{B.34}$$

is a general Redlich–Kwong fluid. Find the mechanical and thermal equations of state for this fluid.

3.1.E The fundamental Helmholtz potential for a fluid is given as

$$f = f_0 - s_0\,(T - T_0) - \frac{a}{v} - RT \log\left(\frac{v}{v_0}\right) - \frac{bRT(v_0 - v)\left[3(v + v_0)b - 4vv_0 - 2b^2\right]}{(v - b)^2 v_0^2} + \int_{T_0}^{T}\left(\frac{T' - T}{T'}\right)c_v^{\text{ideal}}(T')dT'. \tag{B.27}$$

Find the thermal and mechanical equations of state for this fluid. The *PVT* relation is called the *Carnahan–Starling generalization* to the van der Waals equation of state.

3.1.F The fundamental Helmholtz potential for a fluid is given as

$$f = f_0 - s_0(T - T_0) - RT \log\left(\frac{v}{v_0}\right) + \left[B_0RT - A_0 - \frac{C_0}{T^2}\right]\frac{1}{v} + \frac{bRT - a}{2}\left(\frac{1}{v^2}\right) + \frac{a\alpha}{5}\left(\frac{1}{v^5}\right) - \frac{C}{T^2}\left[\left(\frac{1}{\gamma} + \frac{1}{2v^2}\right)\exp\left(-\frac{\gamma}{v^2}\right) - \left(\frac{1}{\gamma}\right)\exp\left(-\frac{\gamma}{v_0^2}\right)\right] + \int_{T_0}^{T}\left(\frac{T' - T}{T'}\right)c_v^{\text{ideal}}(T')dT'. \tag{3.63}$$

Find the thermal and mechanical equations of state for this fluid. The *PVT* relation is called the *Benedict–Webb–Rubin equation of state*.

3.1.G The fundamental Helmholtz potential for a fluid is given as

$$f = f_0 - s_0(T - T_0) - RT \log\left(\frac{v - b}{v_0}\right)$$
$$- \frac{a(T)}{\gamma - \beta} \log\left[\left(\frac{v + \beta - b}{v + \gamma - b}\right)\right] + \int_{T_0}^{T} \left(\frac{T' - T}{T'}\right) c_v^{\text{ideal}}(T')dT'. \quad \text{(B.58)}$$

Find the thermal and mechanical equations of state for this fluid. The *PVT* relation is called *Martin's generalized cubic equation of state*.

3.1.H The fundamental Helmholtz potential for a fluid is given as

$$f = f_0 - s_0(T - T_0) - RT \log\left(\frac{v - b}{v_0}\right)$$
$$+ \frac{a(T)}{2\sqrt{2}b} \log\left[\left(\frac{v + b(1 + \sqrt{2})}{v + b(1 - \sqrt{2})}\right)\right] + \int_{T_0}^{T} \left(\frac{T' - T}{T'}\right) c_v^{\text{ideal}}(T')dT'. \quad \text{(B.48)}$$

Find the thermal and mechanical equations of state for this fluid. The *PVT* relation is called the *Peng–Robinson equation of state*. Since v_0 is the specific volume for which the fluid is ideal, it is much larger than b, and the second ratio that appears inside the logarithm may be safely approximated as 1.

3.1.I Show that the fundamental Gibbs relation

$$G = Ng_0 + [(c + 1)NR - Ns_0](T - T_0) + NRT \log\left[\left(\frac{T_0}{T}\right)^{c+1}\left(\frac{P}{P_0}\right)\right]$$

is that of an ideal gas, where c, R, s_0, T_0, and P_0 are constants. Describe what each of these constants denotes.

3.1.J Find the generalized potential $\tilde{U}(S, P, \mu)$ that has S, P, and μ as canonical independent variables. Write its differential.

3.1.K Similarly to Example 2.6.6, we wish to use a gas as a refrigerator, except this time the gas is real instead of ideal. We use the van der Waals equation of state for oxygen as our gas. The gas is in a container that has a piston so that it can exchange work with its environment. All the walls are adiabatic except for one, which is isothermal. However, we also have some insulation, so we can make that wall adiabatic at will. The gas begins with temperature $T_1 = 50\,°\text{C}$ and pressure $P_1 = 10\,\text{atm}$. In the first step, step a, we compress the gas adiabatically (i.e. with the insulation covering the diathermal wall) to temperature $T_2 = 100\,°\text{C}$. Next, during step b, we isobarically cool the gas to $T_3 = 50\,°\text{C}$. Step c then isothermally expands the gas back to the first state. See Figure 3.18.

We wish to find the amount of cooling performed per step, per mole of gas. To aid in the calculation, you might perform the following steps.

- Using the expressions in Appendix B for the van der Waals fluid, find the constants for the PvT equation of state, a and b. From the fundamental relation, find the equations of state, $s = s(T, v)$ and $u = u(T, v)$, for the general van der Waals fluid.

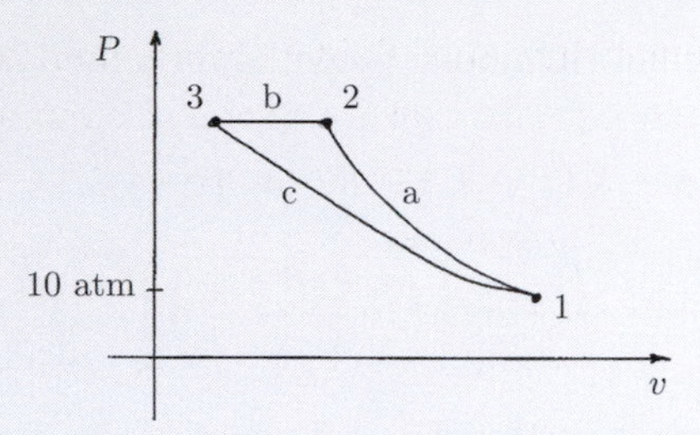

Figure 3.18 A three-step process to use a real gas as a refrigerator. In state 1, the gas is at temperature $T_1 = 50\,°\text{C}$ and pressure $P_1 = 10\,\text{atm}$. In step a, the gas is compressed adiabatically to temperature $T_2 = 100\,°\text{C}$; in step b, it is cooled isobarically to $T_3 = 50\,°\text{C}$; and in step c, it is expanded isothermally back to its original state.

- Construct two tables to be filled in. The first table should have boxes for the pressure, temperature, and specific volume at each of the points 1, 2, and 3. The second table should have boxes for Δu, Q, and W for each of the steps a, b, and c. Fill in the boxes from the problem statement.
- Now begin filling in each of the boxes from the necessary calculations. For example, you could use the PvT equation of state to find the specific volume v_1.
  Then, step a is assumed adiabatic and reversible, so it is isentropic. Use your entropy equation of state above to find the specific volume v_2. From T_2 and v_2, you should be able to find the pressure P_2. From your u equation of state, you can also find the change in internal energy in step a, Δu_a. From an energy balance, you can find Q_a.
- Continue filling out the tables in this way. When you have finished, calculate the total work necessary and the amount of cooling that can be accomplished. Which steps require work, and which steps provide cooling?

As a bonus, you could also run the machine backwards and use heat to generate work. In that case, how much work is generated per amount of heat added to the system?

3.1.L Use the thermodynamics diagram for ammonia (Figure 5.10 on p. 176) to estimate the change in chemical potential for ammonia at constant $T = 140\,°\text{C}$, but with the pressure changing from 20 to 1 MPa.

3.2.A Prove that a system in contact with both a pressure and a temperature reservoir reaches equilibrium when $dG_{sys} = 0$.

3.2.B Use the result of Example 3.2.1 to find the quasi-static work required to compress one mole of oxygen isothermally at 170 K from 2 to 0.8 l using a Peng–Robinson equation of state. Can you think of another way to find this result?

3.2.C Using the results of Example 3.2.1, find the work necessary to compress reversibly and isothermally one mole of ethane at 330 K from 8 to 1 l. Use the Benedict–Webb–Rubin equation of state.

3.2.D Using Figure 5.10 on p. 176, find the work necessary to compress ammonia isothermally from a density of 2 kg/m^3 to 250 kg/m^3 at 340 °C.

3.2.E Using the results of Example 3.2.1, find the work necessary to compress reversibly and isothermally one mole of DuPont refrigerant HCFC-123 at 330 K from 200 to 105 l. Apparently, this fluid is well described by a modified Benedict–Webb–Rubin equation of state. Details of the appropriate equation of state can be found at http://www.dupont.com/suva/na/usa/literature/pdf/h47753.pdf. This document also contains tables of thermodynamic properties, which may be used to avoid explicit use of the equation of state altogether.

3.2.F Using Figure 5.10 on p. 176, find the work necessary to compress 5 l of ammonia adiabatically from a density of 2 kg/m^3 and pressure of 250 kPa to a density of 60 kg/m^3.

3.3.A Derive the differentials for the generalized potentials H and G.

3.3.B Derive the remaining Maxwell relations, Eqns. (3.56), (3.58), and (3.59).

3.3.C In Example 3.1.2, we found $S(T, V)$ in Eqn. (3.17) for a general ideal gas. Verify that the third Maxwell relation is satisfied for the general ideal gas. Verify the other three relations.

3.3.D In this chapter we performed a Legendre transform on the internal energy in the variables S and V. One may also perform transforms in the variable N, or on the entropy. By performing a Legendre transform on U in the mole number, find the definition of the generalized potential Ψ with canonical independent variables (T, V, μ). Find the differential for this variable, and show that

$$\left(\frac{\partial \Psi}{\partial \mu}\right)_{T,V} = -N. \tag{3.121}$$

This generalized potential frequently arises in statistical mechanics, but is often the negative of the one defined here.

3.3.E Prove the Maxwell-like relation

$$\left(\frac{\partial \mu_i}{\partial P}\right)_{T,\{N_j\}} = \left(\frac{\partial V}{\partial N_i}\right)_{T,P,\{N_{j\neq i}\}}. \tag{3.122}$$

3.3.F A methanol fuel cell relies on the reaction

$$CH_3OH + \frac{3}{2}O_2 \rightarrow CO_2 + 2H_2O.$$

Find the maximum available work as a function of temperature for this fuel cell, assuming that it works reversibly. Also find the maximum work available from a heat engine. Do your results agree with Table 3.3?

3.5.A Find the coefficient of thermal expansion and the heat capacity for the simple van der Waals fluid.

3.5.B Find the isothermal compressibility and heat capacity predicted by the Redlich–Kwong equation of state in Section B.5 of Appendix B.

3.5.C Find the isothermal compressibility and heat capacity predicted by the Peng–Robinson equation of state in Section B.6 of Appendix B.

3.5.D It is reported that the isothermal compressibility at the boiling point (which is reported as $-252.77\,^\circ$C) for hydrogen is $-50.3\,\text{MPa}^{-1}$. How does this value compare with the prediction by the van der Waals equation of state? Note that the constants for the van der Waals fluid may be found from critical data by application of Eqn. (4.22).

3.5.E Find α as predicted by the Peng–Robinson equation of state.

3.5.F Find α as predicted by the Redlich–Kwong model.

3.5.G Use the Soave–Redlich–Kwong model (see Section B.5 in Appendix B) to estimate the constant-pressure heat capacity of water at $25\,^\circ$C and 2 bar.

3.5.H Using the Peng–Robinson model, derive an expression for the constant-volume heat capacity. Plot $(c_v - c_v^{\text{ideal}})/c_v^{\text{ideal}}$ for pure nitrogen as a function of pressure at room temperature. Your pressure range should be 1–100 atm. The parameters for the Peng–Robinson model can be found from the critical-point properties using the equations given in Appendix B.

3.6.A What is the constant-volume heat capacity for a simple ideal gas?

3.6.B What is $(\partial U/\partial T)_P$ for a general ideal gas?

3.6.C Reduce the following derivatives in terms of standard quantities:
(1) $(\partial S/\partial V)_P$, (2) $(\partial V/\partial T)_S$, (3) $(\partial H/\partial V)_T$, (4) $(\partial V/\partial P)_H$, (5) $(\partial V/\partial S)_H$.

3.6.D Find an expression that relates the **isentropic compressibility** κ_S to the three second-order derivatives α, κ_T, and c_P:

$$\kappa_S := -\frac{1}{v}\left(\frac{\partial v}{\partial P}\right)_{S,N}. \tag{3.123}$$

What is κ_S for a simple ideal gas?

3.6.E Find the change in chemical potential for steam at $400\,^\circ$C that is compressed from 0.5 to 2 MPa. Explain why one cannot use the steam tables to find the change in chemical potential for steam undergoing *temperature* changes. *Hint:* look at the entry for the entropy of saturated liquid at $0\,^\circ$C. Does it agree with the Nernst postulate? How could this problem be fixed, so that one could use the steam tables?

3.6.F Express the Joule–Thomson coefficient

$$\left(\frac{\partial T}{\partial P}\right)_{H,N}, \quad \text{Joule–Thomson coefficient,} \tag{3.124}$$

in terms of measurable quantities.

3.6.G The Joule–Thompson coefficient defined in the previous problem is useful in flow through valves. As is shown in Section 5.2, if one assumes that flow across a valve is adiabatic, it is also isenthalpic. Hence, the change in temperature as the pressure drops across the valve can be found from the Joule–Thompson coefficient. Find the change in temperature when the pressure of ammonia drops from 200 MPa and $180\,^\circ$C to a pressure of 50 MPa in two ways: (1) using the thermodynamic diagram (Figure 5.10 on p. 176) and (2) using the Peng–Robinson model.

3.6.H (1) Find $(\partial T/\partial P)_{S,N}$ as a function of temperature and volume for a general ideal gas.

(2) Plot $(\partial T/\partial P)_{S,N}$ for oxygen at room temperature as a function of pressure between 1 and 3 atm.

(3) Integrate your expression from (1) to find an expression that relates temperature to pressure. Note that the expression is not explicit in temperature (i.e. it is not possible to write an equation of the form $T = T(P)$). However, it can be made explicit in pressure: $P = P(T)$. Can you figure out a way to plot the temperature as a function of pressure during the reversible, adiabatic compression of part (2)?

3.6.I Using the result of Exercise 3.5.G and thermodynamic manipulations, find the constant-pressure heat capacity of liquid water at the same temperature and pressure.

3.6.J Find a relationship between the constant-tension heat capacity, and the constant-length heat capacity and other measurable quantities.

3.7.A Find the mechanical and thermal equations of state for the rubber band with the fundamental Helmholtz relation

$$F = f_0 L_0 - s_0 L_0 (T - T_0) - \frac{1}{2} N_c RT \log\left[1 - \left(\frac{L - L_0}{L_1 - L_0}\right)^2\right] + L_0 \int_{T_0}^{T} \left(\frac{T' - T}{T'}\right) c_L(T') dT'. \tag{3.125}$$

The quantities N_c, L_0, and L_1 are constants. c_L is some known function of T.

3.7.B Another statistical-mechanical model for a single strand of DNA (see Exercise 2.9.E) is called the **inverse Langevin force law** with Gibbs potential

$$G(T, \mathcal{T}) = g_0 L_0 - (T - T_0) s_0 L_0 - L_0 \int_{T_0}^{T} \left(\frac{T - T'}{T'}\right) c_{\mathcal{T}}^0(T') dT' - N_K k_B T \log\left[\frac{k_B T}{\mathcal{T} a_K} \sinh\left(\frac{\mathcal{T} a_K}{k_B T}\right)\right], \tag{3.126}$$

where, as before, N_K is the number of segments, a_K is the length of a segment, k_B is Boltzmann's constant, and we have introduced the zero-tension heat capacity for the chain $C_{\mathcal{T}}^0(T')$. How well does this model describe the data given in Table 2.1 on p. 50? Be sure to read the comments about persistence length versus Kuhn step length in the solution to Example 3.7.1.

3.7.C Here we derive a method to measure the contributions of entropy and internal energy to the elasticity $\mathcal{T}$. For isothermal stretching, we may write

$$\mathcal{T} = \left(\frac{\partial F}{\partial L}\right)_T = \left(\frac{\partial U}{\partial L}\right)_T - T\left(\frac{\partial S}{\partial L}\right)_T, \tag{3.127}$$

where the second line is obtained from the definition for F. We then seek a way to estimate the contribution from each of the two terms on the right-hand side of this expression.

Table 3.6 Tension divided by cross-sectional area versus percentage extension (Ext.) ($100 \times (L-L_0)/L_0$) for several temperatures as measured by Anthony, Caston, and Guth. These data are for reversible, isothermal extensions.

$T = 10\,°C$		$T = 30\,°C$		$T = 50\,°C$		$T = 70\,°C$	
Ext. (%)	Stress (kg/cm^2)	Ext. (%)	Stress (kg/cm^2)	Ext. (%)	Stress (kg/cm^2)	Ext. (%)	Stress (kg/cm^2)
3.20	0.484	2.52	0.419	2.29	0.428	1.83	0.372
6.17	0.800	5.49	0.781	5.49	0.809	5.03	0.772
13.5	1.35	12.3	1.38	11.7	1.40	11.2	1.41
22.2	2.01	21.3	2.07	21.0	2.14	19.9	2.22
38.2	3.00	37.5	3.11	37.5	3.25	37.5	3.43
72.9	4.52	71.8	4.77	71.3	5.01	70.2	5.27

First, derive a generalized Maxwell relation to relate $(\partial S/\partial L)_T$ to measurable quantities $(T, L, \mathcal{T})$. Then, use the generalized differential for U to obtain $(\partial U/\partial L)_T$ in terms of measurables. By "generalized," we mean here where we have replaced the usual variables (S, V, N) with the new set appropriate for a rubber band (S, L). Finally, explain how measurements of tension at different values of T but fixed L can be used to estimate the contributions of entropy and internal energy to the elasticity at a given L.

Use the data in Table 3.6, which contains tension versus extension length data for a cross-linked rubber, to construct a plot of $(\partial U/\partial L)_T$ and $-T(\partial S/\partial L)_T$ (divided by area) versus $(L-L_0)/L_0$ for rubber. To a good approximation, you may assume that the volume is constant during stretch, so that the area $A = A_0 L_0/L^2$, where A_0 is the cross-sectional area in the unstretched state.

3.7.D Table 3.7 shows the amounts of oxygen and nitrogen that adsorb onto solid zeolite as a function of pressure at 293 K. In order to design a pressure-swing adsorber, it is necessary to find the Langmuir adsorption function $\chi(T)$ at this temperature for each species. Using these data, find the maximum amount of each species that can be adsorbed per gram of zeolite, ρ_{max}, and $\chi(T = 293\text{ K})$. Find an estimate for the size of an oxygen or nitrogen molecule, and use this size to approximate the amount of surface area per mass of zeolite.

3.7.E The Brunauer–Emmett–Teller (BET) model for multilayer chemical adsorption. If we generalize the statistical-mechanical Langmuir model for adsorption to allow for more than one molecule being adsorbed on a single site, we find the following expression for the generalized potential:

$$\Psi = -M_s RT \log\left[\frac{c + (c-1)q_{\text{site}}(T)\exp[\mu/(RT)]}{c - q_{\text{site}}(T)\exp[\mu/(RT)]}\right], \tag{3.128}$$

where c is a dimensionless constant of order 100, and $q_{\text{site}}(T)$ is the single-site partition function as described in Section 3.7. The other constants have the same physical meaning as in the Langmuir model. The generalized potential Ψ is introduced in Exercise 3.3.D, and one might find the result of that problem useful here.

Table 3.7 Isothermal adsorption amount versus pressure for oxygen and nitrogen on a surface of 5-Å zeolite at 293 K. The amount adsorbed, ρ, is given in mmol of adsorbent per gram of bulk zeolite. Data from [112].

Nitrogen		Oxygen	
P (bar)	ρ (mmol adsorbed/g solid)	P (bar)	ρ (mmol adsorbed/g solid)
0.130	0.0710	0.217	0.0410
0.391	0.146	0.5	0.0626
0.957	0.310	0.783	0.0993
1.348	0.396	1.044	0.127
1.696	0.487	1.413	0.162
2.174	0.567	1.761	0.201
2.674	0.657	2.174	0.240
3.304	0.752	2.609	0.279
3.935	0.838	3.022	0.317
4.5	0.909	3.457	0.356
5.174	0.976	4.022	0.410
5.891	1.05	4.609	0.456
6.717	1.121	5.261	0.518
7.565	1.18	6	0.575
		7	0.646
		7.87	0.711
		8.67	0.771

The constant c is related to the difference in energy between adding the first molecule to the site and adding the second, or higher molecule. Hence, in the limit that $c \to \infty$, your answer should reduce to the Langmuir adsorption isotherm. You can use this limit to check your result. Plot $\theta := N/M_s$ vs. $\chi(T)P$ for $c = 160$ for this model. Can you interpret the shape of the plot?

3.7.F Colloidal systems consist of particles in a liquid that do not dissolve, but stay suspended [58]. Milk is such a system, with proteins and fats suspended in water. Many colloidal systems appear "milky." Paints are colloidal suspensions of clay particles, and many foods are also colloids. It is typically desirable to keep the particles suspended as long as possible. However, some particles prefer to aggregate into bigger particles, which then fall out of solution. To avoid this problem, some colloids are *polymerically stabilized*, by dissolving polymer chains in the liquid, which adsorb onto the colloidal particles (see Figure 3.19). These chains typically adsorb only partially, and leave tails and loops sticking out into the solvent. These polymer brushes "push" the particles apart and prevent agglomeration.[11]

[11] The polymer chains have more entropy when their loops and tails can move around. Pushing two particles together decreases the amount of room the loops have.

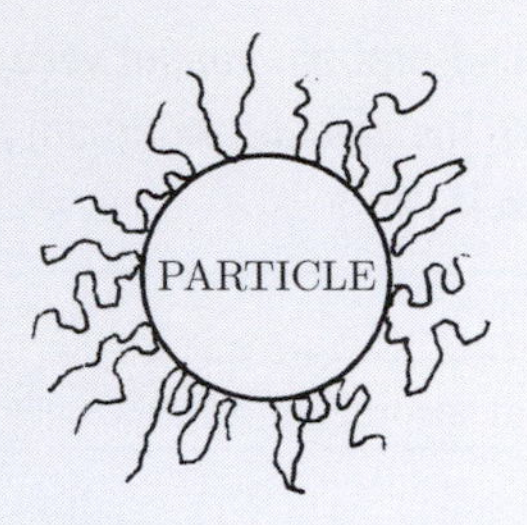

Figure 3.19 A sketch of a spherical particle in a solvent with polymer chains adsorbed. The loops and tails of the polymer can prevent the particle from agglomerating with other particles, making the suspension more stable.

A simple way to understand the phenomenon of adsorption is to model the process by a Langmuir adsorption isotherm. However, the polymer is no longer in the ideal-gas phase, but rather in solution. Here we assume that the concentration of the polymer in the bulk phase C_{eq} plays the role of pressure for the ideal gas, so that $\mu \cong \mu^{\circ}(T) + RT \log C_{\text{eq}}$.

An additional complication that arises for polymer adsorption is that the area occupied by a polymer chain when it is adsorbed can change with temperature, depending on how much of the chain adsorbs, and how much of the chain forms loops.

The temperature dependence of adsorption can exhibit rich behavior; however, it can be understood by a simple free-energy argument: to adsorb, the chain minimizes free energy. Adsorption gives the chain more order, and hence less entropy, so ΔS_{ads} of adsorption is negative. Adsorption requires, then, that ΔH_{ads} is also negative, so that $\Delta G_{\text{ads}} = \Delta H_{\text{ads}} - T\,\Delta S_{\text{ads}}$ of adsorption is negative and adsorption is favorable. As the temperature is changed, the quality of solvent can change, and the entropy of adsorption can be sensitive to temperature. Hence, the temperature dependence can be rather complicated.

Table 3.8 gives the mass of polymer adsorbed per area for ethyl(hydroxyethyl) cellulose on dehydroxylated silica particles, reported in [70]. Fit these data to the Langmuir adsorption equation to find the maximum adsorption and the adsorption parameter χ as functions of temperature. Can you explain the trends that you obtain? Can you say anything about the change in entropy of adsorption if you assume that the energy of adsorption is nearly constant?

3.7.G Pressure-swing adsorption uses the fact that some molecules preferentially adsorb onto the surface. If one begins with a gas phase that has equal numbers of moles of two species, and the pressure is raised, the preferentially adsorbed species will be the first to fill the adsorption sites, and will be more strongly depleted from the gas phase. If this procedure is repeated, on the depleted gas phase, the two species can be separated.

If the gas above the surface is ideal, the chemical potential of each species in the mixture is just the chemical potential of that pure species at the same temperature and partial pressure,

$$\mu_i(T, P, x_i) = \mu_i^{\circ}(T) + RT \log(x_i P), \quad \text{ideal gas.} \tag{3.129}$$

Table 3.8 Adsorption data for ethyl(hydroxyethyl)cellulose on dehydroxylated silica particles in ultra-high-purity water, taken from [70]. Ads. Amt., adsorbed amount.

$T = 18\,°\mathrm{C}$		$T = 25\,°\mathrm{C}$		$T = 37\,°\mathrm{C}$	
C_{eq} (ppm)	Ads. Amt. (mg/m^2)	C_{eq} (ppm)	Ads. Amt. (mg/m^2)	C_{eq} (ppm)	Ads. Amt. (mg/m^2)
1	0.616			2.03	1.19
2	0.811	1	0.790	10.1	1.51
7.41	1.14	6.05	1.14	34.1	1.71
13.7	1.1	10.7	1.11	40.3	1.66
17.3	1.05	31.7	1.34	56.5	1.94
53.7	1.17	40.6	1.28	62.3	1.88
58.6	1.12	75.4	1.39	101	1.99
97	1.22	88.9	1.29	147	2.01
104	1.17	124	1.41	158	1.95
109	1.12				
154	1.17				

However, since each site can hold only one molecule, the free energy for two types of adsorbed species is slightly different [44, p. 426],

$$\begin{aligned}\frac{F(T, M_S, N_1, N_2)}{RT} &= M_S \log\left(1 - \frac{N_1 + N_2}{M_S}\right) + N_1 \log\left(\frac{N_1}{M_S - N_1 - N_2}\right)\\ &\quad + N_2 \log\left(\frac{N_2}{M_S - N_1 - N_2}\right) - N_1 \log\left[q_1(T)\right] - N_2 \log\left[q_2(T)\right],\end{aligned} \tag{3.130}$$

where $q_i^{\mathrm{site}}(T)$, $i = 1, 2$, describes the energy of interaction between the adsorption site and molecule i.

Using this fundamental relation, and assuming that the gas phase is ideal, find the fractions of species 1, $\theta_1 := N_1/M_S$, and species 2, $\theta_2 := N_2/M_S$, adsorbed as functions of temperature, pressure, and mole fraction in the gas phase.

3.7.H Construct a fundamental relation for a rubber band, assuming (1) that the volume is constant as a function of stretch, (2) that the tension is linear in temperature, (3) that the tension is given by the data in Table 3.9, and (4) that the constant-length heat capacity is well described by a cubic function of temperature: $C_L = A_0 + A_1 T + A_2 T^2 + A_3 T^3$.

3.7.I The binding of protein to DNA could also be modeled using a pseudo-reaction expression,

$$K = \frac{c_{\mathrm{B}}}{c_{\mathrm{p}} c_{\mathrm{D}}}. \tag{3.131}$$

Using a similar species balance, as was done in Example 3.7.3, find values for K at the two cAMP concentrations. How good are the fits to the data? Does K show similar trends to χ with cAMP concentration?

Table 3.9 Elongational stress data for uniaxial stretch of cross-linked rubber. A rubber band with rest length L_0 is stretched, and the stress is measured as a function of the length L. Note that the stress is the tension divided by the cross-sectional area of the rubber. Data from Treloar [134].

Extension, $(L - L_0)/L_0$	Stress ($\mathrm{kg_f/cm^2}$)
0.0544	0.4329
0.1096	0.928
0.169	1.383
0.2687	2.054
0.432	3.040
0.7717	4.691
1.282	6.748
1.928	9.0542
2.560	11.56
3.210	14.78
3.704	18.76

3.7.J Here we consider a simplified model for the binding of oxygen to hemoglobin in the bloodstream. Hemoglobin (Figure 3.20) is responsible for oxygen transport in mammals, and makes up a substantial proportion by mass of blood. In humans hemoglobin has four globular proteins, each containing an iron ion, such that the complex can transport four oxygen molecules. Later we will use statistical mechanics to derive the fundamental relation for a simple model able to carry r molecules to be

$$\Psi(M, \mu, T) = -MRT \log\left[\frac{1}{M}\sum_{i=0}^{r} q_i(T) \exp\left(\frac{\mu i}{RT}\right)\right], \tag{3.132}$$

where M is the number of moles of hemoglobin molecules, T is the temperature, μ is the chemical potential of oxygen in the blood stream, and the $\{q_i(T)\}, i = 0, \ldots, r$ are functions of temperature only. Later we will see that the $\{q_i(T)\}$ are what are called "single-molecule partition functions," where i is the number of oxygens bound to each hemoglobin. Derive the ratio $\theta := N/(rM)$ as a function of the temperature and μ.

If the oxygen in the gas phase is assumed ideal, derive an expression for θ as a function of the partial pressure of oxygen in the gas phase.

Later, we will see that, if the binding energies of all oxygen molecules on the same hemoglobin do not interact, then we have just two functions to describe the partition functions: $q_0(T)$ and $q_1(T)$. Depending on the assumed symmetry of the hemoglobin complex [75], we could write, for example, $q_2 = 2q_1^2$, $q_3 = q_1^3$, and $q_4 = q_1^4$. Using the assumption of non-interacting oxygen molecules, is it possible to fit the data of Imai, given in Table 3.10? Do you obtain a significantly better fit to the data if you fit $\{q_i(T)/q_0(T)\}, i = 1, 2, 3, 4$?

3.7.K A Mooney plot is often used to represent the tension of a rubber band under stretch [134]. Here one plots the tension $\mathcal{T}$ divided by the stretch-dependence of the tension

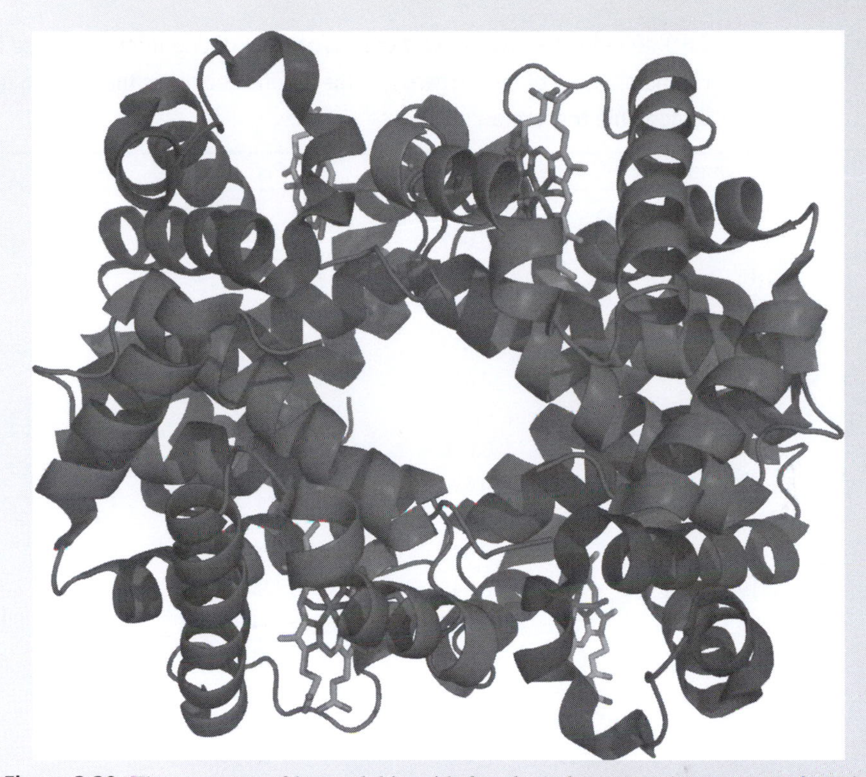

Figure 3.20 The structure of hemoglobin with four bound oxygens (oxygen not shown). From the Protein database (`http://www.rcsb.org/pdb/cgi/explore.cgi?pdbId=1GZX`).

predicted by an ideal rubber band versus the inverse of the stretch ratio $\lambda := L/L_0$. In this plot of $\mathcal{T}/[2(\lambda^2 - 1/\lambda)]$ vs. $1/\lambda$, one observes a significant region where the data fall on a straight line with intercept C_1 and slope C_2. Find the fundamental Helmholtz relation implied by this observation.

3.7.L Consider a model polymer chain of N independent steps, each step of length a_K. For a free, unconfined chain in a solvent, the Helmholtz potential F of the chain as a function of end-to-end length z can be well approximated by

$$F_{\text{free}}(T, z, N) = \frac{3k_B T z^2}{2Na_K^2}, \tag{3.133}$$

where T is the temperature, $k_B = R/\tilde{N}_A$ is Boltzmann's constant, $\tilde{N}_A$ is Avogadro's number, and R is the ideal-gas constant. See Figure 3.21.

On the other hand, if the chain is tethered at one end to a solid, flat surface, a simple statistical-mechanical argument estimates the Helmholtz potential to be

$$F_{\text{teth}}(T, z, r, N) = \frac{3k_B T\left(z^2 + r^2\right)}{2Na_K^2} - k_B T \log\left(\frac{3z}{Na_K}\right), \tag{3.134}$$

Table 3.10 Hemoglobin oxygen-uptake data of Imai, dating from 1990. The first column is the pressure of the oxygen, and the second column is the fraction of sites in the hemoglobin that are occupied.

P_{O_2} (mm Hg)	θ	P_{O_2} (mm Hg)	θ
0.37	0.007	8.72	0.537
0.37	0.007	9.69	0.621
0.46	0.009	10.77	0.700
0.46	0.009	10.77	0.700
0.58	0.011	11.96	0.768
0.72	0.013	13.43	0.826
0.90	0.016	14.77	0.870
1.14	0.020	16.41	0.903
1.14	0.020	18.43	0.929
1.41	0.025	20.49	0.947
1.75	0.032	22.77	0.961
2.19	0.043	28.13	0.977
2.68	0.060	34.75	0.985
3.38	0.085	43.39	0.991
3.38	0.085	53.61	0.992
3.75	0.104	66.93	0.996
4.17	0.129	82.70	0.996
4.64	0.160	102.17	0.998
5.15	0.197	126.24	0.998
5.73	0.246	155.97	1.000
6.36	0.307		
7.07	0.376		
7.85	0.454		

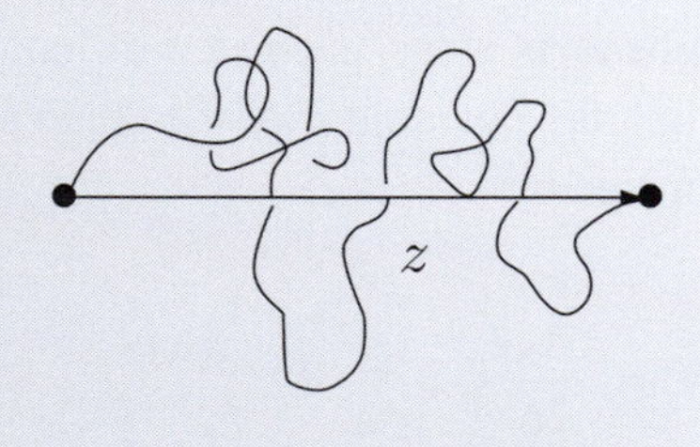

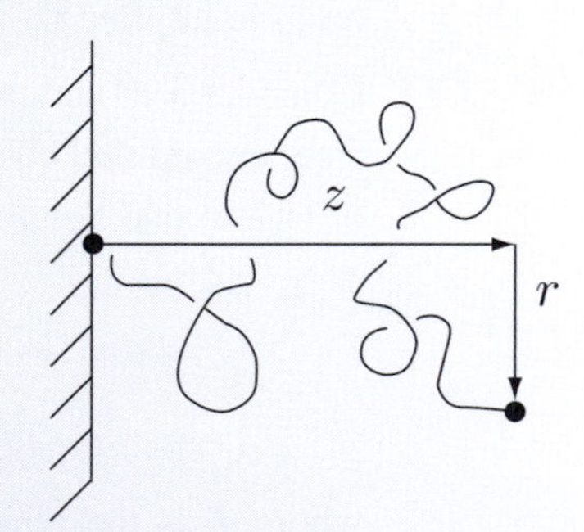

Figure 3.21 Sketches of an untethered polymer chain (left) and a tethered polymer chain (right), to illustrate the position variables r and z of the end-to-end vector.

where z is the perpendicular distance of the untethered end from the flat surface. The variable r is the radial distance from the line that runs perpendicular to the surface, and passes through the tether, as shown on the right in Figure 3.21.

1. If the chain is pulled away from the flat surface by a distance z, find the tension $\mathcal{T}_{\text{teth}}$ in the tethered chain. Find the similar tension $\mathcal{T}_{\text{free}}$ in the untethered chain which would be required in order to pull its ends a distance z apart.
2. Using your results, make a sketch of the dimensionless tension $\mathcal{T} a_K/(k_B T)$, versus the dimensionless length $z/(N_K a_K)$ at a fixed temperature. Make the sketch as quantitative as possible.
3. Explain where the approximations in the two models must break down, because the results are no longer physical.
4. How much work does it take to stretch the tethered chain from $z = \sqrt{N/3}a_K$ to $z = Na_K/3$ isothermally?

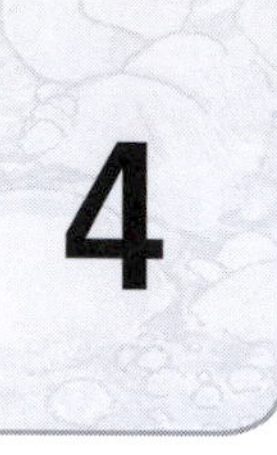

4 First applications of thermodynamics

Now that we have a reasonably complete structure of thermodynamics, we can tackle more complicated problems. In the following section we introduce the concepts of local and global stability, and show how local stability puts restrictions on second-order derivatives. In Section 4.2 we see that application of local stability criteria to the proposed fundamental relations leads to predictions of **spinodal curves**, which indicate when substances will spontaneously change state. In Section 4.2.2 the application of global stability to a van der Waals fluid leads to predictions of vapor saturation curves and liquid saturation curves. These curves are sometimes called **binodal curves** (or **coexistence curves**), and can be predicted from *PVT* equations of state alone. We then show how thermodynamic diagrams useful for refrigeration, or power-cycle design, can be constructed from *PVT* relations in Section 4.4. Section 4.4.2 shows generically how one can make predictions of differences in thermodynamic quantities from any of the equations of state shown in Appendix B or from experimental data.

4.1 STABILITY CRITERIA

What happens if you take gaseous nitrogen and compress it, keeping the internal energy constant by removing heat? Eventually, the nitrogen begins to condense in the container, and you have a mixture of liquid and gaseous nitrogen. If you isolate this system and wait for a long time, then you see that the system is indeed at a stable equilibrium. From Section 2.8 we know that these two phases have the same temperature and pressure. How can that be? Why is some of the nitrogen happy to stay gaseous, while the rest saw fit to condense into a liquid? The two phases have the same temperature and pressure, yet they have different densities. If we compress the fluid more, we find that these two phases are always present and have constant specific volume – further compression changes only the relative amounts of each phase. Why does the fluid not change the density of one or both phases? The answer to these questions comes from the concept of **stability**.

Because a real system is made up of discrete particles such as atoms and molecules, the density of a fluid in some region of the system is always fluctuating. If we divide a system into two equal-sized halves, each subdivision will have the same number of molecules *on average*. However, at any instant of time, one side may have several more molecules than the other. This would be a fluctuation in density. Similarly, there will be fluctuations in energy. It is possible that the entropy of the total system grows from one of these fluctuations. If so, we say that the system is **unstable** – as required by the fourth postulate. Since the fluctuations are unconstrained, any fluctuation that increases the entropy will continue to grow, according to the fourth postulate.

If the entropy grows for any size fluctuation, no matter how small, we say that the system is **locally unstable**. If a single-molecule imbalance raises the entropy, the system is locally unstable. However, if a fluctuation requires a certain finite size in order for it to find the state of higher entropy, we call the system **globally unstable**. Maybe the difference in densities of the two sides needs to be larger than some minimum number. If the entropy never grows for any size of fluctuation, then the system is **stable**.

The first section of this chapter derives the *stability criteria* for simple systems. The second section applies these criteria to predict the vapor–liquid equilibrium of pure substances using PvT equations of state.

4.1.1 Entropy

Recall that the second law states that the unconstrained variables of a system arrange themselves so as to maximize the entropy. We put N moles of a fluid in a rigid container of volume V, and insulate it so that its internal energy stays fixed at U. Now we conceptually split this container into two parts, such that there now exist two subsystems in the container with values $(U^{(1)}, V^{(1)}, N^{(1)})$ and $(U^{(2)}, V^{(2)}, N^{(2)})$ for the internal energy, volume, and mole number of the two subsystems. See Figure 4.1. The physical setup does not constrain the fluid to any particular values for these variables, provided that U, V, and N are conserved. Namely, the isolated system requires

$$U^{(1)} + U^{(2)} = U, \quad V^{(1)} + V^{(2)} = V, \quad N^{(1)} + N^{(2)} = N. \tag{4.1}$$

The system will be homogeneous, that is $U^{(1)}/N^{(1)} = U^{(2)}/N^{(2)} = U/N$ and $V^{(1)}/N^{(1)} = V^{(2)}/N^{(2)} = V/N$, only if the entropy is thus maximized. The system will become inhomogeneous if

$$S(U, V, N) < S(U^{(1)}, V^{(1)}, N^{(1)}) + S(U^{(2)}, V^{(2)}, N^{(2)}), \quad \text{globally unstable.} \tag{4.2}$$

We say that the system is unstable because such a system is susceptible to splitting, according to the postulates of thermodynamics. We shall explain the qualifier "global" shortly.

For simplification, we consider for the moment that the only quantity that fluctuates between the two phases is the volume. In other words, we consider the special (perhaps unphysical) case in which

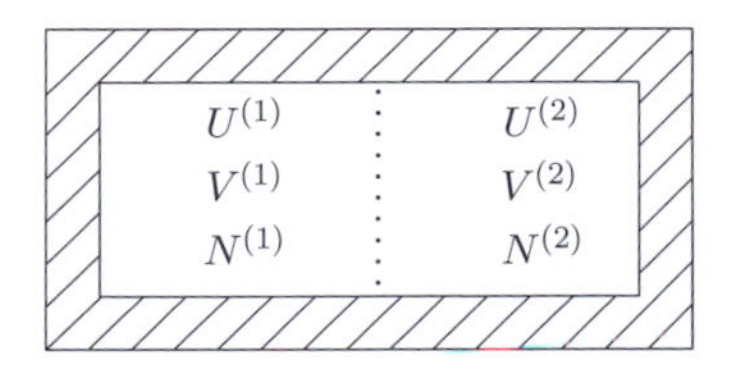

Figure 4.1 We consider an isolated fluid with total energy, volume, and mole number (U, V, N). We then divide the fluid into two compartments with an imaginary partition, shown by the dotted line, and ask when it is thermodynamically more stable for the two different compartments to have different specific volumes.

$$\begin{aligned} U^{(1)} &= U^{(2)} = \frac{1}{2}U, \\ N^{(1)} &= N^{(2)} = \frac{1}{2}N, \\ V^{(1)} &= \frac{1}{2}(V + \Delta V), \\ V^{(2)} &= \frac{1}{2}(V - \Delta V). \end{aligned} \tag{4.3}$$

Then, the global stability criterion becomes

$$\begin{aligned} S(U, V, N) < S&\left(\frac{1}{2}U, \frac{1}{2}(V + \Delta V), \frac{1}{2}N\right) \\ &+ S\left(\frac{1}{2}U, \frac{1}{2}(V - \Delta V), \frac{1}{2}N\right), \quad \text{globally unstable.} \end{aligned} \tag{4.4}$$

Or, if we use the extensive property of the entropy,

$$S(U, V, N) < \frac{1}{2}S(U, V + \Delta V, N) + \frac{1}{2}S(U, V - \Delta V, N), \quad \text{globally unstable.} \tag{4.5}$$

As a specific example, consider the entropy of a van der Waals gas at low internal energy as shown in Figure 4.2. We constrain the volume at V, then the system might begin with entropy S_i. However, in this state the system is not globally stable because it can split into two different phases to increase its overall entropy while keeping its overall volume and internal energy constant. It does this by some of the fluid condensing to a high-density phase with volume $V - \Delta V$, whereas the rest expands to a low-density phase with volume $V + \Delta V$.

As well as having two different volumes, the two phases also have two different entropies, namely $S(V - \Delta V)$ and $S(V + \Delta V)$.[1] The resulting composite entropy is $S_I = \frac{1}{2}(S(V + \Delta V) + S(V - \Delta V))$, also shown in Figure 4.2, which is greater than $S_i = S(V)$. Hence, a homogeneous phase can become unstable. In fact, all volumes between $V - \Delta V$ and $V + \Delta V$ are globally unstable.

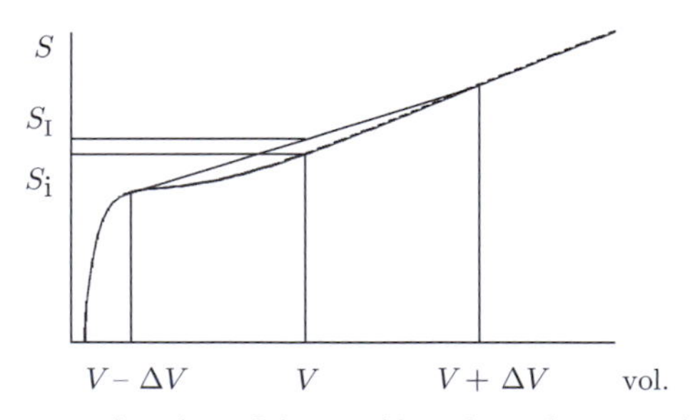

Figure 4.2 The specific entropy as a function of the specific volume for a van der Waals gas at low (constant) internal energy. For volumes near V, the system is locally unstable. See Eqn. (4.6).

[1] In a real system, these two phases would have the same temperature and pressure (not U), so they cannot be shown on this curve, which assumes constant internal energy. We will see in Section 4.2.2 that it is more natural to use the Helmholtz potential to describe phase transitions of pure components instead of entropy.

We explore the consequences of global instability in greater detail in Section 4.2. For the moment we consider the weaker criterion of **local stability**. A system is locally unstable when Eqn. (4.5) holds in the limit $\Delta V \to 0$. If we perform a second-order Taylor-series expansion (see Section A.1 of Appendix A) on both terms on the right-hand side of Eqn. (4.5) around (U, V, N), we obtain to lowest order in ΔV

$$\left(\frac{\partial^2 S}{\partial V^2}\right)_{U,N} > 0, \quad \text{locally unstable.} \tag{4.6}$$

The region of *local* instability in Figure 4.2 is much smaller than (and lies entirely within) the region of *global* instability.

We can create similar local stability criteria by examining changes in internal energy, or simultaneous changes in both volume and internal energy. Similarly to what we saw before, the criteria for global instability are

$$S(U,V,N) < \frac{1}{2}S(U+\Delta U,V,N) + \frac{1}{2}S(U-\Delta U,V,N), \quad \text{globally unstable,}$$
$$S(U,V,N) < \frac{1}{2}S(U+\Delta U,V+\Delta V,N) + \frac{1}{2}S(U-\Delta U,V-\Delta V,N).$$

Expanding the terms on the right-hand side of each equation in a Taylor-series expansion yields the criteria for local instability after some manipulation [21, p. 207]. We can then write the criteria for local stability as

$$\left(\frac{\partial^2 S}{\partial U^2}\right)_{V,N} < 0, \quad \text{locally stable,} \tag{4.7}$$

$$\left(\frac{\partial^2 S}{\partial V^2}\right)_{U,N}\left(\frac{\partial^2 S}{\partial U^2}\right)_{V,N} - \left(\frac{\partial^2 S}{\partial U\,\partial V}\right)_N^2 > 0, \quad \text{locally stable.} \tag{4.8}$$

Again, these criteria for local stability are less restrictive than the criteria for global stability.

EXAMPLE 4.1.1 What restriction is placed on second-order derivatives by the local stability criterion on entropy for fluctuations in energy, Eqn. (4.7)?

Solution. We note that we can write the left-hand side of Eqn. (4.7) as

$$\begin{aligned}\left(\frac{\partial^2 S}{\partial U^2}\right)_{V,N} &= \left(\frac{\partial}{\partial U}\left(\frac{\partial S}{\partial U}\right)_V\right)_V \\ &= -\frac{1}{T^2}\left(\frac{\partial T}{\partial U}\right)_V \\ &= -\frac{1}{T^2(\partial U/\partial T)_V} \\ &= -\frac{1}{T^3(\partial S/\partial T)_V},\end{aligned}$$

where the second line follows from the definition of temperature, and the rest are standard thermodynamic manipulations shown in Section 3.6. Note that the denominator in the fraction of the last line is simply $T^2 C_V$, where C_V is called the **constant-volume heat capacity**. Thus, the local stability criterion Eqn. (4.7) then says that the constant-volume heat capacity must be positive

$$C_V := T\left(\frac{\partial S}{\partial T}\right)_V > 0, \quad \text{locally stable,} \tag{4.9}$$

otherwise the system will split into two phases, and each phase will have positive constant-volume heat capacity. □

Note that the requirement found in this example is independent of the specific model for fundamental energy used to describe the system – it is a result strictly of the postulates of thermodynamics. Interpretation of the other local stability relations is more easily handled using the generalized potentials in Section 4.1.3, which we shall cover shortly.

4.1.2 Internal energy

We may also look at stability criteria when other variables are manipulated instead of (U, V). For example, if we manipulate entropy and volume, it is natural to consider the internal energy as the dependent variable. In Section 3.2 we found that an equilibrium state is attained when internal energy is minimized at constant entropy. If we make local stability arguments for U analogous to those we made for S, we find that local stability requires

$$\left.\begin{aligned} \left(\partial^2 U/\partial S^2\right)_{V,N} &> 0 \\ \left(\partial^2 U/\partial V^2\right)_{S,N} &> 0 \\ \left(\partial^2 U/\partial V^2\right)_{S,N}\left(\partial^2 U/\partial S^2\right)_{V,N} - \left(\partial^2 U/\partial S\partial V\right)_N^2 &> 0 \end{aligned}\right\} \quad \text{locally stable.} \tag{4.10}$$

Manipulation of the first two derivatives leads to the requirements that

$$C_V := T\left(\frac{\partial S}{\partial T}\right)_V > 0, \quad \kappa_S := -\frac{1}{V}\left(\frac{\partial V}{\partial P}\right)_S > 0, \quad \text{local stability,} \tag{4.11}$$

where κ_S is the **isentropic compressibility**.

4.1.3 Generalized potentials

Not surprisingly, the generalized potentials also satisfy extremum conditions at equilibrium. Using the properties of Legendre transforms, it is possible to show (also see Exercises 4.1.C and 4.1.D) that the conditions for local stability are

$$\left.\begin{aligned} \left(\partial^2 H/\partial S^2\right)_{P,N} > 0, \quad \left(\partial^2 H/\partial P^2\right)_{S,N} < 0 \\ \left(\partial^2 F/\partial T^2\right)_{V,N} < 0, \quad \left(\partial^2 F/\partial V^2\right)_{T,N} > 0 \\ \left(\partial^2 G/\partial T^2\right)_{P,N} < 0, \quad \left(\partial^2 G/\partial P^2\right)_{T,N} < 0 \end{aligned}\right\} \quad \text{locally stable.} \tag{4.12}$$

EXAMPLE 4.1.2 Prove the second stability criterion for H.

Solution. From Table 3.1 we know that

$$V = \left(\frac{\partial H}{\partial P}\right)_S .$$

Taking the derivative of each side with respect to P holding S constant yields

$$\left(\frac{\partial V}{\partial P}\right)_S = \left(\frac{\partial^2 H}{\partial P^2}\right)_S .$$

However, from the definition of pressure, we also know that

$$\left(\frac{\partial P}{\partial V}\right)_S = -\left(\frac{\partial^2 U}{\partial V^2}\right)_S ,$$

whence

$$\left(\frac{\partial^2 H}{\partial P^2}\right)_S = -1 \Big/ \left(\frac{\partial^2 U}{\partial V^2}\right)_S .$$

Hence, the stability for enthalpy against fluctuations in pressure has sign opposite to that for internal energy against fluctuations in volume. Combining this equation with the stability criterion for U, Eqn. (4.10), yields the second relation in the first line of Eqn. (4.12). □

Note that, in all cases, regions of stability against fluctuations in an extensive variable are concave up, and those in intensive variables are concave down. Manipulation of the second stability criterion for F leads to the requirement (Exercise 4.1.B) that

$$\kappa_T > 0, \tag{4.13}$$

which is an important criterion in constructing phase diagrams for pure-component systems, as will be seen in Section 4.2.

In Example 3.6.3 we found a relationship between the constant-volume heat capacity and the constant-pressure heat capacity. Using Eqn. (3.86) and the stability condition Eqn. (4.11), we find that

$$C_P = C_V + \frac{V\alpha^2 T}{\kappa_T} > C_V > 0. \tag{4.14}$$

Similarly, using the techniques of Section 3.6 and Eqn. (4.11), we find

$$\kappa_T = \frac{C_P}{C_V}\kappa_S > \kappa_S > 0. \tag{4.15}$$

The physical interpretation of these results is straightforward. Increasing the temperature of a system at constant pressure or volume necessarily increases its entropy, which means that heat must be added. Also, increasing the pressure of a system at either constant temperature or constant entropy necessarily decreases its volume. The results are summarized in Table 4.1.

Table 4.1 Restrictions on second-order derivatives from stability considerations.

Restrictions
$C_P > C_V > 0$
$\kappa_T > \kappa_S > 0$

Finally, we can now prove the fourth relation for temperature that we promised on p. 33. From the differential for internal energy, we can prove that

$$\left(\frac{\partial U}{\partial T}\right)_{V,N} = C_V > 0.$$

Since the constant-volume heat capacity must be positive by virtue of stability, raising the temperature at constant volume necessarily increases the internal energy of a substance.

4.2 SINGLE-COMPONENT VAPOR–LIQUID EQUILIBRIUM

Sometimes we have little information about the fundamental relation for a substance, but we know a great deal about its mechanical equation of state. For example, we may know its *PVT* behavior over a wide range of temperatures and pressures, or we may use statistical-mechanical arguments or atomistic simulations to derive such an expression. In this section we show how such information is sufficient to understand the phase stability of single-component systems. By studying the phase stability, we can use equations of state to predict the vapor saturation line and liquid saturation line for pure-component systems (the **binodal curve**). An equation of state that describes both the liquid phase and the vapor phase of a substance is called an equation of state for a **fluid**.

Rather than give the general principles involved in such manipulations, we begin with an example of the van der Waals fluid. The procedure used to reveal the phase behavior of a van der Waals fluid is general for any fluid, and the general procedure is outlined in Section 4.2.3.

4.2.1 The spinodal curve of a van der Waals fluid

The ideal-gas model assumes that molecules consist of point particles that do not interact with each other. From statistical mechanics, it is known that the pressure can be expressed as a sum of repulsive and attractive intermolecular contributions. A simple expression for the repulsive contributions can be obtained by analyzing the behavior of the molar volume in the limit of infinite pressure; the ideal-gas equation of state predicts that the volume should vanish. At high pressures, however, we expect most fluids to solidify; a reasonable model for a fluid should therefore reach an asymptotic volume beyond which it can no longer be compressed. That deficiency can be corrected by assuming that the *repulsive* part of the pressure is given by

$$P = \frac{RT}{v - b}, \tag{4.16}$$

where b can be interpreted as the molar volume that the fluid would have in the limit of infinite pressure. At constant temperature, Eqn. (4.16) predicts a monotonic decrease of the volume with pressure. As v approaches b, the pressure diverges.

We also know that atoms contain electrons and protons, which interact with each other according to Coulomb's law; assuming that interactions between different atoms are pairwise additive, the pressure of a fluid is expected to decrease (below the purely repulsive value) by an amount proportional to the square of the density. By considering both attractive and repulsive contributions, a simple mechanical equation of state can be written in the form

$$P = \frac{RT}{v - b} - \frac{a}{v^2}. \tag{4.17}$$

Equation (4.17) is the familiar van der Waals equation of state, Eqn. (2.35). It was first proposed by Johannes Diderik van der Waals in his doctoral thesis more than a hundred years ago [142]. In spite of its simplicity, this equation is able to capture qualitatively the rich and sometimes complex phase behavior of pure fluids and their mixtures. Therefore, although it is often not sufficiently accurate, the van der Waals fluid may be used as an illustrative example of the techniques used to predict phase behavior for pure fluids.

In Figure 4.2 we see that the van der Waals fluid is locally and globally unstable for some values of U and V, since κ_T is negative in these regions. Hence, the fluid can split into two different phases in order to maximize the entropy. We now explore these phase changes in greater detail. This exploration is done most easily by examining the pressure as a function of volume at constant temperature, Eqn. (4.17), as shown in Figure 4.3 for several temperatures. We find that the pressure decreases monotonically with increasing volume at high temperatures. This trend is consistent with our local stability criterion, Eqn. (4.13).

However, at low temperatures the pressure undergoes oscillatory behavior. When the slope of one of these curves is positive, then the equation of state predicts that κ_T is negative – an

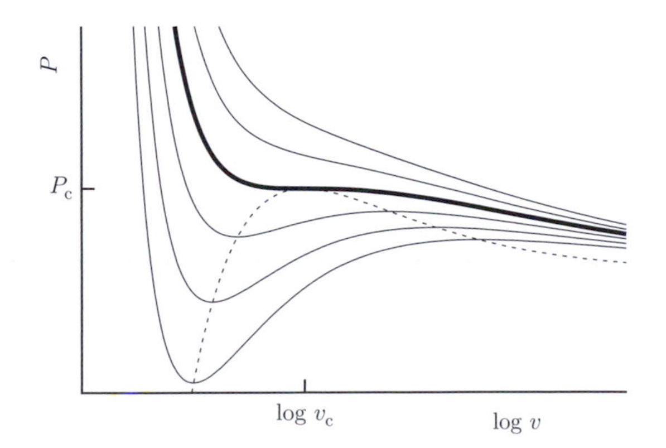

Figure 4.3 Pressure as a function of specific volume for a van der Waals gas at several temperatures (solid lines). At high temperatures, the pressure decreases monotonically with volume, indicating stability. At low temperatures, we see that $(\partial P/\partial v)_T$ becomes positive in some regions, indicating local instability. The envelope of local instability is within the spinodal curve (dashed line). The bold line is the critical isotherm.

indication of local instability according to Eqn. (4.13). The dashed line in Figure 4.3 shows the region of local instability where κ_T is negative, and is called the **spinodal curve**.

There is a unique point in this plot that corresponds to the maximum in the spinodal curve called the **critical point**.[2] The isotherm that passes through this point is called the **critical isotherm**, with **critical temperature** T_c. Correspondingly, there is a **critical pressure** P_c, and a **critical volume** v_c. When the temperature is above T_c, the volume decreases monotonically with pressure; the system is a so-called **super-critical fluid**. Super-critical fluids are neither gases nor liquids; when compressed isothermally, their density changes smoothly, and they undergo no phase change. One can see all densities between liquid and gas. Fluids near the critical point have some very interesting properties, which we wait until Section 6.7 to discuss. For the moment we are interested in temperatures below T_c, where the fluid can be either a gas or a liquid, depending on the pressure and volume.

First, we find the critical variables for the van der Waals fluid of Figure 4.3. The criterion for local instability is $(\partial P/\partial v)_T > 0$. The spinodal curve is where the system goes from stable to unstable. Hence, we can find the spinodal point at a given volume by finding where $(\partial P/\partial v)_T = 0$ or $\kappa_T \to \infty$ from Eqn. (4.17). When we set the derivative to zero and solve for the spinodal temperature, we obtain

$$RT_s(v) = \frac{2a(v-b)^2}{v^3}, \tag{4.18}$$

where $T_s(v)$ is the spinodal temperature as a function of volume. Upon inserting the expression for the spinodal temperature Eqn. (4.18) into the van der Waals equation of state, Eqn. (4.17), we obtain an expression for the spinodal curve:

$$P_s(v) = \frac{a}{v^2}\left(1 - \frac{2b}{v}\right). \tag{4.19}$$

This is the equation used to plot the dashed line in Figure 4.3. We can now find the critical point by finding the maximum of this curve. Taking the derivative of each side of Eqn. (4.19) with respect to volume, we obtain

$$\frac{dP_s}{dv} = \frac{2a}{v^3}\left(\frac{3b}{v} - 1\right). \tag{4.20}$$

Clearly, the derivative is zero when the volume is $3b$, to give the critical volume

$$v_c = 3b, \qquad \text{van der Waals fluid.} \tag{4.21}$$

The critical pressure is found from the spinodal curve, Eqn. (4.19), and the critical temperature from the spinodal temperature, Eqn. (4.18), at the critical volume

$$\left.\begin{aligned} P_c &= P_s(v_c) = a/(27b^2), \\ T_c &= T_s(v_c) = 8a/(27Rb). \end{aligned}\right\} \quad \text{van der Waals fluid.} \tag{4.22}$$

[2] According to Sengers and Levelt [122], James Clerk Maxwell and Diederek Johannes Korteweg credit Arthur Cayley as having been the first to define the critical point.

Using these relations, we can eliminate constants in the van der Waals equation of state Eqn. (4.17) to express it in reduced form as

$$P_r = \frac{8T_r}{3v_r - 1} - \frac{3}{v_r^2}, \quad \text{van der Waals fluid,} \tag{4.23}$$

where we have defined the **reduced variables**

$$T_r := \frac{T}{T_c}, \quad \text{reduced temperature,} \tag{4.24}$$

$$P_r := \frac{P}{P_c}, \quad \text{reduced pressure,} \tag{4.25}$$

$$v_r := \frac{v}{v_c}, \quad \text{reduced volume.} \tag{4.26}$$

We have now completely characterized the region of local instability for the van der Waals fluid. Also, by introducing reduced variables, we have a universal expression for the *PVT* behavior of all fluids, insofar as they are described by the van der Waals fluid. There are many such equations of state that predict such universal behavior for fluids, called **corresponding-states** models. This means that, if you plot the *PVT* data for many fluids in terms of reduced quantities, the data for different materials should lie on top of one another. Such behavior is observed approximately for a number of simple chemical species, even though they do not obey well the van der Waals equation of state. The relationships for some fluids, like water, with special molecular associations (called **hydrogen bonding**) do not fit with those of the other, simple, fluids.

Note that the van der Waals fluid predicts a universal value of the critical compressibility factor,

$$z_c := \frac{P_c v_c}{RT_c} = \frac{3}{8}, \quad \text{van der Waals fluid,} \tag{4.27}$$

where the **compressibility factor** z is defined for a non-ideal fluid as

$$z := \frac{PV}{NRT}. \tag{4.28}$$

However, we know experimentally that fluids exhibit a range of critical compressibility factors typically below ≈ 0.3. Hence, the van der Waals fluid is not capable of capturing the *PVT* relations of a real fluid accurately near the critical region.

The region of global instability is larger than, and entirely encompasses, the region of local instability; the curve demarcating the line of global instability is called the **binodal curve** (not shown in Figure 4.3). In the locally unstable region, the material will spontaneously undergo phase separation without external influence. However, in the region that is globally unstable, but locally stable, a system free of external perturbation may remain as a single phase for long periods of time. For example, there have been news reports about the dangers of heating water in a microwave oven. Water placed in a ceramic mug and heated sufficiently long in the oven (an isobaric process) can be put inside the binodal curve, but outside the spinodal curve. If the mug is not perturbed, the water may stay liquid, although this state is only metastable. When the oven user removes the mug and adds a teabag or spoon, or just perturbs the water, it may suddenly and rapidly boil, dousing the holder of the mug with scalding water.

Analogously, a fluid may sometimes be cooled very slowly and carefully below its freezing point without crystallizing. However, when the container is tapped from the outside, the entire fluid may crystallize extremely rapidly. Such a **supersaturated** liquid is an example of a substance that is inside the binodal curve, but outside the spinodal curve. Once the temperature has been lowered sufficiently to bring the substance to the spinodal curve, crystallization will occur spontaneously, even if the material is kept vibration-free.

These properties are exploited in engineering practice when producing crystals. If a system is kept between the binodal and spinodal curves, crystallization begins at a single point, or at very few points, in the liquid. However, if the system is quenched to a point inside the spinodal region, crystallization is nucleated at many points. In the latter case, many tiny crystals form. However, in the former situation it is possible to produce very large crystals of high purity. Such large crystals are necessary for X-ray crystallography to study conformational properties of proteins, for example.

4.2.2 The binodal (or coexistence) curve of a van der Waals fluid

Consider the sub-critical isotherm of a van der Waals vapor shown in Figure 4.4. The thermodynamic state of the vapor would be on a point in the lower right-hand corner of the figure, and would have low density; in other words, the fluid would be a gas. Now begin to increase the pressure gradually, holding the temperature constant. The evolution of the thermodynamic state of the system may be represented by a point moving slowly along the sub-critical isotherm from right to left. Before reaching the region of local instability, marked by the spinodal curve, the point will first reach the *binodal curve*, which envelops the region of global stability. Physically, this corresponds to the point where the fluid begins to condense; most of the fluid is still a gas, but a tiny amount becomes liquid.

Since these two phases are in thermal, mechanical, and chemical contact, they must satisfy the equilibrium conditions found in Chapter 2, namely

$$T_\mathrm{v} = T_\mathrm{l},$$
$$P_\mathrm{v} = P_\mathrm{l},$$
$$\mu_\mathrm{v} = \mu_\mathrm{l},$$

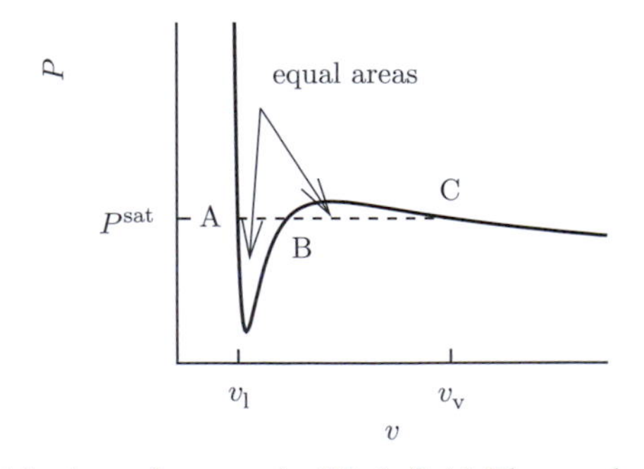

Figure 4.4 A single sub-critical isotherm for a van der Waals fluid. The equal-area construction determines the location of the saturation pressure, the liquid density, and the vapor density.

where the subscripts v and l indicate vapor and liquid, respectively. These are Eqns. (2.69)–(2.71) rewritten for vapor and liquid phases in contact. The pressure at which the liquid is in equilibrium with the vapor is called the **saturation pressure**, $P^{sat} = P_v = P_l$, and is unique for a given temperature.

Hence, the liquid phase in equilibrium with the vapor phase must lie on the same isotherm as the vapor, and must have the same pressure. A curve of constant pressure on this plot is a horizontal line. However, many horizontal lines may intersect with our sub-critical isotherm at three points. Which of these horizontal lines is the correct one is determined by the fact that the chemical potentials of the two phases must also be equal. We can find the location of the binodal curve from just the mechanical equation of state and the equivalence of the chemical potentials in the following way. The result is surprising: when the areas shown in Figure 4.4 are equal, the saturation pressure is found.

Recall that the chemical potential of a pure substance is just the specific Gibbs free energy, Eqn. (3.21). We can find the change in Gibbs free energy (or chemical potential) as we change volume along the isotherm. At constant temperature the differential for the Gibbs potential, Eqn. (3.51), becomes

$$\left(\frac{\partial \mu}{\partial v}\right)_{T,N} \equiv \left(\frac{\partial g}{\partial v}\right)_{T,N} = v\left(\frac{\partial P}{\partial v}\right)_{T,N}, \tag{4.29}$$

which also follows from the Gibbs–Duhem relation. If we integrate each side at constant temperature from v_v to v_l, we obtain

$$\begin{aligned} \mu_l - \mu_v &= \int_{v_v}^{v_l} v\left(\frac{\partial P}{\partial v}\right)_{T,N} dv \\ 0 &= P^{sat}(v_l - v_v) - \int_{v_v}^{v_l} P\, dv \\ 0 &= \int_{v_v}^{v_l} \left[P^{sat}(T) - P(T,v)\right] dv. \end{aligned} \tag{4.30}$$

We integrated by parts to obtain the second line, and used the fact that P^{sat} is independent of volume to obtain the third. Since the left-hand side is zero, the result has a straightforward graphical representation. We search for a horizontal line $P = P^{sat}$ that intersects the isotherm at three points: in the locally unstable region (B), at the liquid volume (A), and at the vapor volume (C); see Figure 4.4. This line creates two closed areas: one below P^{sat}, between points A and B, and one above P^{sat}, between points B and C. In order for the integral above to be zero, these two areas must be equal.[3] The horizontal line that yields two equal areas is then $P = P^{sat}$.

If we find $v_v(P^{sat})$, $v_l(P^{sat})$, and P^{sat} for all temperatures, we can construct the binodal curve, which is shown as the dashed line in Figure 4.5. Whenever the *total specific volume* and temperature of a system lie inside this region, there is not a single phase with this specific volume. Instead, the system has two phases, whose *composite* specific volume lies inside the vapor–liquid region. As we change the volume of the system isothermally, the relative amounts of the

[3] This criterion is sometimes called the Maxwell equal-area rule, after James Clerk Maxwell, who first derived it in 1875.

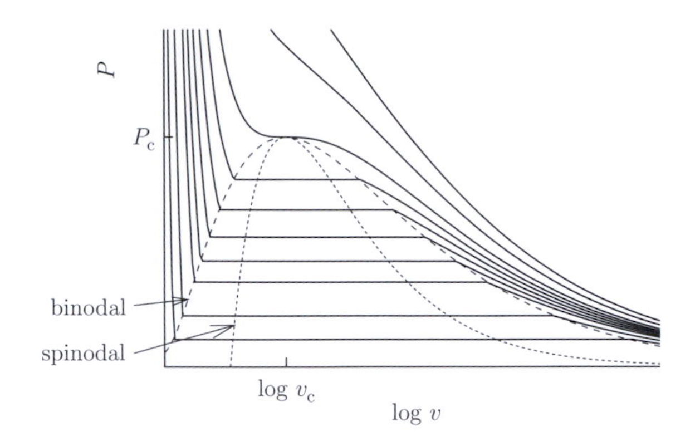

Figure 4.5 The binodal curve for a van der Waals fluid. The isotherms predicted by the equation of state are not drawn inside the vapor–liquid region, where they are globally unstable. Instead, horizontal tie lines connecting the vapor and liquid phases that are in equilibrium with one another are drawn. Note that the volume is plotted on a logarithmic scale in order to make the shape of the vapor–liquid region clearer.

two phases change, but the composite system moves horizontally along the drawn tie line (the dashed line in Figure 4.4).

Let us derive the equations that determine the binodal curve for the van der Waals fluid. Since they are in mechanical contact, the two phases must have identical pressures, the so-called saturation pressure P^{sat}. Using the van der Waals equation of state, Eqn. (2.35), we may write

$$P^{\text{sat}} = \frac{RT}{v_{\text{v}} - b} - \frac{a}{v_{\text{v}}^2} \tag{4.31}$$

and

$$P^{\text{sat}} = \frac{RT}{v_{\text{l}} - b} - \frac{a}{v_{\text{l}}^2}. \tag{4.32}$$

By inserting Eqn. (4.17) into the integral in Eqn. (4.30), we obtain the equal-area construction for a van der Waals fluid:

$$\begin{aligned} 0 &= \int_{v_{\text{v}}}^{v_{\text{l}}} \left(P^{\text{sat}} - P\right) dv \\ &= P^{\text{sat}}(v_{\text{l}} - v_{\text{v}}) + RT \log\left(\frac{v_{\text{v}} - b}{v_{\text{l}} - b}\right) + a\left(\frac{1}{v_{\text{v}}} - \frac{1}{v_{\text{l}}}\right). \end{aligned} \tag{4.33}$$

For a given temperature, these represent three equations, with three unknowns (P^{sat}, v_{v}, v_{l}). Since the last equation is nonlinear in the volumes, these equations cannot be solved to obtain explicit expressions for the unknowns. When they are solved numerically at sub-critical temperatures, the dashed line in Figure 4.5 results. The equation of state is cubic in volume, and may be handled using the formulation given in Section A.7 of Appendix A. This method is the easiest numerical method to find P^{sat}.

A faster way to find the saturation pressure *graphically* is to construct a parametric plot. Figure 4.6 shows the pressure isotherm for a van der Waals fluid. Shown on the same figure is a chemical-potential isotherm. From Eqn. (B.16) in Appendix B, we can find the chemical potential:

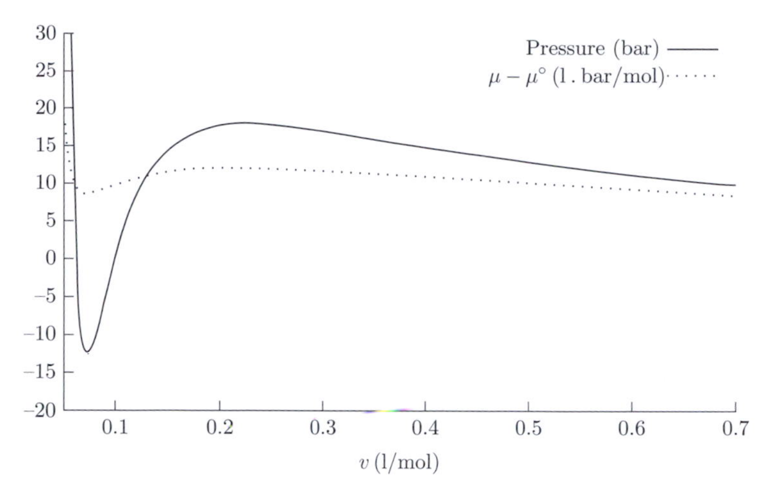

Figure 4.6 The pressure and chemical potential for a van der Waals fluid as a function of specific volume at constant temperature. For this fluid, $a = 1.366\,\mathrm{l}^2 \cdot \mathrm{bar/mol}^2$, and $b = 0.038\,57\,\mathrm{l/bar \cdot mol}$. Note that both the pressure and the chemical potential are non-monotonic.

$$\begin{aligned}\mu &= \left(\frac{\partial F}{\partial N}\right)_{T,V} \\ &= f_0 - s_0(T - T_0) - RT\log\left(\frac{v-b}{v_0}\right) + \frac{RTv}{v-b} - \frac{2a}{v} + \int_{T_0}^{T}\left(\frac{T'-T}{T'}\right)c_v^{\mathrm{ideal}}(T')dT' \\ &= \mu^\circ(T) - RT\log(v-b) + \frac{RTv}{v-b} - \frac{2a}{v}.\end{aligned} \tag{4.34}$$

We have separated out the bits on the right-hand side that depend on temperature only (and not volume), since these are constant for the isotherm. Hence, for convenience we have defined

$$\mu^\circ(T) := f_0 - s_0(T - T_0) + RT\log(v_0) + \int_{T_0}^{T}\left(\frac{T'-T}{T'}\right)c_v^{\mathrm{ideal}}(T')dT'. \tag{4.35}$$

Now, instead of using the equal-area rule, we seek the two thermodynamic states for which the temperature, pressure, and chemical potential are the same. Since the curves in Figure 4.6 are isotherms, the temperature criterion is already satisfied. Equal pressure is a horizontal line on this plot, as is equal chemical potential. It is easier to find this location by constructing a plot of chemical potential versus pressure. Such a plot is shown in Figure 4.7. *This figure is the easiest graphical method by which to obtain the saturation pressure.*

If the apparent specific volume V/N of a system lies between the specific liquid volume v_l and the specific vapor volume v_v, then two phases are present. We can find the relative amount of the phases using the **lever rule**. If x_l is the fraction of liquid in the mixture, the total volume of the system can be written

$$V = v_\mathrm{v}N_\mathrm{v} + v_\mathrm{l}N_\mathrm{l}, \tag{4.36}$$

where N_l and N_v are the numbers of moles of liquid and vapor, respectively. If we replace the number of moles of vapor with $N_\mathrm{v} = N - N_\mathrm{l}$, and divide each side by N, we find

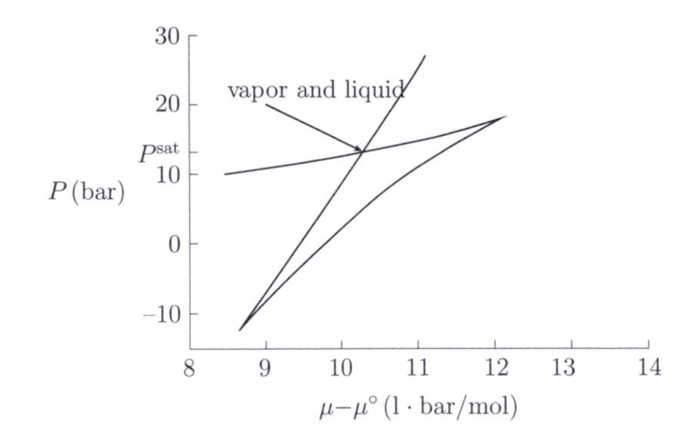

Figure 4.7 Pressure vs. chemical potential for a van der Waals fluid at constant temperature, with specific volume varying parametrically. This is the same fluid as shown in Figure 4.6. Because those curves are non-monotonic, the curve here doubles back on itself. At that point, there are two different thermodynamic states that have the same temperature, pressure, and chemical potential. These must be the liquid and vapor states. From this plot we can read that the saturation pressure is approximately 13 bar.

$$x_{\text{l}} = \frac{v_{\text{v}} - V/N}{v_{\text{v}} - v_{\text{l}}}, \tag{4.37}$$

which has a straightforward graphical representation as a ratio of two lengths along a tie line in Figure 4.5.

In conclusion, we find that the fluid obeys the van der Waals equation of state outside the binodal curve, but follows the straight tie lines constructed inside the binodal curve showing a mixture of vapor and liquid, as drawn in Figure 4.5. The behavior is qualitatively, but not quantitatively, correct. Better success can be had using the Peng–Robinson equation of state, or the Soave modification to the Redlich–Kwong equation of state (see Appendix B).

4.2.3 The general formulation

Given a *PVT* equation of state in the form $P = P(v, T)$, we can find the spinodal curve P_{s}, saturation pressure P^{sat}, vapor and liquid volumes v_{v} and v_{l}, for any temperature, and the critical constants by application of the following relations.

The spinodal curve is determined by solution of the relation

$$\left(\frac{\partial P}{\partial v}\right)_T = 0. \tag{4.38}$$

The critical point is determined by the simultaneous satisfaction of Eqn. (4.38) and the relation

$$\left(\frac{\partial^2 P}{\partial v^2}\right)_T = 0. \tag{4.39}$$

The vapor saturation line, the liquid saturation line, and the saturation pressure (the binodal curve) are found as functions of temperature by the simultaneous solution of the three equations

$$P^{\text{sat}}(T) = P(v_{\text{l}}, T), \tag{4.40}$$

$$P^{\text{sat}}(T) = P(v_{\text{v}}, T), \tag{4.41}$$

$$0 = \int_{v_{\text{v}}}^{v_{\text{l}}} \left[P^{\text{sat}}(T) - P(v,T)\right] dv. \tag{4.42}$$

The last relation may be found graphically by use of the equal-area construction illustrated in Figure 4.4, or by a plot of the sort shown in Figure 4.7.

4.2.4 Approximations based on the Clapeyron equation

The procedures considered above for predicting $P^{\text{sat}}(T)$ can be very accurate and apply over the entire two-phase region, but require some numerical effort. Methods for predicting the saturation pressure over small ranges of temperature also exist. However, these do not give the full binodal curve. Instead, they are based on approximations to the exact Clapeyron equation. The Clapeyron equation can be derived from the equivalence of chemical potentials in the two phases:

$$\begin{aligned} \mu^{\text{liq}}[T, P^{\text{sat}}] &= \mu^{\text{vap}}[T, P^{\text{sat}}], \\ g^{\text{liq}}[T, P^{\text{sat}}] &= g^{\text{vap}}[T, P^{\text{sat}}]. \end{aligned} \tag{4.43}$$

The second line follows from the equivalence between chemical potential and specific Gibbs free energy. If we take the differential of each side of Eqn. (4.43), and use the differential for g, Eqn. (3.51), we obtain

$$\begin{aligned} dg^{\text{liq}}[T, P^{\text{sat}}] &= dg^{\text{vap}}[T, P^{\text{sat}}]. \\ -s^{\text{liq}}\, dT + v^{\text{liq}}\, dP^{\text{sat}} &= -s^{\text{vap}}\, dT + v^{\text{vap}}\, dP^{\text{sat}}. \end{aligned} \tag{4.44}$$

Note that in this differential we are staying on the saturation curve, since we have equated chemical potentials. Therefore, the temperature and pressure are both changing, but each must be the same in both phases. In fact, P must be the saturation pressure. We can rewrite this expression as a differential equation,

$$\begin{aligned} \frac{dP^{\text{sat}}}{dT} &= \frac{s^{\text{vap}} - s^{\text{liq}}}{v^{\text{vap}} - v^{\text{liq}}} \\ &= \frac{h^{\text{vap}} - h^{\text{liq}}}{T(v^{\text{vap}} - v^{\text{liq}})}, \end{aligned} \tag{4.45}$$

which is the **Clapeyron equation**. The second line follows from the Legendre transform expression $g = h - Ts$. This expression is exact, and, in fact, holds for the coexistence curve for any two phases, not just liquid and vapor.

We can derive our first useful expression by noting that the specific volume of the vapor is much greater than that for the liquid, and that it may be approximated (not particularly well) as an ideal gas:

$$v^{\text{vap}} - v^{\text{liq}} \cong v^{\text{vap}} \cong \frac{RT}{P^{\text{sat}}}. \tag{4.46}$$

When this approximation is inserted into the Clapeyron equation, we obtain the **Clausius–Clapeyron equation**

$$\frac{d \log P^{\text{sat}}}{dT} \cong \frac{\Delta h^{\text{vap}}}{RT^2}, \quad \text{Clausius–Clapeyron equation,} \tag{4.47}$$

where $\Delta h^{\text{vap}} := h^{\text{vap}} - h^{\text{liq}}$ is the heat of vaporization, so called because it is the amount of heat necessary to vaporize a liquid at constant pressure. This surprising relation between two rather different quantities can be used to estimate the change in saturation pressure with temperature. We first integrate each side of Eqn. (4.47) from T_0 to T. If the temperature range is not too large, the heat of vaporization is not strongly temperature-dependent, so we can make the approximation

$$\begin{aligned} \log\left(\frac{P^{\text{sat}}(T)}{P^{\text{sat}}(T_0)}\right) &= \int_{T_0}^{T} \frac{\Delta h^{\text{vap}}}{RT'^2}\, dT' \\ &\cong \frac{\Delta h^{\text{vap}}}{R}\left(\frac{1}{T_0} - \frac{1}{T}\right), \end{aligned} \tag{4.48}$$

which may be solved for $P^{\text{sat}}(T)$. The form of this equation suggests that another approximation is valid over somewhat larger regions. This third approximation, called the **Antoine equation**, has three adjustable parameters, A, B, and C, which have been tabulated for many substances,

$$\log_{10} P^{\text{sat}}(T) \cong A - \frac{B}{T + C}. \tag{4.49}$$

We will see in Chapters 8 and 9 that such approximations are useful for predicting solubility, or vapor–liquid equilibrium of mixtures. Values for the parameters A, B, and C are given in Table D.4 in Appendix D for several substances.

Alternatively, one can use the predictions of P^{sat} from a PvT equation of state to make a universal relation. For the van der Waals model, after some numerical effort, we find

$$\log\left(\frac{P_{\text{r}}^{\text{sat}}[T_{\text{r}}]}{P_{\text{c}}}\right) \cong B(1 - T_{\text{r}}) + C(1 - T_{\text{r}})^2 + D(1 - T_{\text{r}})^3, \tag{B.24$'$}$$

where the van der Waals model predicts $B = -4.081\,21$, $C = -2.046\,25$, $D = -7.413\,66$, and the result is valid for reduced temperatures between 0.625 and 1. Similar empirical relations that are based on other models are also given in Appendix B.

4.3 CRYSTALLIZATION OF SOLIDS

Many other possible phase changes exist besides the vapor-to-liquid phase transition. The most commonly known are solid-to-liquid (melting), solid-to-vapor (sublimation), and their opposites (freezing and condensation). Here we consider briefly the transitions between solid and liquid.

For vapor–liquid transitions, the density changes dramatically. However, for liquid–solid transitions, density changes are typically rather small, and sometimes even negative. We understood vapor–liquid transition in terms of a balance between attractive and repulsive forces of

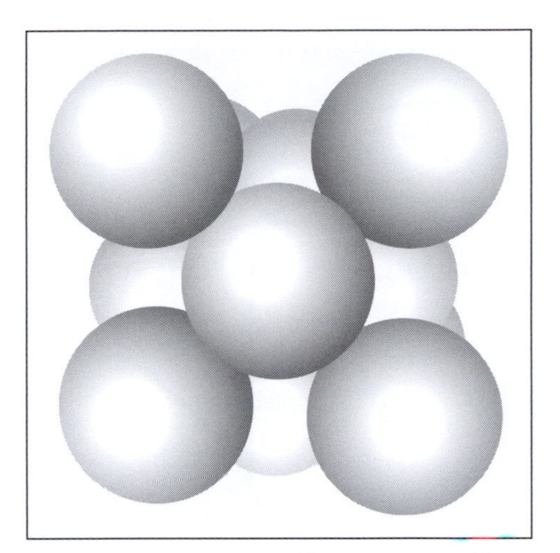

Figure 4.8 A sketch of the positions of gold atoms making up a crystalline solid. This picture shows cubic close packing, determined by X-ray diffraction.

atoms. Attraction is of long range, and repulsion is short range. These two forces are accounted for very crudely in the van der Waals equation of state through the parameters a and b, respectively. Recall that the phase change comes from lowering the Gibbs free energy, G, which is $H - TS$. The denser liquid has lower energy, but also lower entropy. The less dense vapor has higher energy, but also higher entropy. These two phases can become balanced. For the solid–liquid transition, these simple density arguments are insufficient, since the density is nearly constant.

The solid crystalline state arises because of *order* differences. The molten state is very disorderly compared with the regularly arrayed crystal (see Figure 4.8). Therefore, the entropy change of freezing is very negative. So, if the entropy change is negative, and therefore unfavorable, the energy change must be even more negative. This can be possible when the atoms on the molecules prefer to have specific neighbors, for example. Or, it can be that the temperature has become so low that the entropic contribution to G is too low. For water, the —OH groups like to arrange themselves in very specific ways, energetically speaking. However, these arrangements are not packed very efficiently. Therefore, the ordered state of water is less dense than liquid water – *ice floats*. This particularly strong affinity for a certain relative orientation of —OH bonds on adjacent water molecules is called **hydrogen bonding**.

Figure 4.9 shows the particularly rich phase diagram exhibited by water. At higher pressures, several different distinct solid phases exist. Many of these phases have been found only very recently, and researchers continue to study the thermodynamic behavior of this important substance [119].

4.4 THERMODYNAMIC DIAGRAMS

Several types of thermodynamic diagrams are useful for engineering design. For example, in power-plant analysis, diagrams of temperature vs. entropy are commonly used, whereas

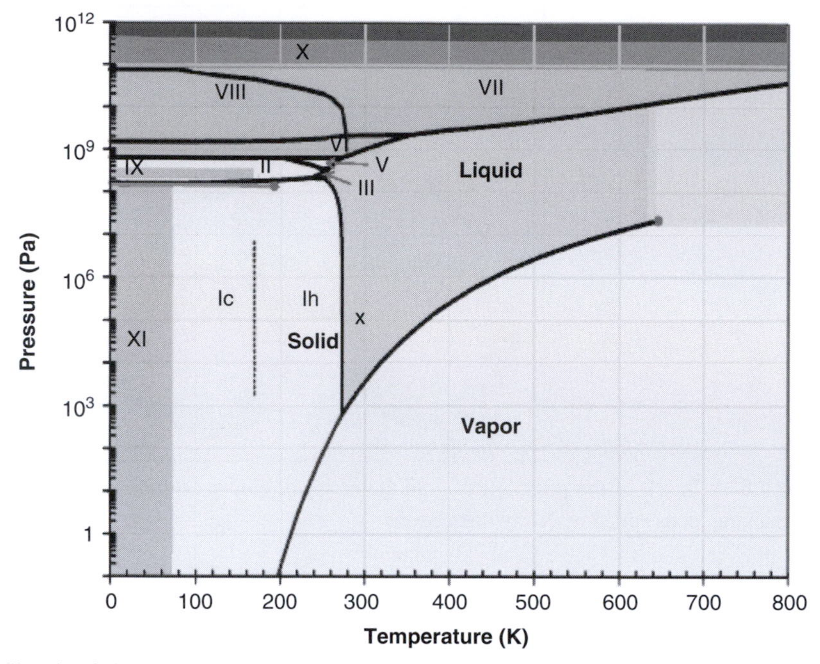

Figure 4.9 Sketch of the phase diagram of water, showing the known solid and fluid phases. Courtesy of Martin Chaplain; reproduced with permission.

pressure vs. enthalpy plots are useful in refrigeration design. For example, Figure 5.10 on p. 176 makes designing a refrigeration cycle simpler.

We saw in the previous section that PvT equations of state can be used to predict phase changes in fluids. In order to construct TS and PH diagrams, we also need a thermal equation of state. However, given a mechanical equation of state, a thermal equation of state may not be chosen arbitrarily; rather, it must be consistent with the fact that both equations of state are derivable from a single fundamental relation, as we did in Examples 2.6.1, 2.9.1, 3.1.2, and 3.1.3.

Even when the underlying fundamental relation is *not* known, we can still construct consistency criteria for the equations of state. For example, consider the Maxwell-like relation that follows from the analytic property of entropy:

$$\begin{aligned} \frac{\partial s(u,v)}{\partial v\, \partial u} &= \frac{\partial s(u,v)}{\partial u\, \partial v}, \\ \left(\frac{\partial}{\partial v}\frac{1}{T}\right)_u &= \left(\frac{\partial}{\partial u}\frac{P}{T}\right)_v . \end{aligned} \tag{4.50}$$

The second line follows from the alternative definitions for temperature and pressure, Eqns. (2.30) and (2.31). If we use the van der Waals equation of state to evaluate the right-hand side, we obtain

$$-\frac{1}{T^2}\left(\frac{\partial T}{\partial v}\right)_u = \frac{a}{T^2 v^2}\left(\frac{\partial T}{\partial u}\right)_v, \qquad \text{van der Waals fluid.}$$

On multiplying each side by $T^2(\partial u/\partial T)_v$ and using Eqn. (A.20) from Appendix A, we obtain

$$\left(\frac{\partial u}{\partial v}\right)_T = \frac{a}{v^2}, \quad \text{van der Waals fluid.} \tag{4.51}$$

Note that the right-hand sides of Eqns. (4.50) and (4.51) are determined by the *PVT* equation of state, whereas the left-hand sides are determined by a thermal equation of state, $u = u(T, v)$. This result shows that it is not correct to use a thermal equation of state that depends only on temperature with a van der Waals mechanical equation of state. Note that we used the *PVT* relation only, without the fundamental relation, to derive Eqn. (4.51). We now illustrate two different (but equivalent) methods for obtaining thermodynamic diagrams.

4.4.1 Construction of fundamental relations from two equations of state for single-component systems

So far we have accepted "given" fundamental relations from which we construct equations of state. However, one typically analyzes experimental data to arrive at an equation of state: pressure measured at several temperatures and volumes, and heat capacity as a function of temperature. The very first example of Chapter 2 made exactly such a connection from experiment to equation of state. Here we illustrate how one can construct the fundamental Helmholtz relation from just these two equations of state.[4]

Say we measure $C_P^{\text{ideal}}(T)$ and $P(T, v)$ for a range of temperatures and specific volumes. Then, since our independent variables are T and v, the natural potential is the Helmholtz potential. Recall that the differential for the Helmholtz potential allows us to write $P = -(\partial f/\partial v)_T$. Hence for the van der Waals equation of state we may write

$$\left(\frac{\partial f}{\partial v}\right)_T = -\frac{RT}{v-b} + \frac{a}{v^2}. \tag{4.52}$$

Integrating both sides over specific volume from an *ideal* reference volume v_0 to arbitrary v yields

$$f(T, v) - f[T, v_0] = -RT \log\left(\frac{v-b}{v_0-b}\right) - a\left(\frac{1}{v} - \frac{1}{v_0}\right). \tag{4.53}$$

We need to find $f[T, v_0]$, which is determined by the thermal equation of state. Recalling that the fluid is an ideal gas at v_0, we write

$$\begin{aligned} f[T, v_0] &= f[T_0, v_0] + \int_{T_0}^{T} \left(\frac{\partial f^{\text{ideal}}}{\partial T}\right)_v \Bigg|_{v=v_0, T=T'} dT' \\ &= f_0 - \int_{T_0}^{T} s^{\text{ideal}}[T', v_0] dT' \\ &= f_0 - \int_{T_0}^{T} \left\{ s_0 + \int_{T_0}^{T'} \left(\frac{\partial s^{\text{ideal}}}{\partial T''}\right)_{v_0} dT'' \right\} dT' \\ &= f_0 - s_0(T - T_0) - \int_{T_0}^{T} \int_{T_0}^{T'} \frac{c_v^{\text{ideal}}[T'']}{T''} dT''\, dT', \end{aligned} \tag{4.54}$$

[4] The first example of deriving a fundamental relation from equations of state was given for a rubber band in Section 3.7.1.

where the second line follows from the differential for F, and the fourth line follows from the definition of the constant-volume heat capacity. We can simplify the double integral a bit by changing the order of integration:

$$
\begin{aligned}
f[T, v_0] - f_0 + (T - T_0)s_0 &= -\int_{T_0}^{T}\int_{T_0}^{T''} \frac{c_v^{\text{ideal}}(T')}{T'}\, dT'\, dT'' \\
&= -\int_{T_0}^{T}\int_{T'}^{T} \frac{c_v^{\text{ideal}}(T')}{T'}\, dT''\, dT' \\
&= -\int_{T_0}^{T} \left(\frac{T - T'}{T'}\right) c_v^{\text{ideal}}(T') dT'. \qquad (4.55)
\end{aligned}
$$

The second line follows from a change in the order of integration, and the third from performing the integration over T''. We have now found a very convenient form for the fundamental Helmholtz relation of a general van der Waals fluid:

$$
\begin{aligned}
f = f_0 - s_0(T - T_0) - RT\log\left(\frac{v - b}{v_0 - b}\right) - a\left(\frac{1}{v} - \frac{1}{v_0}\right) \\
+ \int_{T_0}^{T} \left(\frac{T' - T}{T'}\right) c_v^{\text{ideal}}(T') dT', \quad \text{general van der Waals fluid.} \qquad (4.56)
\end{aligned}
$$

Since v_0 is an ideal volume, we could further simplify this equation using $v_0 \gg b$, and a/v_0 makes a negligible contribution to the pressure. We may now construct all possible thermodynamic diagrams for this fluid.

For example, we may find an equation of state for the entropy by taking the derivative of each side of Eqn. (4.56) with respect to T,

$$
s(T, v) = s_0 + R\log\left(\frac{v - b}{v_0 - b}\right) + \int_{T_0}^{T} \frac{c_v^{\text{ideal}}(T')}{T'} dT'. \qquad (4.57)
$$

Also, the thermal equation of state may be found from the definition of f, Eqn. (3.10),

$$
\begin{aligned}
u(T, v) &= f(T, v) + Ts(T, v) \\
&= u_0 - a\left(\frac{1}{v} - \frac{1}{v_0}\right) + \int_{T_0}^{T} c_v^{\text{ideal}}(T') dT'. \qquad (4.58)
\end{aligned}
$$

It is worth considering the form of Eqn. (4.58). We see that the internal energy depends both on temperature and on volume. Only in the ideal limit $v = v_0$ (or $a = 0$) does the internal energy depend on temperature only. Whenever one uses a volume-independent heat-capacity expression alone to find changes in internal energy or enthalpy, one is necessarily assuming ideality, and the heat capacities in Eqns. (4.58) and (4.56) are *ideal heat capacities*. If the system is non-ideal, either the fundamental Helmholtz potential should be found, or the appropriate *PVT* relation should be used to find the residual properties discussed in Section 4.4.2.

For a general pressure-explicit equation of state, using the steps above we find

$$
f = f_0 - s_0(T - T_0) + RT\int_{v}^{v_0} \frac{z(T, v)}{v} dv + \int_{T_0}^{T} \left(\frac{T' - T}{T'}\right) c_v^{\text{ideal}}(T') dT', \qquad (4.59)
$$

where z is the compressibility factor, $z := P_v/(RT)$. Now we see from where many of the fundamental relations in Appendix Section B come. Also, using relations derived from statistical mechanics, one can even get c_v^{ideal} from infrared-spectroscopy measurements (see Exercise 6.3.C). Alternatively, a calorimeter can be used.

EXAMPLE 4.4.1 Construct a *TS* diagram using a simple van der Waals fluid.

Solution. If the van der Waals fluid is simple, then $c_v^{\text{ideal}} = cR$ is constant, and Eqn. (4.57) is

$$s = -\left(\frac{\partial f}{\partial T}\right)_v = s_0 + R\log\left[\left(\frac{T}{T_0}\right)^c\left(\frac{v-b}{v_0}\right)\right], \quad \text{simple van der Waals fluid.} \tag{4.60}$$

We may also solve the van der Waals equation of state to obtain an explicit expression for $T = T(P, v)$:

$$T = \frac{v-b}{Rv^2}(a + Pv^2). \tag{4.61}$$

Using Eqns. (4.61) and (4.60), we can plot T as a function of S at fixed P, by changing v parametrically. For example, on a spreadsheet, we fix the pressure for a single value of P, and make a column of values for v, between, say $2b$ and $20b$. In the second column, we can then calculate values for T using Eqn. (4.61). In the third column, we find values for s using Eqn. (4.60). We can then plot an isobar from the second and third columns.

However, we know from our considerations in Section 4.2.2 that, for $P < P_c$ and some regions of v, the van der Waals fluid is globally unstable and undergoes phase separation. In these regions, there are liquid and vapor phases that are in equilibrium with each other; hence, the two phases have the same temperature and pressure, but may have different volumes, entropies, enthalpies, etc. Inside this region, the fluid no longer obeys the equation of

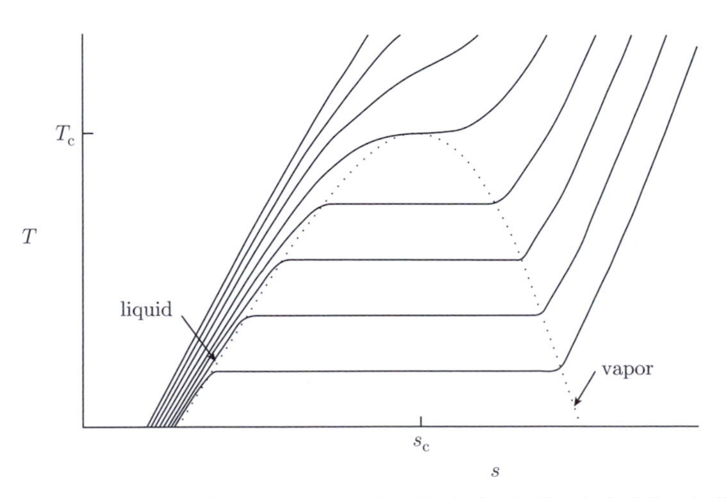

Figure 4.10 Temperature vs. entropy for a simple van der Waals fluid. The dashed line indicates the vapor–liquid region. The solid lines are isobars. Outside the vapor–liquid region, the curves are determined by the equations of state. Inside, the fluid separates into a liquid phase and a vapor phase connected by horizontal tie lines.

state. Hence, we do not draw the isobars through this region, but instead connect the points of saturated vapor in equilibrium with saturated liquid using tie lines. Therefore, when constructing the columns for sub-critical isobars, we must vary the values for the volume in a more careful way. Namely, we systematically vary the values for v until $v = v^{\text{liq}}$. At this point, $P = P^{\text{sat}}$. In the next row, we put $v = v^{\text{vap}}$, and again $P = P^{\text{sat}}$. In the following rows we again vary v systematically with values greater than $v = v^{\text{vap}}$. Note that the binodal region is always determined by the procedure in Section 4.2, and cannot be found directly from $s(T, P)$ curves.

If we plot T as a function of s at constant pressure, the curves in Figure 4.10 result. □

The dashed line in Figure 4.10 encloses the vapor–liquid region. The entropy at the critical point is found by evaluating s in Eqn. (4.60) at T_c and v_c given in Eqns. (4.21) and (4.22).

Similarly, a *PH* diagram for a van der Waals fluid may also be constructed. The mechanical equation of state for the van der Waals fluid is Eqn. (4.17),

$$P = \frac{RT}{v - b} - \frac{a}{v^2}.$$

From its definition, the enthalpy for a simple van der Waals fluid is found from Eqn. (2.34) to be

$$\begin{aligned} h(T, v) &:= u(T, v) + P(T, v)v \\ &= RT\left(c + \frac{v}{v - b}\right) - \frac{2a}{v}. \end{aligned} \tag{4.62}$$

Hence, we can plot P as a function of h parametrically in v, when we hold the temperature constant. Again, there are regions of T and v where two phases exist, and the equation of state does not apply. Using Eqns. (4.17) and (4.62), we can create Figure 4.11.

We have seen that the van der Waals equation of state does a reasonably good job of predicting the trends of real fluids, at least qualitatively. However, for engineering practice we typically require more accurate equations of state. There is no shortage of more realistic equations of

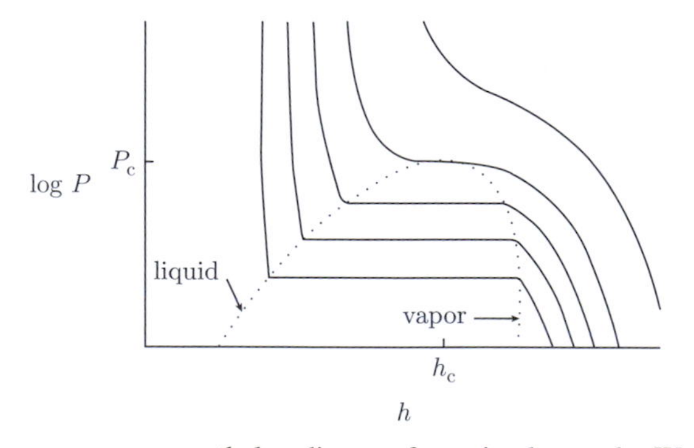

Figure 4.11 Isotherms on a pressure vs. enthalpy diagram for a simple van der Waals fluid. The dashed line indicates the vapor–liquid region. The solid lines are isotherms. Outside the vapor–liquid region, the curves are determined by the equations of state. Inside, the fluid separates into a liquid phase and a vapor phase connected by horizontal tie lines.

state being used. In Appendix B we summarize briefly just a few that are found empirically to describe many materials over reasonable ranges of T and P.

Many useful thermodynamic diagrams have been constructed using this method with more realistic equations of state. For example, a PH diagram for ammonia, which is useful for designing freezers, is shown in Figure 5.10 on p. 176. This diagram was constructed from a more complicated equation of state, but the procedure is the same.

4.4.2 Residual properties

From the fundamental Helmholtz relations found in the last section, we could, in principle, find all possible thermodynamic properties at any thermodynamic state. However, it is customary to use **residual properties** to estimate changes in thermodynamic properties. The two methods are equivalent.

Consider the TP thermodynamic-state plane shown in Figure 4.12. We wish to find the difference in enthalpy between the two states (T_f, P_f) and (T_i, P_i),

$$\Delta H \equiv H[T_f, P_f] - H[T_i, P_i]. \tag{4.63}$$

To a first approximation, we could assume that the change in enthalpy is that found for an ideal gas for the same states,

$$\Delta H^{\text{ideal}} \equiv H^{\text{ideal}}[T_f, P_f] - H^{\text{ideal}}[T_i, P_i]. \tag{4.64}$$

However, for many states of interest, such an approximation is quite poor, and we need to find a correction to this value – for example, when at least one of the states is liquid. Hence, using just the ideal-gas approximation for an isothermal pressure change is very risky.

An accurate estimate of the change in a thermodynamic property can be found from the following hypothetical path. First, we change the pressure at constant temperature from (T_i, P_i) to $(T_i, 0)$. At the latter pressure, the system behaves like an ideal gas. Then, we change the pressure isothermally back to its original value P_i, but assuming that the system is ideal for the entire path.

Third, we change the temperature and pressure in the ideal state from (T_i, P_i) to (T_f, P_f). This third step is given by Eqn. (4.64). The last two steps of the hypothetical path are similar to the first two, except that the final pressure is P_f instead of P_i. This path is also shown in Figure 4.12. The path is written mathematically as

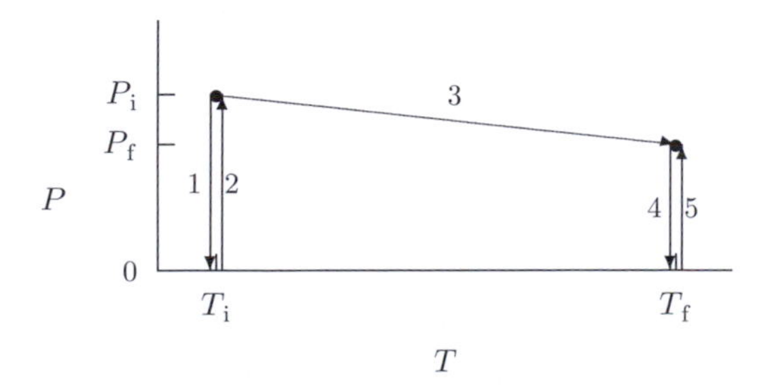

Figure 4.12 The hypothetical path used to find changes in thermodynamic properties of real quantities. The vertical arrows are the hypothetical path used to construct residual properties.

$$\begin{aligned}\Delta H &= \underbrace{[H[T_f,P_f]-H[T_f,0]]}_{\textbf{Step 5}} + \underbrace{\left[H^{\text{ideal}}[T_f,0]-H^{\text{ideal}}[T_f,P_f]\right]}_{\textbf{Step 4}} \\ &\quad + \underbrace{\left[H^{\text{ideal}}[T_f,P_f]-H^{\text{ideal}}[T_i,P_i]\right]}_{\textbf{Step 3}} + \underbrace{\left[H^{\text{ideal}}[T_i,P_i]-H^{\text{ideal}}[T_i,0]\right]}_{\textbf{Step 2}} \\ &\quad + \underbrace{[H[T_i,0]-H[T_i,P_i]]}_{\textbf{Step 1}} \\ &= H^{\text{R}}[T_f,P_f] + \Delta H^{\text{ideal}} - H^{\text{R}}[T_i,P_i]. \end{aligned} \tag{4.65}$$

Note that we have assumed that $H(T,0) = H^{\text{ideal}}(T,0)$. In the second line we have introduced the **residual property**, which is defined as

$$M^{\text{R}}(T,P) := M(T,P) - M^{\text{ideal}}(T,P), \tag{4.66}$$

the difference between the real property M (e.g., H, U, S, F, or G) and its value *at the same temperature and pressure if the system were an ideal gas*. We see from Eqn. (4.65) that the residual properties are corrections to changes found assuming an ideal gas.

Using the first two steps of the hypothetical path shown in Figure 4.12, we can find the residual properties from an equation of state. Assume that we are given a volume-explicit PVT equation of state $z = z(T,P)$, where $z := Pv/(RT)$ is called the **compressibility factor**. We can find the residual Helmholtz potential from the following manipulation:

$$\begin{aligned} F^{\text{R}}(T,P) &:= F(T,P) - F^{\text{ideal}}(T,P) \\ &= [F(T,P)-F(T,0)] - \left[F^{\text{ideal}}(T,P)-F^{\text{ideal}}(T,0)\right] \\ &= \int_0^P \left[\left(\frac{\partial F}{\partial P}\right)_T - \left(\frac{\partial F^{\text{ideal}}}{\partial P}\right)_T\right] dP \\ &= -\int_0^P \left[P\left(\frac{\partial V}{\partial P}\right)_T - P\left(\frac{\partial V^{\text{ideal}}}{\partial P}\right)_T\right] dP \\ &= -\int_0^P \left[NRT\left(\frac{\partial z}{\partial P}\right)_T - \frac{zNRT}{P} + \frac{NRT}{P}\right] dP \\ &= -NRT(z-1) + NRT\int_0^P (z-1)\frac{dP}{P}. \end{aligned}$$

We used the differential for F to obtain the fourth line; that $V = zNRT/P$ and $V^{\text{ideal}} = NRT/P$ to obtain the fifth; and the fundamental theorem of calculus to obtain the third and sixth lines. If we perform similar manipulations on the other potentials and entropy, we obtain the following relations:

$$\frac{G^{\text{R}}(T,P,\{N_i\})}{NRT} = \int_0^P [z(T,P',\{N_i\})-1]\frac{dP'}{P'}, \tag{4.67}$$

$$\frac{H^{\text{R}}(T,P,\{N_i\})}{NRT} = -T\int_0^P \left(\frac{\partial z}{\partial T}\right)_{P',\{N_i\}} \frac{dP'}{P'}, \tag{4.68}$$

$$\frac{U^{R}(T,P,\{N_i\})}{NRT} = \frac{H^{R}}{NRT} - (z-1)\,, \tag{4.69}$$

$$\frac{F^{R}(T,P,\{N_i\})}{NRT} = \frac{G^{R}}{NRT} - (z-1)\,, \tag{4.70}$$

$$\frac{S^{R}(T,P,\{N_i\})}{NR} = \frac{H^{R}-G^{R}}{NRT}. \tag{4.71}$$

These equations are useful whenever an equation of state is available that is explicit in volume. There is no reason to memorize them.

EXAMPLE 4.4.2 Calculate the residual volume of carbon dioxide at 50 °C and 7 bar using a virial equation of state truncated after the second term.

Solution. The residual molar volume of a fluid is given by

$$v^{R} = v - v^{\text{ideal}}. \tag{4.72}$$

For simplicity, we use the volume-explicit form of the virial equation of state truncated after the second term. The actual volume of the system is therefore given by

$$v = \frac{RT}{P} + B. \tag{4.73}$$

The residual volume is therefore

$$v^{R} = \frac{RT}{P} + B - \frac{RT}{P} = B. \tag{4.74}$$

This result is interesting in that it shows that the residual volume of a virial gas is equal to the virial coefficient itself; it therefore depends on temperature but not on pressure!

To estimate B for CO_2 we can use one of the correlations discussed in Chapter 6, or the Pitzer correlation given in Section B.2 of Appendix B. The critical pressure and temperature of CO_2 are 73.8 bar and 304.1 K, respectively. The acentric factor is $\omega_{CO_2} = 0.239$. At 50 °C, the residual molar volume of CO_2 is therefore estimated to be -102.5 cm^3/mol. This residual volume is negative, indicating that at 50 °C the real fluid is more compact, or denser than the corresponding ideal gas. This was to be expected, considering that, at low to modest temperatures, attractive forces between the molecules will tend to bring them closer together. □

EXAMPLE 4.4.3 Find an expression for the residual enthalpy using a pressure-expansion virial equation of state (see Section B.2).

Solution. The virial equation of state can be written

$$z(T,P) = 1 + B'(T)P + C'(T)P^2 + \cdots .$$

We see from Eqn. (4.68) that we need the partial derivative of z with respect to T at constant P. So, if we take the derivative of each side of this equation with respect to T holding P constant, we obtain

$$\left(\frac{\partial z}{\partial T}\right)_P = P\frac{dB'}{dT} + P^2\frac{dC'}{dT} + \cdots.$$

Now we multiply each side of this equation by $(-T/P)$, and integrate over P from 0 to P to obtain

$$\frac{H^{\mathrm{R}}}{NRT} = -T\left[P\frac{dB'}{dT} + \frac{P^2}{2}\frac{dC'}{dT} + \cdots\right],$$

using Eqn. (4.68). □

Virial expansions are very accurate when restricted to gases that are not very dense. For dense gases or liquids, we typically have available equations of state that are explicit in pressure, rather than volume. Hence, Eqns. (4.67)–(4.71) cannot be used for these PVT relations. Therefore, we require integrals over volume instead of pressure for pressure-explicit equations of state. We can use arguments analogous to those above to obtain the appropriate integrals; however, a subtle, but important, difference should be noted.

The definition for residual properties, Eqn. (4.66), requires that we find the property in the ideal state *at the same temperature and pressure, not the same volume*. Hence, to find the residual Helmholtz potential at a given temperature and volume, we write

$$F^{\mathrm{R}}(T, V) = F(T, V) - F^{\mathrm{ideal}}\left(T, \frac{NRT}{P(T, V)}\right).$$

Note that the volume specified for the ideal property is given by NRT/P, not by V. The P in this expression must be found from the non-ideal equation of state $P(T, V)$. To find these values we integrate from $V = \infty$ to V, since the system behaves ideally at infinite volume,

$$\begin{aligned} F^{\mathrm{R}}(T, V) &= \int_\infty^V \left(\frac{\partial F}{\partial V}\right)_T dV - \int_\infty^{NRT/P} \left(\frac{\partial F^{\mathrm{ideal}}}{\partial V}\right)_T dV \\ &= -\int_\infty^V P\, dV + \int_\infty^{NRT/P} P^{\mathrm{ideal}}\, dV, \quad \text{constant } T. \end{aligned}$$

We used the differential for F to obtain the second line. Now we break up the second integral into two pieces.

$$\begin{aligned} F^{\mathrm{R}}(T, V) &= -\int_\infty^V P\, dV + \int_\infty^V P^{\mathrm{ideal}}\, dV + \int_V^{NRT/P} P^{\mathrm{ideal}}\, dV, \\ &= -\int_\infty^V \left[P - P^{\mathrm{ideal}}\right] dV + \int_V^{NRT/P} P^{\mathrm{ideal}}\, dV, \\ &= -\int_\infty^V \left[\frac{zNRT}{V} - \frac{NRT}{V}\right] dV + \int_V^{NRT/P} \frac{NRT}{V}\, dV \\ &= NRT\int_V^\infty (z[T, V] - 1)\frac{dV}{V} - NRT \log z. \end{aligned}$$

To obtain the third line, we used the ideal-gas law and the definition for the compressibility factor z. If we perform similar manipulations for other quantities, we obtain the following useful relations:

$$\frac{F^{\mathrm{R}}(T, V, \{N_i\})}{NRT} = \int_V^\infty \left[z(T, V', \{N_i\}) - 1\right] \frac{dV'}{V'} - \log z, \tag{4.75}$$

$$\frac{U^{\mathrm{R}}(T, V, \{N_i\})}{NRT} = -T \int_V^\infty \left(\frac{\partial z}{\partial T}\right)_{V', \{N_i\}} \frac{dV'}{V'}, \tag{4.76}$$

$$\frac{G^{\mathrm{R}}(T, V, \{N_i\})}{NRT} = \frac{F^{\mathrm{R}}}{NRT} + (z - 1), \tag{4.77}$$

$$\frac{H^{\mathrm{R}}(T, V, \{N_i\})}{NRT} = \frac{U^{\mathrm{R}}}{NRT} + (z - 1), \tag{4.78}$$

$$\frac{S^{\mathrm{R}}(T, V, \{N_i\})}{NR} = \frac{U^{\mathrm{R}} - F^{\mathrm{R}}}{NRT}. \tag{4.79}$$

It is easier to obtain these results by a straightforward change of variable of integration for the integrals showing up in Eqns. (4.67)–(4.71). Either way, these equations need not be memorized.

EXAMPLE 4.4.4 Find the change in specific residual internal energy predicted by the Redlich–Kwong equation of state for methyl chloride for an initial state $P_0 = 70\,\mathrm{bar}$, $T_0 = 460\,\mathrm{K}$ and a final state $P_\mathrm{f} = 50\,\mathrm{bar}$, and $T_\mathrm{f} = 450\,\mathrm{K}$.

Solution. The Redlich–Kwong equation of state, Eqn. (B.33), can be written as

$$z(T, v) = \frac{v}{v - b} - \frac{a}{RT^{3/2}(v + b)},$$

which is a pressure-explicit PVT relation. Hence, we insert this equation into Eqn. (4.76) to obtain

$$\begin{aligned} u^{\mathrm{R}} &= -\frac{3a}{2\sqrt{T}} \int_v^\infty \frac{dv}{v(v + b)} \\ &= \frac{3a}{2b\sqrt{T}} \log\left(\frac{v}{v + b}\right). \end{aligned} \tag{4.80}$$

To estimate the parameters for the Redlich–Kwong equation of state, we use the predictions for the critical constants, Eqns. (B.38) and (B.39). For methyl chloride, $T_\mathrm{c} = 416\,\mathrm{K}$, and $P_\mathrm{c} = 67\,\mathrm{bar}$, according to the NIST Chemistry Web Book. Hence, Eqns. (B.38) and (B.39) yield

$$a \approx 159\,\frac{\mathrm{l}^2 \cdot \mathrm{bar} \cdot \sqrt{\mathrm{K}}}{\mathrm{mol}^2} = 1.59 \times 10^4\,\frac{\mathrm{l} \cdot \mathrm{J} \cdot \sqrt{\mathrm{K}}}{\mathrm{mol}^2},$$

$$b \approx 0.0447\,\frac{\mathrm{l}}{\mathrm{mol}}.$$

We can use these relations to find the residual properties at the beginning and final states. The final volume is found by using the formulation for the roots of a cubic equation, which is given in Section A.7 of Appendix A. We find that there are single real roots, and

$$v_{\mathrm{f}} \approx 0.566\,\frac{\mathrm{l}}{\mathrm{mol}},$$

$$v_0 \approx 0.363\,\frac{\mathrm{l}}{\mathrm{mol}}.$$

Inserting these values into Eqn. (4.80) yields

$$u_{\mathrm{f}}^{\mathrm{R}} \approx -1.92\,\frac{\mathrm{kJ}}{\mathrm{mol}},$$

$$u_0^{\mathrm{R}} \approx -2.896\,\frac{\mathrm{kJ}}{\mathrm{mol}}.$$

These are the corrections to the ideal internal-energy change. We recall from Chapter 2 that the internal energy of an ideal gas depends on temperature only, so we write

$$\Delta u^{\mathrm{ideal}} = \int_{T_0}^{T_{\mathrm{f}}} c_v^{\mathrm{ideal}}[T]dT, \tag{4.81}$$

$$= \int_{T_0}^{T_{\mathrm{f}}} [c_P^{\mathrm{ideal}}[T] - R]dT, \tag{4.82}$$

where we used the relation derived earlier, Eqn. (3.85). Especially see the text following this equation. We can find an expression for the constant-pressure ideal heat capacity from the NIST Web Book also, which suggests

$$c_P^{\mathrm{ideal}} = A_0 + A_1 T + A_2 T^2 + A_3 T^3 + \frac{A_{-2}}{T^2}, \tag{4.83}$$

where $A_0 = 3.524\,690$, $A_1 = 136.9277 \times 10^{-3}$, $A_2 = -82.141\,96 \times 10^{-6}$, $A_3 = 20.227\,97 \times 10^{-9}$, and $A_{-2} = 0.278\,032 \times 10^6$. This yields the heat capacity in J/(mol K), if the temperature is in kelvins. If we insert this relation into Eqn. (4.82), we can integrate to find

$$\Delta u^{\mathrm{ideal}} \approx -437\ \mathrm{J/mol}. \tag{4.84}$$

Therefore, the total internal-energy change is

$$\Delta u = u_{\mathrm{f}}^{\mathrm{R}} + \Delta u^{\mathrm{ideal}} - u_i^{\mathrm{R}} \tag{4.85}$$

$$\approx 1.42\ \mathrm{kJ/mol}. \tag{4.86}$$

In other words, we find that the estimate for the change in internal energy is opposite in sign of that predicted for the ideal change alone. □

It is also possible to create fundamental relations for F or G using residual properties. See Exercises 4.4.L and 4.4.M for more details.

Summary

In this chapter we introduced the following ideas.

- *Local stability*, to show that there are restrictions on values for second-order derivatives: $C_P > C_V > 0$ and $\kappa_T > \kappa_S > 0$.
- *Global stability*, to show how a PvT relation can predict boiling and condensation of fluids upon heating and cooling.
- *Binodal and spinodal curves.* Given any PvT equation of state, the student should be able to construct numerically a vapor–liquid phase diagram like that shown in Figure 4.5.
- *Residual properties*, to generate thermodynamic diagrams, or thermodynamic properties of real substances. Given a PvT equation of state, the student should be able to generate an expression for any residual property. From this residual property, and an expression for the ideal heat capacity $c_v^{\text{ideal}}(T)$, the student should also be able to construct any sort of thermodynamic diagram. Examples are TS diagrams, Figure 4.10, and PH diagrams, Figure 4.11.

We also learned how to construct thermodynamic diagrams from fundamental relations, or equations of state. In the two-phase region of these diagrams, we used the lever rule. As approximations to the binodal curve, we also derived expressions that are based on the Clapeyron equation, such as the Antoine equation. Alternatives to the Antoine equation have been fit to the PVT equations of state given in Appendix B.

As examples, we used the van der Waals fluid to predict the vapor–liquid region of a fluid; we found the chemical potential of water from steam tables; and we used the Redlich–Kwong equation of state to find energy changes in methyl chloride.

Exercises

4.1.A Prove that the local stability criteria on internal energy place restrictions on the constant-volume heat capacity and the isentropic compressibility, Eqn. (4.11).

4.1.B Prove that the local stability criteria on F yield restrictions on the isothermal compressibility, Eqn. (4.13).

4.1.C Following Example 4.1.2, prove the following stability criterion for the generalized potential:

$$\left(\frac{\partial^2 F}{\partial T^2}\right)_{V,N} < 0.$$

4.1.D Prove the remaining local stability criteria (compared with the above exercise) on the generalized potentials, Eqns. (4.12), in the following way. Begin by noticing that all of the second-order derivatives of the generalized potential can be written as first-order derivatives. For example,

$$\left(\frac{\partial^2 F}{\partial V^2}\right)_T = \left(\frac{\partial}{\partial V}\left(\frac{\partial F}{\partial V}\right)_T\right)_T = -\left(\frac{\partial P}{\partial V}\right)_T.$$

Then use Jacobian transformations to write this first-order derivative in terms of other first-order derivatives that use S and V as independent variables. Finally, replace P and T with their definitions to arrive at second-order derivatives in U, for which the local stability criteria are known.

4.1.E Prove Eqn. (4.15).

4.2.A Derive the reduced form for the van der Waals equation (4.23) using the expressions for the critical constants.

4.2.B A third graphical representation, which is mathematically equivalent to the equal-area rule, Eqn. (4.30), may be found from the Helmholtz potential. The two phases must have equal temperature, but are globally unstable against fluctuations in volume. Since it has T and v as canonical independent variables, f is the natural choice for global instability of pure-component systems. Figure 4.13 shows the Helmholtz potential at a constant, sub-critical temperature as a function of volume for the simple van der Waals fluid. We see that there is a portion of the curve that is globally unstable against volume fluctuations, since it is concave from below. Whenever the total volume of the system is constrained to be between v_l and v_v, the system can separate into these two phases, thereby lowering its effective Helmholtz potential to lie on the dashed line connecting these phases. This construction requires complete thermodynamic information, whereas our earlier consideration required only an equation of state.

The mathematical criterion for the binodal instability can be found from the equation for the slope of the line in Figure 4.13:

$$\left(\frac{\partial f}{\partial v}\right)_T\bigg|_{v_l} = \left(\frac{\partial f}{\partial v}\right)_T\bigg|_{v_v} = \frac{f(T, v_v) - f(T, v_l)}{v_v - v_l}.$$

Recognizing from its differential that $(\partial f/\partial v)_T = -P$, derive Eqn. (4.30).

4.2.C Estimate the parameters of the van der Waals equation of state for the following substances using their critical constants. Using a spreadsheet (this is probably the simplest method) or a similar packaged program, or by writing a computer code, find a numerical solution for a prediction of the binodal curve. Because the equations are nonlinear, an iterative approach is necessary. For example, you could perform the following steps.

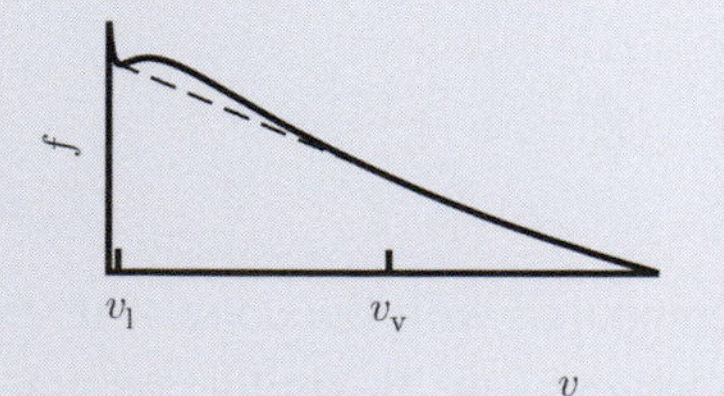

Figure 4.13 The Helmholtz potential at constant, sub-critical temperature for a van der Waals fluid as a function of volume. The straight dashed line connects the liquid and vapor volumes with a slope of $-P^{\mathrm{sat}}$.

1. Pick a sub-critical temperature, and plot the pressure vs. volume relationship predicted by the equation of state. You should see a local minimum and a local maximum, indicating instability.
2. Find the chemical potential vs. volume relationship as well, and construct a plot of the chemical potential vs. pressure. Where the curve crosses over itself indicates the saturation pressure.
3. Find the three volumes that intersect with the equation of state at the saturation pressure. For a cubic equation of state, these values can be found analytically using the expressions in Section A.7 of Appendix A. The largest root is the vapor volume, and the smallest root is the liquid volume. The root between these two is the unstable one.
4. As a check, calculate the two areas between these two curves, from the equal area formula, Eqn. (4.30), to verify that this is the correct saturation pressure and that these are the correct volumes. Record the values for the temperature, saturation pressure, liquid volume, and vapor volume.
5. Pick a new sub-critical temperature and repeat the procedure. Once you have several values for the saturation pressure and liquid volumes, you can plot the left-hand side of the binodal curve. The vapor volumes provide the right-hand side, and these meet at the critical point.

Compare your predictions with the tabulated (experimentally determined) values for the normal boiling points of nitrogen, oxygen, methanol, and carbon monoxide.

Check: your binodal curve should be well approximated in reduced form by the expression

$$\frac{P^{\mathrm{sat}}}{P_{\mathrm{c}}} = \frac{0.1458}{v_{\mathrm{r}}^4} - \frac{0.419}{v_{\mathrm{r}}^3} - \frac{0.6203}{v_{\mathrm{r}}^2} + \frac{1.9266}{v_{\mathrm{r}}} - 0.0341.$$

4.2.D Derive an equation for the spinodal curve predicted by the Redlich–Kwong equation of state. Plot the spinodal curve, and several isotherms for reasonable values for the parameters. Derive the nonlinear equation that determines the dimensionless critical volume v_{c}/b, and find an approximate value for it.

Show that

$$v_{\mathrm{c}}/b \approx 3.847\,32,$$

$$P_{\mathrm{c}} \approx 0.029\,894\,4 \left(\frac{a^2 R}{b^5}\right)^{1/3},$$

$$T_{\mathrm{c}} \approx 0.345\,04 \left(\frac{a}{Rb}\right)^{2/3},$$

and, hence, Eqns. (B.38) and (B.39).

4.2.E Do Exercise 4.2.C using a Redlich–Kwong fluid.

Check: Your binodal curve should be well approximated in reduced form by the expression

$$\frac{P^{\mathrm{sat}}}{P_{\mathrm{c}}} = -\frac{0.0133}{v_{\mathrm{r}}^4} + \frac{0.3215}{v_{\mathrm{r}}^3} - \frac{1.6131}{v_{\mathrm{r}}^2} + \frac{2.3194}{v_{\mathrm{r}}} - 0.0152.$$

4.2.F Do Exercise 4.2.C using a Peng–Robinson fluid.

Check: Your binodal curve should be well approximated in reduced form by the expression

$$\frac{P^{\text{sat}}}{P_c} = -\frac{0.0266}{v_r^4} + \frac{0.3747}{v_r^3} - \frac{1.6451}{v_r^2} + \frac{2.3075}{v_r} - 0.0049$$

over a significant range of v_r near the critical point for oxygen.

4.2.G Do Exercise 4.2.D using the Schmidt–Wenzel form of Martin's generalized cubic equation of state.

4.2.H Do Exercise 4.2.D using the Peng–Robinson equation of state. Show that

$$\frac{v_c}{b} = \left[4 + 2\sqrt{2}\right]^{1/3} + \left[4 - 2\sqrt{2}\right]^{1/3} + 1$$
$$P_c \approx 0.013\,236\,6\frac{a_0}{b^2}$$
$$T_c \approx 0.170\,144\,4\frac{a_0}{bR}.$$

and, hence, Eqns. (B.49) and (B.50).

4.2.I Do Exercise 4.2.D using the Carnahan–Starling equation of state. Show that

$$v_c/b \cong 7.666\,13,$$
$$P_c \cong 0.004\,416\,81\frac{a}{b^2},$$
$$T_c \cong 0.094\,328\,7\frac{a}{bR},$$

and, hence, Eqns. (B.28).

4.2.J Find the critical point for Martin's cubic equation of state when $a = a_0\sqrt{T}$, $\beta = 2b$, and $\gamma = b$.

4.2.K Do Exercise 4.2.C for the Carnahan–Starling equation of state. Note that this problem is more difficult than for cubic equations of state, since the equation of state is a fifth-order polynomial in specific volume. It will be necessary to find all the roots, and determine which are physically reasonable for a given T and P^{sat}. One trick to find all the roots is to use a standard root-finding algorithm (such as that found in a spreadsheet), and use synthetic division to reduce the polynomial to fourth order. Then, the second root can be found by the algorithm, and the procedure is repeated.

Check: your binodal curve should be well approximated in reduced form by the expression

$$\frac{P^{\text{sat}}}{P_c} = \frac{0.0089}{v_r^6} - \frac{0.1102}{v_r^5} + \frac{0.4831}{v_r^4} - \frac{0.6908}{v_r^3} - \frac{0.6825}{v_r^2} + \frac{2.0051}{v_r} - 0.0110$$

over a significant range of v_r near the critical point.

4.2.L Do Exercise 4.2.C using the Schmidt–Wenzel form of Martin's generalized cubic equations of state for oxygen.

Check: your binodal curve should be well approximated in reduced form by the expression

$$\frac{P^{\mathrm{sat}}}{P_{\mathrm{c}}} = -\frac{0.0970}{v_{\mathrm{r}}^4} + \frac{0.9129}{v_{\mathrm{r}}^3} - \frac{3.2786}{v_{\mathrm{r}}^2} + \frac{4.2369}{v_{\mathrm{r}}} + 0.0443$$

over a significant range of v_{r} *near the critical point, when you use a value of 0.487 (that for ethylene glycol) for the acentric factor.*

4.2.M Do Exercise 4.2.C using the Anderko–Pitzer equation of state for propane. Compare your results with those in Table B.1 in Appendix B. Note that this problem is a little harder than the cubic equations, since five roots are possible, and there exists no analytic method to find them. One straightforward way to find all the roots is to first find one root v_1 by iteration (e.g. *goal seek* in Excel). Then, divide the original polynomial by $(v - v_1)$ using synthetic division to find a fourth-order polynomial. Again search for a root to this equation, divide synthetically once more by $(v - v_2)$, and then use the analytic expressions for the resulting cubic equation.

(*Hint:* it is easiest to perform the entire calculation in reduced form to obtain $P_{\mathrm{r}}^{\mathrm{sat}}$ vs. v_{r}.)

4.2.N Use the Benedict–Webb–Rubin model to construct binodal and spinodal curves for one of the fluids listed in Table B.3 in Appendix B.

4.2.O Find an expression for the spinodal curve of the simplified Beattie–Bridgeman model

$$P = \frac{RT}{v^2}\left[v + B_0\left(1 - \frac{a}{v}\right)\right] - \frac{A_0}{v^2}\left(1 - \frac{a}{v}\right). \tag{4.87}$$

4.2.P The fundamental relation for a fluid suggested by Patel and Teja [132] is

$$f = f_0 - s_0(T - T_0) - RT\log\left(\frac{v-b}{v_0}\right) + \frac{a(T)}{A}\log\left(\frac{2v+b+c-A}{2v+b+c+A}\right) + \int_{T_0}^{T}\left(\frac{T'-T}{T'}\right)c_v^{\mathrm{ideal}}(T')dT',$$

where $A := \sqrt{b^2 + 6bc + c^2}$, a is a function of temperature, and b and c are constants.

1. Find a PVT equation of state for this fluid.
2. Find a $U = U(T, V)$ equation of state for this fluid.
3. Find an expression for κ_T for this fluid.
4. Find $P^{\mathrm{spinodal}}(v)$ for this fluid, when $a(T)$ is assumed constant. Also find an equation to determine the critical volume, and find a numerical approximation for it. Also determine the critical pressure and temperature.

4.2.Q Figure 4.14 shows the pressure as a function of the specific volume for a single isotherm of a fluid. Estimate the spinodal and binodal volumes at this temperature, and the saturation pressure.

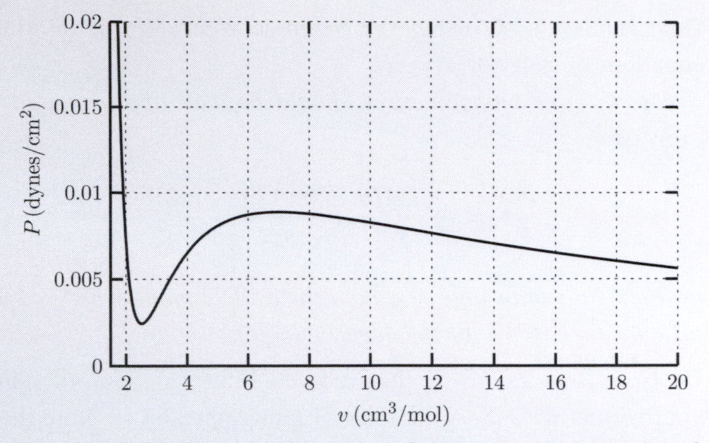

Figure 4.14 Pressure vs. specific volume as predicted for a fluid by some PvT equation of state.

4.2.R Since a PvT equation of state predicts the binodal curve, it could also be used to predict the parameters in the Antoine equation, over some range of temperatures. If the equation of state uses only two parameters, the results can be put into reduced form (reduced saturation pressure P_r^{sat} as a function of reduced temperature T_r). Therefore, the Antoine equation could also be put into reduced form. First, using the result of Exercise 4.2.D, show that the Redlich–Kwong equation of state can be put into the reduced form

$$P_r = \frac{3T_r}{v_r - 0.259\,921} - \frac{3.8473\,2}{\sqrt{T_r}v_r(v_r + 0.259\,921)}. \tag{4.88}$$

If we use the results from Exercise 4.2.E,

$$P_r^{sat} = -\frac{0.0133}{v_r^4} + \frac{0.3215}{v_r^3} - \frac{1.6131}{v_r^2} + \frac{2.3194}{v_r} - 0.0152,$$

then we have two equations with three unknowns, v_r, T_r, and P_r^{sat}. Therefore, it is possible to find the reduced saturation pressure for a given reduced temperature. Using these equations, find the saturation pressure of ammonia at 100 °C. Compare your estimate with that given by the Antoine equation.

4.2.S The Redlich–Kwong model for fluids has a PvT equation of state

$$P = \frac{RT}{v - b} - \frac{a}{\sqrt{T}v(v + b)},$$

with parameters a and b, which can be estimated from the critical properties. Figure 4.15 shows such a sub-critical isotherm for benzene. Using **only** the Redlich–Kwong equation of state and Figure 4.15, estimate the following quantities at the temperature shown, 200 °C:

1. the saturation pressure P^{sat};
2. the liquid and vapor specific volumes, v^v and v^l; and
3. the heat of vaporization, $\Delta h^{vap}(T) = h^v(T) - h^l(T)$.

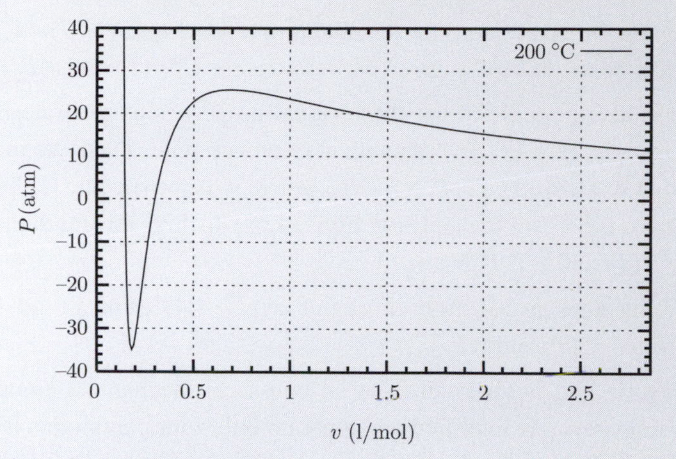

Figure 4.15 An isotherm for benzene, $T = 200\,°\text{C}$, as predicted by the Redlich–Kwong equation of state.

4.4.A Create a *TS* diagram for nitrogen using the van der Waals equation of state given in Appendix B.

4.4.B Create a *PH* diagram for nitrogen using the van der Waals equations of state given in Appendix B.

4.4.C We can find a generalization to Eqn. (4.51) that is valid for any equation of state $z = z(T, v)$. Insert the relation $P = z(T, v)RT/v$ into Eqn. (4.50) and use manipulations similar to those used to obtain Eqn. (4.51) to find

$$\left(\frac{\partial u}{\partial v}\right)_T = \frac{RT^2}{v}\left(\frac{\partial z}{\partial T}\right)_v .$$

Compare this result with Eqn. (4.76), which was used to find the residual internal energy.

4.4.D Create a *Ts* diagram for nitrogen using the Redlich–Kwong equation of state given in Section B.5 of Appendix B.

4.4.E Create a *Ts* diagram for nitrogen using the Carnahan–Starling equation of state given in Section B.4 of Appendix B.

4.4.F Create a *Ts* diagram for nitrogen using a Peng–Robinson fluid.

4.4.G Derive Eqns. (4.67)–(4.71).

4.4.H Derive Eqns. (4.76)–(4.79).

4.4.I Oxygen at 22 bar, 298 K is being pumped along a pipe, and passes through a valve, which causes the pressure to drop to 15 bar. Using the Peng–Robinson model, and assuming that such a **throttling process** is isenthalpic, find the temperature downstream of the valve.

4.4.J Use the Anderko–Pitzer equation of state to create expressions for the residual properties. Write a spreadsheet or computer program to plot values for these quantities as functions of reduced temperature, and reduced volume; two plots for each quantity will be necessary in order to include the acentric factor (e.g. $h = h_0(T_r, P_r) + \omega h_1(T_r, P_r)$).

4.4.K Using the expressions for residual properties, Eqns. (4.67)–(4.71), and its definition, find an expression for C_P, given expressions for $C_P^{\text{ideal}}(T)$ and $z(T, P, \{N_i\})$.

4.4.L It is also possible to use the residual properties to generate fundamental relations. For example, pick (T_0, P_0) as an ideal reference state. Then, we may use an equation similar to (4.65) to find $G(T, P, N) - Ng_0$, where g_0 is the specific Gibbs potential at the reference state. Find ΔG^{ideal}, and use Eqn. (4.67) to find the fundamental relation for G from $c_v^{\text{ideal}}(T)$ and $z(T, P, N)$.

4.4.M Using steps analogous to those in Exercise 4.4.L, find a fundamental relation $F(T, V, N)$ from $c_v^{\text{ideal}}(T)$ and $z(T, V, N)$.

4.4.N Beattie and Bridgeman have an empirical mechanical equation of state with several parameters. A simplified version with only four parameters is

$$P = \frac{RT}{v^2}\left[v + B_0\left(1 - \frac{a}{v}\right)\right] - \frac{A_0}{v^2}\left(1 - \frac{a}{v}\right), \tag{4.89}$$

where a, b, A_0, and B_0 are all constants.

Derive a fundamental relation for this fluid.

4.4.O Estimate the minimum amount of work necessary to pump up my bicycle tire sufficiently slowly that the process is isothermal, by using the following steps.

1. The tire has a width of 1.5 in, and a diameter of 22 in. Estimate the volume of air in the tire.
2. When full, the tire is at 7 bar. Assuming the air is all nitrogen at room temperature, use the Peng–Robinson model to estimate the number of moles of gas in the tire.
3. What is the volume of the air at the same temperature, but at ambient pressure?
4. Now calculate the minimum work necessary to compress this air to 7 bar isothermally. You should remember that isothermal work is equivalent to the change in a certain thermodynamic quantity.

4.4.P We here wish to do the same calculation as Exercise 4.4.O, except when the process is adiabatic instead of isothermal. This calculation is more involved, because reversible, adiabatic compression of the gas is isentropic, not isothermal. Therefore, the final temperature (after pumping adiabatically, but before the tire has cooled down) is not known. However, we do want the pressure in the tire to be 7 bar at ambient temperature (after the tire has cooled down), so the number of moles is still the same. Now estimate the final temperature and pressure in the tire assuming adiabatic pumping. Also calculate the minimum total work necessary. Which process requires more work, the adiabatic or the isothermal pumping? Why?

5 Application to process design: flow systems

Traditional application of thermodynamics to engineering problems involves processes that are flowing. For example, an engineer might design a refrigerator in which a refrigerant is pumped in a continuous cycle through a coil of tubing, or a generator where steam is pumped in a power cycle through several pieces of equipment. Hence, engineering has placed a lot of emphasis on balances in flowing systems. Such flow systems involve *time* as an independent variable. However, thermodynamics applies only to equilibrium states, and the introduction of time is strictly forbidden. The study of time-dependent processes actually falls within the domains of *transport phenomena* and *non-equilibrium thermodynamics*.

Whereas the field of transport phenomena is relatively well advanced and well understood, non-equilibrium thermodynamics is a developing field of research, and the fundamental postulates are by no means agreed upon [11, 94].

We will restrict ourselves here to the simplest of such time-dependent systems. Namely, we will assume that our system is in a **local state of equilibrium**. Such an assumption allows us to use the quantities derived for equilibrium systems as local variables that depend upon position and time. This simplification is usually applicable whenever the *local* response time of a system is much smaller than the time scale of the whole process. In this way, we can simplify many engineering *flow* problems to equivalent *equilibrium* thermodynamics problems.

For example, consider water flowing through a heat exchanger. If we follow a small packet of water as it flows through the system, it will experience work, through pressure forces, and heat fluxes from neighboring bits of fluid. The water can respond very quickly to such changes, and reach equilibrium within one microsecond of a change. The time rate for imposition of such changes is much longer, perhaps never faster than one millisecond for our example. Hence, the fluid element "feels" as if it is always in a heat bath at the local temperature and in a pressure bath at the local pressure. Also, it is "unaware" that other fluid elements elsewhere in the exchanger are at different temperatures and pressures.

On the other hand, such an approximation is *not* valid for a polymer melt flowing through a piece of processing equipment. Polymer melts are composed of long chains that rearrange by Brownian motion on relatively long time scales, perhaps several seconds. Hence, as the polymer fluid element travels through the equipment it can easily be subjected to rapid changes in pressure and temperature to which it cannot quickly respond, so local equilibrium is not recovered.

Also, there are manufacturing processes in which reactions take place on surfaces – such as in semiconductor manufacturing. In these reactions, a plasma is created above the surface. For example, a plasma may consist of ions and electrons in a low-density, gas-like state subjected to an oscillating electromagnetic field. Because the mass of the electrons is so much smaller than that of the ions, the electrons are more strongly affected by the electromagnetic field. Hence, the ions and the electrons, although occupying the same space, need not be in thermal equilibrium with one another, and temperature has no meaning in this situation. Assuming local equilibrium would be a poor approximation for these two systems.

To derive the governing equations for flow systems rigorously, we will need to draw upon knowledge gleaned from transport phenomena. Unfortunately, the reader may be unfamiliar with this subject. Therefore, we show the details of transport phenomena in Appendix C for the interested reader, and instead go straight to the necessary *macroscopic balances* in Section 5.1, where these equations are first applied. We strongly recommend, however, that the student skim Appendix C and return to it, when appropriate, during a course on transport phenomena. The method of attack to obtain the important equations is outlined in Figure 5.1 on p. 153.

We apply these macroscopic balances to single pieces of equipment in Section 5.1. Then, in Section 5.2, we consider power and refrigeration cycles. Note in these problems that *flowing* systems are reduced to simpler *equilibrium* problems of thermodynamics.

5.1 MACROSCOPIC MASS, ENERGY, AND ENTROPY BALANCES

The macroscopic balances useful for industrial applications may be derived from the microscopic balances introduced in Appendix C. The details are largely omitted here, because excellent derivations may be found in textbooks on transport phenomena [e.g. 14, pp. 198–205, 221–223, 454–458, 667, and 728–729], and really belong in that domain. The derivations are not exceedingly complex, however, and rely only on the Gauss divergence theorem and a generalization of the Leibniz formula, which are given in Sections A.6 and A.5 of Appendix A, respectively.

If we integrate the continuity equation, Eqn. (C.5), over the volume of our system of interest, we obtain the **macroscopic mass balance**,

$$\frac{d}{dt}M_{\text{tot}} + \Delta\dot{m} = 0, \tag{5.1}$$

where M_{tot} is the total mass of the system, $\dot{m}$ is the mass flow rate of an inlet or exit stream, and

$$\Delta(\ldots) := \sum_{\text{exits}}(\ldots) - \sum_{\text{inlets}}(\ldots) \tag{5.2}$$

is the difference in values between all the exit and inlet streams.

Similarly, an integration over the volume of interest of the microscopic energy balance, Eqn. (C.14), yields the **macroscopic energy balance**,

$$\frac{d}{dt}E_{\text{tot}} + \Delta\left[\frac{h}{M_{\text{w}}} + \frac{\langle|\vec{v}|^3\rangle}{2\langle|\vec{v}|\rangle} + \phi_{\text{pot}}\right]\dot{m} = \dot{Q} + \dot{W}_{\text{s}}, \tag{5.3}$$

where E_{tot} is the sum total of internal, kinetic, and potential energy in the volume, $\langle|\vec{v}|\rangle$ is the average velocity of the stream, ϕ_{pot} is the potential energy of the stream (height times gravity), $\dot{Q}$ is the rate at which heat flows into the volume from outside, and $\dot{W}_{\text{s}}$ is the rate at which the environment does "shaft" work on the volume through moving boundaries.

A few important things to note about this equation are as follows.

- It is necessary to make the local-equilibrium approximation (to use Eqn. (C.15)) in deriving this equation.
- A few minor assumptions have been made: the extra stress $\vec{\vec{\tau}}$ contribution to work at the inlets and exits is neglected, and the velocity field is in the direction of the unit normal at the flow streams [12].
- The heat flow $\dot{Q}$ contains all forms of heat flow, not just conduction [14, Eqn. (18.4-2)]. If the boundaries are impermeable to matter, then conduction is sufficient. However, if part of the boundary of the macroscopic control volume is semi-permeable, the heat flux would include conduction and transfer from molar flux

$$\vec{q} = -k\vec{\nabla}T + \sum_{k=1}^{m} \bar{h}_k \vec{J}_k, \tag{5.4}$$

where $\bar{h}_k := (\partial H/\partial N_k)_{T,P,\{N_{j\neq k}\}}$ is the partial molar enthalpy,[1] k is the **thermal conductivity** of the fluid, and $\vec{J}_k$ is the molar flux of species k relative to the macroscopic velocity $\vec{v}$. The first term is called **Fourier's law**.
- The "shaft-work" term, $\dot{W}_s$, arises from the motion of solid boundaries that can do work on the system. However, an "electrical-work" term can also arise if electrons flow across the boundary making an electrical current (e.g. as in a fuel cell).
- The ratio of velocity averages is usually approximated in turbulent flow as simply $\langle|\vec{v}|^2\rangle = \dot{m}/(\rho A_s)$, where A_s is the cross-sectional area of the stream. When the flow is laminar, the term is usually neglected.
- The potential energy of a stream is typically just the height of the stream above some arbitrarily chosen plane times the gravitational constant. This term is often neglected.
- This equation is commonly used for energy balances in chemical processes [38, Eqn. (7.4-15)]; [63, Eqn. (5.14)], and in unit operations [14, Eqn. (15.1-3)].
- The total energy can be written as the sum of internal, kinetic, and potential energies if we assume "local equilibrium:"

$$E_{\text{tot}} := U_{\text{tot}} + K_{\text{tot}} + \Phi_{\text{tot}}, \tag{5.5}$$

where

$$\begin{aligned} U_{\text{tot}} &:= \iiint\limits_{V(t)} \frac{u(\vec{r},t)}{v(\vec{r},t)}\, dV(t) \\ K_{\text{tot}} &:= \iiint\limits_{V(t)} \frac{M_w(\vec{r},t)|\vec{v}(\vec{r},t)|^2}{2v(\vec{r},t)}\, dV(t) \\ \Phi_{\text{tot}} &:= \iiint\limits_{V(t)} \frac{M_w(\vec{r},t)\phi_{\text{pot}}}{v(\vec{r},t)}\, dV(t) \end{aligned} \tag{5.6}$$

For many unsteady-state applications of interest, the kinetic and potential contributions are neglected, and the internal energy is assumed to be uniform throughout the volume.

[1] For more details on partial molar properties, see Section 8.3.

To estimate the importance of terms, it is typical to use *dimensional analysis*. Details of this procedure are outlined in [14, Sections 3.7, 11.5, and 19.5] for the microscopic equations. Examples of applications to microscopic balances can be found in [124].

Finally, the **macroscopic entropy balance** can be obtained by integrating the *microscopic* entropy balance, Eqn. (C.26), over the volume of interest:

$$\frac{d}{dt}S_{\text{tot}} + \Delta\left[\frac{\dot{m}s}{M_{\text{w}}}\right] = \iint\limits_{A_{\text{S}}(t)} \vec{\boldsymbol{n}} \cdot \left(\frac{k\,\vec{\nabla}T}{T} - \sum_i \bar{s}_i \vec{\boldsymbol{J}}_i\right) dA_{\text{s}} + \dot{\Sigma}, \tag{5.7}$$

where the differential area dA_{s} has unit normal vector $\vec{\boldsymbol{n}}$ pointing out of the volume. The first term inside the integral on the right-hand side of Eqn. (5.7) is the heat flux (energy/area/time) divided by temperature, and the second is the molar flux (moles/area/time) $\vec{\boldsymbol{J}}_i$ times the partial molar entropy $\bar{s}_i := (\partial S/\partial N_i)_{T,P,\{N_{j\neq i}\}}$. Both terms provide entropy entering the system through the infinitesimal area dA_{s} at the boundaries.

It is typically assumed that the contribution from the molar flux is negligible. However, the term may not be neglected when there is sufficient molar flux across the control volume, as may happen with mass flow across a membrane. When the contribution from the molar flux is neglected, we may replace the integral on the right-hand side with $\iint_{A_{\text{S}}(t)} d\dot{Q}/T$.

The last term on the right-hand side represents the total entropy generation in the system, which is the sum of contributions from heat flow, mass flow, stress work, and chemical reactions:

$$\begin{aligned} \dot{\Sigma} &:= \dot{\Sigma}_{\text{hf}} + \dot{\Sigma}_{\text{mf}} + \dot{\Sigma}_{\text{sw}} + \dot{\Sigma}_{\text{r}}, \\ \dot{\Sigma}_\kappa &:= \iiint\limits_{V(t)} \sigma_\kappa \, dV(t), \qquad \kappa = \text{hf, mf, sw, r.} \end{aligned} \tag{5.8}$$

Definitions of the rates of entropy generation per unit volume σ_κ are given by Eqns. (C.29)–(C.32). Calculation of these quantities requires detailed information about the process inside the control volume that is not typically available. However, adherence to the second law requires that each of these terms be non-negative. We give below an example in which the contribution of each of these terms is calculated explicitly. Often we must content ourselves with just the inequality

$$\dot{\Sigma} \geq 0. \tag{5.9}$$

The total entropy of the system is defined as

$$S_{\text{tot}} := \iiint\limits_{V(t)} \frac{s(\vec{\boldsymbol{r}}, t)}{v(\vec{\boldsymbol{r}}, t)} \, dV(t). \tag{5.10}$$

All of the above equations are exact within the assumption of local equilibrium. Typically, the rigor and usefulness of an equation are inversely proportional to one another. To become useful, the equations require some assumptions and simplifications. For applications in the rest of this chapter, we will use a simplified set of equations. The simplified mass, energy, and entropy balances are

$$\frac{dM_{\text{tot}}}{dt} + \Delta\dot{m} = 0, \tag{5.11}$$

$$\frac{dU_{\text{tot}}}{dt} + \Delta\left[\frac{\dot{m}h}{M_{\text{w}}}\right] = \sum_i \dot{Q}_i + \dot{W}_{\text{s}}, \tag{5.12}$$

$$\frac{dS_{\text{tot}}}{dt} + \Delta\left[\frac{\dot{m}s}{M_{\text{w}}}\right] = \sum_i \frac{\dot{Q}_i}{T_i} + \dot{\Sigma}. \tag{5.13}$$

The summations in the energy and entropy equations allow one to apply the equations to a system with more than one heat flow, if these are applied at different temperatures. Whereas the first equation is still exact, the energy balance neglects all contributions from kinetic and potential energies. The entropy balance assumes that any heat flow occurs at a single temperature.

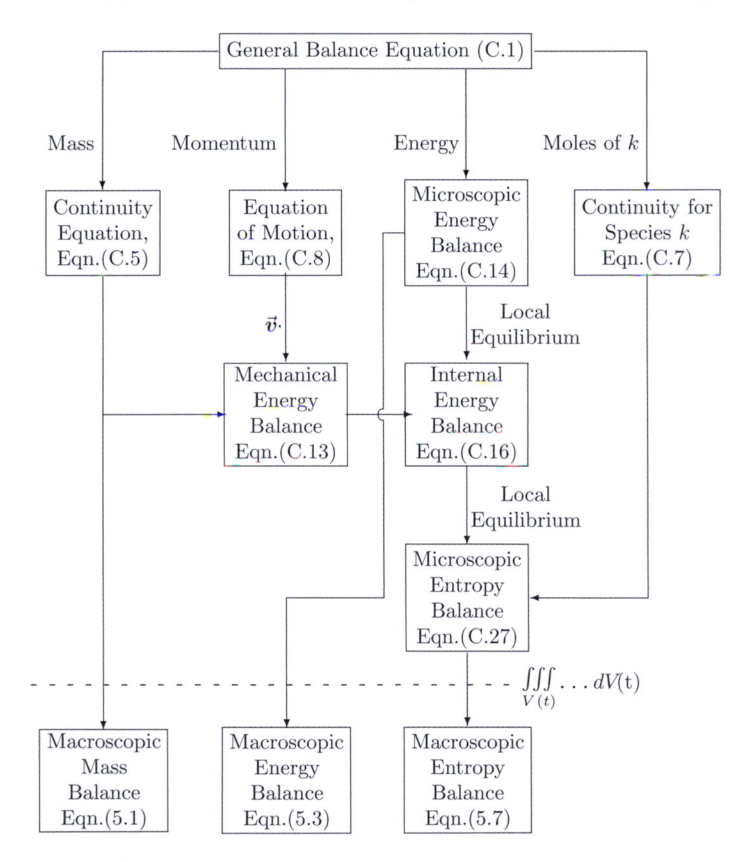

Figure 5.1 A schematic outline of the derivation of the macroscopic balance equations. Derivation of the macroscopic energy and mass balances is straightforward: a microscopic balance on these quantities is integrated over the macroscopic volume. However, the microscopic entropy balance is derived from the others. The microscopic energy balance contains both kinetic and internal energy. The kinetic energy can be eliminated, however, using the equation of motion, leaving a microscopic internal energy balance. Using the differential for internal energy $dU = T\,dS - P\,dV + \sum \mu_k\,dN_k$, this balance can be converted to a balance involving S, V, and N_k. The last two can be eliminated using the continuity equations, leaving a microscopic entropy balance. This microscopic entropy balance is then integrated over the macroscopic volume.

The macroscopic balances are the major goals of this section, and represent the tools used in many industrial applications. Because they contain some assumptions, it is important to keep in mind the paths of their derivations. These paths are sketched schematically in Figure 5.1. The balances above also use the list of assumptions begun on p. 151. The following subsections contain examples that utilize these simplified macroscopic balances.

5.1.1 The throttling process

A throttling process occurs when a fluid passes through a valve or opening and experiences a pressure drop. In commercial processes, it is important to be able to predict from the resulting temperature and pressure changes whether the fluid evaporates. Such two-phase flow can cause significant corrosion that leads to hazardous conditions. Also, a throttle is used in some refrigeration cycles to cause a fluid to evaporate, so that it absorbs heat from the environment.

We assume that kinetic and potential energies can be neglected in the steady-state flow through the valve. We also approximate the system as being adiabatic. Since there is no shaft work, the macroscopic energy balance Eqn. (5.3) becomes

$$\Delta h = h(T_\mathrm{f}, P_\mathrm{f}) - h(T_\mathrm{i}, P_\mathrm{i}) = 0. \tag{5.14}$$

Typically, we know the upstream conditions of the fluid $(T_\mathrm{i}, P_\mathrm{i})$. From straightforward fluid-mechanics calculations of the sort we do in transport phenomena using Bernoulli's equation, we can estimate the pressure drop through the valve, P_f. The energy balance above can then be used to find the downstream thermodynamic conditions.

In the following example we show how a flow process in a butane lighter can be treated as a throttling process.

EXAMPLE 5.1.1 On a day when the temperature is 25 °C, you pick up a cigarette lighter and push the valve without striking the flint. The butane exiting the lighter seems slightly cooler than the vapor–liquid mixture in the lighter. Why does it feel cooler? What is its thermodynamic state upon exit?

Solution. The fluid in the lighter is a mixture of vapor and liquid in equilibrium. It is not isolated from the environment, so its initial temperature T_i must also be 25 °C, which fixes the thermodynamic state for a two-phase, pure-component system. After a little searching in the NIST Chemistry Web Book (see Section D.1), we find that the vapor pressure of n-butane is approximately 0.244 MPa at this temperature. Hence, the initial pressure P_i is known. The final pressure is just $P_\mathrm{f} = P_\mathrm{atm} = 1$ atm. We also notice, since the lighter is made of colored, transparent plastic that there is a small tube attached to the valve that leads down to the bottom of the liquid. Hence, we have liquid butane at a known temperature and pressure entering the valve, fixing the entering enthalpy.

Recall the typical P–H diagram for a pure substance shown in Figure 4.11 on p. 134. The entering state of the fluid lies on the liquid (left) side of the binodal curve. On passing through the valve, the fluid must stay on a vertical line of constant H, but P decreases, as sketched in

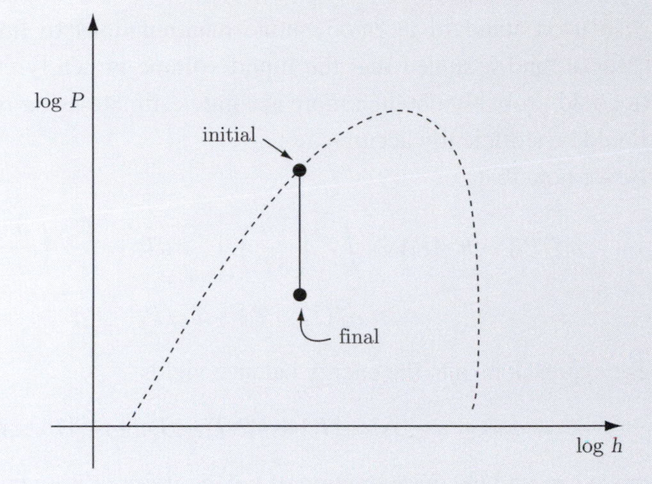

Figure 5.2 A sketch of the throttling process on a P–H diagram.

Figure 5.2. Thus, the fluid moves from the liquid saturation line to a point directly below, *inside the coexistence region*. Hence, the exiting fluid could be a vapor–liquid mixture.

The final enthalpy is also fixed by the macroscopic energy balance for the throttling process, Eqn. (5.14). Assuming for the moment that the exiting stream is then a mixture of liquid and vapor, it is convenient to write the energy balance as

$$\begin{aligned} 0 = \Delta h &= h_{\mathrm{f}} - h_{\mathrm{i}} \\ &= w^{\mathrm{f}}_{\mathrm{vap}} h^{\mathrm{sat}}_{\mathrm{vap}}[P_{\mathrm{f}}] + (1 - w^{\mathrm{f}}_{\mathrm{vap}}) h^{\mathrm{sat}}_{\mathrm{liq}}[P_{\mathrm{f}}] - h^{\mathrm{sat}}_{\mathrm{liq}}[P_{\mathrm{i}}] \\ &= w^{\mathrm{f}}_{\mathrm{vap}} (h^{\mathrm{sat}}_{\mathrm{vap}}[P_{\mathrm{f}}] - h^{\mathrm{sat}}_{\mathrm{liq}}[P_{\mathrm{i}}]) + (1 - w^{\mathrm{f}}_{\mathrm{vap}})(h^{\mathrm{sat}}_{\mathrm{liq}}[P_{\mathrm{f}}] - h^{\mathrm{sat}}_{\mathrm{liq}}[P_{\mathrm{i}}]), \end{aligned} \tag{5.15}$$

where $w^{\mathrm{f}}_{\mathrm{vap}}$ is the mass fraction of vapor in the exiting fluid, and $h^{\mathrm{sat}}_{\mathrm{vap}}[P_{\mathrm{f}}]$ and $h^{\mathrm{sat}}_{\mathrm{liq}}[P_{\mathrm{f}}]$ are the saturated vapor and liquid specific enthalpies at the exiting pressure. We use mass fraction instead of mole fraction, because the specific-enthalpy data we have are per mass instead of per mole. This equation assumes that we have a vapor–liquid mixture exiting the lighter, which we will need to check at the end. If it is a mixture, then the exiting temperature T_{f} is the boiling point of n-butane, 272.7 K, and the remaining unknown is the mass fraction of vapor. To find the first enthalpy change above, we use the path

$$\begin{aligned} h^{\mathrm{sat}}_{\mathrm{vap}}[P_{\mathrm{f}}] - h^{\mathrm{sat}}_{\mathrm{liq}}[P_{\mathrm{i}}] &= h^{\mathrm{sat}}_{\mathrm{vap}}[P_{\mathrm{f}}] - h^{\mathrm{sat}}_{\mathrm{liq}}[P_{\mathrm{f}}] + h^{\mathrm{sat}}_{\mathrm{liq}}[P_{\mathrm{f}}] - h_{\mathrm{liq}}[T_{\mathrm{i}}, P_{\mathrm{f}}] + h_{\mathrm{liq}}[T_{\mathrm{i}}, P_{\mathrm{f}}] - h_{\mathrm{liq}}[T_{\mathrm{i}}, P_{\mathrm{i}}] \\ &= \Delta h_{\mathrm{vap}}[T_{\mathrm{f}}] + \int_{T_{\mathrm{i}}}^{T_{\mathrm{f}}} \left(\frac{\partial h}{\partial T}\right)_P \bigg|_{P=P_{\mathrm{f}}} dT + \int_{P_{\mathrm{i}}}^{P_{\mathrm{f}}} \left(\frac{\partial h}{\partial P}\right)_T \bigg|_{T=T_{\mathrm{i}}} dP \\ &\cong \Delta h_{\mathrm{vap}}[T_{\mathrm{f}}] + c_P^{\mathrm{liq}}(T_{\mathrm{f}} - T_{\mathrm{i}}) + v_{\mathrm{liq}}(P_{\mathrm{f}} - P_{\mathrm{i}}). \end{aligned} \tag{5.16}$$

To obtain the second line, we have used the definition of the heat of vaporization, $\Delta h_{\mathrm{vap}}[T_{\mathrm{f}}] := h^{\mathrm{sat}}_{\mathrm{vap}}[P_{\mathrm{f}}] - h^{\mathrm{sat}}_{\mathrm{liq}}[P_{\mathrm{f}}]$. To arrive at the last line, we have used the definition for the specific constant-pressure heat capacity c_P, and assumed that it is constant over this small temperature change.

We have also used standard thermodynamic manipulations to find that $(\partial h/\partial P)_T = v - T(\partial v/\partial T)_P \approx v$, and assumed that the liquid volume is nearly constant over this pressure change. We could probably obtain a more accurate estimate using residual properties, but this estimate should be sufficiently accurate.

Similarly, we note that

$$h_{\rm liq}^{\rm sat}[P_{\rm f}] - h_{\rm liq}^{\rm sat}[P_{\rm i}] = \int_{T_{\rm i}}^{T_{\rm f}} \left(\frac{\partial h}{\partial T}\right)_P\bigg|_{P=P_{\rm f}} dT + \int_{P_{\rm i}}^{P_{\rm f}} \left(\frac{\partial h}{\partial P}\right)_T\bigg|_{T=T_{\rm i}} dP$$
$$\approx c_P^{\rm liq}(T_{\rm f} - T_{\rm i}) + v_{\rm liq}(P_{\rm f} - P_{\rm i}). \qquad (5.17)$$

Putting these expressions into the energy balance yields

$$0 = w_{\rm vap}^{\rm f}\,\Delta h_{\rm vap}[T_{\rm f}] + c_P^{\rm liq}(T_{\rm f} - T_{\rm i}) + v_{\rm liq}(P_{\rm f} - P_{\rm i}), \qquad (5.18)$$

from which we can find the mass fraction of butane vapor exiting the lighter, $w_{\rm vap}^{\rm f}$. After a little more searching in the NIST Chemistry Web Book, we find that $c_P^{\rm liq} \approx 132.4$ J/mol · K, $v_{\rm liq} \approx 0.01$ l/mol, and $\Delta h_{\rm vap}[T_{\rm f} = 272.7\,{\rm K}] = 22.44$ kJ/mol, from which we calculate $w_{\rm vap}^{\rm f} \approx 0.15$. Because the mass fraction of vapor is between zero and one, our assumption about a mixture of vapor and liquid is correct.

This is a surprising answer. However, a simple experiment with a lighter and a thermocouple[2] verifies that the temperature is at the boiling point of n-butane at 1 atm. Although there is a greater mass fraction of liquid, the saturated specific vapor volume is approximately 22 l/mol, 2000 times greater than that of the liquid. Hence, the volume fraction of vapor is approximately 0.997, and the liquid is not easily detected. □

5.1.2 Specifications for a turbine generator

A turbine contains blades attached to a shaft that are turned by an expanding gas – something like blowing on the propeller of a model plane, forcing it to turn. It is possible to estimate the power output from the turbine, and the thermodynamic condition of the exit stream from the temperature, pressure, and mass flow rate of the entrance stream, and the exit pressure.

The turbine operates in a steady-state manner, and we neglect changes in kinetic and potential energy and approximate the process as being adiabatic. Hence, the macroscopic energy balance Eqn. (5.3) simplifies to

$$\Delta[\dot{m}h/M_{\rm w}] = \dot{W}_{\rm s}. \qquad (5.19)$$

The steady-state mass balance requires that the mass flow rates in and out are equal: $\dot{m}_{\rm in} = \dot{m}_{\rm out} = \dot{m}$. Hence, our energy balance can be written

$$\dot{m}(h[T_{\rm f}, P_{\rm f}]/M_{\rm w} - h[T_{\rm i}, P_{\rm i}]/M_{\rm w}) = \dot{W}_{\rm s}. \qquad (5.20)$$

Since we know the entering pressure and temperature, we can use the steam tables to find the enthalpy of the inlet stream. However, we cannot find the exit enthalpy, because we do not

[2] Make certain that it is the liquid butane entering the tube, not the vapor.

know the exit temperature T_f. Hence, we currently have only one equation, but two unknowns: T_f and $\dot{m}$.

A second equation is provided by the macroscopic entropy balance Eqn. (5.7), which, under our adiabatic approximation and steady-state operation, becomes

$$\Delta[\dot{m}s/M_w] = \dot{\Sigma}. \tag{5.21}$$

However, here we have introduced a third unknown, namely $\dot{\Sigma}$, and the situation is not improved. On the other hand, it is conventional to neglect the entropy generation in the turbine, as a first approximation. Hence, we arrive at the approximation $s_{in} = s_{out}$. We can then use this relation to find the exit temperature. This last approximation, as well as the neglect of heat losses, will make our estimate the maximum possible power.

EXAMPLE 5.1.2 Steam is used to drive a turbine. The steam entering the turbine is at a pressure of $P_i = 2\,\text{MPa}$ and a temperature of $T_i = 350\,°\text{C}$, and exits at a pressure of $P_f = 30\,\text{kPa}$. What flow rate of steam is necessary if the turbine is to provide 12 MW of power?

Solution. From the steam tables, we find that

$$\begin{aligned} h[T_i, P_i]/M_w &= 3138.6\ \text{kJ/kg}, \\ s[T_i, P_i]/M_w &= 6.9596\ \text{kJ/kg}\cdot\text{K}. \end{aligned} \tag{5.22}$$

Again using the steam tables, we seek a temperature that corresponds to a pressure of $P_f = 30\,\text{kPa}$, and a specific entropy of $s[T_f, P_f]/M_w = s[T_i, P_i]/M_w = 6.9596\,\text{kJ/kg}\cdot\text{K}$. On inspection, we find that this specific entropy lies between the specific entropy of the liquid ($s^{liq}[30\,\text{kPa}]/M_w = 0.9441\,\text{kJ/kg}\cdot\text{K}$) and that of the saturated vapor ($s^{vap}[30\,\text{kPa}]/M_w = 7.7695\,\text{kJ/kg}\cdot\text{K}$) at this pressure. Hence, the exit stream must be a mixture of liquid and vapor. If w^{vap} is the mass fraction of vapor in the exit stream, then we can write

$$s[T_f, P_f]/M_w = w^{vap}s^{vap}[P_f]/M_w + (1 - w^{vap})s^{liq}[P_f]/M_w. \tag{5.23}$$

Solving for the mass fraction, we find $w^{vap} \approx 0.881$. We have now completely specified the thermodynamic state of the exit stream, and the exit specific enthalpy can be found:

$$\begin{aligned} h[T_f, P_f]/M_w &= w^{vap}h^{vap}[P_f]/M_w + (1 - w^{vap})h^{liq}[P_f]/M_w \\ &\approx 2348.2\,\text{kJ/kg}. \end{aligned} \tag{5.24}$$

We then solve our energy balance equation for the mass flow rate:

$$\dot{m} = \frac{\dot{W}_s M_w}{\Delta h} \approx 15.2\,\text{kg/s}. \tag{5.25}$$

This is a very large flow rate of steam. Note that we have made some rather extreme assumptions in order to arrive at the answer, namely we assumed that there is no heat loss from the turbine and no entropy generation; we also made a couple of minor assumptions, namely that the kinetic and potential energies are negligible.

If we had included entropy generation in the entropy balance, we would have found a higher exit entropy, which would have required a higher exit temperature T_f. The higher exit temperature would, in turn, have led to a higher value for the exit enthalpy and a smaller change in enthalpy Δh, and a larger mass flow rate would be required in order to obtain the same shaft work $\dot{W}_s$ from the turbine.

On the other hand, if we had included the effects of heat loss from the turbine, the exiting stream would be predicted to have a *decreased* entropy and a decreased enthalpy drop. The conventional *ad hoc* method to account for irreversibility in the turbine is to assume an efficiency factor η, which is the ratio of predicted change in enthalpy divided by the isentropic enthalpy change. □

5.1.3 Work requirements for a pump

EXAMPLE 5.1.3 Water passes through a pump that raises the pressure from P_i to P_f at a mass flow rate of $\dot{m}$. Find the rate of work required in order to operate the pump.

Solution. To a first approximation, we assume that water is incompressible, so that the change in kinetic energy is negligible. We also neglect any change in potential energy. Hence, the energy balance Eqn. (5.3) becomes

$$\Delta[\dot{m}h/M_w] = \dot{W}_s. \tag{5.26}$$

If we neglect entropy and heat fluxes at the boundaries of the pump, then the system is isentropic, and we can manipulate the above equation in the following manner:

$$\begin{aligned}\dot{W}_s &= \left(\frac{\dot{m}}{M_w}\right)(\Delta h)_s \\ &= \left(\frac{\dot{m}}{M_w}\right)\int_{P_i}^{P_f}\left(\frac{\partial h}{\partial P}\right)_s dP \\ &= \left(\frac{\dot{m}}{M_w}\right)\int_{P_i}^{P_f} v\, dP \\ &\approx \left(\frac{\dot{m}}{M_w}\right) v(P_f - P_i).\end{aligned} \tag{5.27}$$

The third line is obtained from standard thermodynamic manipulations, and the last line is obtained by assuming incompressibility of the fluid. The pressure drop $(P_f - P_i)$ required to obtain a given mass flow rate is a problem best tackled using the techniques of transport phenomena. □

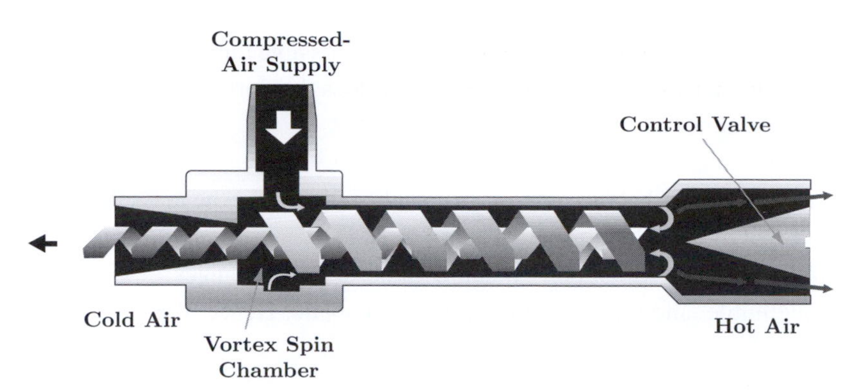

Figure 5.3 A schematic diagram of a commercially available Ranque–Hilsch vortex tube. Drawing courtesy of EXAIR Corporation.

5.1.4 The Ranque–Hilsch vortex tube

The Ranque–Hilsch vortex tube (Figure 5.3)[3] is a device that delivers hot and cold streams of air from a single initial stream at room temperature. The instrument might seem to violate the second law of thermodynamics, but a careful thermodynamic analysis shows otherwise.

There are many different designs of the tube, including some sold for commercial applications,[4] and plans for home experimentation. However, all designs work on a similar principle: compressed air is injected into a circular tube tangentially, such that it swirls inside the tube at a high velocity. The air in the center is ejected at one end of the tube as the cold stream, and the air circulating near the walls is expelled as the hot stream from the end of the tube either opposite to, or the same as, the cold stream. The transport mechanism describing how the tube works is still contested [52]. The fact that we are nonetheless able to predict the device's feasibility attests to the power of thermodynamics.

To analyze the device, it is necessary to perform mass, energy, and entropy balances. If the input conditions are known ($\dot{m}_\text{i}, T_\text{i}, P_\text{i}$), and the outlet pressures, and one exit mass flow rate are also known, then the unknowns are the other exit mass flow rate and the two outlet temperatures. If the entropy generation is also unknown, then we are left with four unknowns, but only three equations. The irreversibility inequality in entropy generation provides a range of possible solutions, with the reversible case providing bounds to the range.

In the following example we consider a thermodynamic analysis of a simplified Ranque–Hilsch vortex tube with a simple ideal gas.

EXAMPLE 5.1.4 An ideal gas enters a *Ranque–Hilsch vortex tube* at a pressure P_0 (with $P_0 > P_\text{atm}$), room temperature T_atm, and mass flow rate $\dot{m}_0$, as shown in Figure 5.3. The gas exiting at the left side of the tube has temperature $T^\text{c} < T_\text{atm}$, and the gas that exits the right side has temperature $T^\text{h} > T_\text{atm}$; both exit streams are at room pressure P_atm, and have the

[3] For the original work, see Ranque's text [106]; interest in the device was revived by Hilsch [62].
[4] See http://www.exair.com/.

same mass flow rate $\dot{m}$. What is the maximum possible temperature difference between the two exiting streams $\Delta T_{\max} = T^{\mathrm{h}}_{\max} - T^{\mathrm{c}}_{\min}$? Why does the vortex tube not violate the second law of thermodynamics?

Solution. A mass balance on the vortex tube tells us that the mass flow rate of each exit stream is half that of the inlet: $\frac{1}{2}\dot{m}_0 = \dot{m}^{\mathrm{c}} = \dot{m}^{\mathrm{h}} \equiv \dot{m}$.

To a first approximation, we can assume that the changes in kinetic and potential energy are negligible, and that the vortex tube is adiabatic. Hence, the steady-state energy balance is written

$$\begin{aligned} 0 &= \Delta(\dot{m}h) \\ &= h[T^{\mathrm{h}}, P_{\mathrm{atm}}]\dot{m}^{\mathrm{h}} + h[T^{\mathrm{c}}, P_{\mathrm{atm}}]\dot{m}^{\mathrm{c}} - h[T_{\mathrm{atm}}, P_0]\dot{m}_0 \\ &= \left(h[T^{\mathrm{h}}, P_{\mathrm{atm}}] + h[T^{\mathrm{c}}, P_{\mathrm{atm}}] - 2h[T_{\mathrm{atm}}, P_0]\right)\dot{m}. \end{aligned} \tag{5.28}$$

We used the results from the mass balance to obtain the last line. If we had an equation of state for $h(T, P)$, we would now have one equation, but two unknowns: T^{c} and T^{h}. Hence, we need another equation, which is provided by the entropy balance. Again neglecting entropy fluxes at the solid boundaries from heat flows, we obtain for the steady-state entropy balance

$$s[T^{\mathrm{h}}, P_{\mathrm{atm}}] + s[T^{\mathrm{c}}, P_{\mathrm{atm}}] - 2s[T_{\mathrm{atm}}, P_0] = \frac{\dot{\Sigma}}{\dot{m}}. \tag{5.29}$$

Our second equation has introduced another unknown, $\dot{\Sigma}$, whose estimation requires detailed information about the fluid inside the vortex tube. Since we do not have this detailed information, it might appear that the situation is not improved. However, we do know from the second law of thermodynamics that

$$\dot{\Sigma} \geq 0. \tag{5.30}$$

We are able to calculate the reversible case $\dot{\Sigma} = 0$ as a limit. For the moment, let us keep the term around, and remember that its value must be non-negative.

We can now introduce the necessary equations of state for the ideal gas. If we also assume that the gas is simple, then

$$h = u + Pv = cRT + RT = (c+1)RT, \quad \text{simple ideal gas,} \tag{5.31}$$

and the energy balance (5.28) becomes

$$2T_{\mathrm{atm}} = T^{\mathrm{h}} + T^{\mathrm{c}}. \tag{5.32}$$

The fundamental entropy relation for a simple ideal gas is given in Eqn. (2.39) of Chapter 2,

$$s = s_0 + R\log\left[\left(\frac{u}{u_0}\right)^c \left(\frac{v}{v_0}\right)\right], \quad \text{simple ideal gas.} \tag{5.33}$$

We may find the more useful equation of state $s = s(T, P)$ from this fundamental relation using the equations of state $u = cRT$ and $Pv = RT$:

$$s = s_0 + R \log \left[\left(\frac{cRT}{u_0} \right)^c \left(\frac{RT}{Pv_0} \right) \right], \quad \text{simple ideal gas.} \tag{5.34}$$

Using this equation of state in the entropy balance Eqn. (5.29) yields, after a little algebraic manipulation,

$$\log \left[\left(\frac{T^h T^c}{T_{atm}^2} \right)^{c+1} \left(\frac{P_0}{P_{atm}} \right)^2 \right] = \frac{\dot{\Sigma}}{\dot{m}R}. \tag{5.35}$$

If we use the result from our energy balance, Eqn. (5.32), to eliminate T^c from this equation, we find that

$$0 = \left(\frac{T^h}{T_{atm}} \right)^2 - 2 \left(\frac{T^h}{T_{atm}} \right) + \left(\frac{P_{atm}}{P_0} \right)^{2/(c+1)} \exp \left[\frac{\dot{\Sigma}}{(c+1)\dot{m}R} \right]. \tag{5.36}$$

This is a quadratic equation in T^h/T_{atm} with two roots. One of the roots gives T^h less than T_{atm}, which is clearly the cold stream. Hence, we find

$$\frac{T^h}{T_{atm}} = 1 + \sqrt{1 - (P_{atm}/P_0)^{2/(c+1)} \exp \left[\frac{\dot{\Sigma}}{(c+1)\dot{m}R} \right]}. \tag{5.37}$$

There are two interesting things to note from this solution. First, since the entropy-generation term $\dot{\Sigma}$ must be positive, we see that the maximum value that can be reached by the hot stream is for $\dot{\Sigma} = 0$. Therefore,

$$T_{max}^h = T_{atm} \left[1 + \sqrt{1 - (P_{atm}/P_0)^{2/(c+1)}} \right]. \tag{5.38}$$

The temperature of the cold stream can be found from this relation and Eqn. (5.32). Thus, we find that the maximum temperature difference predicted by a simple ideal gas with no entropy generation is

$$\Delta T_{max} = 2T_{atm} \sqrt{1 - (P_{atm}/P_0)^{2/(c+1)}}. \tag{5.39}$$

If $P_0 \cong 2$ atm, and $c = 5/2$ (which is appropriate for a diatomic gas, like nitrogen, at sufficiently high temperatures), then this expression gives $\Delta T_{max}/T_{atm} \cong 1.14$. For $T_{atm} \cong 298$ K, this gives a cold stream of 128 K!

The second observation about Eqn. (5.37) is that there is a maximum possible entropy generation of

$$\dot{\Sigma}_{max} = 2\dot{m}R \log \left(\frac{P_0}{P_{atm}} \right). \tag{5.40}$$

This limit in entropy generation exists because we have already specified the entrance and exit pressures. When the entropy generation is at a maximum, the radical in Eqn. (5.37) is zero, and both exit streams have the same temperature as the inlet stream, T_{atm}.

Note that throughout the development we have kept careful track of entropy generation. Yet, we find that it is possible to generate a stream of hot air and a stream of cold air from a stream at room temperature. This sounds like we are getting "something for nothing," which may violate our intuition about the second law of thermodynamics. However, the vortex tube exploits the fact that a gas at high pressure has lower entropy than the same gas at low pressure. Since a pressure drop is required in order to drive the gas through the system, the entropy of the gas rises. The vortex tube exploits this rise in entropy to drive the process of temperature separation, and the second law is not violated. □

Real vortex tubes are more complicated than this analysis suggests. First, the mass flow rate is not necessarily split evenly between the cold and hot exit streams. Secondly, there can be substantial friction losses in the tube, leading to large entropy generation. Perhaps most importantly, the velocity of the inlet stream is extremely large, even near the speed of sound [52]. Therefore, not only should kinetic energy be taken into account, but also the assumption of local equilibrium might not be valid, violating the derivation of the macroscopic balances, Eqns. (5.3) and (5.7).

5.1.5 Fuel cells

In Examples 3.5.3 and 3.5.4 we did a closed-system analysis of a fuel cell. To perform the analysis we somewhat artificially considered the reactants and products as part of the system. We return here to the fuel cell, but now as an open system in which reactants flow in, products (and unreacted reactants) flow out, and electrical work is produced. Depending upon the entropy of reaction, heat may flow in or out of the cell.

We first perform an energy balance on the cell using Eqn. (5.3). If we assume a steady state and neglect kinetic and potential energy, we obtain

$$\Delta\left[\frac{h\dot{m}}{M_{\mathrm{w}}}\right] = \dot{Q} + \dot{W}_{\mathrm{elec}}. \tag{5.41}$$

We have also assumed here that the general shaft work can be replaced by a term for electrical work. If the fuel cell operates isothermally, then the unreacted reactants do not contribute to the term on the left-hand side, because their enthalpy is unchanged. Hence, $\dot{m} \to \dot{m}_{\mathrm{rxn}}$ is the mass flow rate of products in the exit stream, and only the reacted reactants at the entrance side. We neglect the mass of the electrons in the flow.

If the temperature and pressure of the fuel cell are known, then the left-hand side can be found. In order to find the available electrical work, the heat flow must be known, which we find from the open entropy balance. Again assuming a steady state and isothermal conditions, and neglecting any molar flux not in the entrance and exit streams, simplifies Eqn. (5.7) to

$$\Delta\left[\frac{\dot{m}_{\mathrm{rxn}}s}{M_{\mathrm{w}}}\right] = \frac{\dot{Q}}{T} + \dot{\Sigma}. \tag{5.42}$$

We find that the heat flow into the system is

$$\dot{Q} = \dot{m}_{\mathrm{rxn}}T\frac{\Delta s_{\mathrm{rxn}}}{M_{\mathrm{w}}} - T\dot{\Sigma}. \tag{5.43}$$

If the cell operates reversibly ($\dot{\Sigma} = 0$), then this result is the same as that obtained for the closed system. Namely, the heat flow into (or out of) the fuel cell depends on the sign of ΔS_{rxn}.

If $\Delta S_{\text{rxn}} < 0$, then heat necessarily flows out of the cell. Otherwise, heat is required to enter the cell in order to maintain isothermal conditions. Any irreversibility in the operation (arising from temperature gradients, resistance, etc.) mitigates the amount of necessary added heat.

If we insert $\dot{Q}$ from Eqn. (5.43) into Eqn. (5.41), we obtain

$$\dot{W}_{\text{elec}} = \Delta\left[\frac{\Delta g_{\text{rxn}}\,\dot{m}_{\text{rxn}}}{M_{\text{w}}}\right] + T\dot{\Sigma}. \tag{5.44}$$

Hence, for reversible operation, we get the same result as that found in the closed analysis – the available work is minus the Gibbs free energy of reaction. If the entropy of reaction is positive, an additional cost is incurred from keeping the fuel cell heated, according to Eqn. (5.43). For negative entropy of reaction, no heating is necessary, and the maximum work available is when no entropy is generated, $\dot{\Sigma} = 0$. Therefore, we get from Eqn. (5.47)

$$\dot{W}_{\text{elec}}^{\text{max}} = \Delta\left[\frac{\Delta g_{\text{rxn}}\,\dot{m}_{\text{rxn}}}{M_{\text{w}}}\right], \qquad \Delta S_{\text{rxn}} < 0. \tag{5.45}$$

On the other hand, if the entropy of reaction is positive, irreversibility mitigates the need for external heating, as we see from Eqn. (5.43). In our earlier, closed-system analysis, we used combustion to heat the cell and finish the analysis. Here we assume that irreversibility completely obviates the need for heating to obtain the same result. If no heating is necessary, then Eqn. (5.43) says that the entropy generation must be

$$T\dot{\Sigma} = \frac{\dot{m}_{\text{rxn}}T\,\Delta s_{\text{rxn}}}{M_{\text{w}}}, \qquad \Delta S_{\text{rxn}} > 0. \tag{5.46}$$

Then, the maximum electrical work available according to Eqn. (5.47) is

$$\dot{W}_{\text{elec}}^{\text{max}} = \Delta\left[\frac{\Delta h_{\text{rxn}}\,\dot{m}_{\text{rxn}}}{M_{\text{w}}}\right], \qquad \Delta S_{\text{rxn}} > 0. \tag{5.47}$$

Although the analysis was a little different, these results are completely analogous to the results found for a closed system.

Some operations of the fuel cell leave some fraction of the reactants unreacted. However, the rate of reaction can be found from measuring the electrical current produced, and knowledge of the number of electrons exchanged in the reaction. If the rate of reaction is $\dot{R}_{\text{rxn}}$, then we can write

$$\dot{R}_{\text{rxn}} = \frac{i}{N_{\text{e}}\mathcal{F}}, \tag{5.48}$$

where i is the current, N_{e} is the number of electrons transferred in the reaction, and $\mathcal{F} = \tilde{N}_{\text{A}}e = 9.6492 \times 10^4$ C/mol $= 230\,60$ cal/(mol · eV) $= 2.8025 \times 10^{14}$ esu/mol is **Faraday's constant**, where e is the electron charge. Here C is a coulomb, esu is an electrostatic unit, and eV is an electron volt.

Therefore, from data for the current and power (or, equivalently, voltage), we can find the efficiency of a fuel cell, defined as

$$\eta := \frac{\dot{W}_{\text{elec}}}{\dot{W}_{\text{elec}}^{\text{theoretical}}}, \tag{5.49}$$

from the measured electrical work.

EXAMPLE 5.1.5 The overall reaction in the direct methanol fuel cell is

$$CH_3OH + \frac{3}{2}O_2 \rightarrow CO_2 + H_2O.$$

The redox reaction that corresponds to oxidation of methanol at the anode of the fuel cell is [6]

$$CH_3OH + H_2O \rightarrow CO_2 + 6H^+ + 6e^-.$$

A direct methanol fuel cell is operated at ambient pressure and 110 °C. The methanol concentration is 1 M and the flow rate is 360 ml/min. The electrode has a geometrical surface area of 225 cm^2. The average cell performance is given in Table 5.1. What is the efficiency of the direct methanol fuel cell described above as a function of the current density?

Solution. For this reaction, the entropy of reaction is negative, so the maximum theoretical electrical work available is equal to the change in enthalpy for the reaction. Hence, Eqn. (5.49) becomes for the methanol fuel cell

$$\eta = \frac{\dot{W}_{\text{elec}} M_{\text{w}}}{R\,\Delta h_{\text{rxn}}} \tag{5.50}$$

Since we know the area of the electrodes, the data in Table 5.1 are sufficient for us to calculate the power, which is the voltage times the current density times the area. From the current

Table 5.1 Data taken on a methanol fuel cell, from [20]. The results represent the average performance for a single cell.

Cell potential (V)	Current density (mA/cm^2)	Cell potential (V)	Current density (A/cm^2)
0.578	14	0.405	281
0.552	35	0.393	305
0.535	60	0.380	326
0.520	78	0.364	350
0.505	107	0.349	371
0.492	129	0.337	396
0.479	152	0.317	416
0.466	176	0.298	438
0.457	196	0.274	461
0.440	218	0.252	485
0.431	239	0.226	506
0.418	263	0.173	558

Table 5.2 The total current, reaction rate, and power production for the fuel cell whose data are given in Table 5.1

Voltage potential (V)	Current (mA)	Reaction rate (mol/s)	Power (W)
0.779	0	0	0
0.578	3.15	5.441×10^{-6}	1.8207
0.552	7.875	1.360×10^{-5}	4.347
0.535	13.5	2.332×10^{-5}	7.223
0.52	17.55	3.032×10^{-5}	9.126
0.505	24.08	4.159×10^{-5}	12.158
0.492	29.03	5.014×10^{-5}	14.280
0.479	34.2	5.908×10^{-5}	16.382
0.466	39.6	6.840×10^{-5}	18.454
0.457	44.1	7.618×10^{-5}	20.154
0.44	49.05	8.473×10^{-5}	21.582
0.431	53.78	9.289×10^{-5}	23.177
0.418	59.18	0.000 102 22	24.735
0.405	63.23	0.000 109 2	25.606
0.393	68.63	0.000 118 5	26.970
0.38	73.35	0.000 126 7	27.873
0.364	78.75	0.000 136 0	28.665
0.349	83.48	0.000 144 2	29.133
0.337	89.1	0.000 153 9	30.027
0.317	93.6	0.000 161 7	29.671
0.298	98.55	0.000 170 2	29.368
0.274	103.73	0.000 179 2	28.421
0.252	109.13	0.000 188 5	27.500
0.226	113.85	0.000 196 7	25.730
0.173	125.55	0.000 216 9	21.720

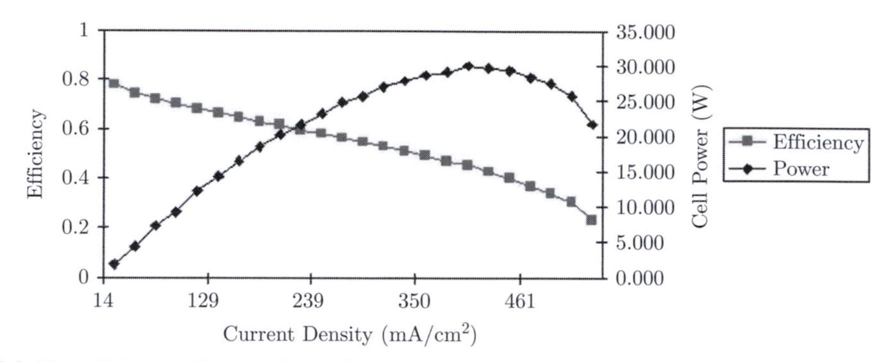

Figure 5.4 The efficiency of the methanol fuel cell as a function of the cell voltage from the data shown in Table 5.1.

density and the area, we can also find the reaction rate using Eqn. (5.48). These results are shown in Table 5.2.

Finally, to calculate the theoretical maximum, we find the heat of reaction to be −429.559 kJ/mol, using the same procedure as in Example 3.5.3 and the data in [25], or the NIST Web Book.

Hence, we can find the efficiency, which is plotted in Figure 5.4.

The reduction in the efficiency is believed to be caused by the irreversible energy losses that occur as a result of electrode polarization [6, Figure 2-2]. In other words, as the current density increases, the cell voltage decreases. This decrease in cell voltage causes a peak in the cell power as shown in Figure 5.4. Also, note how much steeper the curve is in the high-current-density region. Any fluctuations in the load (which are very common in electrical systems) for this cell would cause substantial changes in the power delivery, which may cause sporadic power outages in the system. For this reason, fuel-cell designers always operate on the up-slope, far from the peak power. This operation translates into an efficiency of approximately 57%, which is lower than the efficiency of the hydrogen fuel cell; however, it is still much better than the reasonable operating range of a heat engine, which will be discussed in the next section. □

5.2 CYCLES

Using the work of a pump, we can make heat flow from a cold region to a hot region, as we do every day with refrigerators, air conditioners, and heat pumps. These appliances pump a fluid in a cycle, where it undergoes a time-periodic expansion and contraction, with heat losses and heat gains. Similarly, a fluid, for example steam, is pumped in a cycle to produce shaft work from a heat source, such as a boiler. These cycles take in heat from a hot reservoir, and expel it to a cold reservoir, converting some to shaft work. Here we consider the thermodynamic analysis and design of such cycles.

5.2.1 The Carnot cycle

In the cycles that follow, we analyze the thermodynamic state of a fluid as it is pumped through devices such as heaters, compressors, and turbines. However, we first consider a fluid that is stationary, but undergoes work and heat flows that change its thermodynamic state. We consider this idealized system for a few reasons. First, it is of historic interest, since it was developed by one of the very early pioneers in thermodynamics, the French engineer Sadi Carnot, who was born in 1796 in Palais du Petit-Luxembourg. Secondly, it is a simple ideal cycle that is easily analyzed, and provides a simple criterion for the maximum efficiency that can be extracted from any work cycle. Lastly, we use it to discuss how irreversibility corrupts the maximum amount of work available from a real cycle.

The Carnot cycle is performed on a fluid in a container that contains a frictionless piston that allows reversible work to be done on the fluid, or by the fluid on the environment. It also has a diathermal wall that allows heat to pass to heat reservoirs. The thermodynamic states that the fluid reaches are shown in Figure 5.5.

The diathermal wall of the container is first placed in contact with the hot thermal reservoir, which has temperature T_h. Using the piston to manipulate the volume, we change the entropy of the system to S_l. Hence, the fluid begins at point A on the $T-S$ diagram in Figure 5.5. The fluid is then allowed to expand reversibly and isothermally by pushing on the piston until the entropy increases to S_h. During this step A $\rightarrow$ B, the fluid draws in heat from the hot reservoir, and does work on its environment.

In the next step, the fluid is expanded further, but this time reversibly and adiabatically. As we learned in Section 2.6, such a process is isentropic. Since the system is doing work on the environment, its internal energy decreases during this step. Hence, the temperature of the fluid drops until it is at temperature T_c. This is step B $\rightarrow$ C in Figure 5.5.

Once the fluid is at the lower temperature T_c, we place the diathermal wall of the container in contact with the cold reservoir at the same temperature. Now we compress the fluid isothermally and reversibly back to its original entropy S_l. During this step C $\rightarrow$ D, the fluid expels heat to the cold thermal reservoir, and requires that we do work.

To complete the cycle, the fluid is compressed adiabatically and reversibly back to its initial state A. Summarizing, we find that the fluid delivers work during steps A $\rightarrow$ B and B $\rightarrow$ C, and requires work during steps C $\rightarrow$ D and D $\rightarrow$ A; it takes in heat during step A $\rightarrow$ B, and expels heat during step C $\rightarrow$ D.

We can find the total work delivered (the work delivered minus the work required) from an energy balance on the fluid following all four steps. Since the fluid ends at the same

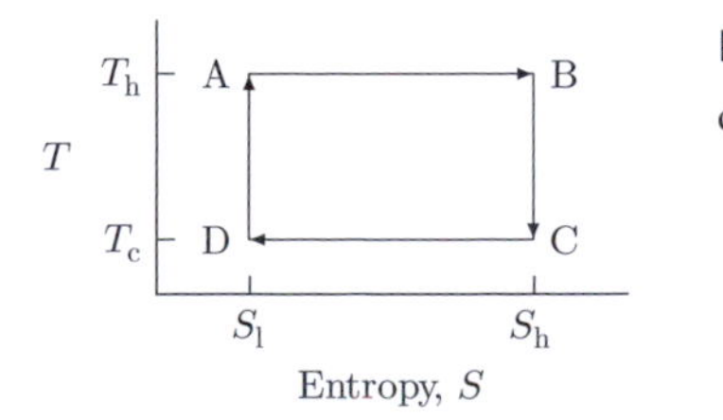

Figure 5.5 The thermodynamic states reached by a fluid in a Carnot cycle.

thermodynamic state as that at which it started, its total change in internal energy is zero. An energy balance says that the total heat delivered to the fluid must equal the total work delivered. During the isothermal steps, we may write

$$\begin{aligned} dU &= dU \\ đQ_{\text{rev}} + đW_{\text{rev}} &= T\,dS - P\,dV \\ đQ_{\text{rev}} &= T\,dS, \end{aligned} \tag{5.51}$$

where we used the definition of heat (2.1) and the differential for U (2.27) to obtain the second line, and the fact that $đW_{\text{rev}} = -P\,dV$ to obtain the third. If we integrate this equation over the two isothermal steps in the cycle, we obtain

$$Q_{\text{AB}} = T_{\text{h}}\,\Delta S, \quad Q_{\text{CD}} = -T_{\text{c}}\,\Delta S, \tag{5.52}$$

where $\Delta S := S_{\text{h}} - S_{\text{l}}$. Hence, the total work ($-W$) delivered *by the fluid to its environment* is $-W = Q_{\text{AB}} + Q_{\text{CD}} = (T_{\text{h}} - T_{\text{c}})\Delta S$, which is just the area of the box ABCD. The **coefficient of performance** ε_{p} is defined as the total work delivered divided by the heat taken from the hot reservoir, Q_{AB}. Hence, for the reversible Carnot cycle, we find

$$\begin{aligned} \varepsilon_{\text{p}} &:= \frac{W}{Q_{\text{AB}}} \\ &= 1 - \frac{T_{\text{c}}}{T_{\text{h}}}, \quad \text{Carnot cycle.} \end{aligned} \tag{5.53}$$

Since every step was performed reversibly, no entropy was created during the process, and this is the best performance that could ever be obtained by any cycle that absorbs heat from a thermal reservoir at temperature T_{h} and expels heat to a thermal reservoir at T_{c}.

In order to attain this efficiency, however, note that the isothermal steps require that the fluid and the reservoirs be at the same temperature during those steps. Such heat transfer takes an infinite amount of time. The adiabatic steps must also be done infinitely slowly in order for them to be reversible. If you would like to deliver work in a finite amount of time, you will need to place the diathermal wall of the container in contact with a reservoir that is hotter than T_{h} and another one that is colder than T_{c}. However, this heat flux will generate entropy (see Section C.4 in Appendix C), and the coefficient of performance will be less than that given by Eqn. (5.53).

In our description above, we assumed that the system was a fluid whose entropy was changed by manipulations of the volume. However, we could just as easily use, for example, a rubber band, and manipulate the entropy by manipulation of the length. In fact, some clever designs have been made of rubber-band engines that can be built with simple materials [128]. Other systems could also be used, such as in recent attempts to create motors on a nanometer length scale [8].

Finally, we note that the cycle could be run in reverse, so that the system acts as a *refrigerator*, taking in heat from the cold reservoir, and expelling it to the hot reservoir. It is straightforward to find the coefficient of performance ε_{r} for the refrigeration cycle as well.

It is defined as the heat absorbed at T_c divided by the net work required to run the system. By reversing the direction of the cycle, we simply change the signs of the heat transfer in the isothermal steps. Hence, we find

$$\varepsilon_r := \frac{Q_{CD}}{W} = \frac{T_c}{T_h - T_c}, \quad \text{Carnot cycle.} \tag{5.54}$$

Again, this is the best performance that can be expected from any refrigeration cycle that removes heat from a reservoir at temperature T_c and expels it at T_h. Exercise 2.9.D illustrates how a rubber band can also be made into a crude refrigerator.

5.2.2 The Rankine power cycle

The Carnot cycle operates in a sort of "batch" process with a quiescent fluid. However, engineers usually prefer continuous processes because of efficiency. Hence, most cycles utilize a flowing fluid. For these processes, we follow a parcel of fluid around the cycle and analyze its thermodynamic state at each point using the macroscopic balances.

The Rankine cycle uses steam in a power plant to generate shaft work. The Rankine cycle is shown schematically in Figure 5.6. Liquid water at point 1 is fed to a boiler where it becomes supersaturated steam, 2. The steam is fed through a turbine where it undergoes (in the ideal case) isentropic expansion to drive turbine blades on a shaft. The fluid that exits the turbine is usually a mixture of steam and water condensation, 3. This liquid–vapor mixture is then fed to a condenser, so that the resulting liquid, 4, can be pumped back to the boiler. The boiler requires heat flow $\dot{Q}_b$, perhaps from oil combustion, and the pump requires work $\dot{W}_p$, or some other form of energy. Waste heat $-\dot{Q}_c$ is generated in the condenser, which may be used elsewhere in the plant, or to preheat the liquid flowing from the pump to the boiler. The turbine generates the usable shaft work $\dot{W}_s$.

Thermodynamic analysis of the process is most easily seen on a *TS* diagram of the sort derived in Figure 4.10. If we follow a bit of steam as it travels through the cycle, it experiences time-periodic changes in its thermodynamic state. For example, at point 1 it is a liquid at high pressure (from the pump), but low temperature; whereas, after it has passed through the boiler,

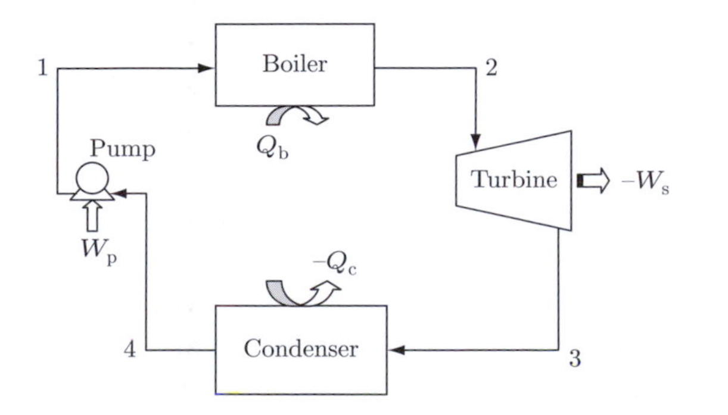

Figure 5.6 A schematic representation of the Rankine cycle using steam for power generation.

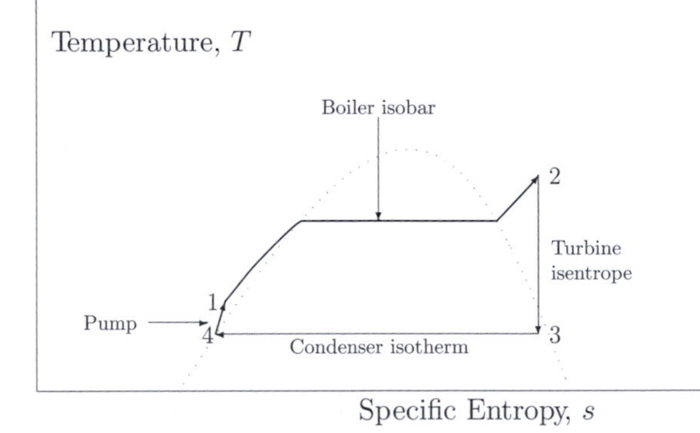

Figure 5.7 The thermodynamic path taken by a "characteristic" volume of steam as it travels through a Rankine cycle to generate work. The numbers correspond to the points in the cycle shown in Figure 5.6, and the dashed line is the binodal curve demarcating the two-phase region. The full $T\!-\!S$ diagram is shown in Figure 4.10.

it is superheated steam. We follow the thermodynamic state of the steam in the $T\!-\!S$ diagram shown in Figure 5.7.

As the water passes through the boiler, it is heated, boils, and is then superheated, all at constant pressure. Hence, the curve connecting points 1 and 2 is an isobar. We approximate the flow through the turbine to be isentropic (as was done in Example 5.1.2), so the line connecting points 2 and 3 must be vertical. On exiting the turbine, the steam is a mixture of liquid and vapor, so its state now falls inside the two-phase region of the diagram, point 3. The vapor is completely condensed in the aptly named condenser to be the saturated liquid of point 4. The pump returns the liquid back to its original pressure, point 1.

Design of the cycle is accomplished by putting together the parts as considered in the previous section. The following example also illustrates the usefulness of constructing two thermodynamic tables to determine the heat requirement of the boiler, work requirement for the pump, and available shaft work from the turbine. In the past, such design calculations were typically performed by reading the values for entropy, enthalpy, and so on off such a plot as Figure 5.7. However, it is more convenient now to use software. Here we will use the steam tables to obtain these values and illustrate the methodology.

EXAMPLE 5.2.1 We wish to design a power plant utilizing a steam-driven turbine generator. The turbine is designed to operate optimally with input steam at 525 °C and 9 MPa. The exhaust from the turbine is estimated to be approximately 20 kPa. Find the operating conditions necessary for all the equipment in the cycle. Estimate the coefficient of performance (work produced/heat required) for the cycle. If a total power generation of 10 MW is required, what is the flow rate of steam?

Solution. Design of the plant is facilitated by the construction of two tables: one describing the thermodynamic state of the steam at the four points and another describing the change in enthalpy, work done, and heat added. The first is shown in Table 5.3, with values added from the problem statement in shaded boxes. The remaining values will be found below.

Some values in Table 5.3 can be filled in rather quickly. We typically neglect the pressure drops in the boiler and condenser, so we know that $P_1 = P_2$ and $P_4 = P_3$. Also, we design the condenser to create saturated liquid: $w_{vap,4} = w_{vap,1} = 0$.

It is also useful to construct a table for the changes in thermodynamic quantities, as is shown in Table 5.4. The values in shaded boxes are typical assumptions for the equipment pieces. The remaining values will be determined below.

We now proceed stepwise to fill in the necessary values in the tables. Notably, we need to find the work required for the pump $\dot{W}_p$, the heat input required for the boiler $\dot{Q}_b$, and the shaft work obtained from the turbine $-\dot{W}_s$.

Since the fluid is a single-component system, only two thermodynamic variables are necessary in order to determine the state at any point. Hence, once any two values (excluding the mass fraction) have been filled in for a row in Table 5.3, the others may be found from an appropriate equation of state. For steam, the steam tables in Section D.4 of Appendix D may be used. Note that, if two phases are present (when the vapor mass fraction is not zero or one), only a single entry in the row is necessary in order to determine the thermodynamic state.

Table 5.3 A table of thermodynamic states in a Rankine cycle: temperature, pressure, specific entropy, specific enthalpy, and mass fraction of vapor. Note that the specific quantities are per mass instead of per mole, as is conventional. The boxed values are known from the problem statement, whereas others are found by application of the macroscopic balance equations.

State	T (°C)	P (kPa)	s/M_w (kJ/kg · K)	h/M_w (kJ/kg)	w_{vap}
1		9000		260.59	0
2	525	9000	6.7388	3448.7	1
3	60.09	20	6.7388	2219.8	0.8346
4	60.09	20	0.8321	251.5	0

Table 5.4 A table of changes in thermodynamic quantities for equipment in a Rankine cycle. The boxed values are typical assumptions for the Rankine cycle: adiabatic and isentropic turbine, and an adiabatic pump. No mechanical work is done either on the boiler or on the condenser.

Step	$\Delta h/M_w$ (kJ/kg)	$\dot{W}_s/\dot{m}$ (kJ/kg)	$\dot{Q}/\dot{m}$ (kJ/kg)
Boiler: 1 → 2	3188.1	0	3188.1
Turbine: 2 → 3	−1228.9	−1228.9	0
Condenser: 3 → 4	−1968.3	0	−1968.3
Pump: 4 → 1	9.1344	9.1344	0

- From the steam tables we may find the specific entropy and enthalpy for row 2: $s_2/M_w = 6.7388\,\text{kJ/kg}\cdot\text{K}$ and $h_2/M_w = 3448.7\,\text{kJ/kg}$, which we add to Table 5.3. We also note from Table 5.4 that the turbine is assumed isentropic, so we can immediately add $s_3/M_w = s_2/M_w = 6.7388\,\text{kJ/kg}\cdot\text{K}$ to Table 5.3.
- Since we know the pressure and entropy of point 3, we may now determine, using the steam tables, the remaining quantities in this row. When we check the tables, we find that at $P_3 = 20\,\text{kPa}$, the entropy at point 3 falls between the values of entropy for the vapor and the liquid. Hence, the water must be a mixture of liquid and vapor, as is shown in Figure 5.7. From the steam tables, we have $s_{\text{vap}}(20\,\text{kPa})/M_w = 7.9094\,\text{kJ/kg}\cdot\text{K}$, and $s_{\text{liq}}(20\,\text{kPa})/M_w = 0.8321\,\text{kJ/kg}\cdot\text{K}$. Hence, we can find the vapor mass fraction, from

$$s_3 = w_{\text{vap}} s_{\text{vap}} + (1 - w_{\text{vap}}) s_{\text{liq}},$$

  to be $w_{\text{vap}} = 0.8346$. Using an analogous equation for enthalpy, we can also find $h_3/M_w = 2219.8\,\text{kJ/kg}$. Using the tables for saturated steam, we estimate the temperature at this point to be $60.09\,^\circ\text{C}$.
- We note that the values for enthalpy in Table 5.3 can be used to find the change in enthalpy $2 \to 3$ for Table 5.4: $\Delta h_{2\to3}/M_w = h_3/M_w - h_2/M_w = -1228.9\,\text{kJ/kg}$. A macroscopic energy balance on the turbine yields

$$\frac{\dot{W}_s}{\dot{m}} = \Delta h_{2\to3}/M_w,$$

  which allows us to fill in the shaft work delivered by the turbine.
- We can also find the thermodynamic state at 4, which is the saturated liquid from point 3. Hence, we can immediately write down that $T_4 = 60.09\,^\circ\text{C}$, $P_4 = 20\,\text{kPa}$, $s_4/M_w = s_{\text{liq}}(20\,\text{kPa})/M_w = 0.8321\,\text{kJ/kg}\cdot\text{K}$, and $h_4/M_w = h_{\text{liq}}(20\,\text{kPa})/M_w = 251.5\,\text{kJ/kg}$.
- Again, we use the enthalpies from rows 3 and 4 of the first table to obtain the change in enthalpy in the second table: $\Delta h_{3\to4}/M_w = h_4/M_w - h_3/M_w = -1968.3\,\text{kJ/kg}$. An energy balance on the condenser yields

$$\frac{\dot{Q}_c}{\dot{m}} = \Delta h_{3\to4}/M_w,$$

  from which we fill in the condenser duty in Table 5.4.
- We find the pump specifications as we did in Example 5.1.3,

$$\frac{\dot{W}_p}{\dot{m}} \approx v_4(P_1 - P_4)/M_w.$$

  Hence, we find that $\dot{W}_p/\dot{m} \approx 9.1344\,\text{kJ/kg}$. An energy balance on the pump yields

$$\dot{W}_p/\dot{m} = \Delta h_{4\to1}/M_w,$$

  which allows us to fill in the change in enthalpy, and $h_1/M_w = 260.6\,\text{kJ/kg}$.
- Finally, an energy balance yields the heat duty on the boiler,

$$\frac{\dot{Q}_b}{\dot{m}} = \Delta h_{1\to2}/M_w = h_2/M_w - h_1/M_w,$$

  from which we find $\dot{Q}_b/\dot{m} = 3188.1\,\text{kJ/kg}$.

With the second table completed, we can now find the coefficient of performance for the cycle ε_c, which is the ratio of available work to the heat duty of the boiler,

$$\varepsilon_c := \frac{-\dot{W}_s - \dot{W}_p}{\dot{Q}_b}, \tag{5.55}$$

which yields 38.3% efficiency for this cycle. We can compare this with the maximum efficiency that would be attained by a Carnot cycle that operates between the temperature output by the boiler and the temperature of the condenser. Using Eqn. (5.53), we find that the maximum coefficient of performance for this example is

$$\varepsilon_{\text{Carnot}} = 1 - \frac{T_3}{T_2} = 0.582, \tag{5.56}$$

or 58.2% efficiency.

We find the necessary flow rate of steam, from the desired $\dot{W}_s = -10\,\text{MW} = 10\,000\,\text{kJ/s}$ and the calculated $\dot{W}_s/\dot{m} = -1228.9$, to be $\dot{m} = 8.14\,\text{kg/s}$. □

A real cycle operates less efficiently than this, however, because of heat losses in the turbine, entropy generation in the turbine and pump, and the fact that the heat reservoirs must have different temperatures from the condenser and boiler in order to drive heat flows.

5.2.3 The refrigeration cycle

Refrigeration has the opposite goal to that of a power cycle. Instead of adding heat to the cycle in order to extract work, we add work in order to extract heat. Here we consider an evaporation cycle that uses a single-component fluid. When we manipulate the pressure of the fluid, we can cause it to condense to give off heat in one part of the cycle, and to evaporate in another part of the cycle in order to extract heat from, say, a refrigerator. Work is required in order to compress the fluid and condense it. The cycle is shown schematically in Figure 5.8. In this cycle a valve (or throttle) is used to cause the pressure drop and evaporate the fluid. The changes in the state of the fluid can be illustrated on a P–H diagram such as is shown in Figure 5.9. The sketch shows that we assume an isentropic compressor, an isobaric evaporator and condenser, and an isenthalpic throttle.

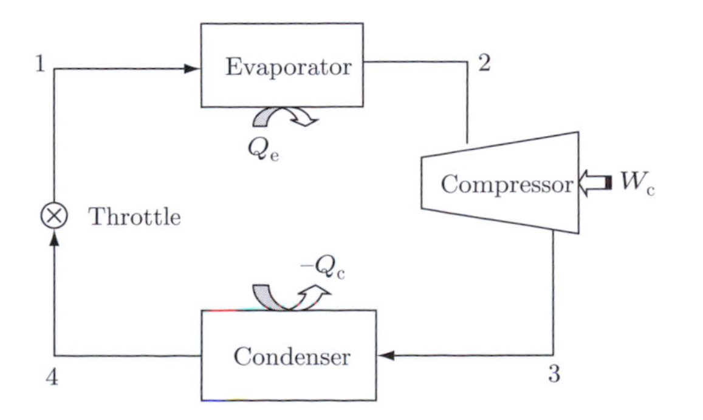

Figure 5.8 A schematic representation of a typical refrigeration cycle that utilizes evaporation.

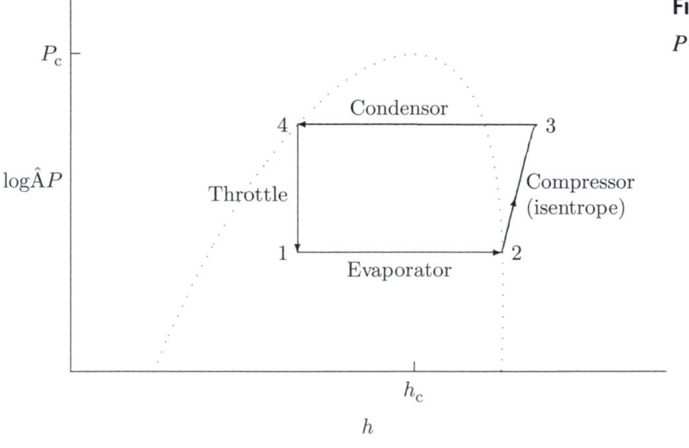

Figure 5.9 The refrigeration cycle on a P–H diagram.

In the following example we use ammonia to refrigerate a freezer to $-15\,°\text{C}$. As is customary in refrigeration, we use a P–H diagram to aid in our design. The diagram shown in Figure 5.10 is a standard one used by engineers to aid in design. It is derived from an equation of state that is more complicated than the ones considered so far. However, there are no new ideas introduced in the construction of the diagram, just more-involved algebra. The same principles as those discussed in Section 4.4 are used.

EXAMPLE 5.2.2 Design an evaporation cycle using ammonia to cool a freezer to $-15\,°\text{C}$ and expel the heat to the atmosphere. To achieve this, we assume that we can operate the coils of the evaporator at $-20\,°\text{C}$ and the condenser at $80\,°\text{C}$. Find the coefficient of performance, and compare it with the Carnot cycle. If heat-transfer calculations estimate that the refrigerator will need to remove 200 BTU/h, what is the volume flow rate of ammonia, and what size is necessary for the compressor motor?

Solution. We have designed the evaporator to operate at $-20\,°\text{C}$. Hence, the fluid at point 2 on our cycle in Figure 5.8 is saturated vapor at $-20\,°\text{C}$. From Figure 5.10, we see that

$$\begin{aligned} P_2 &\approx 0.18\,\text{MPa}, \\ h_2 &\approx 475\,\text{kJ/kg}, \\ s_2 &\approx 10.6\,\text{kJ/kg}\cdot\text{K}. \end{aligned} \tag{5.57}$$

We now assume that the fluid passes isentropically through the compressor, which must compress the fluid to enter the condenser. We have specified that the condenser operates at $80\,°\text{C}$. Hence, the compressor must condense the vapor to P^{sat} at $80\,°\text{C}$. Again, from Figure 5.10, we see that $P^{\text{sat}}(80\,°\text{C}) \approx 4.0\,\text{MPa}$. Therefore, point 3 in the cycle must occur at an entropy given by s_2 and this pressure. From Figure 5.10 we find that this is

$$
\begin{aligned}
P_3 &\approx 4.0\,\text{MPa}, \\
s_3 &\approx 10.6\,\text{kJ/kg}\cdot\text{K}, \\
h_3 &\approx 975\,\text{kJ/kg}.
\end{aligned}
\tag{5.58}
$$

At point 3, the fluid is supersaturated vapor. This vapor is condensed to a saturated liquid in the condenser. Hence, the condenser operates at the saturation pressure, 4.0 MPa. From Figure 5.10, we find that

$$
\begin{aligned}
P_4 &\approx 4.0\,\text{MPa}, \\
h_4 &\approx -375\,\text{kJ/kg}, \\
s_4 &\approx 6.9\,\text{kJ/kg}\cdot\text{K}.
\end{aligned}
\tag{5.59}
$$

In Section 5.1.1 we considered a throttling process to be isenthalpic. Hence, we estimate the enthalpy of point 1 to be the same as that at point 4, $h_1 = h_4 \approx -375$ kJ/kg. We also know that the evaporator operates isobarically, since it contains a mixture of liquid in vapor at equilibrium: $P_1 = P_2 \approx 0.18$ MPa. These two values determine the location on the figure, and we can write down all of the thermodynamic quantities. We find that the point lies inside the binodal region. Hence, the fluid is here a mixture of liquid and vapor:

$$
\begin{aligned}
h_1 = h_4 &\approx -375\,\text{kJ/kg}, \\
P_1 = P_2 &\approx 0.18\,\text{MPa}.
\end{aligned}
\tag{5.60}
$$

We can find the mass fraction of vapor x_{vap} from this enthalpy by noting

$$
h_1 = w^{\text{vap}} h_1^{\text{vap}} + (1 - w^{\text{vap}}) h_1^{\text{liq}}. \tag{5.61}
$$

We find from Figure 5.10 that

$$
\begin{aligned}
h_1^{\text{vap}} = h_2 &\approx 475\,\text{kJ/kg}, \\
h_1^{\text{liq}} &\approx -850\,\text{kJ/kg}.
\end{aligned}
\tag{5.62}
$$

Hence, we find $w^{\text{vap}} \approx 0.358$ from the last three equations.

From the thermodynamic states at all of the points, we can now find the work and heat in the different components. From the energy balance on the compressor, we find that

$$
\frac{\dot{W}_{\text{c}}}{\dot{m}} = \frac{\Delta h_{2\to 3}}{M_{\text{w}}} = \frac{h_3 - h_2}{M_{\text{w}}} \approx 500\,\text{kJ/kg}. \tag{5.63}
$$

From an energy balance on the evaporator we find

$$
\frac{\dot{Q}_{\text{e}}}{\dot{m}} = \frac{\Delta h_{1\to 2}}{M_{\text{w}}} = \frac{h_2 - h_1}{M_{\text{w}}} \approx 850\,\text{kJ/kg}. \tag{5.64}
$$

Our design requires that the heat-transfer rate be 200 BTU/h. Hence, we find that the mass flow rate of the fluid should be

$$
\dot{m} = \frac{\dot{Q}_{\text{e}}}{\dot{Q}_{\text{e}}/\dot{m}} \approx \left(\frac{200\,\text{BTU/h}}{850\,\text{kJ/kg}}\right)(1.055\,\text{kJ/BTU}) = 250\,\text{g/h}. \tag{5.65}
$$

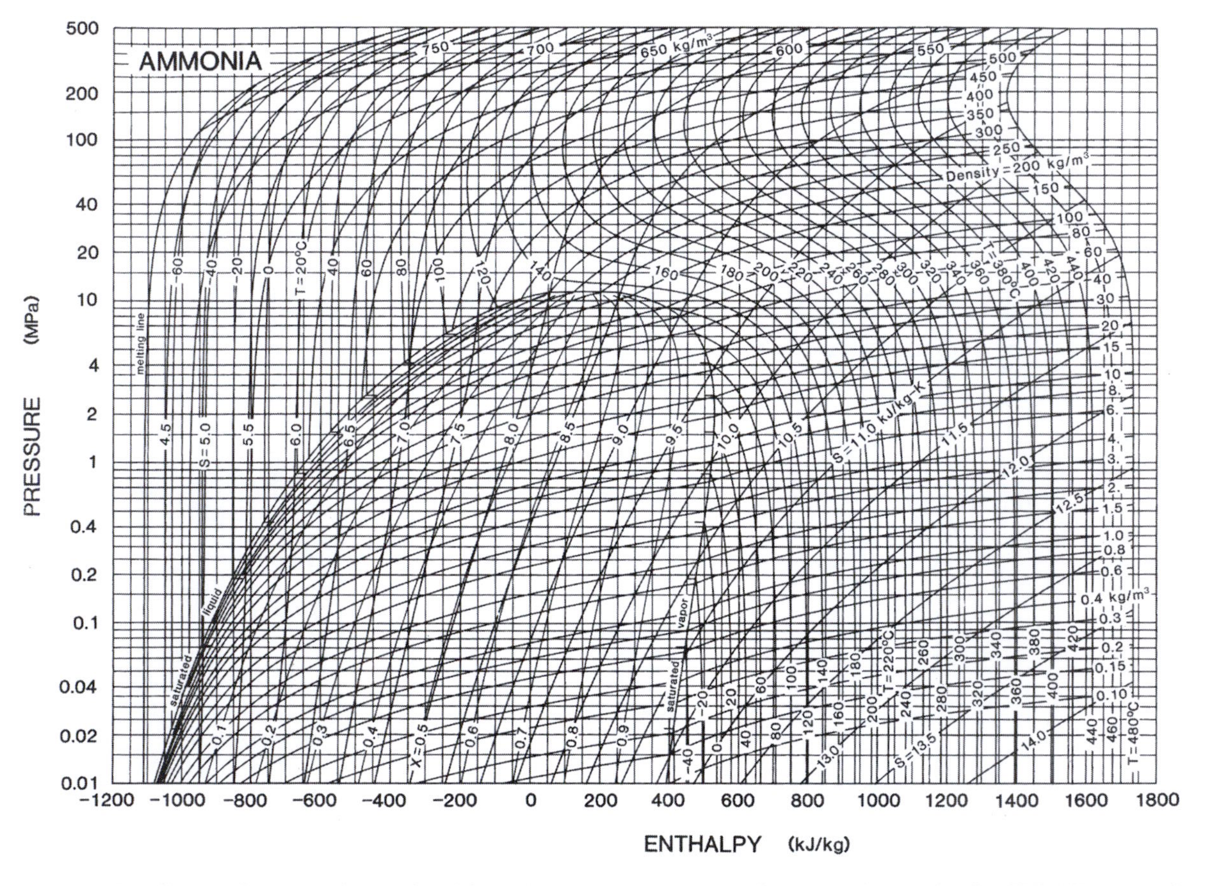

Figure 5.10 The pressure–enthalpy diagram for ammonia, reprinted from the 1997 ASHRAE Handbook-Fundamentals. ©ASHRAE, www.ashrae.org.

Using our energy balance on the compressor, we find that the pump in the compressor uses

$$\dot{W}_c = \dot{m}\left(\frac{\dot{W}_c}{\dot{m}}\right) \approx (0.25\,\text{kg/h})(500\,\text{kJ/kg}) = 125\,\text{kJ/h}. \tag{5.66}$$

We can calculate the coefficient of performance for refrigeration, which is the ratio of heat removed to work required:

$$\varepsilon_r = \frac{\dot{Q}_e}{\dot{W}_c} \approx 1.7, \tag{5.67}$$

which can be compared with the Carnot performance, Eqn. (5.54),

$$\varepsilon_{\text{Carnot}} = \frac{T_c}{T_h - T_c} = 2.53, \tag{5.68}$$

where we have taken the cold and hot reservoirs to be at −20 °C and 80 °C, respectively, instead of at the temperatures of the freezers and ambient. □

For very large systems, it can be more economical to replace the throttle with a turbine to get some work back out of the fluid at this step. The extracted work can then be used to help run the compressor and lower operating costs. This changes the thermodynamic analysis in step $4 \rightarrow 1$. Instead of being isenthalpic, the step is isentropic.

Summary

In this chapter we applied the ideas of thermodynamics to flowing systems when one can safely assume *local equilibrium*. By applying this assumption to tiny volume elements, we derived the *microscopic balance equations* for density, energy, velocity, and entropy in Appendix C. By integrating these equations over a larger control volume we derived the *macroscopic balance equations* in Section 5.1. These macroscopic balance equations are then used to find approximate solutions to large flowing systems. After completion of this chapter, the student should be able to analyze the thermodynamic limitations of any flowing system that is well approximated as being in local equilibrium.

Most of the chapter is devoted to solving specific problems. Those considered are

- the flow of butane through a valve (the throttling process),
- work in a steam-driven turbine,
- the work requirements for a pump,
- the thermodynamic analysis of an ideal gas in a Ranque–Hilsch vortex tube,
- the available electrical work in a fuel cell,
- the thermodynamic analysis of a Carnot cycle, the Rankine power cycle using steam, and a refrigeration cycle using ammonia.

Exercises

5.1.A You want to power a 60-W light bulb with a methanol-fuel-cell stack of 225 cm^2 connected in series. Use the data provided in Table 5.1, and possibly the results from Example 5.1.5, to design a fuel-cell stack to power it. Do you want your fuel cells to be stacked in series (keeping the current constant, but changing the voltage), or in parallel (keeping the voltage constant, but changing the current). Aside from giving the necessary quantitative answers (e.g. the number of cells, operating voltage, etc.), explain in detail how you chose the optimal operating conditions for the cells. For example, at what efficiency would your stack operate?

5.1.B Consider the turbine in Example 5.1.2.

(1) Suppose that the entropy generation is neglected, but that the heat loss from the turbine $-\dot{Q}$ is assumed to be 10% of the generated work. Further assume that the heat loss occurs completely at the inlet temperature. Calculate the mass flow rate and exit conditions of the steam. Is the turbine more or less efficient than the adiabatic case?

(2) What are the exit conditions and flow rates if you assume that the entropy generation is equivalent to the entropy loss along the sides of the turbine?

(3) Finally, assume that the turbine operates adiabatically, but that the entropy generation inside the turbine is approximately $0.10\dot{W}_s/T_i$. Now find the exit conditions of the steam and the required mass flow rates. How good is the ideal estimate, and how important is it to estimate accurately the entropy generation?

5.1.C Steam at temperature $T_i = 400\,°C$ and pressure $P_i = 9.8\,MPa$ is fed into a vortex tube. If the hot and cold streams have the same mass flow rates and pressure $P_f = 2.0\,MPa$, what range of temperatures is thermodynamically permissible for the exit streams?

5.1.D Nitrogen at temperature $T_i = 400\,°C$ and pressure $P_i = 9.8\,MPa$ is fed into a vortex tube. If the hot and cold streams have the same mass flow rates and pressure $P_f = 2.0\,MPa$, what range of temperatures is thermodynamically permissible for the exit streams? Use the Peng–Robinson equation of state for your estimate.

5.1.E R. Hilsch suggested that the efficiency of the Ranque–Hilsch vortex tube be measured by taking the ratio of the temperature drop in the cold stream $T_0 - T_c$ to the temperature drop that would occur for an isentropic expansion of the gas from (T_0, P_0) to T_{atm}. What is the efficiency of the Ranque–Hilsch vortex tube considered in Example 5.1.4?

5.1.F Your bicycle tire is pumped to 65 psi and the temperature is 25 °C. When you open the valve, what is the temperature of the air coming out, if you assume it is all nitrogen?

5.1.G In Example 5.1.1, we held the lighter so that liquid butane entered the valve. If we hold the lighter so that vapor enters the valve, what would the temperature of the exiting stream be? Use the Peng–Robinson equation of state for the vapor inside the lighter.

5.1.H A vortex tube using compressed air is operated, and data on the inlet and outlet streams are gathered. Table 5.5 shows the properties of these streams. We wish to analyze the

Table 5.5 Measured quantities for a vortex tube with adjustable position of inlet valve. The entries, from left to right, are the inlet pressure, the position of the inlet valve, the temperature of the exiting hot stream, the temperature of the exiting cold stream, the temperature of the inlet stream, the volumetric flow rate in, and the volumetric flow rate out.

Inlet P (psig)	Valve position (degrees)	T_h (°C)	T_c (°C)	T_i (°C)	Volumetric flow in (l/min)	Volumetric flow out (l/min)
16.8	0	31.8	28.8	30.2	14.05	12.5
16.8	30	34.2	23.0	30.2	15.1	11.1
16.8	60	36.2	22.8	30.2	15.3	8.4
45	0	43.6	16.6	30.6	33.2	30
45	30	46.6	7.6	30.8	33.4	24.9
45	60	49.4	12.2	30.8	33.8	20
65	0	39.4	3.8	30.2	48.8	41.4
65	30	45.2	−3.6	30.6	49.6	33.5
65	60	50.4	4.2	30.2	50.1	26.1

performance of the tube assuming that air is an ideal gas under these conditions and that its properties may be estimated as those of a mixture of 80% nitrogen and 20% oxygen. To perform the analysis, first find the mass flow rates of the streams given, and then estimate the mass flow rate of the remaining stream. Then perform an energy balance on the system. Is it operated adiabatically? Next perform an entropy balance and determine whether the entropy generation is zero, positive, or negative. Make plots of $\dot{Q}/\dot{m}$ versus valve position, and entropy generation per mass versus valve position, each for several inlet pressures. Is there any correlation between entropy generation and the temperature of the cold stream? What role does heat flow play?

5.2.A Using a steam-driven Rankine cycle, design a power plant to deliver 1.5 MW of power. Assume that the turbine operates with its input steam at 475 °C and 5 MPa, and that the exit stream is at 25 kPa. What is the coefficient of performance?

5.2.B Design a steam power plant using the Rankine cycle. Assume that the turbine operates with entering and exiting pressures of 7.25 MPa and 5 kPa, respectively. The entering temperature of the steam is 425 °C. Find the operating conditions of all the components in the cycle. Find the coefficient of performance, and compare it with the Carnot efficiency. If a total power generation of 625 kW is required from the cycle, what is the mass flow rate of steam necessary?

5.2.C Find the coefficient of performance, evaporator duty, and compressor size if the throttle in Example 5.2.2 is replaced by a turbine. Assume that the turbine is used to help power the compressor.

5.2.D Design a steam power plant using the Rankine cycle assuming that the turbine operates with an entering pressure of 6.1 MPa, an entering temperature of 510 °C, and an exiting pressure of 1.5 kPa. Find the operating conditions of the fluid everywhere in the cycle, the coefficient of performance, and compare it with a Carnot cycle operating at the same temperatures. If 310 kW of power is required, what flow rate of steam is necessary?

5.2.E We wish to use a refrigerant from DuPont (HCFC-123) to design a refrigeration cycle. The thermodynamic properties of the fluid can be found online. Design a refrigeration cycle to operate between 40 and 140 °F. While the pump is running, the refrigerator should provide 1000 BTU/h cooling capacity. What flow rate of refrigerant is necessary, and what size pump is needed?

5.2.F Analyze the Carnot cycle in reverse (i.e. as a refrigerator) to obtain the coefficient of performance expression given in Eqn. (5.54).

6 Statistical mechanics

In this chapter, we make the connection between molecules and thermodynamics. Until this chapter, we have dealt only with macroscopic quantities, and the influence of molecular details has been discussed only qualitatively. Here, we make quantitative connections between molecules and much of the thermodynamics we have already seen. For example, we derive the fundamental relations for ideal gases (simple and general), molecular adsorption on solid surfaces, and the elasticity of polymer chains. The procedure is general, and may be used to derive fundamental relations for more complex systems, to relate macroscopic experimental results to molecular interactions, or to exploit computer simulations. Since statistical mechanics deals with more detailed information than does classical thermodynamics, we can also find how real systems fluctuate in time when at equilibrium. We will see that these fluctuations are important for all systems, and can even dominate the behavior of fluids near a critical point (or spinodal curve) or that of small systems.

This chapter is not essential for understanding most of the rest of the book. However, the remaining chapters will also have a few sections dealing with the statistical mechanics of mixtures and polymers, and the material in this chapter is essential for understanding those topics. Statistical mechanics is both beautiful and powerful, but often difficult to grasp the first time you see it. However, we believe that statistical mechanics, like thermodynamics, rewards the student who visits it repeatedly. Therefore, we include a short introduction to the subject here, and encourage the student to try and capture a rough idea of how it works. If the details are baffling at first, try instead to capture what goes into the modeling, what assumptions are made, and what comes out at the end.

Most introductory texts on statistical mechanics start from a quantum-mechanical description of matter [59, 61]. Relatively few authors have adopted a purely classical introduction to the subject [21, 144]. In order to present a self-contained discussion of statistical mechanics, here we follow the approach of Callen [21] and that of Landau and Lifshitz [78], who introduce the subject without invoking quantum states of matter.

6.1 ENSEMBLE AND TIME AVERAGES

Most of the systems of interest in this text consist of collections of a large number of molecules (of order 10^{23}). These molecules are moving incessantly, and yet the macroscopic equilibrium thermodynamic properties are stable in that they neither change with time nor become heterogeneous. The reason for this is that, in simple fluids, molecular motion occurs on time scales that are much shorter than those associated with macroscopic measurements; that is, our

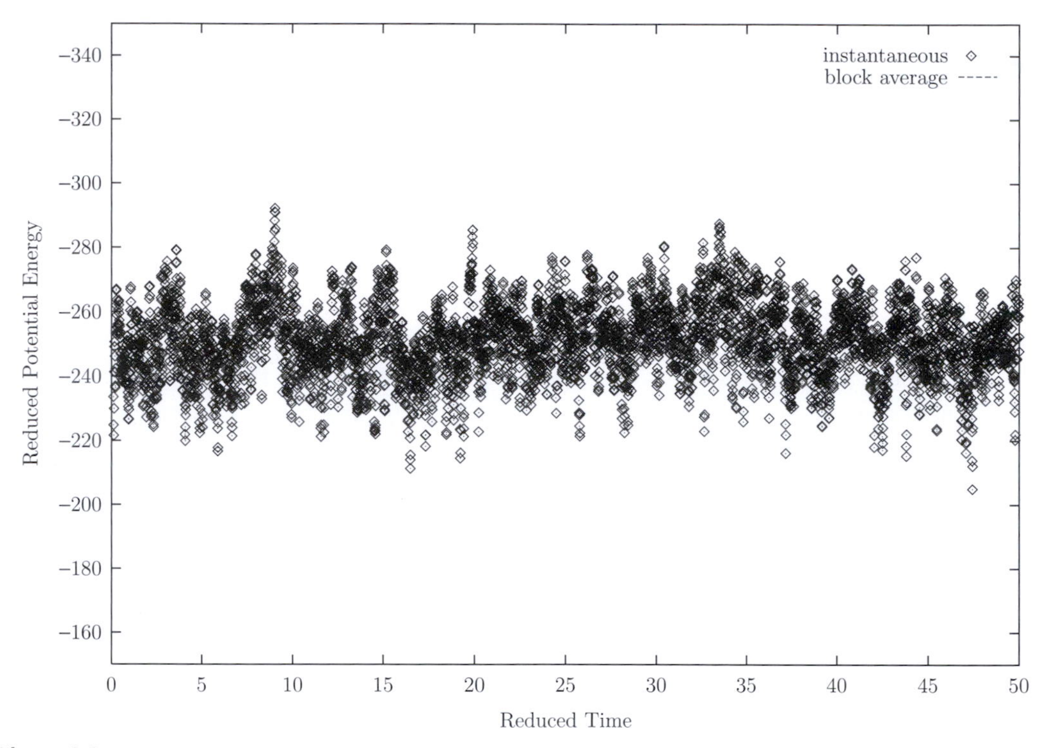

Figure 6.1 The instantaneous energy of a Lennard-Jones fluid as a function of reduced time. The points show results of a molecular-dynamics simulation.

macroscopic probes are insensitive to fast processes and measure only "average" properties of a collection of molecules. This observation suggests that the behavior of a collection of molecules could be analyzed using statistical methods, thereby allowing us to extract average quantities and the magnitude of fluctuations around the mean.

Figure 6.1 shows the instantaneous internal energy of a collection of just 200 "Lennard-Jones" molecules (a good model for the behavior of simple fluids such as argon or methane) as a function of reduced time. The energy undergoes rapid fluctuations about a mean value. The time scale for these fluctuations is relatively fast (on the order of 10^{-15} s); the whole abscissa comprises events that happened within just a fraction of a nanosecond. As the number of molecules increases, the size of the fluctuations decreases relative to the mean energy, roughly as the square root of the inverse of the number of molecules. A macroscopic measurement technique for the energy would be unable to resolve these fast fluctuations and would merely sample an average internal energy value. The methods of statistical mechanics allow us to do precisely that. *The statistical properties of collections or ensembles of molecules are studied in order to make predictions about the average value of a thermodynamic property and the likelihood of observing deviations from that average.*

Following our development of classical thermodynamics presented in previous chapters, we begin by introducing a postulate. That postulate states that, for an isolated system, all microscopic states having the same volume, number of molecules, and energy are equally likely to occur.

Statistical postulate I. For an isolated system consisting of N molecules, having volume V and energy U, all microscopic states consistent with those constraints occur with the same probability.

Let's consider an example. Think of a collection of N molecules of a noble gas (e.g. argon) confined to a small container with volume V. Furthermore, we can assume that the container is well insulated, so as to prevent interactions with the exterior, thereby maintaining a constant energy. At any non-zero temperature, the molecules in the container are constantly moving and exchanging energy with one another, but the total energy must somehow remain constant. This is equivalent to saying that there are many distinct arrangements of the molecules, or microscopic configurations, which have the same energy. The postulate stated above is intuitively attractive in that it simply says that all of these microstates have the same chance of occurring.

In *classical mechanics*, of course, if molecules are free to sample all of the volume of the container, it would be difficult to determine the number of microstates consistent with a distinct value of the energy. From *quantum mechanics*, however, we know that only distinct states are possible, and such a problem does not arise. Here we wish to keep things simple, and exploit the work of Gibbs, which used purely classical ideas. To get around the problem, Gibbs determined the number of microstates comprised within a certain, narrow energy range. To quantify this idea, we introduce a quantity $\Omega(N, V, U)$ that represents the number of distinct microscopic states of the system having number of molecules N, volume V, and energy U in the range of energy between U and $U+\delta U$. As we shall see later, the results of our analysis are not sensitive to the value of δU, provided that we satisfy the constraints $0 < \delta U \ll U$. If we use p_j to denote the probability of observing a distinct microstate j of a system having N, V, and energy between U and $U + \delta U$ at equilibrium, we can write

$$p_j = \frac{1}{\Omega(N, V, U)}, \tag{6.1}$$

since the probability must be normalized to sum up to unity. A system for which N, V, and U are constant is called a **microcanonical ensemble**. The quantity $\Omega(N, V, U)$ is called the **microcanonical partition function**.

To make a connection with thermodynamics, we introduce the second and final postulate of statistical mechanics. It seems intuitively plausible that the entropy of a system would depend on the number of available microstates – a measure of disorder. On the other hand, if we double the size of a system, the number of microstates grows quadratically. Since entropy must grow only linearly with size (it is extensive), Boltzmann introduced the following postulate.

Statistical postulate II.

$$S = k_B \log \Omega(U, V, N). \tag{6.2}$$

The constant k_B appearing in Eqn. (6.2) is called **Boltzmann's constant**. An analysis of experimental data reveals that its value is approximately 1.381×10^{-16} erg/K, which, not coincidentally, is the ideal-gas constant divided by Avogadro's number. The presence of k_B merely gives entropy the correct units; that is, the units which are used in classical thermodynamics. We will see later why this is a sensible definition, and we will show that it is consistent with nature, and our earlier postulates. As articulated by Richard Feynman in his *Lectures on Physics* [40], "...entropy is just the logarithm of the number of ways of internally arranging a system while having it look the same from the outside." Isn't that great? To perform statistical mechanics you just have to be able to count! Unfortunately, we shall see that counting is not as easy as it sounds, because the numbers are so large. For example, if the entropy is on the order of the ideal-gas constant, then $\log \Omega \sim \tilde{N}_A$, Avogadro's number. Hence, the number of microstates is approximately $\exp(10^{23})$.

6.2 THE CANONICAL ENSEMBLE

The purpose of the statistical-mechanical formalism presented in this text is ultimately that of providing a useful framework in which to analyze the structure and properties of molecular-level systems. This framework would of course be used to interpret experimental data, and in that respect the microcanonical ensemble is not the most convenient to work with.

As we considered in earlier chapters, experiments are often conducted at constant temperature, volume, and number of molecules. An ensemble in which these three variables are kept constant is called a **canonical ensemble**. To derive a partition function for this ensemble, we can go through the following thought experiment. Consider a subsystem, or group of molecules, immersed in a giant thermal reservoir whose energy is U_{bath}. The energy of the subsystem fluctuates through interactions with the thermal bath, but the energy of the total system remains constant. If, for simplicity, we assume that the energy of the subsystem adopts discrete values[1] U_j, then we have

$$U_{\text{total}} = U_{\text{bath}} + U_j, \quad \text{constant,} \tag{6.3}$$

where the total energy is constant because the combined system is isolated. We now wish to determine the probability that the subsystem has energy U_j. The entire system is isolated, so the probability $p_{j,k}$ that the subsystem is in state j and the bath is in state k is found from the postulate

$$p_{j,k} = \frac{1}{\Omega_{\text{total}}[U_{\text{total}}]}. \tag{6.4}$$

To find the probability p_j that the subsystem is in state j and the bath is in any other state, we sum this probability over all bath states:

$$\begin{aligned} p_j &= \sum_k p_{j,k} \\ &= \sum_k \frac{1}{\Omega_{\text{total}}[U_{\text{total}}]} \\ &= \frac{\Omega_{\text{bath}}[U_{\text{total}} - U_j]}{\Omega_{\text{total}}[U_{\text{total}}]}. \end{aligned} \tag{6.5}$$

[1] Actually, from quantum mechanics we know that real systems can adopt only discrete values for energy.

We used the property of probabilities to write the first line, and Eqn. (6.4) to obtain the second. To obtain the last line we used the fact that Ω_{total} is independent of the particular energy distribution between subsystem and reservoir, and the definition of Ω_{bath}. Namely, $\Omega[U_{\text{total}}]$ is the total number of distinct configurations of the combined system, and $\Omega_{\text{bath}}[U_{\text{total}} - U_j]$ is the number of configurations available to the reservoir when the subsystem has energy U_j. We have suppressed arguments involving the volume and mole numbers in the bath and subsystem, because these are held constant. The quantity p_j is nothing other than the fraction of the total number of configurations of the total system in which the subsystem is in a configuration having energy U_j.

The probability p_j can be related to the entropy of the bath and that of the total system through Eqn. (6.2). Upon substitution of that expression into Eqn. (6.5) we have

$$p_j = \exp\left[\frac{S_{\text{bath}}[U_{\text{total}} - U_j] - S_{\text{total}}[U_{\text{total}}]}{k_B}\right]. \tag{6.6}$$

The total entropy of the system consists of the entropy of the subsystem and that of the reservoir. At equilibrium, the energy of the subsystem immersed in the reservoir undergoes rapid fluctuations around some average value, U_{avg}. In order to determine the entropy of the subsystem we resort to thermodynamics and calculate the entropy corresponding to that average value of the energy. Following that reasoning, we write

$$S_{\text{total}}[U_{\text{total}}] = S[U_j] + S_{\text{bath}}[U_{\text{total}} - U_j], \tag{6.7}$$

since entropy is extensive. According to the definition of a heat reservoir, the fluctuations in energy exchanged with the subsystem are infinitesimally small. Following Callen [21, p. 350], the energy of the reservoir can then be expanded about its average value to give

$$\begin{aligned} S_{\text{bath}}[U_{\text{total}} - U_j] &\cong S_{\text{bath}}[U_{\text{total}} - U_{\text{avg}}] + (U_{\text{avg}} - U_j)\left(\frac{\partial S_{\text{bath}}}{\partial U}\right)_{V_{\text{bath}}, N_{\text{bath}}}\Bigg|_{U = U_{\text{tot}} - U_{\text{avg}}} \\ &= S_{\text{bath}}[U_{\text{total}} - U_{\text{avg}}] + \frac{U_{\text{avg}} - U_j}{T}. \end{aligned} \tag{6.8}$$

Upon substitution of Eqn. (6.8) into Eqn. (6.6) the probability p_j becomes

$$p_j = \exp\left[\frac{S_{\text{bath}}[U_{\text{total}} - U_{\text{avg}}] - S_{\text{total}}[U_{\text{total}}]}{k_B}\right]\exp\left[\frac{U_{\text{avg}} - U_j}{k_B T}\right]. \tag{6.9}$$

We can now use the fact that entropy is extensive to eliminate the entropies of the combined system and the bath above to obtain

$$p_j = \exp\left[\frac{U_{\text{avg}} - TS[U_{\text{avg}}]}{k_B T}\right]\exp\left[-\frac{U_j}{k_B T}\right]. \tag{6.10}$$

The quantity $U_{\text{avg}} - TS[U_{\text{avg}}]$ corresponds to the Helmholtz free energy of the subsystem, F. Note what we have accomplished: the bath influences this last expression *only through the temperature* T.

We can now find an expression for the free energy; the sum over all possible configurations of the probability p_j must be unity:

$$\sum_j p_j = \exp\left(\frac{F}{k_B T}\right) \sum_j \exp\left(-\frac{U_j}{k_B T}\right) = 1. \tag{6.11}$$

The free energy of the system is therefore given by

$$\frac{F}{k_B T} = -\log Q(T, V, N), \tag{6.12}$$

where the quantity Q is called the **canonical-ensemble partition function**, and is given by

$$Q(T, V, N) := \sum_j \exp\left(-\frac{U_j}{k_B T}\right). \tag{6.13}$$

The partition function plays a central role in statistical mechanics. It should not be confused with heat; which quantity is meant by Q should be clear from the context. We are unhappy when books confuse notation, but this choice of notation for the partition function is so ubiquitous in the literature (or sometimes the notation Z is used) that we are willing to make an exception here. Also, problems in which both the partition function and heat flow appear at the same time practically never arise.

We see from Eqn. (6.12) that knowledge of $Q(T, V, N)$ yields a fundamental relation. Equation (6.13) shows us how to find it. Similarly, we can see that $\Omega(U, V, N)$ also yields a fundamental relation from Postulate II. Hence, it is called the **microcanonical partition function**. The bulk of statistical mechanics is an attempt to estimate partition functions for real systems.

From Eqns. (6.10), (6.12), and (6.13) we also see that the probability of the system being in microstate j is

$$p_j = \frac{1}{Q} \exp\left(-\frac{U_j}{k_B T}\right). \tag{6.14}$$

It is interesting to find the entropy in terms of the probabilities. Recall from the differential for F that

$$\begin{aligned} S &= -\left(\frac{\partial F}{\partial T}\right)_{V,N} \\ &= \left(\frac{\partial}{\partial T} k_B T \log Q\right)_{V,N} \\ &= k_B \log Q + \frac{k_B T}{Q} \sum_j \frac{U_j}{k_B T^2} \exp\left(-\frac{U_j}{k_B T}\right), \end{aligned} \tag{6.15}$$

where we used Eqn. (6.12) to obtain the second line, and the definition of the canonical partition function, Eqn (6.13), to obtain the third. Now we use our result for the probability, Eqn. (6.14), to simplify the second term on the right-hand side:

$$\begin{aligned} S &= k_B \log Q + \sum_j \frac{U_j}{T} p_j, \\ &= k_B \sum_j p_j \left[\log Q + \frac{U_j}{k_B T} \right], \\ &= -k_B \sum_j p_j \log p_j. \end{aligned} \tag{6.16}$$

To obtain the second line, we used the fact that the probability is normalized: $\sum_j p_j = 1$. The third line used again our expression for p_j.

What is interesting about this result is that it also works for the microcanonical ensemble. In other words, if we insert Eqn. (6.1) into Eqn. (6.16), we obtain the second statistical-mechanical postulate, Eqn. (6.2). In fact, the result above plays an important role not just in thermodynamics, but also in non-equilibrium thermodynamics [94], information theory [71], applied mathematics [103], and process control.

6.3 IDEAL GASES

6.3.1 A simple ideal gas

To illustrate the use of statistical mechanics and the canonical ensemble, in this section we derive the partition function, fundamental relation, and equations of state for a simple, monatomic ideal gas. It is easiest to solve this problem using the canonical ensemble, so we consider a large thermal bath in which a subsystem is immersed. The number of molecules N and volume V of the subsystem are constant, and the thermal bath holds the temperature fixed. Whereas T, V, and N are constant, the instantaneous energy of the subsystem fluctuates through interactions with the thermal bath. The total, instantaneous energy of the subsystem consists of a sum of kinetic and potential energy terms,

$$U(\vec{\boldsymbol{r}}^{\tilde{N}}, \vec{\boldsymbol{p}}^{\tilde{N}}) = U_{\text{kin}}(\vec{\boldsymbol{p}}^{\tilde{N}}) + U_{\text{pot}}(\vec{\boldsymbol{r}}^{\tilde{N}}). \tag{6.17}$$

The specific value of the energy corresponding to a particular "configuration" of the subsystem depends on the positions of the atoms and their momenta, where $\vec{\boldsymbol{r}}^{\tilde{N}} := \vec{\boldsymbol{r}}_1, \vec{\boldsymbol{r}}_2, \ldots, \vec{\boldsymbol{r}}_{\tilde{N}}$ and $\vec{\boldsymbol{p}}^{\tilde{N}} := \vec{\boldsymbol{p}}_1, \vec{\boldsymbol{p}}_2, \ldots, \vec{\boldsymbol{p}}_{\tilde{N}}$ are used to denote collectively the position coordinates and the momenta of all the $\tilde{N}$ molecules in the system, respectively.

Following the discussion of the previous section, we know that the energy of the subsystem should be distributed according to the following probability distribution:

$$p(\vec{\boldsymbol{r}}^{\tilde{N}}, \vec{\boldsymbol{p}}^{\tilde{N}}) = \frac{1}{Q} \exp\left[-\frac{U(\vec{\boldsymbol{r}}^{\tilde{N}}, \vec{\boldsymbol{p}}^{\tilde{N}})}{k_B T} \right]. \tag{6.18}$$

As we saw in the previous section, the partition function Q is obtained by summing the exponential terms corresponding to all of the distinct configurations that the subsystem can adopt. For a large number of molecules in a continuum, the energies of different configurations are going to be closely spaced, and it is therefore appropriate to replace the summation appearing in Eqn. (6.13) by an integral over all of the coordinates of all the molecules in the subsystem,

$$Q = \frac{1}{(2\pi\hbar)^{3\tilde{N}}} \int \int_V \exp\left(-\frac{U[\vec{r}^{\tilde{N}}, \vec{p}^{\tilde{N}}]}{k_B T}\right) d\vec{r}^{\tilde{N}}\, d\vec{p}^{\tilde{N}}. \tag{6.19}$$

A factor of $1/\hbar^{3\tilde{N}}$, where $\hbar \cong 1.054\,571\,6 \times 10^{-34}$ J·s is the reduced Planck constant, has been included in Eqn. (6.19) to reconcile the classical arguments presented here with quantum mechanics; as can be seen from Eqn. (6.12), its only impact is to shift the reference state for the free energy by a constant amount proportional to $-3\tilde{N} \log \hbar$. Intuitively, one could think of the presence of Planck's constant as a way of introducing Heisenberg's uncertainty principle: we cannot subdivide $dr\,dp$ any finer than $2\pi\hbar$ in the integral.

We have also introduced the number of molecules $\tilde{N} := N\tilde{N}_A$, where N is the usual mole number and $\tilde{N}_A$ is Avogadro's number. We have used the shorthand that $\int \ldots d\vec{p}^{\tilde{N}}$ represents $3\tilde{N}$ integrals – each molecule has three components of momentum. Although that represents a lot of integrals, we will soon see that it is really just the same integral done many times.

By definition, the molecules of an ideal gas are point particles that do not interact with each other. The internal energy of an ideal gas is therefore purely kinetic, and the integral over the position coordinates of the molecules appearing in Eqn. (6.19) can be simplified considerably to give a volume factor V for each molecule:

$$\begin{aligned}
Q &\sim \frac{1}{(2\pi\hbar)^{3\tilde{N}}} \int_{-\infty}^{-\infty} \int_V \exp\left(-\frac{U[\vec{r}^{\tilde{N}}, \vec{p}^{\tilde{N}}]}{k_B T}\right) d\vec{r}^{\tilde{N}}\, d\vec{p}^{\tilde{N}} \\
&\sim \frac{1}{(2\pi\hbar)^{3\tilde{N}}} \int_V \exp\left(-\frac{U_{\text{pot}}[\vec{r}^{\tilde{N}}]}{k_B T}\right) d\vec{r}^{\tilde{N}} \int_{-\infty}^{-\infty} \exp\left(-\frac{U_{\text{kin}}[\vec{p}^{\tilde{N}}]}{k_B T}\right) d\vec{p}^{\tilde{N}} \\
&\sim \frac{1}{(2\pi\hbar)^{3\tilde{N}}} \int_V d\vec{r}^{\tilde{N}} \int_{\infty}^{\infty} \exp\left(-\frac{U_{\text{kin}}[\vec{p}^{\tilde{N}}]}{k_B T}\right) d\vec{p}^{\tilde{N}}, \quad \text{dilute, monatomic gases} \\
&\sim \frac{V^{\tilde{N}}}{(2\pi\hbar)^{3\tilde{N}}} \int_{-\infty}^{+\infty} \exp\left(-\frac{U_{\text{kin}}[\vec{p}^{\tilde{N}}]}{k_B T}\right) d\vec{p}^{\tilde{N}} \\
&\sim \frac{V^{\tilde{N}}}{(2\pi\hbar)^{3\tilde{N}}} \left[\int_{-\infty}^{+\infty} \exp\left(-\frac{U_{\text{kin}}^{\text{mol}}[\vec{p}]}{k_B T}\right) d\vec{p}\right]^{\tilde{N}}.
\end{aligned} \tag{6.20}$$

An important assumption has been made in order to pass from the second to the third line above. Namely, we assume that the particles do not interact with one another, so that $U_{\text{pot}} = 0$. In real gases this is attained by making the density low. Also, we have neglected any potential interactions between atoms within a molecule. Therefore, the derivation is restricted to monatomic, low-density gases. To see how we went from the third to the fourth line it is instructive to consider the special case in which our volume is a rectangular box with sides of length L_x, L_y, and L_z:

$$\begin{aligned}
\int d\vec{r}^{\tilde{N}} &= \int \dots \int d\vec{r}_1 \dots d\vec{r}_{\tilde{N}} \\
&= \int d\vec{r}_1 \dots \int d\vec{r}_{\tilde{N}} \\
&= \left(\int d\vec{r}\right)^{\tilde{N}} \\
&= \left(\int_0^{L_z}\int_0^{L_y}\int_0^{L_x} dx\, dy\, dz\right)^{\tilde{N}} \\
&= \left(L_x L_y L_z\right)^{\tilde{N}} \\
&= V^{\tilde{N}}.
\end{aligned} \tag{6.21}$$

The momenta of all the molecules in the gas are independent of each other: $U_{\text{kin}}[\vec{p}^{\tilde{N}}] = \sum_{i=1}^{\tilde{N}} U_{\text{kin}}^{\text{mol}}[\vec{p}_i]$; it therefore suffices to perform the integration over the momentum of one molecule and then factorize the result. The integral for one molecule is then

$$\begin{aligned}
\int_{-\infty}^{+\infty} \exp\left(-\frac{U_{\text{kin}}^{\text{mol}}[\vec{p}]}{k_{\text{B}}T}\right) d\vec{p} &= \int_{-\infty}^{\infty}\int_{-\infty}^{\infty}\int_{-\infty}^{\infty} \exp\left[-\frac{p_x{}^2 + p_y{}^2 + p_z{}^2}{2mk_{\text{B}}T}\right] dp_x\, dp_y\, dp_z \\
&= \left[2\pi m k_{\text{B}} T\right]^{3/2}.
\end{aligned} \tag{6.22}$$

With these results, the canonical-ensemble partition function becomes

$$Q \sim \left[V\left(\frac{mk_{\text{B}}T}{2\pi\hbar^2}\right)^{3/2}\right]^{\tilde{N}}. \tag{6.23}$$

The group of variables appearing in Eqn. (6.23) is generally referred to as the **de Broglie thermal wavelength**, Λ, defined by

$$\Lambda := \frac{\sqrt{2\pi}\,\hbar}{(mk_{\text{B}}T)^{1/2}}. \tag{6.24}$$

Our discussion so far has assumed that all molecules are *distinguishable*. In reality, however, molecules are indistinguishable from each other (they don't have tags or labels to identify them). Our expression for the partition function therefore overestimates the number of arrangements of the molecules; this can be corrected by dividing the right-hand side of Eqn. (6.23) by $\tilde{N}!$, which is the number of permutations that we can have amongst the $\tilde{N}$ molecules (see Section A.8.1 of Appendix A). With this correction, the free energy of the ideal gas becomes

$$\begin{aligned}
\frac{F}{k_{\text{B}}T} &= -\log Q \\
&= -\log\left\{\frac{1}{\tilde{N}!}\left[\frac{V}{\Lambda^3}\right]^{\tilde{N}}\right\} \\
&\approx -\tilde{N}\log\left[\frac{V}{\tilde{N}\Lambda^3}\right] - \tilde{N},
\end{aligned} \tag{6.25}$$

where the natural logarithm of $\tilde{N}!$ was approximated using the fact that $\log N! \approx N\log N - N$ for large N (this is the **Stirling approximation** [1, p. 257]).

Equation (6.25) is a fundamental relation, so we can use it to find equations of state. For example, the entropy is

$$S = -\left(\frac{\partial F}{\partial T}\right)_{V,\tilde{N}}$$
$$= \tilde{N}k_B + \tilde{N}k_B \log\left[\frac{V}{\tilde{N}\Lambda^3}\right] + \frac{3}{2}\tilde{N}k_B, \tag{6.26}$$

and the thermal equation of state is given by

$$U = F + TS = \frac{3}{2}NRT, \tag{6.27}$$

where we used the relation $R = k_B\tilde{N}_A$. The mechanical equation of state is also obtained from

$$P = -\left(\frac{\partial F}{\partial V}\right)_{N,T}$$
$$= \frac{\tilde{N}k_B T}{V} = \frac{NRT}{V}. \tag{6.28}$$

These are the equations of state introduced in Chapter 2 for a simple ideal gas, since $\tilde{N}k_B = NR$.

Also note how the contribution to the partition function from each molecule is independent of the contributions from the others, Eqn. (6.20). This factorization of the partition function into contributions from energetically independent parts of the system – in this case non-interacting molecules – is a general principle. In fact, the principle is used repeatedly in statistical mechanics, and we use it again in the next example for a general ideal gas.

We conclude this section by pointing out that statistical mechanics also provides a direct route to correct Equation (6.28) for intermolecular interactions (or deviations from ideal-gas behavior). This is illustrated in Problem 6.3D, and in chapters 7, 8 and 9.

6.3.2 A general ideal gas

We now show that a *general* ideal gas is made of molecules that do not interact with one another, but may have atoms within the molecule that do interact. In other words, the molecules may have bond vibrational energies, rotational energies, etc., and are no longer just point particles. Therefore, we can write the energy of just one molecule of type 1 as

$$E^1_{\text{mol}}(\vec{\boldsymbol{R}}) + \frac{\vec{\boldsymbol{p}}\cdot\vec{\boldsymbol{p}}}{2m_1}, \tag{6.29}$$

where $\vec{\boldsymbol{R}}$ is some vector of dimension $2d$ describing the internal conformations and momenta of the polyatomic molecule, and tells us about bond lengths, rotation angles, etc. For example, maybe R_1 is the length of a covalent bond, R_2 is the angle between successive bonds, and R_3 is a *trans–gauche* rotation angle. The momentum of the center of mass of the molecule is denoted by $\vec{\boldsymbol{p}}$, and the mass of the molecule of type 1 is m_1.

We place one such molecule in a box of volume V. The partition function Q_1 for our single polyatomic molecule in a box is then, from the definition of Q,

$$Q_1 = \frac{1}{(2\pi\hbar)^{3+d}}\int\int\int \exp\left[-\frac{E^1_{\text{mol}}(\vec{\boldsymbol{R}})}{k_B T} - \frac{\vec{\boldsymbol{p}}\cdot\vec{\boldsymbol{p}}}{2m_1 k_B T}\right] d\vec{\boldsymbol{r}}\, d\vec{\boldsymbol{p}}\, d\vec{\boldsymbol{R}}. \tag{6.30}$$

The two innermost integrals have already been found for the simple ideal gas, where $\vec{r}$ is the position of the center of mass of the molecule. The number of internal degrees of freedom of the molecule is written as d here. The outermost integral is the contribution from internal degrees of freedom of the molecule. Hence, we can write

$$\begin{aligned} Q_1 &= \frac{1}{(2\pi\hbar)^{3+d}} \int d\vec{r} \times \int \exp\left[-\frac{\vec{p}\cdot\vec{p}}{2m_1 k_B T}\right] d\vec{p} \times \int \exp\left[-\frac{E^1_{\text{mol}}(\vec{R})}{k_B T}\right] d\vec{R} \\ &= \frac{V}{\Lambda^3} Q_1^{\text{int}}(T), \end{aligned} \tag{6.31}$$

where

$$Q_1^{\text{int}}(T) := \frac{1}{(2\pi\hbar)^d} \int \exp\left[-\frac{E^1_{\text{mol}}(\vec{R})}{k_B T}\right] d\vec{R} \tag{6.32}$$

is the contribution to the partition function from the internal degrees of freedom for a molecule of type 1. Note that it depends only on temperature, and not on volume.

Now we add a second molecule of type 1 to our box. The system is dilute, so the molecules can be assumed not to interact, and the total energy of the system is now

$$E^1_{\text{mol}}(\vec{R}_1) + \frac{\vec{p}_1\cdot\vec{p}_1}{2m_1} + E^1_{\text{mol}}(\vec{R}_2) + \frac{\vec{p}_2\cdot\vec{p}_2}{2m_1}. \tag{6.33}$$

We go through the same steps as before, and exploit the fact that the nested integrals just become products again, to find the two-molecules-in-a-box partition function

$$Q_2 = \frac{V^2}{2!\Lambda^6} Q_1^{\text{int}}(T) Q_1^{\text{int}}(T). \tag{6.34}$$

Since the two molecules are indistinguishable, we divided by a factor of 2! to avoid overcounting. It is now straightforward to generalize this result by adding more and more molecules to the box, as long as they do not interact significantly. If we add to the box $\tilde{N}$ molecules, all of the same type, then the partition function becomes

$$Q = \frac{V^{\tilde{N}}}{\tilde{N}!\Lambda^{3\tilde{N}}} \left[Q_1^{\text{int}}(T)\right]^{\tilde{N}}. \tag{6.35}$$

Compare this result with that for non-interacting, monatomic molecules, Eqn. (6.23). The only change is the addition of the internal contribution to the partition function, Q_1^{int}, which depends only upon temperature.

Therefore, the fundamental relation for a box of $\tilde{N}$ identical, non-interacting particles is

$$\frac{F}{k_B T} = -\log Q \tag{6.36}$$

$$\approx -\tilde{N} \log\left[\frac{V}{\tilde{N}\Lambda^3}\right] - \tilde{N} - \tilde{N}\log Q_1^{\text{int}}(T). \tag{6.37}$$

We can obtain the mechanical equation of state by taking the derivative of each side of this equation with respect to V. Since $Q_1^{\text{int}}(T)$ is independent of V, we again obtain the ideal-gas law:

$$\frac{P}{k_B T} = -\left(\frac{\partial}{\partial V}\frac{F}{k_B T}\right)_{T,\tilde{N}} = \frac{\tilde{N}}{V}. \tag{6.38}$$

However, the thermal equation of state is different,

$$\frac{U}{NRT} = -T\left(\frac{\partial}{\partial T}\frac{F}{NRT}\right)_{V,\tilde{N}}$$
$$= \frac{3}{2} + \frac{d\log Q_1^{\mathrm{int}}(T)}{dT}. \tag{6.39}$$

As a result, we see that a simple ideal gas is a dilute gas of monatomic molecules that rarely see each other, and a general ideal gas includes dilute polyatomic molecules. The more degrees of freedom within the molecule, the greater $Q_1^{\mathrm{int}}(T)$ is. In fact, infrared spectroscopy can be used to measure the energetics available to the molecular vibrations, and these energetics can in turn be used to predict the heat capacity to great accuracy [101].

Finally, we mention a general result found by the same procedure. If we have r kinds of non-interacting molecules in our system, we can generally write the partition function as a product over each molecule,

$$Q = \prod_{i=1}^{r} \frac{Q_i^{\mathrm{mol}}(T,V)^{\tilde{N}_i}}{\tilde{N}_i!}, \tag{6.40}$$

where r is the number of types of molecules, $\tilde{N}_i$ is the number of indistinguishable molecules of type i, and $Q_i^{\mathrm{mol}}(T,V)$ is the partition function for a molecule of type i, such as that given in Eqns. (6.31) and (6.32).

6.4 LANGMUIR ADSORPTION

We consider here the statistical-mechanical derivation of the fundamental relation used in Section 3.7.3 for the adsorption of molecules on a solid surface. Recall that, in Langmuir adsorption, we assume that there are $\tilde{M}_s$ distinct adsorption sites, with $\tilde{N} \leq \tilde{M}_s$ non-interacting, identical molecules adsorbed.

Since the molecules are non-interacting, the energy of our system, E_{tot}, can be written as

$$E_{\mathrm{tot}}(\vec{r}^{\tilde{N}}, \vec{p}^{\tilde{N}}) = \sum_{i=1}^{\tilde{N}} E_{\mathrm{mol}}(\vec{r}_i, \vec{p}_i), \tag{6.41}$$

where E_{mol} is the energy of interaction between the adsorption site and the molecule. The position of molecule i, relative to its site, is given by $\vec{r}_i$, and $\vec{p}_i$ is the molecule's momentum. If the molecule has internal degrees of freedom, they could also be added to E_{mol}, but we ignore this for the time being. Note, however, that there is no interaction between molecules adsorbed near each other.

Recall that the partition function is a sum over the possible microstates of the system, Eqn. (6.13). For our example here, this sum is a sum not just over the energies that a molecule can have on a site, but also the sum $\sum_{\mathrm{dist}}$ over the ways in which these molecules can be distributed among the sites. As before, we replace the sums over energies of the sites by integrals. Therefore, we write the partition function as

$$Q(T, \tilde{M}_s, \tilde{N}) = \sum_{\text{dist}} \int \dots \int \exp\left[-\frac{E_{\text{tot}}(\vec{r}^{\tilde{N}}, \vec{p}^{\tilde{N}})}{k_B T}\right] \frac{d\vec{r}^{\tilde{N}}\, d\vec{p}^{\tilde{N}}}{(2\pi\hbar)^{3\tilde{N}}}. \tag{6.42}$$

The sum is over the ways of distributing the molecules on the adsorption sites, and the integrals are over the possible energy states that the system might have. Now we use Eqn. (6.41) to replace the total energy and simplify the integral. As before, the nested integrals can be written as products, so we write

$$\begin{aligned} Q(T, \tilde{M}_s, \tilde{N}) &= \sum_{\text{dist}} \int \dots \int \exp\left[-\frac{\sum_{i=1}^{\tilde{N}} E_{\text{mol}}(\vec{r}_i, \vec{p}_i)}{k_B T}\right] \frac{d\vec{r}^{\tilde{N}}\, d\vec{p}^{\tilde{N}}}{(2\pi\hbar)^{3\tilde{N}}} s \\ &= \sum_{\text{dist}} \int \dots \int \left(\prod_{i=1}^{\tilde{N}} \exp\left[-\frac{E_{\text{mol}}(\vec{r}_i, \vec{p}_i)}{k_B T}\right]\right) \frac{d\vec{r}^{\tilde{N}}\, d\vec{p}^{\tilde{N}}}{(2\pi\hbar)^{3\tilde{N}}} \\ &= \sum_{\text{dist}} \prod_{i=1}^{\tilde{N}} \int \exp\left[-\frac{E_{\text{mol}}(\vec{r}_i, \vec{p}_i)}{k_B T}\right] \frac{d\vec{r}_i\, d\vec{p}_i}{(2\pi\hbar)^3} \\ &= \sum_{\text{dist}} \prod_{i=1}^{\tilde{N}} q_{\text{site}}(T) \\ &= \sum_{\text{dist}} q_{\text{site}}(T)^{\tilde{N}}, \end{aligned} \tag{6.43}$$

where $q_{\text{site}}(T)$ is the partition function of one molecule on any given site. Since this molecular partition function does not depend upon the arrangement of molecules on sites, we can pull it out of the summation above. So our problem is now reduced to finding the number of ways of distributing the $\tilde{N}$ indistinguishable molecules on the $\tilde{M}_s$ distinguishable adsorption sites.

Finding the number of ways of arranging the molecules on the sites is a classic problem in combinatorics (see Section A.8.2 in Appendix A). The calculation is accomplished in three steps. First, we count the number of ways of adding the molecules one at a time on the sites. This calculation over-counts the number, however, because the molecules are indistinguishable. The second and third steps are meant to calculate the number of over-counts.

We go to each adsorption site in turn, calculate the number of possibilities for that site, and then take the product for each site. For the first site, we have $\tilde{N}$ possibilities for a molecule being present and $\tilde{M}_s - \tilde{N}$ for no molecule being present, so the first site has $\tilde{M}_s$ possibilities. Because we had to make a choice for the first site, the second site has $\tilde{M}_s - 1$ possibilities. The third site has $\tilde{M}_s - 2$, and so on. The total number of possibilities is therefore $\tilde{M}_s \times (\tilde{M}_s - 1) \times \cdots = \tilde{M}_s!$.

Now we correct for the over-counting. Note that, if we swapped the molecules on two occupied sites, the system would not have changed microstates. In a similar way, for a given conformation of occupation, we can calculate the number of ways of rearranging the molecules while keeping the same microstate. We mark the $\tilde{N}$ sites that are occupied, and count the number of ways we could distribute the molecules on these. As before, the number of possibilities for site 1 is $\tilde{N}$, that for site 2 is $\tilde{N} - 1$, etc. Therefore, we have over-counted by $\tilde{N}!$. We have also over-counted the number of unoccupied sites by $(\tilde{M}_s - \tilde{N})!$. So, we obtain

$$Q(T, \tilde{M}_s, \tilde{N}) = \sum_{\text{dist}} q_{\text{site}}(T)^{\tilde{N}}$$

$$= q_{\text{site}}(T)^{\tilde{N}} \sum_{\text{dist}} 1$$

$$= q_{\text{site}}(T)^{\tilde{N}} \frac{\tilde{M}_s!}{(\tilde{M}_s - \tilde{N})!\tilde{N}!}. \tag{6.44}$$

Finally, we can obtain the fundamental relation by taking the logarithm of each side of this equation:

$$\begin{aligned} F(T, \tilde{M}_s, \tilde{N}) &= -k_B T \log Q(T, \tilde{M}_s, \tilde{N}) \\ &= -\tilde{N} k_B T \log q_{\text{site}}(T) + k_B T \log \tilde{N}! - k_B T \log \tilde{M}_s! + k_B T \log(\tilde{M}_s - \tilde{N})! \\ &\cong -NRT \log q_{\text{site}}(T) + NRT \log \tilde{N} - M_s RT \log \tilde{M}_s + (M_s - N)RT \log(\tilde{M}_s - \tilde{N}) \\ &= -NRT \log q_{\text{site}}(T) + NRT \log \left(\frac{N}{M_s - N} \right) + M_s RT \log \left(\frac{M_s - N}{M_s} \right). \end{aligned} \tag{6.45}$$

To obtain the third line we used Stirling's approximation. We also used the fact that $\tilde{N} k_B = NR$, and that any ratio of molecule numbers is equal to the ratio of mole numbers. This is exactly the fundamental relation introduced in Section 3.7.3 to find the Langmuir adsorption isotherm. In that section, an explicit form was also given for the molecular partition function, which was derived from statistical and quantum mechanics assuming that the molecule has only energies from adsorption, vibration parallel to the surface, and vibration perpendicular to the surface [59, Section 7-1].

6.5 THE GRAND CANONICAL ENSEMBLE

Not only do we often manipulate temperature instead of energy in an application or experiment, but also we sometimes manipulate pressure instead of volume. In Chapter 3 we learned that the generalized potential with (T, P, N) as canonical independent variables is the Gibbs free energy G. Just as there is an ensemble for $F(T, V, N)$, so there also exists an ensemble for $G(T, P, N)$. In Section 6.2, we derived the canonical ensemble by considering a subsystem in contact with a heat bath – a small constant-volume chamber connected to a large system through a diathermal wall. Analogously, it is possible to derive an **isobaric, isothermal ensemble** $\Delta(T, P, N)$ using a subsystem in contact with a heat bath and a pressure bath (see Exercise 6.5.A). The results of this exercise are the two relationships

$$\begin{aligned} \frac{G}{k_B T} &= -\log \Delta(T, P, N), \\ \Delta(T, P, N) &:= \int_V Q(T, V, N) \exp\left[-\frac{PV}{k_B T} \right] dV. \end{aligned} \tag{6.46}$$

Similarly, the generalized potential introduced in Exercise 3.3.D, $\Psi := U - TS - \mu N$, has the so-called grand canonical ensemble given by

$$\frac{\Psi}{k_B T} = -\log \Xi(T, V, \mu),$$
$$\Xi(T, V, \mu) := \sum_{\tilde{N}} Q(T, V, \tilde{N}) \exp\left[\frac{\mu N}{k_B T}\right]. \quad (6.47)$$

There are many statistical mechanics problems that are easier to solve in these larger ensembles than in the canonical or microcanonical ensemble.

EXAMPLE 6.5.1 Find the $\Psi(T, \tilde{M}_s, \mu)$ fundamental relation for Langmuir adsorption, and find the corresponding isotherm.

Solution. The solution here assumes that you are familiar with Example 3.7.2 and Section 3.7.3. These should be consulted before considering this example.

In Section 6.4 we found the canonical-ensemble partition function Q, Eqn. (6.44). From the definition for the grand canonical-ensemble partition function, Eqn. (6.47), we can write

$$\begin{aligned}\Xi(T, \tilde{M}_s, \mu) &:= \sum_{\tilde{N}=0}^{\tilde{M}_s} Q(T, \tilde{M}_s, \tilde{N}) \exp\left(\frac{\tilde{N}\mu}{RT}\right) \\ &= \sum_{\tilde{N}=0}^{\tilde{M}_s} \left[Q_{mol}(T) \exp\left(\frac{\mu}{RT}\right)\right]^{\tilde{N}} \frac{\tilde{M}_s!}{(\tilde{M}_s - \tilde{N})!\tilde{N}!}. \end{aligned} \quad (6.48)$$

Note the limits of the sum, which cover all possible values for the number of adsorbed molecules. Now we use the binomial theorem (see Section A.8.1 in Appendix A), which just states that the above summation is what arises when raising a sum to a power:

$$\Xi(T, \tilde{M}_s, \mu) = \left[1 + Q_{mol}(T) \exp\left(\frac{\mu}{RT}\right)\right]^{\tilde{M}_s}. \quad (6.49)$$

It might be easier to see this relation by "going backwards," and seeing how the sum above arises on expanding out the product in Eqn. (6.49). It is now straightforward to find the fundamental relation, and we obtain the isotherm

$$\begin{aligned} N &= -\left(\frac{\partial \Psi}{\partial \mu}\right)_{M_s, T} \\ &= \left(\frac{\partial}{\partial \mu} k_B T \log \Xi\right)_{M_s, T} \end{aligned}$$

$$= M_s RT \left(\frac{\partial}{\partial \mu} \log \left[1 + Q_{\text{mol}}(T) \exp \left(\frac{\mu}{RT} \right) \right] \right)_{M_s, T}$$
$$= M_s \frac{Q_{\text{mol}}(T) \exp[\mu/(RT)]}{1 + Q_{\text{mol}}(T) \exp[\mu/(RT)]}. \tag{6.50}$$

If we insert the chemical potential for a pure ideal gas into the last expression, we obtain exactly the Langmuir adsorption isotherm found in Example 3.7.2.

Since the Langmuir adsorption may be thought of as manipulation of the temperature and chemical potential of the adsorbed species, this ensemble is the more "natural" one for this problem. Hence, derivation of the isotherm from Ψ is easier than it was in Example 3.7.2. □

6.6 AN ELASTIC STRAND

Here we derive a fundamental relation for an elastic polymer strand. This is accomplished in three steps. First, we consider a model that is slightly idealized. We assume that the chain is made up of $\tilde{M}$ independent steps, each with length a_K, but with varying orientation. See Figure 6.2.

Secondly, we write down the canonical-ensemble partition function $Q(T, L, \tilde{M})$, where L is the length of the strand. However, the sums that arise in this derivation have restrictions that make the summation difficult. This difficulty is overcome in the third step by passing to the grand canonical ensemble, whereupon the restrictions disappear.

The elastic strand, or polymer chain, is pictured as having r different kinds of steps. Each step has the same length, but there are different orientations, indicated by i. For example, if L lies along the z axis, then l_i is the component of a step of type i in the z direction. We assume that each step is independent of the others, so we can write that it has a partition function $Q_i^{\text{step}}(T)$.

For a given conformation of chain we might specify, for example, that the first step in the chain is of type 3, the second step is of type 5, etc.,

$$Q(T, L, \tilde{M}) \sim Q_1^{\text{step}}(T)^{\tilde{N}_1} Q_2^{\text{step}}(T)^{\tilde{N}_2} \dots Q_r^{\text{step}}(T)^{\tilde{N}_r}, \tag{6.51}$$

where $\tilde{N}_i$ is the number of steps of type i. The number r of types of steps is somehow determined by quantum mechanics. However, for any given $(L, \tilde{M})$, there are many possible conformations, and the strand partition function requires that we sum over these. We have $\tilde{M}$ steps on the strand,

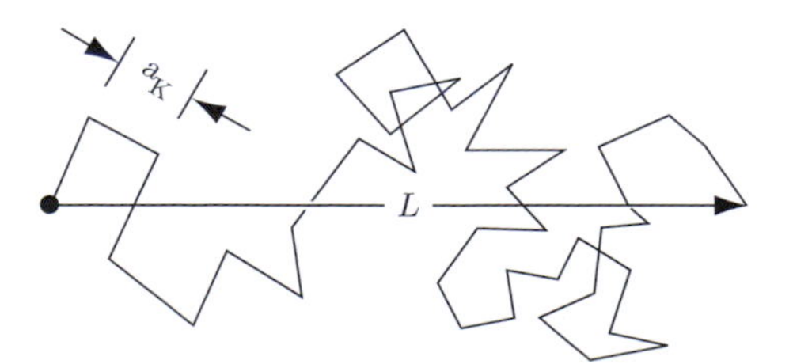

Figure 6.2 A sketch of a polymer strand modeled as a random walk of constant-length steps. Each step has fixed length a_K, but the steps are of varying orientation. The end-to-end length of the strand is L.

which are *distinguishable*, because we know which one is first, which is second, etc. However, the types of steps are themselves indistinguishable. Therefore, the canonical ensemble is

$$
\begin{aligned}
Q(T,L,\tilde{M}) &= \sum_{\tilde{N}_1}\sum_{\tilde{N}_2}\ldots\sum_{\tilde{N}_r}\frac{\tilde{M}!}{\tilde{N}_1!\tilde{N}_2!\ldots\tilde{N}_r!}Q_1^{\text{step}}(T)^{\tilde{N}_1}Q_2^{\text{step}}(T)^{\tilde{N}_2}\ldots Q_r^{\text{step}}(T)^{\tilde{N}_r},\\
&= \sum_{\tilde{\mathbf{N}}}\tilde{M}!\prod_{i=1}^{r}\frac{Q_i^{\text{step}}(T)^{\tilde{N}_i}}{\tilde{N}_i!}, \quad \text{plus restrictions.}
\end{aligned} \tag{6.52}
$$

The second line is just a compact way of writing the first line. The summation above is not as simple as it looks – not that it looks terribly simple – because there are restrictions on the possible values that the $\tilde{N}_i$ might take. First, the total number of steps must be $\tilde{M}$, and, secondly, their lengths in the z direction must add up to L:

$$\sum_{i=1}^{r}\tilde{N}_i = \tilde{M}, \tag{6.53}$$

$$\sum_{i=1}^{r}\tilde{N}_i l_i = L. \tag{6.54}$$

It is worth thinking about Eqns. (6.52)–(6.54), and understanding what each part denotes. The equation is basically a generalization of Eqn. (6.40), which accounts for the fact that the positions of the steps are distinguishable, and that they are connected in a certain way. We can formally include the restrictions by using a **Kronecker delta function**,

$$\delta(i,j) := \begin{cases} 1, & \text{if } i=j,\\ 0, & \text{otherwise,}\end{cases} \tag{6.55}$$

and write (6.52) as

$$Q(T,L,\tilde{M}) = \sum_{\tilde{\mathbf{N}}}\tilde{M}!\,\delta\left(\tilde{M},\sum_{k=1}^{r}\tilde{N}_k\right)\prod_{i=1}^{r}\delta\left(L,\sum_{k=1}^{r}\tilde{N}_k l_k\right)\frac{Q_i^{\text{step}}(T)^{\tilde{N}_i}}{\tilde{N}_i!}. \tag{6.56}$$

Including the restrictions represented by the delta functions $\delta(\,,)$ is difficult, and it is precisely here where passing to the grand canonical ensemble makes the calculation possible. Recall that passing from the canonical ensemble Q to the grand canonical ensemble Δ, Eqn. (6.46), means summing over all possible lengths,[2]

$$
\begin{aligned}
\Delta(T,\mathcal{T},\tilde{M}) &:= \int_L Q(T,L,\tilde{M})\exp\left[\frac{\mathcal{T}L}{k_B T}\right]dL\\
&= \int_L \sum_{\tilde{\mathbf{N}}}\tilde{M}!\,\delta\left(\tilde{M},\sum_{k=1}^{r}\tilde{N}_k\right)\\
&\quad\times\exp\left[\frac{\mathcal{T}L}{k_B T}\right]\prod_{i=1}^{r}\delta\left(L,\sum_{k=1}^{r}\tilde{N}_k l_k\right)\frac{Q_i^{\text{step}}(T)^{\tilde{N}_i}}{\tilde{N}_i!}\,dL.
\end{aligned} \tag{6.57}
$$

[2] Note that we had to make two slight adjustments here. Quantum mechanics dictates that the possible lengths L are not continuous but discrete. Hence, the integral over L should actually be a sum. However, since we ignore these quantum complexities, we use an integral and treat the Kronecker delta function $\delta\left(L,\sum_{k=1}^{r}\tilde{N}_k l_k\right)$ as a Dirac delta function $\delta\left(L-\sum_{k=1}^{r}\tilde{N}_k l_k\right)$.

We can now perform the integration over L by making use of the delta function

$$\begin{aligned}\Delta(T,\mathcal{T},\tilde{M}) &= \sum_{\tilde{\tilde{N}}} \tilde{M}! \left(\exp\left[\frac{\mathcal{T}\sum_{j=1}^{r}\tilde{N}_j l_j}{k_B T}\right]\right) \prod_{i=1}^{r} \delta\left(\tilde{M}, \sum_{k=1}^{r}\tilde{N}_k\right) \frac{Q_i^{\text{step}}(T)^{\tilde{N}_i}}{\tilde{N}_i!} \\ &= \sum_{\tilde{\tilde{N}}} \tilde{M}! \left(\prod_{j=1}^{r}\exp\left[\frac{\mathcal{T}\tilde{N}_j l_j}{k_B T}\right]\right) \prod_{i=1}^{r} \delta\left(\tilde{M}, \sum_{k=1}^{r}\tilde{N}_k\right) \frac{Q_i^{\text{step}}(T)^{\tilde{N}_i}}{\tilde{N}_i!} \\ &= \sum_{\tilde{\tilde{N}}} \tilde{M}! \prod_{i=1}^{r} \delta\left(\tilde{M}, \sum_{k=1}^{r}\tilde{N}_k\right) \frac{\left[Q_i^{\text{step}}(T)\exp[\mathcal{T}l_i/(k_B T)]\right]^{\tilde{N}_i}}{\tilde{N}_i!}. \end{aligned} \tag{6.58}$$

We can exploit the multinomial theorem (see Section A.8.2 of Appendix A) to solve this summation:

$$\Delta(T,\mathcal{T},\tilde{M}) = \left[\sum_{i=1}^{r} Q_i^{\text{step}}(T)\exp\left(\frac{\mathcal{T}l_i}{k_B T}\right)\right]^{\tilde{M}}. \tag{6.59}$$

Next, we assume (1) that there is a large number of kinds of steps, $r \gg 1$; (2) that the difference in their projected lengths l_i is very small from one kind to the next, relative to the longest possible length a_K, $l_{i+1} - l_i \ll a_K$; and (3) that the partition function for a step is independent of its orientation (the environment is isotropic), $Q_1^{\text{step}}(T) = Q_2^{\text{step}}(T) = \ldots = Q_r^{\text{step}}(T) \equiv Q^{\text{step}}(T)$. Then we can approximate the sum as an integral:

$$\begin{aligned}\sum_{i=1}^{r} Q_i^{\text{step}}(T)\exp\left(\frac{\mathcal{T}l_i}{k_B T}\right) &= Q^{\text{step}}(T)\sum_{i=1}^{r}\exp\left(\frac{\mathcal{T}l_i}{k_B T}\right) \\ &\cong Q^{\text{step}}(T)\int_{-a_K}^{+a_K}\exp\left(\frac{\mathcal{T}l}{k_B T}\right) r\frac{dl}{2a_K} \\ &= Q^{\text{step}}(T)\, r\frac{k_B T}{2\mathcal{T}a_K}\exp\left(\frac{\mathcal{T}l}{k_B T}\right)\Bigg|_{l=-a_K}^{+a_K} \\ &= Q^{\text{step}}(T) r\frac{k_B T}{\mathcal{T}a_K}\sinh\left(\frac{\mathcal{T}a_K}{k_B T}\right). \end{aligned} \tag{6.60}$$

When we insert Eqn. (6.60) into Eqn. (6.59) we obtain our final expression for the grand canonical partition function $\Delta(T,\mathcal{T},\tilde{M})$ for a random-walk chain with $\tilde{M}$ steps under tension $\mathcal{T}$,

$$\Delta(T,\mathcal{T},\tilde{M}) = \left[Q^{\text{step}}(T) r\frac{k_B T}{\mathcal{T}a_K}\sinh\left(\frac{\mathcal{T}a_K}{k_B T}\right)\right]^{\tilde{M}}. \tag{6.61}$$

When we insert this result into Eqn. (6.46) we obtain the fundamental relation

$$\frac{G}{\tilde{M}k_B T} = -\log\left[rQ^{\text{step}}(T)\frac{k_B T}{\mathcal{T}a_K}\sinh\left(\frac{\mathcal{T}a_K}{k_B T}\right)\right]. \tag{6.62}$$

This fundamental relation was first introduced in Eqn. (3.126). That expression, however, has replaced the step partition function $Q^{\text{step}}(T)$ in favor of the heat capacity. Taking the appropriate derivative to find the length leads to a tension–length equation of state with the form shown in Figure 6.3, which is called the **inverse Langevin force law**. We can see from this figure that

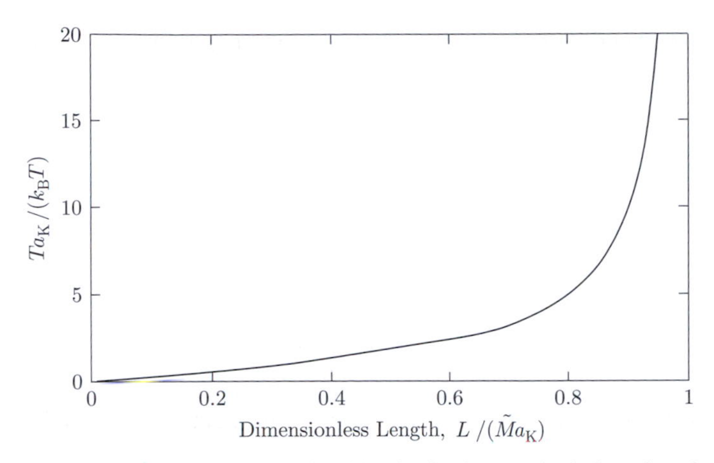

Figure 6.3 The inverse Langevin force law predicted by the fundamental relation given by Eqn. (6.62) for a random-walk chain model for a polymer strand.

the results are qualitatively very reasonable; at low extensions, the tension rises linearly with stretching, and, as the length gets stretched to its maximum, the tension becomes very large.

However, this expression does *not* describe the force–extension relation for the stretching of a DNA segment particularly well. The reason for the discrepancy is our assumption that each step is independent. Even a rather large segment of DNA has an end-to-end length comparable to its persistence length $a_K/2$. Therefore, this derivation needs modification, which has been done by Fixman and Kovacs, as explained in [111]. The (approximate) result is called the *worm-like chain* [81], which was introduced in Eqn. (2.85). If the strand is not greatly extended, the initial part of the force-law curve is linear, with a slope of 3. The resulting approximate expression is called the **Gaussian force law**.

6.7 FLUCTUATIONS

Macroscopic thermodynamic properties, such as energy, temperature, pressure, etc. are averages of fluctuating quantities. Not only does statistical mechanics provide a way to predict these average quantities, but also it can tell us the magnitude of the fluctuations around the averages. Why are fluctuations important? Well, for relatively large systems (say approximately 10^{23} molecules), fluctuations are important near the critical point. Also, for small systems, the fluctuations can become as large as the averages themselves. Therefore, if we are designing tiny labs on a silicon chip, the data we take will inherently have fluctuations, which might make it difficult to estimate model parameters. Biological systems are typically composed of many small subsystems where fluctuations can also dominate. In this section we show how to estimate the size of fluctuations.

To estimate the size of fluctuations, we use the probability of a system being in a particular microstate. For a closed system, the postulates tell us the probability of a system being in a particular microstate. By dividing a closed system into a subsystem and a bath, we also derived the probabilities for particular microstates. These results are summarized in Table 6.1.

Table 6.1 A summary of the probability of being in a given microstate for microcanonical, canonical, and grand canonical ensembles. The fundamental quantity is $\Omega(U, V, N)$, which is the number of microstates available to a system with macroscopic energy U, volume V, and mole number N. The partition functions, Q, Δ, and Ξ, defined in Eqns. (6.13), (6.46), and (6.47), normalize the probabilities to unity.

Fixed variables	Probability of being in microstate j
U, V, N	$p_j = 1/\Omega(U, V, N)$
T, V, N	$p_j(U_j) = \exp[-U_j/(k_B T)]/Q(T, V, N)$
T, P, N	$p_j(U_j, V) = \exp[-U_j/(k_B T)] \exp[-PV/(k_B T)]/\Delta(T, P, N)$
T, V, μ	$p_j(U_j, N) = \exp[-U_j/(k_B T)] \exp[\mu N/(k_B T)]/\Xi(T, V, \mu)$

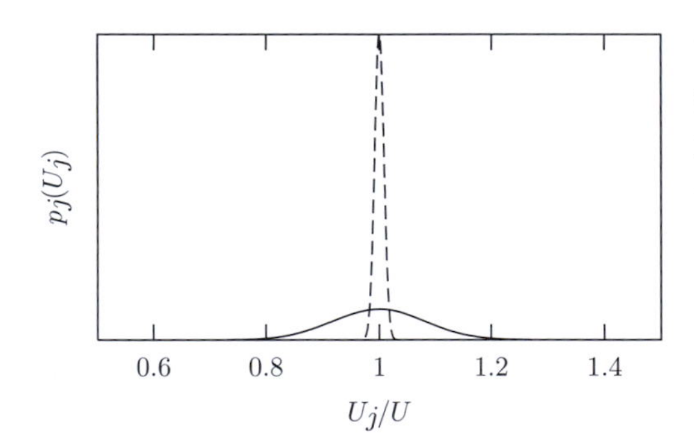

Figure 6.4 The probability $p_j(U_j)$ that an ideal gas at fixed temperature and volume has energy U_j. The solid line is for a system of 100 monatomic molecules, and the dashed line is for 10 000 molecules.

A system, say, held at fixed temperature, pressure, and mole number, has a probability of being in microstate j with energy U_j and pressure P given by $p_j(U_j, V)$ in the third line of the table. Each of these probabilities is normalized, so that summing (or integrating) over all possible microstates yields the value 1.

Any system with fixed T, P, and N will spend time at different energy levels. For very large systems far from the critical point, the system spends nearly all its time very close to the average value $U = \sum_j p_j U_j$, which we call the internal energy in classical thermodynamics. However, for small systems, the energy can spend a significant amount of time away from the average value. For example, Figure 6.4 shows the distribution of energies for an ideal gas at fixed T, V, and $\tilde{N}$, where $\tilde{N} = 100$ (solid line) and $\tilde{N} = 10\,000$ (dashed line). The system with just 100 molecules can easily fluctuate to values 10% above or below the average. From Figure 6.4, we see that the distribution becomes more sharply peaked with a larger system.

Here we derive general equations to estimate the size of the fluctuations, through a single parameter called the variance,

$$\sigma_U^2 := \langle U_j^2 \rangle - \langle U_j \rangle^2, \tag{6.63}$$

where $\langle \ldots \rangle := \sum_j \ldots p_j$ indicates an average. Roughly speaking, the widths of the peaks in Figure 6.4 are approximately σ_U.

First, we find the fluctuations in energy for a canonical ensemble (fixed T, V, and N). We start with the statistical-mechanical definition for the internal energy,

$$U = \sum_j U_j p_j(U_j), \tag{6.64}$$

and take the derivative of each side with respect to temperature:

$$\begin{aligned}
\left(\frac{\partial U}{\partial T}\right)_{V,N} &= \sum_j U_j \left(\frac{\partial p_j}{\partial T}\right)_{V,N}, \\
C_V &= \sum_j U_j \left[\frac{U_j}{k_B T^2} p_j - \frac{p_j}{Q}\left(\frac{\partial Q}{\partial T}\right)_{V,N}\right] \\
&= \frac{\langle U_j^2 \rangle}{k_B T^2} - \frac{\langle U_j \rangle}{Q}\left(\frac{\partial Q}{\partial T}\right)_{V,N} \\
&= \frac{\langle U_j^2 \rangle}{k_B T^2} - \langle U_j \rangle \sum_j \frac{U_j}{k_B T^2} \frac{\exp[-U_j/(k_B T)]}{Q} \\
&= \frac{\langle U_j^2 \rangle}{k_B T^2} - \frac{\langle U_j \rangle^2}{k_B T^2} \\
&= \frac{\sigma_U^2}{k_B T^2}.
\end{aligned} \tag{6.65}$$

To obtain the second line, we used the canonical probability distribution, from Table 6.1, and the definition of the constant-volume heat capacity. To get to the third line, we used the definition of an average. The fourth line arises from the definition of the canonical partition function, from Table 6.2, and the fifth line uses the probability distribution again. The last line comes from the definition for the variance, Eqn. (6.63). The result given in Eqn. (6.65) is remarkable; as molecules collide against one another in a fluid they give rise to fluctuations of the energy. The heat capacity of a fluid is a measure of those fluctuations; Eqn. (6.65) gives a precise correspondence between the variance of such fluctuations and the heat capacity.

Note the size dependence of the final result:

$$\frac{\sigma_U(T,V,N)}{\langle U_i \rangle} = \frac{\sqrt{k_B T^2 C_V}}{\langle U_i \rangle} = \frac{1}{\sqrt{\tilde{N}}} \underbrace{\frac{\sqrt{RT^2 c_V}}{u}}_{\text{size independent}}. \tag{6.66}$$

We introduced the notation here for the variance σ, where the subscript indicates the fluctuating quantity, and the arguments indicate which variables are held constant. The left-hand side here is dimensionless, so it gives the relative magnitude of the fluctuations. Hence, the relative width of the distribution shrinks with the square root of the system size (number of molecules).

EXAMPLE 6.7.1 What are the relative fluctuations of internal energy in 1 μl of a simple ideal gas at standard temperature and pressure?

Solution. A simple ideal gas has average specific internal energy $\langle u \rangle = \frac{3}{2}RT$. Hence, it has specific constant-volume heat capacity $c_v = \frac{3}{2}R$. We can use Eqn. (6.66) to estimate the relative energy fluctuations as

$$\begin{aligned}\frac{\sigma_U}{\langle U \rangle} &= \sqrt{\frac{2}{3\tilde{N}}} \\ &= \sqrt{\frac{2RT}{3PV\tilde{N}_A}} \\ &= \sqrt{\frac{2(298\text{ K})(0.0831\ \ell.\text{ bar/mol K})}{3(1\text{ bar})(10^{-6}\ \ell)(6.023 \times 10^{23}\text{ mol}^{-1})}} \\ &\cong 1.6 \times 10^{-9}. \end{aligned} \tag{6.67}$$

Hence, even for a box of volume only 1 μl, the fluctuations are in the eighth decimal place, and typically insignificant. □

Similarly, one can show that

$$\begin{aligned}\frac{\sigma_N(T, V, \mu)}{\langle N \rangle} &= \sqrt{\frac{k_B T \kappa_T}{V}}, \\ \frac{\sigma_V(T, P, N)}{\langle V \rangle} &= \sqrt{\frac{k_B T \kappa_T}{V}}. \end{aligned} \tag{6.68}$$

The last two results show why fluctuations are large near the critical point. Recall from Section 4.2.1 that the spinodal curve is where $(\partial P/\partial v)_T = 0$, or where $\kappa_T = \infty$. The critical point is where the spinodal curve is at a maximum, so, near this point, the fluctuations in volume (or density) and in mole numbers become extremely large, even for large systems. This result in fact explains why the heat capacity of a fluid diverges as the critical point is approached.

EXAMPLE 6.7.2 Bustamante *et al.* [18] measured the length of a segment of DNA under tension isothermally using an optical trap. At low tensions, they observe large fluctuations in the length, and it takes some time to find the average length. What is the size of the fluctuations in length for an isothermal polymer strand held under tension?

Solution. Using our usual mapping, we can use the results above on volume fluctuations with the substitutions: $V \to L$, $-P \to \mathcal{T}$, and $\tilde{N} \to \tilde{M}$. The length as a function of tension is found from the free energy of the strand, Eqn. (6.62),

$$\frac{\langle L \rangle}{\tilde{M} a_{\mathrm{K}}} = -\frac{1}{\tilde{M} a_{\mathrm{K}}} \left(\frac{\partial G}{\partial \mathcal{T}} \right)_{T,\tilde{M}}$$
$$= \coth\left(\frac{\mathcal{T} a_{\mathrm{K}}}{k_{\mathrm{B}} T} \right) - \frac{k_{\mathrm{B}} T}{\mathcal{T} a_{\mathrm{K}}}, \tag{6.69}$$

which is the inverse Langevin force law plotted in Figure 6.3. By taking the derivative of this expression, and dividing the result by the equation itself, we obtain the isothermal "compressibility"

$$\kappa_T := \frac{1}{\langle L \rangle} \left(\frac{\partial \langle L \rangle}{\partial \mathcal{T}} \right)_{T,\tilde{M}}$$
$$= \frac{1}{\mathcal{T}} \frac{k_{\mathrm{B}} T/(\mathcal{T} a_{\mathrm{K}}) - [\mathcal{T} a_{\mathrm{K}}/(k_{\mathrm{B}} T)] \mathrm{cosech}^2[\mathcal{T} a_{\mathrm{K}}/(k_{\mathrm{B}} T)]}{\coth[\mathcal{T} a_{\mathrm{K}}/(k_{\mathrm{B}} T)] - k_{\mathrm{B}} T/(\mathcal{T} a_{\mathrm{K}})}. \tag{6.70}$$

Now we can use our results from this section, Eqn. (6.68), to find the relative magnitude of the fluctuations in length of the strand:

$$\frac{\sigma_L(T, \mathcal{T}, \tilde{M})}{\langle L \rangle} = \sqrt{\frac{k_{\mathrm{B}} T \kappa_T}{\langle L \rangle}}$$
$$= \frac{1}{\sqrt{\tilde{M}}} \frac{\sqrt{x^{-2} - \mathrm{cosech}^2(x)}}{\coth(x) - x^{-1}}, \tag{6.71}$$

where $x := \mathcal{T} a_{\mathrm{K}}/(k_{\mathrm{B}} T)$ is the dimensionless tension. There are two things that it is worthwhile to note about this result. First, we see again that the relative magnitude depends inversely upon the square root of the number of segments in the chain, which is another specific example of the general result (the simple ideal gas was the first). Secondly, we see that there is a tension-dependent part that can make the fluctuations very large, even for long chains. To illustrate the importance of this second term, we plot it against the dimensionless tension in Figure 6.5.

In accord with observations, we see that the relative fluctuations are much larger for smaller tensions. The result is not surprising, since the average length goes to zero for no tension, but the fluctuations remain finite.

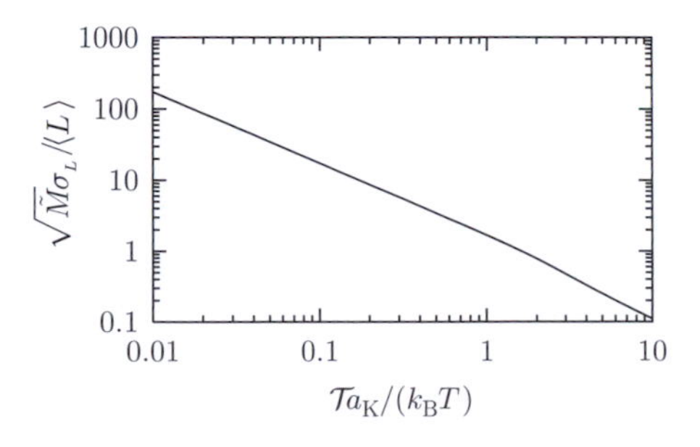

Figure 6.5 The relative magnitude of end-to-end length fluctuations in a polymer strand of $\tilde{M}$ steps under tension $\mathcal{T}$. When the magnitude is multiplied by the square root of the number of steps, as is done here, the result depends only on the (dimensionless) tension. From this plot, we see, for example, that, if the dimensionless tension is one, $\mathcal{T} = k_{\mathrm{B}} T/a_{\mathrm{K}}$, the chain must be longer than 100 segments in order to attain fluctuations of less than 10%. □

Summary

In this chapter we showed how to relate atomic interactions to thermodynamic properties. In theory, the connection arises from two simple statistical-mechanical postulates.

- SMP I: for an isolated system consisting of N molecules having volume V and energy U, all microscopic states consistent with those constraints occur with the same probability.
- SMP II: $S(U,V,N) = k_B \log \Omega(U,V,N)$.

In practice, however, the connection between molecules and thermodynamics is accomplished through *partition functions*. Depending on which set of thermodynamic variables is being held fixed for a system, the appropriate partition function calculates the number of ways in which the molecules or atoms can arrange themselves, with appropriate weighting. The partition functions and their relationships to thermodynamics are summarized in Table 6.2.

Using these definitions for the partition functions, we found fundamental relations for simple ideal gases, general ideal gases, Langmuir adsorption of a gas onto a surface, and the free energy of stretched polymer strands.

We also exploited the fact that statistical mechanics contains information about the entire distribution of quantities to find estimates for the size of fluctuations of real systems. Namely, we found the variance of fluctuations in mole number (grand canonical), volume (grand canonical) and energy (canonical) to be

$$\frac{\sigma_N(T,V,\mu)}{\langle N\rangle} = \sqrt{\frac{k_B T \kappa_T}{V}},$$

$$\frac{\sigma_V(T,P,N)}{\langle V\rangle} = \sqrt{\frac{k_B T \kappa_T}{V}}, \tag{6.68}$$

and

$$\frac{\sigma_U(T,V,N)}{\langle U\rangle} = \frac{\sqrt{k_B T^2 C_V}}{U}. \tag{6.66}$$

Table 6.2 A summary of the ensembles and partition functions used in this text. The fundamental quantity is $\Omega(U,V,N)$, which is the number of microstates available to a system with macroscopic energy U, volume V, and mole number N.

Variable set	Partition function	Thermodynamic relation
U,V,N	$\Omega(U,V,N)$	$S = k_B \log \Omega$
T,V,N	$Q(T,V,N) := \sum_j \exp[(-U_j/(k_B T)]$	$F = -k_B T \log Q$
T,P,N	$\Delta := \int_V Q(T,V,N)\exp[-PV/(k_B T)]dV$	$G = -k_B T \log \Delta$
T,V,μ	$\Xi := \sum_{\tilde{N}} Q(T,V,\tilde{N})\exp[\mu N/(k_B T)]$	$\Psi = -k_B T \log \Xi$

Exercises

6.2.A Consider a simple binomial coefficient of the form

$$\Omega = \frac{(N_1 + N_2)!}{N_1!N_2!}. \tag{6.72}$$

Find the value of N_1 for which Ω is a maximum, for fixed $M = N_1 + N_2$. If we use an asterisk to denote this "maximum" distribution, derive an approximate (first-order) expression in terms of N_1 and N_1^*. Use this expression to show that, for large values of N_1, the binomial coefficient is strongly peaked around N_1^*.

6.3.A Verify that the first line of Eqn. (6.39),

$$\frac{U}{NRT} = -T\left(\frac{\partial}{\partial T}\frac{F}{NRT}\right)_{V,\tilde{N}},$$

is generally true.

6.3.B Show that the canonical partition function derived for a mixture of non-interacting polyatomic gases, Eqn. (6.35), leads to the fundamental relation for a mixture of general ideal gases, Section 7.1.

6.3.C Show that, for an ideal gas comprising $\tilde{N}$ molecules at temperature T, the number of molecules having velocity between v and $v + dv$ is given by

$$4\pi\tilde{N}\left(\frac{m}{2\pi k_{\mathrm{B}}T}\right)^{3/2} \exp\left[-\frac{mv^2}{2k_BT}\right] v^2\, dv, \tag{6.73}$$

where m is the mass of each molecule.

6.3.D Non-ideal gases are often described by a mechanical equation of state of the form

$$\frac{P}{\rho k_{\mathrm{B}}T} = 1 + B\rho + O(\rho^2), \tag{6.74}$$

where B is a so-called second virial coefficient that depends only on temperature. Show that, for a system of spherical molecules in which each pair of molecules interacts through a potential-energy function denoted by $\Gamma(r)$, the second virial coefficient is given by

$$B = 2\pi \int_0^\infty \left\{1 - \exp\left[-\frac{\Gamma(r)}{k_{\mathrm{B}}T}\right]\right\} r^2\, dr, \tag{6.75}$$

where r denotes the distance between the two molecules.

Hint: use a grand canonical ensemble, and begin by showing that the partition function can be written as

$$\Xi = 1 + \sum_{\tilde{N}=1}^{\infty} \frac{Z_N(V,T)}{\tilde{N}!} z^{\tilde{N}}, \tag{6.76}$$

where z is an activity defined as $z = (Q_1/V)e^{\mu/(k_{\mathrm{B}}T)}$, and Q_1 is a single-particle partition function, i.e. $Q_1 := V/\Lambda^3$. Note that the $\tilde{N}$-particle canonical ensemble partition function can be written as $Q_N = (1/\tilde{N}!)(Q_1/V)^{\tilde{N}} Z_N$, where

$$Z_N = \int_V \exp\left[-\frac{U_{\text{pot}}(\vec{\mathbf{r}}^N)}{k_B T}\right] d\vec{\mathbf{r}}^{\bar{N}} \tag{6.77}$$

is the so-called configurational integral and $U_{\text{pot}}(\vec{\mathbf{r}}^N)$ is the potential energy of the system, which in this case is given by $U_{\text{pot}}(\vec{\mathbf{r}}^N) = \frac{1}{2}\sum_i^N \sum_j^N \Gamma_{ij}(r_{ij})$. Next show that the pressure can be written in the following form:

$$\frac{P}{k_B T} = \sum_{j=1}^{\infty} b_j(T) z^j, \tag{6.78}$$

where the $b_j(T)$ are temperature-dependent coefficients given by

$$b_1 = \frac{1}{V} Z_1, \tag{6.79}$$

$$b_2 = \frac{1}{2V}(Z_2 - Z_1^2), \tag{6.80}$$

$$b_3 = \frac{1}{6V}(Z_3 - 3Z_1 Z_2 - 2Z_1^3), \tag{6.81}$$

$$\vdots$$

The series for the pressure can then be inverted by postulating that

$$z = \sum_{i=1}^{\infty} a_i \rho^i, \tag{6.82}$$

substituting z from Eqn. (6.82) into a series for the density, which can be derived from Eqn. (6.78), and matching the coefficients corresponding to each power of z. This procedure should allow you to show that

$$B(T) = -b_2(T) = -\frac{1}{V}(Z_2 - Z_1^2), \tag{6.83}$$

where

$$Z_1 = \int_V d\mathbf{r}_1, \tag{6.84}$$

$$Z_2 = \int \int_V \exp\left[-\frac{\Gamma(r)}{k_B T}\right] d\mathbf{r}_1\, d\mathbf{r}_2. \tag{6.85}$$

Equations (6.83–6.85) are of central importance in applied statistical mechanics in that they provide a direct connection between molecular interactions and macroscopic, measurable thermodynamic properties.

6.3.E Quantum mechanics tells us the energy of a diatomic molecule. There are three contributions: electronic, vibrational, and rotational. If we neglect dissociation of the molecule into two atoms, then the electronic energy ε_0 is a constant. Hence, the microstate of the molecule can be described by two integers ν and K, where

$$E_{\nu K} = \varepsilon_0 + \hbar\omega\left(\nu + \frac{1}{2}\right) + \hbar^2 \frac{K(K+1)}{2I}; \quad \nu, K = 0, 1, 2, \ldots, \tag{6.86}$$

where $E_{\nu K}$ is the energy of the molecule in that microstate, $\hbar$ is Planck's constant, ω is the vibrational frequency (from infrared measurements), and I is the moment of inertia of the molecule. Note that this result applies only for an asymmetric diatomic molecule, and that the rotational energy has degeneracy $2K + 1$, meaning that there are $2K + 1$ distinct microstates with the same rotational energy.

Find the molecule partition function for this asymmetric diatomic molecule. Approximate the sum for the vibrational component as an integral. State explicitly when this approximation is valid. The vibrational part can be summed analytically.

6.3.F Prove that Eqn. (6.40) is correct for a system of two independent molecules with two microstates each.

6.4.A Derive the Langmuir adsorption fundamental relation for the simultaneous adsorption of two types of molecules A and B. In other words, find $F(T, \tilde{M}_s, \tilde{N}_A, \tilde{N}_B)$ for $\tilde{N}_A$ molecules of type A and $\tilde{N}_B$ molecules of type B adsorbed on a surface with $\tilde{M}_s$ sites. Make a detailed comparison between your results and Eqn. (3.130).

6.4.B Consider surface adsorption if the molecules in the Langmuir adsorption have internal degrees of freedom, such that

$$E_{\text{tot}}(\vec{r}^{\tilde{N}}, \vec{p}^{\tilde{N}}, \vec{R}^{\tilde{N}}) = \sum_{i=1}^{\tilde{N}} E_{\text{mol}}(\vec{r}_i, \vec{p}_i, \vec{R}_i), \tag{6.87}$$

where $\vec{R}_i$ is the vector of internal conformations, like the one we used for a general ideal gas (see Eqn. (6.29)). Derive the new fundamental relation.

6.4.C Derive the partition function and fundamental relation for hemoglobin, Eqn. (3.132).

6.5.A Derive the expressions for the grand canonical ensemble, Eqns. (6.46–6.47). Follow the same procedure as in Section 6.2; however, now let molecules exchange between the bath and the subsystem to obtain Ψ, or the volume to obtain G.

6.5.B Derive an expression for the grand canonical ensemble partition function $\Xi(\mu_1, \mu_2, V, T)$ of a two-component system, in which the chemical potentials of components 1 and 2 are held constant, as well as the temperature and volume.

6.5.C Derive an expression for the partition function in the isobaric–isothermal (NPT) ensemble. What is the natural thermodynamic potential for this ensemble? How is this potential related to the partition function? Can you propose any general rules to relate an arbitrary partition function to its "natural" thermodynamic potential?

6.5.D Show that the partition function $\Xi'(N_1, \mu_2, U, P)$ for a two-component system defined by constant U, constant P, constant number of particles for component 1, and constant chemical potential for component 2 is given by

$$\Xi' = \sum_{N_2} \int_V dV \exp\left(\frac{\mu_2 N_2}{k_B T}\right) \exp\left(-\frac{PV}{k_B T}\right). \tag{6.88}$$

Derive the natural thermodynamic potential for this ensemble.

6.6.A A polymer chain is placed in an electric field. Each segment of the chain has a dipole moment, so it has energy $\vec{E} \cdot \vec{u}\alpha$, where $\vec{E}$ is the electric field, and $\vec{u}$ is a unit vector describing the orientation of the segment. Unlike the derivation in Section 6.6, the step partition function now depends on i, with an isotropic part, and an energy from the field

$$Q_i^{\text{step}}(T) = Q_0(T)\exp\left(-\vec{E} \cdot \vec{u}\alpha/(k_\text{B}T)\right).$$

The ends of the chain are pulled apart with tension $\vec{\mathcal{T}}$, which is a vector. Find an expression for the end-to-end vector of the chain $\vec{L}$ as a function of the temperature T, number of segments $\tilde{M}$, tension $\vec{\mathcal{T}}$, and electric field $\vec{E}$. For convenience, put the tension vector $\vec{\mathcal{T}}$ along the z axis so that $\vec{\mathcal{T}} \cdot \vec{l}_i = \mathcal{T} a_\text{K} u_{i,z}$.

6.6.B Show how to obtain Eqn. (3.126) from Eqn. (6.62). What is the heat capacity?

6.6.C Consider a generalization of the Langmuir adsorption model, such that multiple molecules can be adsorbed on a single site. We assume that molecules adsorbed on *adjacent* sites do not interact energetically. However, the molecules on a particular site *do* interact, so we write $q_1(T)$ as the partition function of a site with one molecule adsorbed, $q_2(T)$ for a site with two molecules adsorbed, etc.

1. Show how the canonical partition function for this problem is similar to that for the elastic strand, Eqn. (6.52).
2. Now pass to the grand canonical partition function Ξ, and find an expression for the fraction of adsorbed molecules $\theta := N/M_\text{s}$.
3. From this result, one can derive the Brunauer–Emmett–Teller (BET) model, Eqn. (3.117), by making certain assumptions. One assumes that $q_1 = q_\text{site}(T)$, $q_2 = q_\text{site}(T)/c$, $q_3 = q_\text{site}(T)/c^2$, $q_4 = q_\text{site}(T)/c^3$, and so on. Show how this assumption leads to the BET model. What is c? Note that it is not necessary to approximate the sums as integrals here.

6.6.D In Section 3.7.2, we found the tension–temperature phase diagram for unzipping DNA by pulling. Here we consider just one step in the statistical mechanical derivation of that fundamental relation, Eqn. (3.102). We begin with the canonical partition function $Q(T, L, \tilde{M})$, where T is the temperature, L is the distance between the two ends of the two different strands being pulled, and $\tilde{M}$ is the total number of base pairs on the DNA. Just like in the zipper model, we assume that the base pairs are energetically independent, so that we can assume a partition function $q_\text{a}(T)$ for attached bases and a partition function $q_\text{u}(T)$ for unattached bases. Additionally, n_b consecutive, unattached base pairs can form a Kuhn step having orientation i, with partition function $q_j^{\text{step}}(T)$. So, the canonical partition function can be written

$$Q = \sum_{\tilde{N}=0}^{\tilde{M}} q_\text{a}(T)^{\tilde{N}} q_\text{u}(T)^{\tilde{M}-\tilde{N}} \sum_{I_1} \cdots \sum_{I_r} \left[\frac{2(\tilde{M}-\tilde{N})}{n_b}\right]! \prod_{j=1}^{r} \frac{q_j^{\text{step}}(T)^{I_j}}{I_j!},$$

where $\tilde{N}$ is the number of attached base pairs, which we sum over. This expression also contains a sum over possible orientations of the Kuhn steps, where I_j is the number of steps with orientation j. The summation has the following two restrictions:

$$\sum_j I_j = \frac{2(\tilde{M} - \tilde{N})}{n_b}, \tag{R1}$$

$$\sum_j l_j I_j = L, \tag{R2}$$

where l_j is the length of a step of type j in the direction of L. From this expression for the canonical partition function, derive the fundamental relation, Eqn. (3.102).

6.6.E Actin is a globular protein that is contained in all species. In a cell, actin polymerizes to form filaments that form the cytoplasmic skeleton. Actin can also depolymerize. One can therefore propose a simple model of actin in which actin "monomer" can either attach to a polymer chain (or filament), thereby increasing the molecular weight, or leave the chain, decreasing the length of a filament.

In this problem you are required to find the partition function that describes the polymerization of actin in a cell. Also give an expression for the mean length of the actin polymer, and for the distribution of lengths of actin filaments. Assume that the total number of actin molecules N_{tot} is large. Let E_i denote the energy of a filament with i units, and let L_i denote its length ($L_i = L_0 i$, where L_0 is the length of a single unit and i is the number of units.) The cell can be treated as an isothermal system.

6.6.F Imagine a single polymer chain, oriented more or less in a straight line. Along the backbone of this chain are arranged evenly spaced side groups, each of which acts as a dipole:

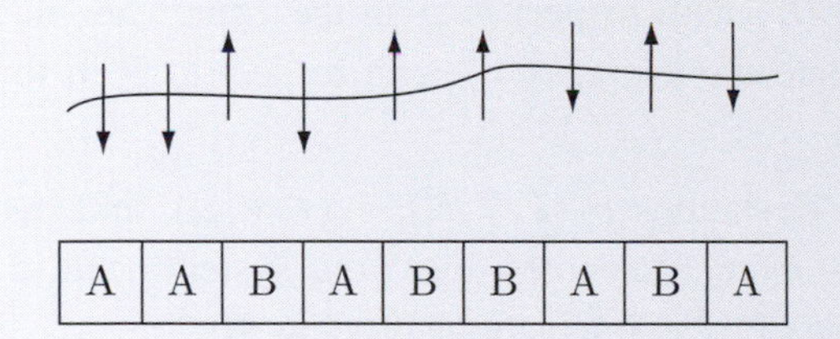

For simplicity we assume that the dipoles can be oriented only up ($s_i = +1$) or down ($s_i = -1$), where s_i describes the orientation of side group i. We have $\tilde{N}$ such side groups. We also assume that only adjacent dipoles interact energetically. The energy contribution from groups i and $i + 1$ is given as

$$-\Gamma s_i s_{i+1},$$

where Γ is a constant. You could think of such a model as a one-dimensional lattice, with two possible species, A and B, as drawn above.

Find the canonical partition function $Q(\tilde{N}, T)$ for this model.

6.6.G We consider here a simplified model for the denaturation of a folded protein. We consider a protein that is divided up into $\tilde{M}$ segments that are either folded (the α state) or unfolded (the β state). The unfolded state is of longer extension, but is energetically less favorable. We consider each segment of the protein to be independent. At any instant, the protein could have $\tilde{M}_\alpha$ folded segments and $\tilde{M}_\beta$ unfolded segments. A folded segment has partition function $q_\alpha(T)$, and an unfolded one has partition function $q_\beta(T)$.

1. If the folded (unfolded) segment has length a_α (a_β), find the length of the protein as a function of temperature and of the tension $\mathcal{T}$ applied on its ends. Plot the fraction of folded segments as a function of the tension. Does this model relate to any of the other statistical-mechanical models considered as examples in this chapter? How?
2. Using the above model it is possible to imagine modeling a muscle tissue made up of many such proteins. Note that the partition function for the unfolded state involves interactions between newly exposed portions of the chain and the solvent. So, by changing the solvent conditions, the partition function could be changed. Design a muscle using this idea.

6.7.A The probability of finding a particular value of the energy U in the canonical ensemble is denoted by $p(U)$. Show how this probability function can be expanded about the average (or most probable) value of U, namely $\langle U\rangle$. Also show that terms beyond those quadratic in the energy are negligible.

6.7.B The partition function for an *NPT* ensemble is given by

$$\Delta = \sum_i \exp(-\beta U_i)\exp(-\beta PV).$$

For this ensemble show that

1. $\langle EV\rangle - \langle U\rangle\langle V\rangle = -(\partial V/\partial\beta)_{P,N} - P\sigma_V^2 = k_B T^2\alpha V - P\sigma_V^2$ and
2. $\langle H^2\rangle - \langle H^2\rangle = k_B T^2 C_P$.

6.7.C Prove the fluctuation formulas given in Eqns. (6.68). The first can be found by differentiating $\langle N\rangle$ with respect to μ in the grand canonical ensemble (T, V, μ). Similarly, differentiate the average $\langle V\rangle$ with respect to P in the (T, P, N) ensemble to find the second.

6.7.D Find the deviations $\sigma_U(T, V, N)$, $\sigma_N(T, V, \mu)$, and $\sigma_V(T, P, N)$ for a monatomic ideal gas. How many molecules does your system need in order to insure that the relative magnitude of these fluctuations is below 10^{-6}?

6.7.E Find the magnitude of the fluctuations in length for a worm-like chain as a function of tension.

7 Molecular interactions

Previous chapters established many relations between various thermodynamic properties of a fluid. We have seen, for example, that the heat capacity of a fluid is always positive. We have also shown that the thermal, mechanical, and chemical equations of state must be mutually consistent (i.e. they must satisfy the Gibbs–Duhem equation). In the previous chapter, we showed how statistical mechanics can be used to derive fundamental thermodynamic relations from simple models of molecular interactions. In this chapter we discuss in greater detail important *molecular* interactions and their relative magnitudes. Specific mathematical expressions taken from physical chemistry are used to estimate when interactions are important and how they might influence thermodynamic quantities. Molecular interactions ultimately determine the behavior of materials and fluids. Hence they have received considerable attention and entire texts have been devoted to their study. Interested readers are referred to [68, 80, 99] for a more complete and thorough discussion.

The equations of state and the transport properties of gases, liquids, and solids are intimately related to the forces between the molecules. The methods of statistical mechanics provide a connection between these forces and measurable thermodynamic properties. An introduction to statistical mechanics was presented in the previous chapter. Here we merely discuss the nature and the origin of intermolecular forces. In the following chapter we will illustrate how everything comes together to generate thermodynamic property predictions from intermolecular interactions. Finally, we note that calculation of intermolecular forces requires knowledge of several fundamental properties of molecules. A good source of information for such properties is provided by the NIST Chemistry WebBook (http://webbook.nist.gov).

7.1 IDEAL GASES

As a starting point, we consider a gas with *no* intermolecular forces and no internal degrees of freedom. Such a fluid is called a simple ideal gas, and was derived in Section 6.3.1. The pressure exerted by N moles of an ideal gas confined to a volume V therefore arises as a result of the sum of the momenta (or kinetic energy) of the particles, and it is given by the familiar equation

$$P = \frac{NRT}{V}. \tag{7.1}$$

The internal energy is also a sum of the contributions of the particles, and is

$$U = cNRT. \tag{7.2}$$

A fundamental entropy expression for the ideal gas was given in Chapter 2 as

$$S = Ns_0 + NR\log\left[\left(\frac{U}{Nu_0}\right)^c \frac{V}{Nv_0}\right], \tag{7.3}$$

where s_0, u_0, and v_0 are constants of integration corresponding to the reference-state molar entropy, energy, and volume of the gas in an ideal state.

Since the molecules of an ideal gas do not interact with each other, the energy of a mixture of ideal gases can be written as

$$U = \left(\sum_i^r N_i c_i\right) RT, \tag{7.4}$$

and an entropy equation of state for a mixture of ideal gases may be written

$$S = \sum_i^r N_i s_{i0} + \sum_i^r (N_i c_i) R \log\left(\frac{T}{T_0}\right) + \sum_i^r N_i R \log\left(\frac{V}{N_i v_0}\right). \tag{7.5}$$

In other words, the entropy of a mixture of ideal gases is given by the sum of the entropies that each component would have if it, alone, were to occupy all of volume V at temperature T. This statement is also known as Gibbs' theorem. From the last two expressions it is straightforward to write down a fundamental Helmholtz relation for a mixture of ideal gases, $F(T, V, \{N_i\}) := U - TS$. Next we consider what happens when molecules *do* interact.

7.2 INTERMOLECULAR INTERACTIONS

7.2.1 The significance of "$k_B T$"

The product of the Boltzmann constant and the temperature, known affectionately as "$k_B T$," is often used as a measure of the strength of an intermolecular interaction. Indeed, it is not uncommon to attend lectures in which entire hypotheses and discussions are cast in terms of multiples of $k_B T$. The thought is that, if an interaction energy exceeds $k_B T$, it will overcome the disorganizing effects of thermal motion. Or, on the other hand, if an interaction energy is many times $k_B T$, thermal motion will rarely overcome it.

The significance of $k_B T$ can be illustrated by analyzing the cohesive energy responsible for condensing a gas into a liquid. This cohesive energy is approximately equal to the latent enthalpy of vaporization, Δh^{vap}. Loosely speaking, the insertion of one molecule into a liquid phase requires that 12 nearest-neighbors be displaced in order to create a cavity. This would necessitate that six nearest-neighbor "bonds" be broken. If the energy of two molecules at near contact is comparable to ϵ, then the energy required in order to carry out this process is 6ϵ.

Experimental data for numerous substances [64, p. 281] suggest the following relation between the enthalpy of vaporization and the boiling temperature T_B:

$$\frac{\Delta h^{\text{vap}}}{T_B} \approx 80 \frac{\text{J}}{\text{K} \cdot \text{mol}} \cong 9.6R. \tag{7.6}$$

Equation (7.6) is known as Trouton's law. The quantity on the right gives $9.6k_B$ per molecule, and the cohesive energy per molecule is equal to $9.6k_B T$ [68]. In other words, molecules will condense if their cohesive energy with other molecules in the liquid is on the order of $9.6k_BT$. We also said that 6ϵ are required to create a cavity in a liquid; we can therefore inter that if the interaction energy between two molecules ϵ exceeds $\frac{3}{2}k_BT$, then it is strong enough to drive a condensation process into a liquid. This simple analysis helps explain why k_BT is used as the "unit" with which one gauges the strength of intermolecular interactions. It is thus useful to keep in mind that $k_BT \cong 4.14 \times 10^{-21}$ J/molecule $\cong 2.5$ kJ/mol at room temperature.

7.2.2 Interactions at long distances

Coulombic forces

The beginning point of the interatomic or intermolecular interactions in which we are interested resides in the attractions or repulsions between charged subatomic particles, namely electrons and protons. The energy of interaction Γ_{ij} between two point charges q_i and q_j *in vacuo* is given by Coulomb's law,

$$\Gamma_{ij} = \frac{q_i q_j}{4\pi\epsilon_0 r}, \quad \text{point charges,} \tag{7.7}$$

where r is the distance between the particles, and ϵ_0 is the dielectric permittivity of vacuum, namely $\epsilon_0 = 8.8542 \times 10^{-12}\ \mathrm{C^2\,J^{-1}\,m^{-1}}$. In general, the force $\vec{F}_{ij}$ between two particles interacting through a potential energy function Γ_{ij} is given by

$$\vec{F}_{ij} = -\nabla\Gamma_{ij}. \tag{7.8}$$

Recall the definition of the vector differential operator: $\nabla := \vec{\delta}_x\partial/\partial x + \vec{\delta}_y\partial/\partial y + \vec{\delta}_z\partial/\partial z$. The magnitude of the force between two charges is therefore given by

$$|\vec{F}_{ij}| = \frac{q_i q_j}{4\pi\epsilon_0 r^2}. \tag{7.9}$$

Two charges of the same sign have a positive energy of interaction, thereby giving rise to a positive, repulsive force between them; two charges of opposite sign have a negative energy of interaction and attract each other. As we shall discuss below, Coulombic or electrostatic interactions are relatively strong and long-ranged. Consider two isolated ions, Na^+ and Cl^-, at contact. The distance between them is the sum of their atomic radii, approximately 0.276 nm. From Eqn. (7.7), their energy of interaction is

$$\Gamma_{ij} = \frac{-(1.602 \times 10^{-19}\ \mathrm{C})^2}{4\pi(8.8542 \times 10^{-12}\ \mathrm{C^2\,J^{-1}\,m^{-1}})(0.276 \times 10^{-9}\ \mathrm{m})} = -8.4 \times 10^{-19}\ \mathrm{J}, \tag{7.10}$$

where $e = 1.602 \times 10^{-19}$ C is the elementary charge. At room temperature, this energy of interaction corresponds to approximately $-200k_BT$. This simple calculation illustrates that the energy of interaction between two ions in vacuum is comparable to that of a covalent bond (several hundred times k_BT) [68]. Furthermore, this calculation shows that energy does not decay to k_BT until the distance between the ions is on the order of 0.1 μm; the range of interaction is indeed long, particularly when we consider that the diameter of an ion is on the order of just a few ångström units! A more realistic calculation of the cohesive energy of an ionic crystal must therefore take into account contributions arising from nearest neighbors, next-nearest neighbors, and so on and so forth. For a cubic lattice such as that formed by NaCl, a central Na^+ ion is surrounded by 6 Cl^- at distance 0.276 nm, by 12 next-nearest Na^+ neighbors at $\sqrt{2}r$, eight Cl^- at $\sqrt{3}r$, etc. The cohesive energy Γ^c of the NaCl crystal is therefore given by

$$\Gamma^c = \frac{q_i q_j}{4\pi\epsilon_0 r}\left(6 - \frac{12}{\sqrt{2}} + \frac{8}{\sqrt{3}} + \cdots\right)$$
$$\cong 1.748\frac{q_i q_j}{4\pi\epsilon_0 r} = -1.46 \times 10^{-18}\,\text{J}. \tag{7.11}$$

This is greater than that predicted for a pair of ions in vacuum by a factor of 1.748; this factor appearing in Eqn. (7.11) is known as **Madelung's constant,**[1] and it assumes different values depending on the type of crystal lattice formed by the ions [68].

Permanent dipoles

Electrostatic or Coulombic forces arise not only between charged particles (e.g. ions), but also between neutral molecules (i.e. molecules that do not exhibit a net charge). Consider here the case of a neutral molecule that possesses two charges of the same magnitude q_i but opposite sign. If the distance between the two charges is denoted by d, that molecule has a **dipole moment** μ_i^{d} given by

$$\mu_i^{\text{d}} = q_i d. \tag{7.12}$$

Molecules such as ethanol, acetone, and water have a relatively large dipole moment, because of the asymmetry in the distribution of charges (electrons and protons) within the molecule. In contrast, symmetric molecules such as methane have a relatively small, usually negligible, permanent dipole moment.

The energy of interaction between two dipolar molecules depends on the distance and the orientation between the dipoles. Figure 7.1 describes schematically our coordinate system for the relative position of the dipoles. If r is used to denote the distance between the dipole centers of mass, and ϕ, θ_i, and θ_j denote the angles between these (see Figure 7.1), then it can be shown [80] that the instantaneous energy of interaction between them is given by

$$\Gamma_{ij} = -\frac{\mu_i^{\text{d}}\mu_j^{\text{d}}}{4\pi\epsilon_0 r^3}(2\cos\theta_i\cos\theta_j - \sin\theta_i\sin\theta_j\cos\phi_{ij}), \quad \text{permanent dipole.} \tag{7.13}$$

A simple analysis of Figure 7.1 suggests that two dipoles attract each other strongly when they are collinear and their "directions" are opposite. They repel each other when they are collinear and their directions are the same. These observations are consistent with Eqn. (7.13); the energy

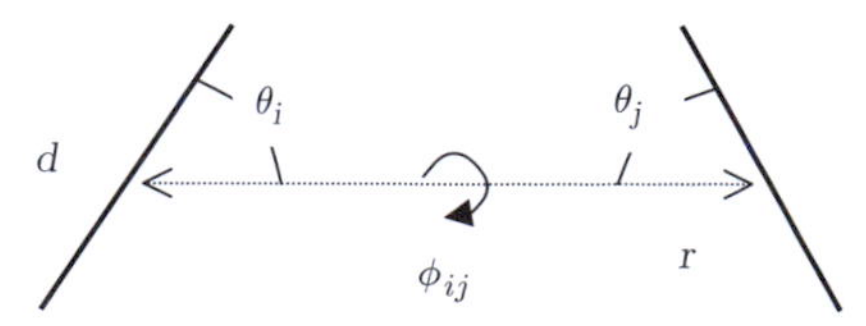

Figure 7.1 The distance between the dipoles is r, and the angles θ and ϕ determine their relative orientation.

[1] Note that this term is a geometrical correction for a specific kind of lattice, in this case NaCl. The term "Madelung's constant" is sometimes used only for this lattice, and sometimes for any general lattice, in which case a number different from 1.748 can arise.

in the first case is $\Gamma_{ij} = -\mu_i^d \mu_j^d/(2\pi\epsilon_0 r^3)$, whereas in the latter case it is $\Gamma_{ij} = \mu_i^d \mu_j^d/(2\pi\epsilon_0 r^3)$. It is interesting to note that, when the origin of the interactions between two neutral, dipolar molecules is electrostatic, their instantaneous energy decays as the third power of the distance between them (as opposed to the inverse power).

In an actual fluid the relative orientation between any two dipolar molecules is not constant, but fluctuates incessantly as the molecules travel through space. It is then useful to calculate an *average* energy of interaction between two such molecules. The average is determined by considering all possible orientations between the dipoles, and weighting more heavily (according to their Boltzmann factor; see Table 6.1) those configurations that have a lower energy. The result of this procedure is an average energy given, to leading order, by [80]

$$\langle \Gamma_{ij} \rangle = -\frac{2(\mu_i^d)^2(\mu_j^d)^2}{3(4\pi\epsilon_0)^2 k_B T r^6} + \cdots . \tag{7.14}$$

It is important to emphasize that Eqn. (7.14) results from a series expansion that is valid only in the limit of large separations ($r \gg d$, where d is the length of the dipole). Several interesting features about Eqn. (7.14) should be highlighted. First, we note that, to leading order, the average energy of interaction is attractive, and it decays as the sixth power of the distance between the dipoles. Secondly, we note that it is inversely proportional to the temperature. Thirdly, we point out that polar interactions depend on the fourth power of the dipole moment; we can anticipate from this result significant differences in the thermodynamic properties of a mixture when one of the components is replaced by a different one of different dipole moment.

To gain an appreciation for the magnitude of dipole–dipole interactions, we can evaluate the energy of interaction between two molecules for a particular configuration. For concreteness, let us consider the case in which the molecules find themselves in a collinear configuration, with the two dipoles pointing in the same direction, as illustrated in Figure 7.2. If we assume that the dipole moment of each molecule is 1 debye (D) (the dipole moment of hydrogen chloride, for example, is 1.05 D) and that the separation between their centers of mass is 3 Å, then their energy of interaction is 7.4×10^{-14} erg $\cong -1.1$ kcal/mol $\cong$ 4.6 kJ/mol, or a little less than $2k_B T$ at room temperature.

Dipole–ion interactions

It is also of interest to consider the interaction energy between an ion and a polar molecule. If q_{ion} denotes the charge of the ion and q_j denotes the charges on the dipole, it can be shown that

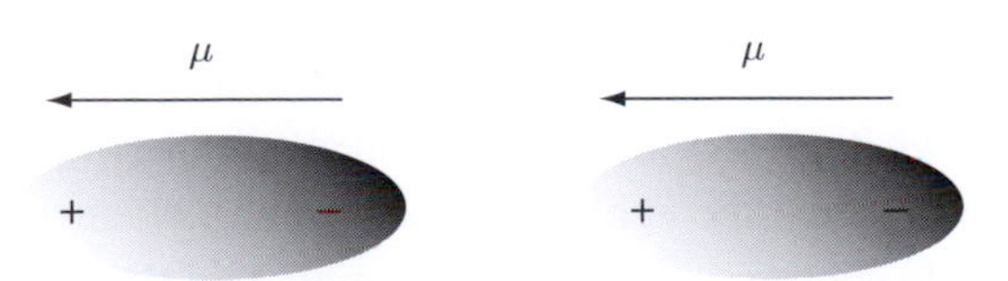

Figure 7.2 While the energy of interaction between two dipoles depends on relative orientation, low-energy configurations such as that depicted above are particularly favorable, and often lead to the formation of "chains" of molecules.

$$\Gamma_{ij} = -\frac{q_{\text{ion}} q_j}{4\pi\epsilon_0} \left(\frac{1}{[(r - \frac{1}{2} d \cos\theta_\text{d})^2 + (\frac{1}{2} d \sin\theta_\text{d})^2]^{1/2}} \right), \tag{7.15}$$

where θ_d is now the angle formed by the projection of the dipole and the line joining the center of mass of the dipole to the ion. As before, r is the distance from the ion to the center of mass of the dipole. For sufficiently large separations, i.e. for $r \gg d$, Eqn. (7.15) can be approximated by

$$\Gamma_{ij} = -\frac{q_{\text{ion}} \mu^\text{d} \cos\theta_\text{d}}{4\pi\epsilon_0 r^2}, \quad \text{dipole–ion interaction, } r \gg d. \tag{7.16}$$

In words, the energy of interaction between an ion and a permanent dipole decays as the square of the distance between them. To estimate the energy which arises between ions and dipoles, we consider the case of Na^+ and a water molecule. The radius of a sodium ion is approximately 0.095 nm, and that of a water molecule is approximately 0.14 nm. The dipole moment of a water molecule is 1.85 D. The maximum interaction energy that arises between Na^+ and water is therefore

$$\begin{aligned} \Gamma &= -\frac{(1.609 \times 10^{-19}\,\text{C})(1.85\,\text{D})(3.336 \times 10^{-30}\,\text{C} \cdot \text{m/D})}{4\pi(8.854 \times 10^{-12}\,\text{C}^2\,\text{J}^{-1}\,\text{m}^{-1})(0.235 \times 10^{-9}\,\text{m})^2} \\ &= -1.6 \times 10^{-19}\,\text{J} = -39 k_\text{B} T, \quad \text{at } T = 300\text{ K}. \end{aligned} \tag{7.17}$$

Note that we used the dipole moment of water in debye (D), from Table 7.2 on p. 218, and the conversion factor to Coulomb-meters. This energy is relatively large, and is responsible for the so-called solvation or hydration of ions in highly polar solvents such as water. A cation such as sodium in aqueous solution at room temperature usually has four or five water molecules "bound" to it; that is, the hydration number of sodium in water is between 4 and 5. While these water molecules are free to exchange with other molecules further away, they do so at a relatively slow rate. The bare radius of a sodium ion is 0.095 nm, but that of the hydrated ion is much larger; the so-called hydration radius of sodium is approximately 0.36 nm. In contrast, the hydration number of a negatively charged chlorine anion is unity.

Molecules can also have quadrupole moments. Some molecules, such as carbon dioxide, have no significant dipole moment but exhibit a large quadrupole moment. A discussion of quadrupole moments is slightly more involved than that of dipole moments, and interested readers are referred to the literature for additional information [80]. Here it suffices to point out that orientation-averaged dipole–quadrupole interactions decay as the eighth power of the distance between them, and quadrupole–quadrupole interactions decay as the tenth power. Such interactions are therefore weaker than those between charges or dipoles, but are nevertheless important in systems that lack charges or dipoles (e.g. CO_2).

Induced dipoles

Our discussion so far has been limited to permanent dipoles. It turns out that most molecules (irrespective of whether they have a permanent dipole or not) can exhibit a transient dipole induced in them through the application of an external electric field. For small electric fields, the induced dipole is proportional to the strength E of the applied field; it is given by

$$\mu^\text{d} = \alpha^\text{p} E, \tag{7.18}$$

where α^{p} is the **polarizability** of the molecule. As its name indicates, the polarizability provides a measure of how easily the electrons of a molecule can be displaced by the application of an external field. Molecules with delocalized electrons, such as benzene and other aromatic compounds, have relatively high polarizabilities. Small, symmetric molecules such as oxygen and nitrogen have small polarizabilities.

Induced dipoles often arise in fluids not through the application of external fields, but through the electric field generated by a nearby polar molecule. In that case, the orientation-averaged energy of interaction between the polar molecule i and the non-polar (but polarizable) molecule j in which a dipole is being induced is given by

$$\langle \Gamma_{ij} \rangle = -\frac{\alpha_j^{\mathrm{p}} (\mu_i^{\mathrm{d}})^2}{(4\pi\epsilon_0)^2 r^6}, \quad \text{permanent dipole–induced dipole.} \tag{7.19}$$

If the two molecules i and j are polar and polarizable, the resulting orientation-averaged energy of interaction is given by

$$\langle \Gamma_{ij} \rangle = -\frac{\alpha_j^{\mathrm{p}} (\mu_i^{\mathrm{d}})^2 + \alpha_i^{\mathrm{p}} (\mu_j^{\mathrm{d}})^2}{(4\pi\epsilon_0)^2 r^6}. \tag{7.20}$$

Note that, in contrast to the average interaction energy between two permanent dipoles in Eqn. (7.14), the average interaction energy between induced dipoles is independent of temperature; the temperature drops out of the calculation when the average is taken. This energy is also attractive.

If we think of a neutral, non-polar molecule as a collection of electrons moving rapidly and chaotically about a nucleus, we realize that such a molecule is only non-polar on average; at any given point in time, it exhibits an *instantaneous* dipole moment due to the fluctuations of the electrons. This instantaneous dipole moment can induce an instantaneous dipole on any nearby molecule, thereby giving rise to an interaction between two induced dipoles. The resulting energy is often called **dispersion energy**. An analysis of the resulting instantaneous interaction between two such molecules in terms of classical mechanics would predict a vanishing energy. That result is incorrect in that we know from experience that neutral, non-polar molecules attract each other to form condensed phases. It is therefore necessary to invoke quantum mechanics and analyze the behavior of the electrons using Schrödinger's equation. A simple analysis of dispersion energy in terms of quantum mechanics can be found in the literature [80]. Here we limit our discussion to the resulting expression for the energy of interaction,

$$\Gamma_{ij} = -\frac{3}{2}\frac{\alpha_i^{\mathrm{p}}\alpha_j^{\mathrm{p}}}{(4\pi\epsilon_0)^2 r^6}\left(\frac{h^2 \nu_{0i}\nu_{0j}}{h\nu_{0i} + h\nu_{0j}}\right), \quad \text{induced dipole–induced dipole,} \tag{7.21}$$

where h is Planck's constant and ν_{0i} is the electronic frequency of the molecule in its lowest energy state. It turns out that the product $\hbar\nu_{0i}$ is well approximated by the *first ionization potential* I_i (the energy required to ionize a molecule from the ground state). Equation (7.21) can therefore be approximated by

$$\Gamma_{ij} \cong -\frac{3}{2}\frac{\alpha_i^{\mathrm{p}}\alpha_j^{\mathrm{p}}}{(4\pi\epsilon_0)^2 r^6}\left(\frac{I_i I_j}{I_i + I_j}\right), \quad \text{induced dipole–induced dipole.} \tag{7.22}$$

Table 7.1 Ionization potentials for various molecules; 1 eV $= 1.60218 \times 10^{-19}$ J. From [158] and [159].

Molecule	I (eV)
He	24.5
Ar	25.2
H_2	15.4
CH_4	13.0
CO	14.1
H_2O	12.6
NH_3	11.5
CCl_4	11
n-C_7H_{16}	10.4

Table 7.2 The dipole moment, polarizability, and attractive energy of interaction for various molecules. The parameter C corresponds to the prefactor of the r^{-6} term in Eqn. (7.26). (T = 300 K). Based on data from [158], [159], [160], [161] and [162].

Molecule	μ^d (debye)	$\alpha^p/(4\pi\epsilon_0)$ (10^{30} m^3)	C (10^{79} J m^6) electrostatic	C (10^{79} J m^6) induction	C (10^{79} J m^6) dispersion
Ar	0	1.63	0	0	50
CH_4	0	2.60	0	0	102
CCl_4	0	10.5	0	0	1460
CO	0.1	1.95	0.002	0.039	64.43
NH_3	1.47	2.1	75.16	9.06	60.94
H_2O	1.85	1.5	188.53	10.27	34.07

Table 7.1 provides values for the first ionization potentials of several molecules; Table 7.2 provides values for the polarizabilities. In general, large molecules are more polarizable than small ones. The opposite is true for ionization potentials. It is also interesting to note that, in contrast to the polarizabilities, the ionization potentials don't change considerably from molecule to molecule.

For small molecules, a typical value of $\alpha^{p}/(4\pi\epsilon_0)$ is around $(0.15\,\text{nm})^3$. A typical ionization potential is $I_i \approx 2 \times 10^{-18}$ J. Two small molecules in contact are separated by approximately 0.3 nm, making their dispersion energy of interaction (at contact) approximately $\Gamma \approx -4.6 \times 10^{-21}$ J, or $k_B T$ at room temperature [68]. This simple numerical calculation should be contrasted with the hand-waving arguments used earlier (in Section 7.2.1) to explain the significance of $k_B T$.

Dispersion interactions such as those described by Eqn. (7.22) arise between virtually all molecules, irrespective of whether they are polar or not, be they asymmetric or symmetric. They always give rise to a net attraction between two molecules, and they are in fact responsible for the occurrence of condensed liquid or solid phases in simple substances that exhibit negligible permanent dipoles or quadrupoles (e.g. noble gases). For example, following [68] we can consider the case of argon, which forms a closely packed solid at low temperatures. Each Ar atom in the solid lattice is surrounded by 12 nearest neighbors, separated by 0.376 nm. This corresponds to six "bonds." If interactions between atoms further away are also considered, this number rises to approximately 7.22 [68]. The molar cohesive energy density of an argon crystal can therefore be estimated to be

$$\begin{aligned} U^{c} &= N_0 \Gamma^{c} \\ &\approx N_0 \times 7.22 \left(\frac{3(\alpha^{p})^2 I}{4(4\pi\epsilon_0)^2 \sigma^6} \right) \\ &= 7.22\, N_0 \frac{3(1.63 \times 10^{-30}\text{m}^3)^2 (2.52 \times 10^{-18} J)}{4(0.376 \times 10^{-9}\text{m})^6} = 7.73\ \text{kJ/mol}, \end{aligned} \tag{7.23}$$

which agrees very well with the experimental enthalpy of melting of argon, namely $\Delta h^{m} = 7.7$ kJ/mol. The agreement between experiment and the simple calculation of Eqn. (7.23) serves to further emphasize that dispersion forces are indeed responsible for the ability of simple, neutral molecules to form condensed phases.

Dispersion interactions are relatively short-ranged; they decay as the sixth power of the separation between the molecules. This helps explain why it is easier to vaporize a simple liquid such as xenon, methane, or a lower alkane than a polar or ionic substance (e.g. water or an ionic crystal), where interactions are much longer-ranged.

In order to gain a physical feel for the magnitude of the interactions between molecules, it is instructive to discuss a few specific examples. Table 7.2 provides dipole moments, polarizabilities, and estimates of the attractive interactions between molecules (i.e. we assume that the attractive energy of interaction is of the form $\Gamma \sim C/r^6$, and we report values for C).

7.2.3 Interactions at short distances

According to Pauli's exclusion principle, distinct electronic clouds cannot overlap each other. At short distances, the electronic clouds of any two molecules begin to overlap, thereby giving rise to a repulsive force between them. A discussion of the precise nature and magnitude of these forces is more elaborate than that of attractive forces, and is therefore left for more specialized texts on intermolecular interactions. Here we merely point out that repulsive forces are extremely short-ranged and are generally described by expressions of the form

$$\Gamma_{ij} = \frac{A_{ij}}{r^n}, \tag{7.24}$$

where n is a positive number, generally taken to be somewhere in the neighborhood of 12, and A_{ij} is a positive constant. An exponential function of the form

$$\Gamma_{ij} = A_{ij} \exp(-\gamma r) \tag{7.25}$$

is also used, where γ and A_{ij} are two constants that determine the steepness of the repulsions and their strength. Theoretical arguments can be used to justify a preference for the use of an exponentially decaying function over that of a simple power law. In practice, however, these two functions lead to comparable descriptions of the thermodynamic properties of fluids [95, 91].

7.2.4 Empirical potential-energy functions

The discussion of intermolecular forces presented in the preceding section can now be used to propose a simple, semi-empirical function to describe interactions between any two molecules of a simple fluid. We discussed how, at large separations, any two molecules of a simple fluid experience interactions between induced dipoles. We also discussed how, at short distances, molecules repel each other according to an inverse power law of the separation between them. This suggests that the total intermolecular energy between two simple, non-polar molecules be described by a function of the form

$$\Gamma_{ij}(r) = \frac{A_{ij}}{r^{12}} - \frac{C_{ij}}{r^6}, \tag{7.26}$$

where the constants A and C are substance-specific. Equation (7.26) can be rewritten in the form

$$\Gamma_{ij}(r) = 4\epsilon \left[\left(\frac{\sigma}{r} \right)^{12} - \left(\frac{\sigma}{r} \right)^{6} \right], \tag{7.27}$$

where the constants σ and ϵ provide a measure of the diameter of the molecules and the energy of interaction between them, respectively. The energy function described by Eqn. (7.27) is known as a **Lennard–Jones potential**. Figure 7.3 provides a schematic representation of the Lennard-Jones potential-energy function. At short distances it rises steeply, and it decays to zero after about four σ. The energy of interaction is zero when the separation between the molecules is σ, and it exhibits a minimum with energy $\Gamma = -\epsilon$ at $r = \sigma/2^{1/6}$.

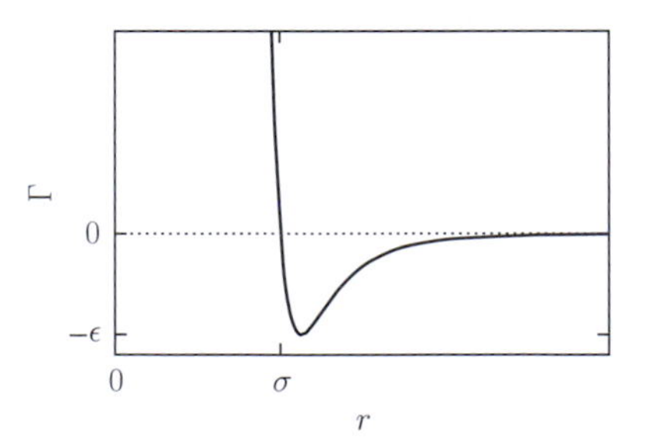

Figure 7.3 The potential-energy function exhibits a minimum with energy ϵ, and it gradually decays to zero after approximately 4σ.

Table 7.3 Lennard-Jones parameters inferred from second-virial-coefficient data.

Molecule	σ (Å)	ϵ/k_B (K)
Ne	2.74	35.7
Ar	3.405	119.75
N_2	3.698	95.05
CH_4	4.010	142.87
CO_2	4.416	192.25
n-C_4H_{10}	7.152	223.74

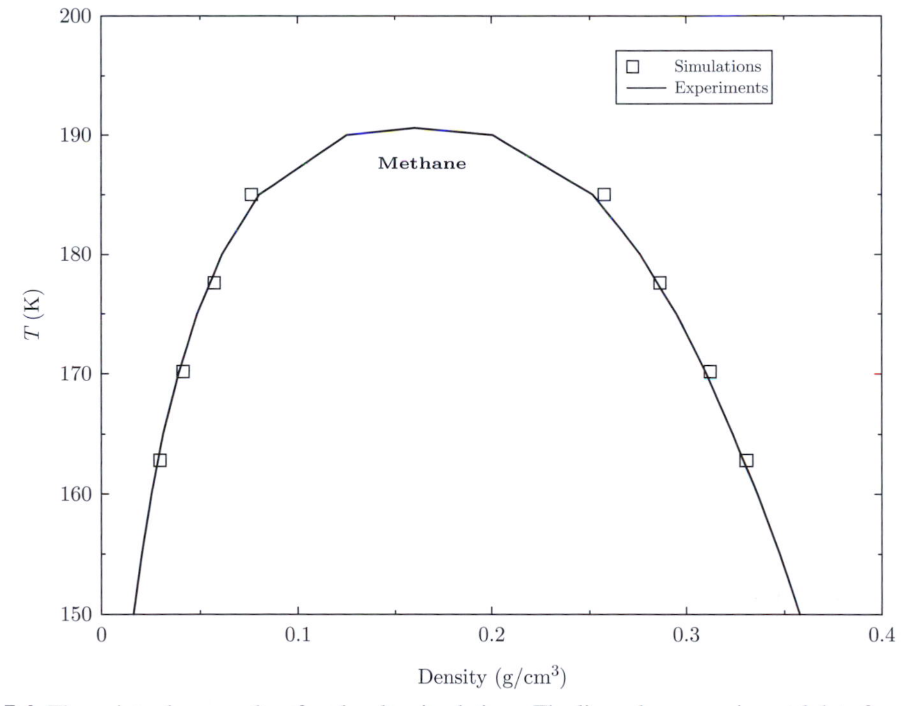

Figure 7.4 The points show results of molecular simulations. The lines show experimental data for methane. The temperature and density have been reduced with respect to the critical temperature and critical density.

Table 7.3 provides values for the Lennard-Jones parameters for several simple fluids. The values of σ are on the order of a few ångström units, and those of ϵ are on the order of a few kcal per mole. Figure 7.4 shows the vapor–liquid coexistence curve corresponding to the Lennard-Jones energy function with parameters for methane. This coexistence curve was determined from molecular simulations using Eqn. (7.27) and assuming that the total potential energy of the fluid consists of a sum of pairwise additive interactions over all pairs of molecules. The agreement with experiment is good, and indicates that the Lennard-Jones function provides a description of simple fluids that is consistent with experiment.

At this point it is instructive to discuss in more detail the connection between a potential-energy function, such as that given by Eqn. (7.27), and macroscopic thermodynamic properties, such as the coexistence curve shown in Figure 7.4. In this chapter we simply assume that there exists a mechanism, via molecular simulations, to establish an unambiguous, numerically exact correspondence between the energy function and such properties. For concreteness, we adopt a Lennard-Jones model in this discussion. That being the case, the thermodynamic properties should depend only on the values of two parameters, namely ϵ and σ, because these are the only two parameters that enter the Lennard-Jones function Eq. (7.27). It follows that, if we were to express the thermodynamic properties of "Lennard-Jonesium" in dimensionless variables, using ϵ and σ to reduce actual properties, we would arrive at *universal* quantities that pertain to any fluid that can be described by a Lennard-Jones function. These dimensionless variables are $T^* = Tk_B/\epsilon$, $\rho^* = \rho\sigma^3$, and $P^* = P\sigma^3/\epsilon$. That is, we can now replot Figure 7.4 in terms of T^* vs. ρ^* to get a unique, universal coexistence curve for "Lennard-Jonesium." If we now wanted to extract results for, say, methane from this curve, we would read out the dimensionless values, T^* and ρ^*, from this figure, and multiply them by ϵ_{CH_4}/k_B and $1/\sigma^3_{CH_4}$ to determine the actual T and ρ. This idea of universality is known as the molecular principle of **corresponding states**. As we will see throughout this text, this principle has far-reaching consequences, particularly in the development of engineering correlations for prediction of thermodynamic properties.

It is also important to recall that we could have anticipated a macroscopic corresponding-states principle from the results of Chapter 4. In that chapter we saw how, by applying thermodynamic stability at the critical point of a pure fluid, we can arrive at a universal coexistence curve (for a given model) by reducing thermodynamic quantities by the corresponding values evaluated at the critical point (i.e. $T_r = T/T_c$, $v_r = v/v_c$, and $P_r = P/P_c$). In the particular case of fluids described by a van der Waals equation of state, we showed that one can write a "universal" equation in the form

$$P_r = \frac{8T_r}{3v_r - 1} + \frac{3}{v_r^2}.$$

This connection between molecular and macroscopic corresponding states suggests that we should be able to establish a connection between the properties of a pure fluid and its Lennard-Jones parameters. That connection can be made through the critical point: for Lennard-Jonesium, the critical temperature and density are $T_c^* \cong 1.32$ and $\rho_c^* \cong 0.29$. If the critical point of an actual fluid of interest is known, it is possible to obtain Lennard-Jones parameters that will capture the experimental behavior in the region near the critical point. For argon, for example, the critical point is at $T_c = 150.8$ K and $\rho_c = 0.013\,35$ mol/cm^3; appropriate Lennard-Jones parameters are therefore $\epsilon/k_B = 114.2$ K and $\sigma = 3.39$ Å.

Note that more anisotropic molecules, such as long alkanes, can be described as a collection of several Lennard-Jones interaction sites connected to each other. For the particular case of a long alkane, each interaction site could, for example, be assumed to represent an individual CH_2 functional group. Figure 7.5 shows results of molecular simulations of a multi-site Lennard-Jones model for alkanes. As that figure illustrates, this simple model describes experimental coexistence curves reasonably well. When molecules have a dipole moment, or when they

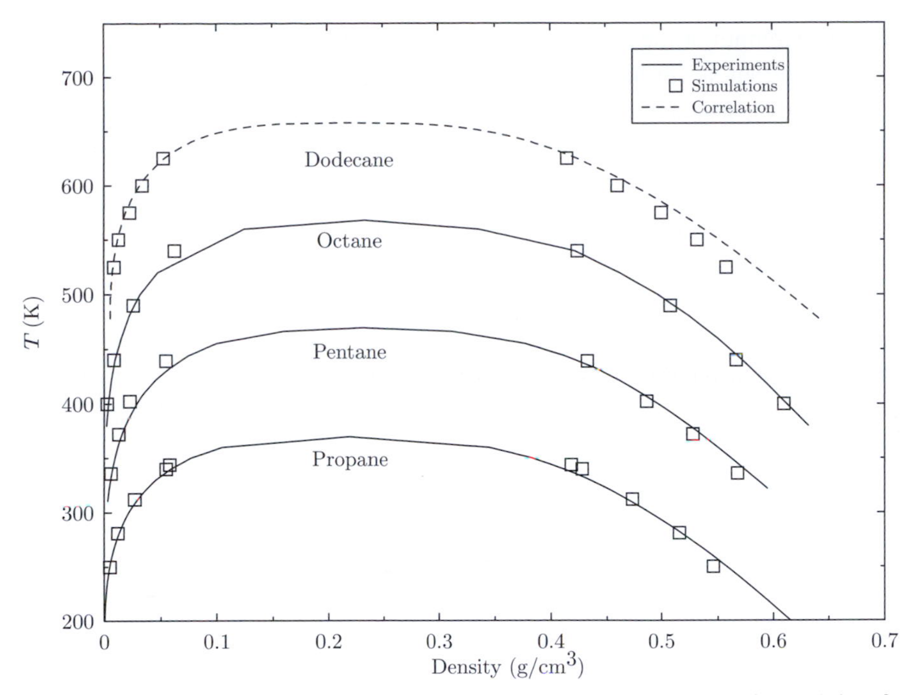

Figure 7.5 The points show results of molecular simulations. The lines show experimental data for various alkanes.

have charges (as ions do), the Lennard-Jones function of Eqn. (7.27) must be supplemented by dipole–dipole interactions, which are proportional to r^{-3}, and by Coulombic interactions, which are proportional to r^{-1}. In that case $\Gamma(r)$ is given by an expression of the form

$$\Gamma(r) = 4\epsilon\left[\left(\frac{\sigma}{r}\right)^{12} - \left(\frac{\sigma}{r}\right)^{6}\right] + \frac{(\mu^{\mathrm{d}})^2}{4\pi\epsilon_0 r^3} + \frac{q^2}{4\pi\epsilon_0 r}, \tag{7.28}$$

where μ^{d} and q denote the dipole moment and the charge of the particles, respectively. While most fluids exhibit dipolar and electrostatic interactions, for the purposes of our discussion here we will assume that Lennard-Jones interactions are sufficient to describe with reasonable accuracy the behavior of simple fluids.

For simplicity, our discussion of the Lennard-Jones energy function has been restricted to pure fluids. In the case of mixtures, the interaction energy between unlike molecules i and j is given by

$$\Gamma_{ij}(r) = 4\epsilon_{ij}\left[\left(\frac{\sigma_{ij}}{r}\right)^{12} - \left(\frac{\sigma_{ij}}{r}\right)^{6}\right]. \tag{7.29}$$

The parameter σ_{ij} defines the length scale for interactions between molecules i and j, and, as discussed earlier, for pure fluids it can be viewed as a measure of the diameter of a molecule. It is therefore logical to assume that diameters are additive and, for mixtures, assume an arithmetic mixing rule for σ_{ij}:

$$\sigma_{ij} = \frac{\sigma_i + \sigma_j}{2}. \tag{7.30}$$

For the strength of the interactions, we can go back to our discussion of intermolecular forces and recall that, according to Eqns. (7.22) and (7.26), the coefficient that governs the strength of attractive, dispersive interactions between two molecules depends on the product of the ionization potentials and the polarizabilities of the molecules. The ionization potentials do not differ too much from one molecule to another. For simplicity, we can assume them to be constant and concentrate our attention on the polarizabilities. In that case (see Exercise 7.4.C), it can be shown that a geometrical mixing rule, i.e.

$$\epsilon_{ij} = (\epsilon_i \epsilon_j)^{1/2}, \tag{7.31}$$

provides a reasonable means of estimating unlike interaction energies from pure-component values of ϵ.

It is important to note that, in general, the expressions given above for $\Gamma(r)$ do not lead to closed-form expressions for the thermodynamic properties of fluids. It is therefore common to assume simplified interaction-energy functions or models that capture some of the essential physics of molecular interactions, but are more tractable from a mathematical point of view. One such model is the hard-sphere fluid, where $\Gamma(r)$ takes the form of a step function. Another example is provided by the square-well fluid, in which intermolecular interactions are assumed to be infinitely large below a certain "overlap" distance, negative and constant for intermediate separations, and zero beyond a certain threshold separation.

7.2.5 Hydrogen bonds

Hydrogen bonds deserve special mention, in that they are 10–40 times stronger than simple dispersion interactions, but considerably weaker than ionic bonds. Hydrogen bonds are particularly relevant in water; they are responsible for its high boiling point and high enthalpy of vaporization, as well as for other anomalies such as the maximum in density that occurs at 1 bar and 4 °C. Note, however, that they also arise in numerous other substances (e.g. methanol, hydrogen fluoride, etc.).

The intramolecular distance of an O—H bond in a water molecule is 0.1 nm. The intermolecular distance between nearest-neighbor oxygen and hydrogen atoms in water is approximately 0.176 nm, which is considerably smaller than the distance that would be expected from simply summing the van der Waals radii of oxygen and hydrogen atoms. While these numbers suggest that hydrogen bonds have a covalent element to them, it is now accepted that their origin is electrostatic.

Hydrogen bonds are highly directional, and arise not only intermolecularly but also intramolecularly. Much of the structure of large biological molecules such as proteins and DNA is in fact a direct manifestation of intramolecular hydrogen bonding.

One particularly intriguing consequence of hydrogen-bonding interactions is the so-called hydrophobic effect. In the case of water, the tendency to form hydrogen bonds between neighboring molecules leads to the formation of a relatively tight tetrahedral network structure. Molecules that are not able to hydrogen bond with water are excluded from the network, as shown in Figure 7.6. A more extensive discussion of the hydrophobic effect is presented later in this chapter.

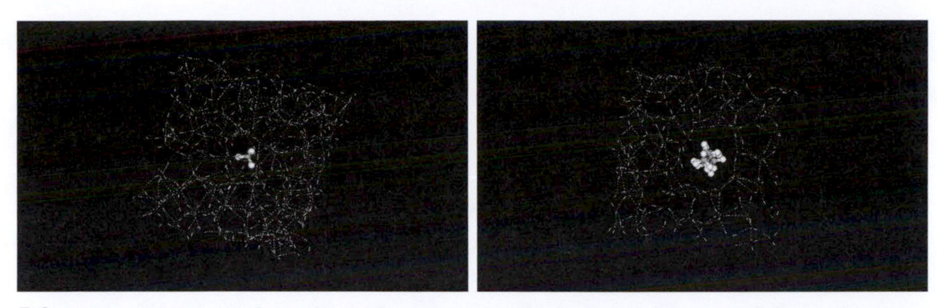

Figure 7.6 Representative configurations of water molecules under ambient conditions around a methane molecule (left) and around a hexane molecule (right). For methane, most water molecules around the solute are able to maintain a tetrahedral hydrogen-bonding structure. Around hexane, however, most of the water molecules must sacrifice some of their hydrogen bonds, thereby leading to a significant restructuring of the hydrogen-bond network.

7.3 MOLECULAR SIMULATIONS

Having established how molecules interact with each other, one of the goals of molecular thermodynamics is to determine how these interactions manifest themselves in the thermodynamic behavior of a fluid. To do that, a connection must be made between molecular interactions and measurable, macroscopic thermodynamic quantities such as the density, pressure, or heat capacity. For two spherical molecules, the interaction energy between them depends on how far apart they are. For molecules with internal degrees of freedom, the interaction energy also depends on their relative orientation and their internal configuration. These quantities change continuously in time as a result of thermal motion. To estimate average, macroscopic thermodynamic quantities, we must therefore have access to millions of distinct configurations of the system and take an average over the "instantaneous" properties corresponding to each of these. These configurations can be easily generated by resorting to what are commonly called "molecular simulations."

For simplicity, it is often assumed that the total energy of a fluid can be decomposed into a sum of pairwise additive interactions. With that assumption, the internal energy U of a fluid can be written in the form

$$U = \frac{1}{2}\left\langle \sum_{i}^{N}\sum_{j}^{N}\Gamma_{ij}(r_{ij})\right\rangle + cNk_{\mathrm{B}}T, \tag{7.32}$$

where the sum is conducted over all the N molecules in the system, and the angle brackets denote an average over many configurations of the system. The first term of Eqn. (7.32) corresponds to potential-energy contributions, which arise from the relative position or arrangement of the molecules. The second term corresponds to kinetic-energy contributions, arising from the velocity or momenta of the molecules; these contributions are responsible for the "simple-ideal-gas" part of any thermodynamic property (see Section 6.3.1).

To generate the configurations required in order to evaluate the average in Eqn. (7.32), one possibility would be to reconstruct a few nanoseconds in the life of a fluid, and to store this information in the form of instantaneous "snapshots" of the system. According to Newton's law of motion, the force acting on a molecule i gives rise to the dynamics

$$\vec{F}_i = m_i \vec{a}_i, \tag{7.33}$$

where m_i is the mass of the molecule and $\vec{a}_i$ is its acceleration. In a many-molecule fluid, the total force acting on molecule i will be given by the sum of the forces $\vec{F}_{ij}$ exerted on it by all the other molecules,

$$\vec{F}_i = \sum_{j \neq i}^{N} \vec{F}_{ij}, \tag{7.34}$$

where the force exerted by molecule j on molecule i is given by

$$\vec{F}_{ij} = -\frac{d\Gamma_{ij}(\vec{r}_{ij})}{d\vec{r}_{ij}}, \tag{7.35}$$

where $\vec{r}_{ij} = \vec{r}_i - \vec{r}_j$. Since we are trying to recreate the trajectory of a collection of molecules as they move about in a fluid, it is natural to write a Taylor-series expansion to estimate the coordinates $\vec{r}_i(t + \delta t)$ of molecule i a short time later, at $t + \delta t$,

$$\begin{aligned} \vec{r}_i(t+\delta t) &= \vec{r}_i(t) + \delta t \left(\frac{\partial \vec{r}_i}{\partial t} \right) + \frac{1}{2}(\delta t)^2 \left(\frac{\partial^2 \vec{r}_i}{\partial t^2} \right) + \cdots \\ &= \vec{r}_i(t) + \delta t\, \vec{v}_i(t) + \frac{1}{2}(\delta t)^2 \vec{a}_i(t) + \cdots \\ &= \vec{r}_i(t) + \delta t\, \vec{v}_i(t) + \frac{1}{2m_i}(\delta t)^2 \vec{F}_i(t) + \cdots , \end{aligned} \tag{7.36}$$

where $\vec{v}_i(t)$ denotes the velocity of molecule i at time t. If the coordinates and velocities of all molecules are known at time t, it is possible to calculate the forces $\vec{F}_i(t)$ at time t; Eqn. (7.36) can then be used to determine $\vec{r}_i(t + \delta t)$ for each molecule i of the system of interest. This process can now be repeated at will, thereby generating a trajectory or "movie" for the system; the total duration of this trajectory will be $\tau = \sum_{n_t} \delta t$, where n_t is the number of "time steps" employed in the integration of Eqn. (7.36).

The integration process outlined above is typically referred to as a molecular-dynamics simulation. It is nowadays typical to use on the order of 1000 molecules for a simulation, and to use a time step of approximately 10^{-15} s. Given the demands of these calculations, typical trajectories seldom exceed a few hundred nanoseconds (the computational time required by these calculations can be on the order of hours to days). For most small-molecule fluids, however, a 1–10-ns simulation, depending on the conditions of temperature and pressure, is sufficient to generate reliable estimates of their thermodynamic properties. Simulations are therefore generally used as computer experiments capable of revealing important insights about a fluid, but they are not used for fast, routine engineering calculations. To do that, we must resort to approximate theoretical treatments, some of which are described in the following sections. Equivalently, one can also use what are called "Monte Carlo simulations," which sample configuration space directly, without generating a time evolution of the system.

7.4 THE VIRIAL EXPANSION

The arguments used above (in Section 4.2.1) to arrive at the van der Waals mechanical equation of state (B.15) in Appendix B are not rigorous. Without a firm understanding of the foundations of van der Waals' equation, it would be difficult to extend it to mixtures, for example.

As alluded to briefly above, it is possible to use molecular simulations to predict the thermodynamic properties of a fluid from knowledge of the potential energy of interaction between molecules (e.g. Eqn. (7.27)). Such simulations, however, are numerically demanding and are not apt for fast, reasonable estimates of the properties of fluids.

For real gases, it is possible to formulate essentially exact equations that relate molecular interactions between pairs of molecules to the macroscopic, thermodynamic properties of the fluid. In this section, we will restrict our attention to real gases at low to moderate pressures.

For an ideal gas the compressibility factor is

$$z := \frac{Pv}{RT} = 1, \quad \text{ideal gas.} \tag{7.37}$$

For a real gas, which is likely to exhibit deviations from ideality, it is reasonable to propose simple, additive corrections to Eqn. (7.37). Our intuition leads us to expect that these corrections will become increasingly important as the density of the gas is increased. A power series in the density (see Section B.2 of Appendix B) satisfies these requirements:

$$\begin{aligned} z &= 1 + B\rho + C\rho^2 + \cdots \\ &= 1 + \frac{B}{v} + \frac{C}{v^2} + \cdots . \end{aligned} \tag{7.38}$$

It turns out that the expression in Eqn. (7.38) is not only physically intuitive, but also supported by experimental data. This point is illustrated in Figure 7.7, where experimental *PVT* data for argon, propane, and butane are shown. When plotted in the form $v(Pv/(RT) - 1)$ vs. $1/v$, data for low to intermediate densities fall on a line; at any given temperature, the intercept with the ordinate axis corresponds to $B(T)$ and the slope to $C(T)$. Equation (7.38) is called a virial expansion, and B and C are the so-called second and third virial coefficients, respectively. As illustrated by Figure 7.7, virial coefficients can be inferred from experimental *PVT* data, which are available for a wide range of fluids.

The virial expansion can also be arrived at on the basis of theoretical considerations. When that calculation is done, it can be shown that the second virial coefficient B is related to $\Gamma(r)$ by

$$B = 2\pi \tilde{N}_{\mathrm{A}} \int_0^\infty \left\{ 1 - \exp\left[-\frac{\Gamma(r)}{k_{\mathrm{B}} T} \right] \right\} r^2 \, dr. \tag{7.39}$$

Derivations of Eqn. (7.39) can be found in the statistical-mechanics literature or in Exercise 6.3.D; here we merely point out that it is exact (within our assumption of pairwise additivity of forces), and has a well-founded origin. The interaction-energy function $\Gamma(r)$ depends only on the distance between two molecules; a closer look at Eqn. (7.39) reveals that virial coefficients depend only on temperature (and not on volume or pressure). That is, given some functional form for $\Gamma(r)$, the integral in Eqn. (7.39) can be generated to yield a value of $B(T)$, which can in turn be used to evaluate the volume of a gas at any given pressure and temperature. Figure 7.8

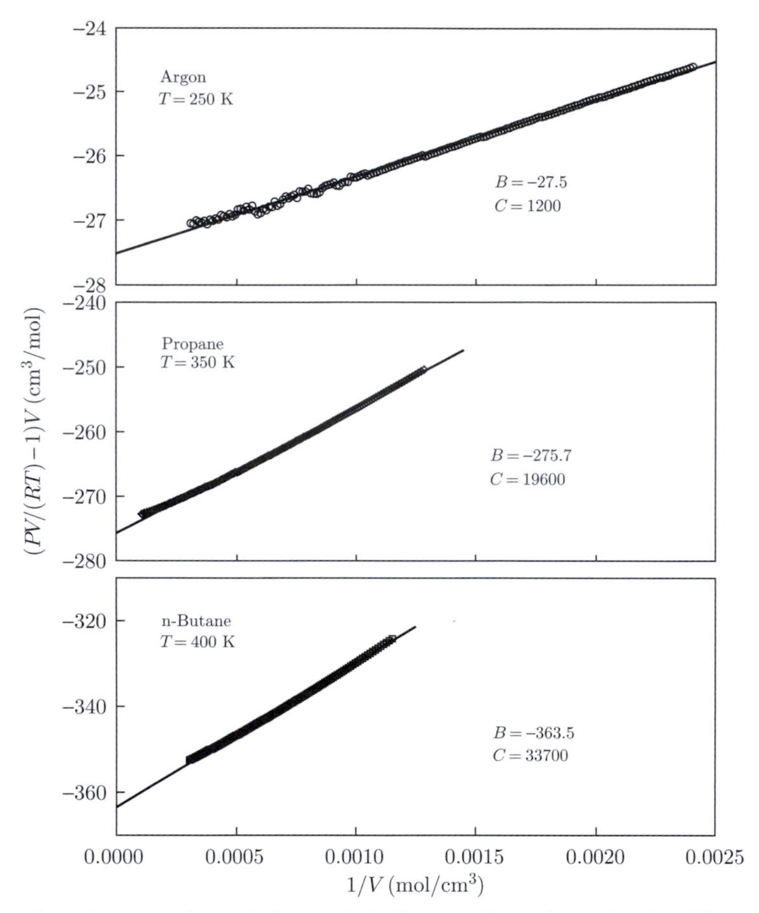

Figure 7.7 The points show experimental data, and the lines are linear fits to the data. The intercept with the ordinate axis corresponds to the second virial coefficient B, and the slope of the line corresponds to the third virial coefficient C.

shows experimental data for the virial coefficients of argon, propane, n-butane, and methyl acetate as functions of temperature.

The lines correspond to Eqn. (7.39). Note that argon is represented by a single, Lennard-Jones spherical molecule; the interaction energy between two argon molecules is therefore of the form Eqn. (7.27). Propane, butane, and methyl acetate, however, are represented by a collection of several spherical interaction sites; the interaction energy $\Gamma(r)$ between two molecules therefore consists of a sum of pair interactions over all sites of the molecule. The agreement between Eqn. (7.39) and experimental data is in general very good. Some small deviations arise at low temperatures, particularly for asymmetric molecules; these are due to the inability of the simple energy functions employed here to describe interaction energies over a wide range of temperature.

Depending on the pressure, a virial expansion can be truncated after the second or third term and produce results of high accuracy. Note, however, that the expansion does not converge for liquid-like densities and is therefore of little use for condensed fluids.

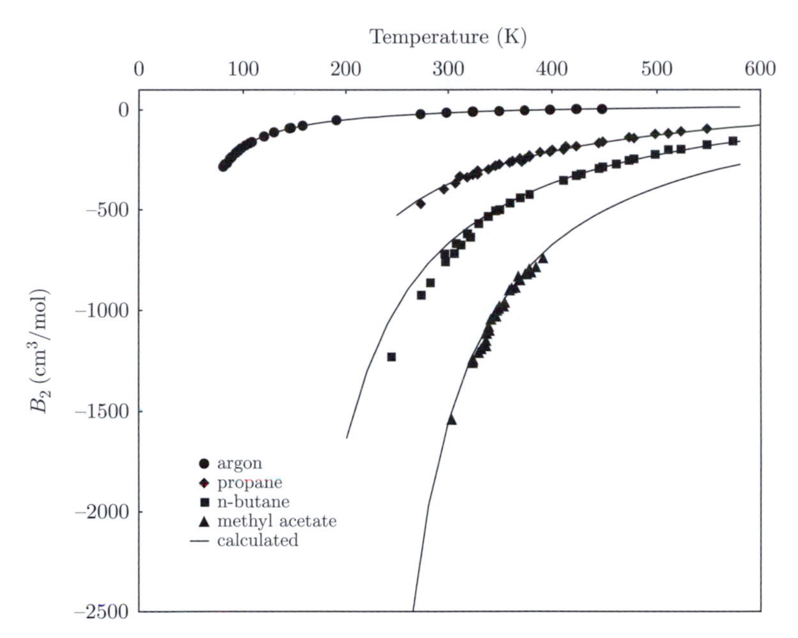

Figure 7.8 The points show experimental data [35], and the lines are calculations using Eqn. (7.39).

For engineering calculations, it is sometimes easier to work with a volume-explicit equation of state, as opposed to a pressure-explicit equation of state. One variant of the virial equation that is often encountered is

$$z = 1 + PB' + C'P^2 + \cdots , \tag{7.40}$$

where the coefficients B' and C' are related to B and C via

$$B' = B/(RT),$$
$$C' = \frac{C - B^2}{(RT)^2}. \tag{7.41}$$

Bear in mind, however, that the pressure-explicit virial equation's convergence is slightly better than that of the volume-explicit equation. Virial coefficients have been measured for a vast number of fluids. A good compilation of experimental data can be found in [35].

In the previous section we introduced the molecular principle of corresponding states, and we made an explicit connection between the experimental critical point of a pure fluid and that of the Lennard-Jones fluid. That connection allowed us to extract values of σ and ϵ from critical-property data. The idea of corresponding states can also be discussed in the context of the virial equation. For simplicity, we consider a virial expansion truncated after the second term. In that case, the PVT behavior of a gas can be described completely using Eqn. (7.38). That implies that the PVT behavior of the gas is just a function of two parameters, namely ϵ and σ, which enter the equation of state through Eqn. (7.39). We can therefore rewrite that expression in reduced units according to

$$B^*(T^*) = \frac{B}{2\pi \tilde{N}_A \sigma^3} = \int_0^\infty \left\{1 - \exp\left[-\frac{\Gamma^*(r^*)}{T^*}\right]\right\} (r^*)^2\, dr^*, \tag{7.42}$$

where Γ^* is used to indicate that the potential-energy function can be written as the product of ϵ times some function of a reduced distance $\Gamma^*(r^*)$. In the case of Lennard-Jonesium, we have

$$\Gamma^*(r^*) = 4\left[\left(\frac{1}{r^*}\right)^{12} - \left(\frac{1}{r^*}\right)^{6}\right].$$

The reduced second virial coefficient B^* can be evaluated as a function of T^* by integrating Eqn. (7.42) numerically. This procedure would lead to a universal figure or chart for a given potential-energy-function model. For the particular case of the Lennard-Jones function, we could read off values of $B^*(T^*)$ from this chart and relate these to experimental $B(T)$ data for a specific fluid, thereby generating estimates of the parameters σ and ϵ.

One important feature of the virial expansion that makes it particularly attractive is that it can be readily extended to mixtures without a need for hand-waving arguments. For a multicomponent mixture, the virial coefficients of a vapor are given by

$$B = \sum_{i=1}^{r}\sum_{j=1}^{r} x_i x_j B_{ij},$$

$$C = \sum_{i=1}^{r}\sum_{j=1}^{r}\sum_{k=1}^{r} x_i x_j x_k C_{ijk}, \tag{7.43}$$

where x_i is the mole fraction of component i, B_{ij} denotes the second virial coefficient for interactions between molecules of type i and type j, and C_{ijk} denotes the third virial coefficient for interactions among molecules of types i, j, and k. The question of how to estimate B_{ij} for two molecules of different chemical species then arises. Equation (7.39) can be used again to determine B from knowledge of an intermolecular potential-energy function for unlike molecules.

7.5 EQUATIONS OF STATE FOR LIQUIDS

As mentioned above, one of the main shortcomings of the virial equation is that it cannot describe the properties of liquid, or high-density, phases. In recent years, new statistical-mechanics-based models have been proposed for such cases. The field of equation-of-state development has evolved very rapidly in the last decade, and new models are published literally every month. The origins of such models, however, are essentially the same and can be summarized in a few lines.

As with the development of the van der Waals model, the pressure of a fluid is separated into repulsive and attractive contributions. It turns out that, for liquid-like densities, the molecular structure of a fluid is dominated by *packing* arrangements. It is therefore particularly important to determine repulsive contributions to the pressure accurately, since they have a profound effect on the system's thermodynamic properties. Developing a quantitative theory for the structure of dense liquids has been one of the main objectives of liquid-state theory for many decades. The advent of molecular simulations has contributed significantly to the refinement of such theories. Providing a detailed account of how such models are developed is beyond the scope of this text. For completeness, however, in this section we describe one of the many successful models of this nature, which has the additional feature of being applicable to mixtures of simple fluids and mixtures containing long, articulated polymer molecules (after minor modifications).

The compressibility factor of a system can be expressed by writing an equation of the form

$$z := \frac{Pv}{RT} = Z^{\text{ideal}} + Z^{\text{hs}} + Z^{\text{disp}}, \tag{7.44}$$

where the superscripts "ideal," "hs," and "disp" are used to denote ideal, hard-sphere, and dispersion contributions, respectively. Ideal contributions to the pressure are given by

$$z^{\text{ideal}} = 1. \tag{7.45}$$

Hard-sphere contributions to the compressibility factor are given by an expression of the form

$$z^{\text{hs}} = \frac{4\eta - 2\eta^2}{(1-\eta^3)}, \tag{7.46}$$

where $\eta := \pi \tilde{N}_{\text{A}} \tilde{\rho} d^{*3}/6$ is the packing fraction of the fluid and d^* is the effective diameter of a hard sphere meant to describe the packing of a molecule in a dense fluid. Here the density is given in units of the number of molecules per unit volume. Equation (7.46) is the so-called "Carnahan–Starling" equation of state, which was derived by fitting polynomials of the density to results of molecular simulations for fluids of hard spheres [22, 88]. Finally, dispersion or attractive contributions to the compressibility factor are given by an expression of the form

$$z^{\text{disp}} = \sum_i \sum_j D_{ij} \left(\frac{u^*}{k_{\text{B}}T}\right)^i \left(\frac{6\eta}{\pi\sqrt{2}}\right)^j, \tag{7.47}$$

where u^* is a characteristic energy of interaction, which is often assumed to be inversely proportional to temperature. The constants D_{ij} are universal parameters obtained by fitting experimental *PVT* data for argon [26]. The form of Eqn. (7.47) was originally proposed by Alder and co-workers from fitting the pressure calculated in molecular simulations of fluids interacting through a square-well potential-energy function [2].

7.6 EXPERIMENTAL MANIFESTATIONS OF INTERMOLECULAR INTERACTIONS

Our discussion so far has focused on the origin of molecular interactions, and on some of the models and methods used to describe them. We now turn our attention to some of the consequences of particular types of interactions. In this section we describe some aspects of the behavior of a number of fluids, and we trace this behavior back to the interactions between molecules.

We begin by discussing steric, or "packing," effects in non-polar fluids. Consider the case of a mixture of small and large molecules, say a solution of a long alkane dissolved in an excess of a small alkane. If the molecules are non-polar, it is reasonable to expect a Lennard-Jones energy function to be adequate for such a mixture. In this case, as mentioned earlier, the long alkane could be described as a collection of several Lennard-Jones interaction sites connected to each other. For such a mixture we can ask ourselves the following question: at saturation, is the density of the pure small-alkane liquid higher or lower than that of the large-alkane liquid? Experimental data for pentane and octane indicate that, at 400 K, the saturated liquid densities of these two alkanes are approximately 0.5 and 0.6 g/cm^3, respectively. These data suggest that octane molecules pack more effectively than do pentane molecules in the liquid phase. This more effective packing of the longer alkanes can be partly attributed to the fact that the "end" (CH_3) groups of the alkane occupy a larger volume than that occupied by the middle CH_2 groups; the latter groups are bound from both sides to other CH_2 groups, thereby occupying less space than the CH_3 groups (which are bound to only one CH_2 group.) Since octane has a smaller concentration of end groups, it exhibits a higher density.

We can now ask the following question: if we add a small amount of the long alkane to the short alkane, will the density of the mixture be higher or lower than that of the pure, small alkane? In this particular case, i.e. for pentane and octane, the answer is relatively straightforward: the density of the resulting mixture can be estimated from a linear combination of the pure-component densities. If, however, the long alkane is relatively long (i.e. if it is a short polyethylene molecule), the long alkane might not be fully miscible with the short alkane. The solution can undergo phase separation into two distinct liquid phases, or, depending on the temperature, pressure, and concentration, the long molecule might drop out of solution and form a semi-crystalline material. The scenario of possibilities for these mixtures is in fact relatively complex, and it can all be related to simple interactions such as those described by the Lennard-Jones function. For mixtures of short and long molecules, the complex thermodynamic behavior that emerges has most of its origins in packing, or steric, effects that come into play when thousands, or millions, of molecules interact with each other. Such packing effects can in turn be quantified through the entropy of the mixture, a point to which we shall return in our discussion of polymer solutions and blends later in this text.

Another class of systems in which packing effects are particularly important is liquid crystals. Liquid crystals are used extensively for a number of applications (e.g. the screens of laptop computers); their ability to diffract light of different wavelengths (and hence exhibit a variety of colors) is related to the way in which molecules pack in the liquid phase. Liquid crystals tend to be elongated, rigid molecules, whose properties can be described with reasonable accuracy by assuming that they have a rod-like or ellipsoidal shape. In the liquid state, at sufficiently low densities, molecules pack in a disordered, isotropic manner. Figure 7.9 shows the instantaneous structure of an isotropic liquid crystal, in which the aspect ratio of the rod-like molecules is 3. As the density (or concentration) is increased, however, the molecules can adopt the structure of an ordered, nematic phase in which molecules assume a well-defined, clear orientation. Figure 7.10 shows the instantaneous structure of such a nematic phase. This preferred orientation can be turned on or off by controlling the density, the temperature, or the strength of an external field. By manipulating molecular orientation, it is possible to control the color of the system and design a display.

We now turn our attention to fluids that exhibit hydrogen bonding. Water, of course, is the hydrogen-bonding fluid *par excellence*, but many other industrially relevant solvents are also

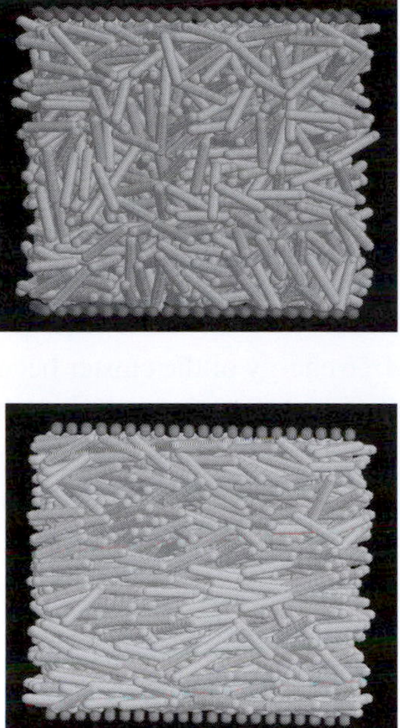

Figure 7.9 A schematic representation of the isotropic phase of a liquid crystal ($\rho = 0.25$). This configuration of the system was generated by means of molecular simulations of rod-like molecules. The system is confined by two surfaces. The distance Z between the surfaces is equivalent to 18 times the smaller dimension of the rods.

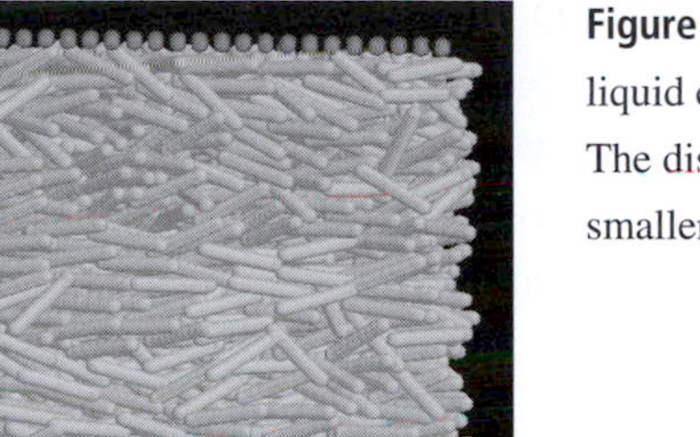

Figure 7.10 A schematic representation of the nematic phase of a liquid crystal ($\rho = 0.30$). The system is confined by two surfaces. The distance Z between the surfaces is equivalent to 18 times the smaller dimension of the rods.

able to hydrogen bond. Hydrogen bonding was discussed briefly in previous sections; in water, a hydrogen bond arises from the electrostatic interaction of the lone electron pairs on the oxygen atom with the electron-deficient hydrogen atoms of other molecules. The strength of a hydrogen bond is approximately an order of magnitude higher than that of simple Lennard-Jones interactions (approximately $15k_BT$). As a result of these stronger interactions water has a high boiling point (relative to non-polar molecules of the same size); more thermal energy is required to break such hydrogen bonds and vaporize the liquid. Also as a result of hydrogen bonding, water is unable to dissolve other substances that do not exhibit such interactions. The solubility of alkanes in water, for example, is remarkably low. As illustrated in Figure 7.6, the energy required to break enough hydrogen bonds to form a "cavity" for a large alkane is too large to allow any appreciable solubility.

The consequences of hydrogen bonding are as diverse as they are intriguing, and are sometimes difficult to predict. To illustrate this point, we discuss briefly the structural rearrangements undergone by water molecules around a simple alkane molecule. Our discussion is based on [24], where a timely discussion of the hydrophobic effect is presented. The presence of a non-polar solute disrupts the hydrogen-bond network of water; the magnitude of this effect depends on the shape and size of the solute molecules. This phenomenon is often referred to as the hydrophobic effect. When the non-polar solute is small, say a methane molecule, the volume that it occupies is less than 5 Å across; water molecules can adopt hydrogen-bonding patterns that go around that solute without a significant loss of energy. Above a critical size, however, the hydrogen-bonding network of water can no longer be maintained, and some bonds are sacrificed. Near a hydrophobic molecule or a hydrophobic surface, the number of hydrogen-bonding possibilities with the rest of the fluid is diminished. To minimize that loss, water tends to move away from the solute, thereby creating an interface around it. At room temperature and ambient pressure, that interface is akin to that which forms between a liquid and a vapor. The cost of maintaining this interface increases linearly with surface area. The solvation free energy of small molecules scales with their volume, but, for solute radii on the order of about 1 nm, a cross-over to a scaling with solute surface area occurs.

In dilute solution, the solvation free energy of n small solute molecules in water is n times the solvation free energy of an individual molecule, and it increases linearly with solute volume. However, as these molecules come together to form a tight cluster, having a surface area greater than approximately 1 nm^2, the solvation free energy increases linearly with surface area. If the number of molecules, n, is large enough, the solvation free energy of the cluster becomes lower than that of the corresponding n individual solute molecules, thereby providing a driving force for cluster assembly.

Thermodynamic arguments can be used to quantify the magnitude of the hydrophobic effect. The Gibbs free-energy difference required to transform a pure solvent (water) into a solvent plus one solute molecule is denoted by $\Delta G_{\text{solvation}}$, and comprises an enthalpic and an entropic component:

$$\Delta G_{\text{solvation}} = \Delta H_{\text{solvation}} - T\,\Delta S_{\text{solvation}}. \tag{7.48}$$

As discussed in Chapter 6 on statistical mechanics, the Gibbs free-energy change of solvation can be expressed by

$$\Delta G_{\text{solvation}} \approx \langle \Delta E_s \rangle, \tag{7.49}$$

where $\langle \Delta E_s \rangle$ represents the average potential-energy change experienced by the solvent upon addition of a solute. The angle brackets in Eqn. (7.49) represent an ensemble average, i.e. an average over multiple realizations of the system (or a time average of the energy experienced by the solute molecule immersed in the solvent). Note that Eqn. (7.49) is approximately valid only when $\Delta E_s/(k_B T)$ is small. A process that is dominated by considerable changes in the number of molecular interactions (e.g. by the breaking or formation of hydrogen bonds), will be dominated by enthalpic contributions; in that case $\Delta G/T$ decreases with increasing temperature. On the other hand, a process that requires the creation of specific hydrogen-bonding patterns is likely to comprise an important entropic element. At room temperature, the process of solvating a hydrophobic solute in water is mostly entropic, and $\Delta G/T$ increases with increasing temperature. At higher temperatures, however, thermal fluctuations are more pronounced and hydrogen bonds are broken and formed with more ease; the entropic contribution to the free energy decreases. This behavior is manifest in the solvation entropy of small alkanes in water; near room temperature $\Delta S_{\text{solvation}}$ is negative, whereas near the boiling point of water it becomes positive.

In addition to effective interactions arising from the creation or destruction of hydrogen bonds, weakly attractive, van der Waals-type interactions between a solute and water molecules can play an important role. Their effect can be particularly pronounced in the case of a relatively large solute, where they can essentially bring water into contact with the hydrophobic surface created by that solute. In other words, while van der Waals forces are much weaker than hydrophobic forces, they can influence the actual position of the interface referred to above [24]. It is of interest to note that the role of van der Waals attractions in the context of the hydrophobic effect continues to be a subject of considerable interest as well as considerable debate. An example is provided in recent work by Huang *et al.* [66], who used computer simulations to show that, when two hydrophobic plates immersed in water are brought together, to a separation less than some critical value, this process induces a spontaneous vaporization of the confined water (between the two plates). A so-called "drying transition" of the plates is observed, followed by collapse of the two surfaces [66]. Interestingly, however, a similar computer experiment using graphite plates by Choudhury and Pettitt [27], in which the attractions between the surfaces are slightly different, does not exhibit the drying transition observed by Huang *et al.*, even at very small separations. This discrepancy can be explained by considering the van der Waals interactions between the plate and the water molecules; the van der Waals interaction parameter for graphite–water interaction is larger than that for paraffin–water interaction, thereby preventing graphite plates from dewetting. As the interaction parameter increases, the critical separation at which drying occurs decreases, and the time scale over which drying occurs increases.

To a first approximation, the interactions that arise in hydrogen-bonding fluids can be described by potential-energy functions of the form given in Eqn. (7.28). We emphasize that this is just a first approximation because water is significantly polarizable and the polarizability has not entered that equation in an explicit manner.

Conclusions

Much of twentieth-century science and engineering aimed at developing an understanding of intermolecular forces. While a brief chapter on intermolecular forces can only provide an introduction to this subject, it is remarkable that one can rationalize, and in many cases predict, the thermodynamic properties of a wide class of fluids from knowledge of several key concepts. Attractive forces can be classified into ionic, dipolar, and induced-dipolar forces. The relative magnitude of these interactions depends on the molecules' overall charge, their dipole moment, and their polarizability.

At a molecular level, one can now resort to molecular simulations on a fairly routine basis to calculate the thermodynamic properties of arbitrary fluids from knowledge of intermolecular interactions. For apolar fluids, such as alkanes, intermolecular interactions that rely on a Lennard-Jones potential-energy function are generally sufficient to achieve quantitative agreement between simulations and experimental data. For polar molecules, more elaborate intermolecular potential functions must be used, and agreement with experiment is more difficult to achieve. This is an active area of research, and new and improved "force fields" continue to be proposed to describe the behavior of polar fluids and solutions, such as those consisting of biological macromolecules dissolved in water.

A wide variety of equations of state is now available to describe the properties of gases and liquids. For gases, the virial equation of state and engineering correlations based on the virial equation can be used to predict the properties of pure systems and multicomponent mixtures. For liquids, equations of state built around the structure of hard-sphere fluids can be used to predict thermodynamic properties. Liquid-phase properties are in general more difficult to quantify than vapor-phase properties, partly as a result of our still incomplete understanding of intermolecular interactions in condensed phases.

Exercises

7.2.A Consider the energy of interaction between two, identical polar molecules. The molecules can be represented as rigid rods of length d, with two charges of opposite sign (q_1 and q_2) at either end of the rod. The distance between the centers of mass of the two molecules is r. (See Figure 7.1.)

(a) Show that, provided that $r \gg d$, the preferred relative orientation between the two molecules is planar, with the two rods parallel to each other. What is the energy of interaction corresponding to that orientation?

(b) What is the most unfavorable orientation that the two molecules can have?

7.2.B Consider a collinear arrangement in which the shortest distance between two identical dipoles is that between the positive charge q_1 of one dipole and the negative charge q_2' of the other dipole. Show that, to leading order, the energy of interaction between the dipoles is given by

$$\Gamma = -\frac{2\mu^{\mathrm{d}}(\mu^{\mathrm{d}})'}{4\pi\epsilon_0 r^3}, \tag{7.50}$$

where μ^{d} and $(\mu^{\mathrm{d}})'$ are the moments of the two dipoles, respectively, and r is the distance between their centers of mass.

Hint: to solve this problem, one can begin by writing an expression for the electrostatic potential energy $\Gamma(P)$ felt at a point P due to the presence of one dipole. If r denotes the distance from P to the center of mass of the dipole, that energy is given by

$$\Gamma(P) = \frac{1}{4\pi\epsilon_0}\left(\frac{q_1}{r_1} + \frac{q_2}{r_2}\right), \tag{7.51}$$

where r_i is the distance between P and charge q_i. Upon introducing an angle θ_d between the main axis of the dipole and the line joining point P to the center of mass of the dipole, Eqn. (7.51) can be written as

$$\Gamma(P) = \frac{1}{4\pi\epsilon_0}\left(\frac{q_1}{\sqrt{r^2 + z_1^2 + 2z_1 r\cos\theta_d}} + \frac{q_2}{\sqrt{r^2 + z_2^2 - 2z_2 r\cos\theta_d}}\right), \tag{7.52}$$

where z_i is the distance between the dipole's center of mass and charge q_i. For $r \gg z_i$, Eqn. (7.52) can be rewritten in powers of z_i/r and expressed in terms of the total charge of the dipole $q = q_1 + q_2$, the dipole moment $\mu^{\mathrm{d}} = q_2 z_2 - q_1 z_1$, the quadrupole moment $\Theta = q_1 z_1^2 + q_2 z_2^2$, etc. By considering that the center of mass of a second dipole is located at point P, one can now determine the energy of interaction between the two dipoles from

$$\Gamma = q_2'\Gamma(q_2') + q_1'\Gamma(q_1'), \tag{7.53}$$

where $\Gamma(q_i')$ represents the electrostatic potential energy experienced at q_i' due to the other (unprimed) dipole.

7.2.C Use the approach proposed in the previous problem to derive Eqn. (7.13).

7.2.D Show that the average energy of interaction between two permanent dipoles i and j at a distance r from each other is given by Eqn. (7.14), namely

$$\langle \Gamma_{ij}(r) \rangle = -\frac{2(\mu_i^{\mathrm{d}})^2(\mu_j^{\mathrm{d}})^2}{(4\pi\epsilon_0)^2 k_{\mathrm{B}} T r^6} + \cdots .$$

Hint: the angle brackets denote an ensemble average in which distinct configurations of a system are weighted according to a Boltzmann factor. The average energy of interaction between two dipoles is therefore given by

$$\langle \Gamma_{ij}(r) \rangle = \frac{\int\int \Gamma_{ij}(\vec{\omega}_i, \vec{\omega}_j)\exp[-\Gamma_{ij}(\vec{\omega}_i, \vec{\omega}_j)/k_{\mathrm{B}}T]d\vec{\omega}_i\, d\vec{\omega}_j}{\int\int \exp[-\Gamma_{ij}(\vec{\omega}_i, \vec{\omega}_j)/k_{\mathrm{B}}T]d\vec{\omega}_i\, d\vec{\omega}_j}, \tag{7.54}$$

where $\vec{\omega}_i$ is used to denote collectively the coordinates of dipole i (i.e. r_i, θ_i, and ϕ_i), and $d\vec{\omega}_i$ can be expressed in spherical coordinates as $r_i^2\, dr_i \sin\theta_i\, d\theta_i\, d\phi_i$.

7.2.E Show that the Lennard-Jones potential-energy function exhibits a minimum having energy ϵ. At what separation or distance (r_{min}) between two molecules does the minimum occur?

7.2.F (a) Two spherically symmetric molecules having charges q_i and q_j and separated by a distance r experience an attractive force between them of magnitude

$$F = \frac{q_i q_j}{r^2}. \tag{7.55}$$

When the molecules are separated by a large distance, they do not interact with each other. Derive an expression for the interaction energy between these molecules as a function of distance.

(b) Two molecules of NH_3 exhibit an interaction energy of approximately -20 kJ/mol at room temperature. The dipole moment of NH_3 is 1.47 D, and the polarizability $\frac{\alpha}{4\pi\epsilon_0}$ is approximately 22.6×10^{-25} cm^3. The ionization potential of NH_3 is $I = 11.5$ eV (1 eV $= 1.6 \times 10^{-19}$ J). The following correlation has been proposed to describe the potential energy of interaction between two NH_3 molecules as a function of temperature:

$$U = \frac{A}{T} - 13.9\,\mathrm{kJ}. \tag{7.56}$$

Using your knowledge of intermolecular interactions, please provide a justification for the form of this empirical expression. What does the constant A represent?

7.2.G Liquid-crystalline systems are often described in terms of the so-called "Gay–Berne" intermolecular energy function [48]. This function describes the energy of interaction between anisotropic, ellipsoidal particles. It is given by an expression similar to Eqn. (7.27), in which the parameters σ and ϵ depend on the relative orientation of the two molecules. If the axial vectors of the molecules are denoted by $\vec{u}_i$ and $\vec{u}_j$, and if $\vec{r}$ is a unit vector between the centers of mass of two molecules, an orientation-dependent size parameter is defined according to

$$\sigma_{\text{GB}}(\vec{u}_i, \vec{u}_j, \vec{r}) = \sigma_0 \left(1 - \frac{1}{2}\chi_a \left[\frac{(\vec{u}_i \cdot \vec{r} + \vec{u}_j \cdot \vec{r})^2}{1 + \chi(\vec{u}_i \cdot \vec{u}_j)} + \frac{(\vec{u}_i \cdot \vec{r} - \vec{u}_j \cdot \vec{r})^2}{1 - \chi(\vec{u}_i \cdot \vec{u}_j)}\right]\right)^{-1/2}, \quad (7.57)$$

The parameter χ_a characterizes the size anisotropy of the molecule, and is given by

$$\chi_a = \frac{\kappa^2 - 1}{\kappa^2 + 1}. \quad (7.58)$$

The parameter σ_0 is the side-by-side contact distance of a pair of rods, and the parameter κ is the aspect ratio (ratio of width to length) of the particles.

The angular dependence of the well depth of the energy is given by

$$\epsilon_{\text{GB}} = (\epsilon')(\epsilon'')^2, \quad (7.59)$$

where

$$\epsilon'(\vec{u}_i, \vec{u}_j, \vec{r}) = \epsilon_0 \left[1 - \chi^2(\vec{u}_i \cdot \vec{u}_j)^2\right]^{-1/2},$$
$$\epsilon''(\vec{u}_i, \vec{u}_j, \vec{r}) = 1 - \frac{1}{2}\chi' \left[\frac{(\vec{u}_i \cdot \vec{r} + \vec{u}_j \cdot \vec{r})^2}{1 + \chi'(\vec{u}_i \cdot \vec{u}_j)} + \frac{(\vec{u}_i \cdot \vec{r} - \vec{u}_j \cdot \vec{r})^2}{1 - \chi'(\vec{u}_i \cdot \vec{u}_j)}\right], \quad (7.60)$$

and

$$\chi' = \frac{\sqrt{\kappa'} - 1}{\sqrt{\kappa'} + 1}. \quad (7.61)$$

In this equation, $\kappa' = \frac{\epsilon_s}{\epsilon_e}$ defines the energetic anisotropy of the particles, i.e. how much stronger the energy of interaction is when the molecules are parallel than when they are oriented head-to-tail or head-to-head. Parameters ϵ_s and ϵ_e represent the minimum of the potential energy when two molecules are side by side or oriented end to end. With the expressions above, the "Gay–Berne" energy of interaction is given by

$$\Gamma_{\text{GB}}(\vec{u}_i, \vec{u}_j, \vec{r}) = 4\epsilon_{\text{GB}} \left[\left(\frac{\sigma_0}{d_{\text{GB}}}\right)^{12} - \left(\frac{\sigma_0}{d_{\text{GB}}}\right)^{6}\right], \quad (7.62)$$

with

$$d_{\text{GB}} = r - \sigma_{\text{GB}} + \sigma_0. \quad (7.63)$$

Plot the energy of interaction between two liquid-crystalline molecules of aspect ratio 3 in the "head-to-tail" or perpendicular orientation and in the "parallel" orientation. Show that, as the aspect ratio of the molecules goes to unity, you recover the Lennard-Jones potential. Please discuss the behavior of the intermolecular energy for very long molecules (polymers) in the limit of infinite aspect ratio. With reference to this energy function, also comment on the energy of interaction between two disk-shaped particles.

7.2.H Consider a Lennard-Jones potential-energy function for the interaction between two spherically symmetric molecules.

(a) Consider a pair of molecules in the gas phase, at some temperature T. Is the average distance between the two molecules the same as r_{min}?

(b) Consider a liquid of spherical molecules interacting via a Lennard-Jones energy function $\Gamma(r)$. For simplicity, assume that the internal energy of the system is given by an expression of the form

$$U = \frac{1}{2}\left\langle \sum_i^N \sum_j^N \Gamma_{ij}(r_{ij}) \right\rangle + cNk_{\text{B}}T, \tag{7.64}$$

where k_{B} is Boltzmann's constant, subscripts i and j are used to denote individual molecules, and N is the number of molecules in the system. Please state clearly the assumptions which went into Eqn. (7.64). Why are there two terms? Why is there a factor of $\frac{1}{2}$ in the first term? What is the meaning of the angular brackets? What is the origin of the second term? At liquid-like densities, what is the average separation between a pair of nearest-neighbor molecules? Is it r_{min}?

7.2.I Consider the energy of interaction between two non-polar molecules. The two molecules are spherical, but they are of different chemical nature.

(a) Show that, if Γ_{ii} is used to denote the attractive energy of interaction between two molecules of the same species (i), it can be written in the form

$$\Gamma_{ii} = k' \frac{\alpha_i^{\text{p}2}}{r^6}, \tag{7.65}$$

where α_i^{p} is the polarizability of molecule i. Please clarify the physical significance of the constant k'.

(b) Show that the energy of interaction between molecules of different species can be approximated by an expression of the form

$$\Gamma_{ij} \approx \sqrt{\Gamma_{ii}\Gamma_{jj}}. \tag{7.66}$$

Please state your assumptions clearly and concisely.

7.2.J Would you expect chain-like molecules to exhibit corresponding-states behavior? Please explain your answer.

7.4.A (a) Use the Lennard-Jones potential-energy function to calculate the second virial coefficient of "Lennard-Jonesium" and construct a plot of this virial coefficient as a function of temperature. To facilitate your analysis, please use reduced (dimensionless) units for the temperature and the distance (i.e. $T^* = Tk_{\text{B}}/\epsilon$, $B^* = B/(2\pi N_{\text{A}}\sigma^3)$, and $r^* = r/\sigma$).

(b) Find experimental data for the virial coefficients of several simple fluids that you might expect to be described accurately by a simple Lennard-Jones energy function (e.g. noble gases, methane, carbon tetrachloride). Use these data to infer "experimental" values for the constants σ and ϵ for those fluids. How well does the Lennard-Jones function describe experiments? Do your parameters follow reasonable trends?

(c) For the fluids considered in part (b), derive values for the Lennard-Jones parameters from critical-property data. Compare these parameters with those obtained in (b) using second-virial-coefficient data. Are the parameters identical? If not, why are they different?

7.4.B The energy of interaction between two Lennard-Jones particles of different species i and j can be estimated by using the so-called Lorentz–Berthelot combining rules, in which the ϵ_{ij} and σ_{ij} cross-interaction parameters are given by

$$\epsilon_{ij} = (\epsilon_i \epsilon_j)^{1/2}, \tag{7.67}$$

$$\sigma_{ij} = \frac{\sigma_i + \sigma_j}{2}. \tag{7.68}$$

(a) On the basis of the calculations of part (a) of Exercise 7.4.A, propose a simple and accurate correlation for $B^*(T^*)$. Express this correlation in terms of σ and ϵ for a pure fluid. Assuming the validity of the Lorentz–Berthelot combining rules, write a correlation for the cross-interaction second virial coefficient in a binary mixture, in terms of pure-component Lennard-Jones parameters.

(b) The following correlation has been proposed in the literature [82] as a means to estimate the second virial coefficients of simple, non-polar fluids consisting of quasi-spherical molecules (e.g. methane):

$$\frac{B}{v_c} = 0.430 - 0.886\left(\frac{T}{T_c}\right)^{-1} - 0.694\left(\frac{T}{T_c}\right)^{-2}. \tag{7.69}$$

Assuming that the interactions between such molecules can be described by a Lennard-Jones potential-energy function, rewrite Eqn. (7.69) in terms of the Lennard-Jones parameters σ and ϵ. Does the resulting equation describe the expression derived in part (a) of Exercise 7.4.A with reasonable accuracy?

(c) If you now wanted to develop a correlation for cross-interaction second virial coefficients in terms of pure-component critical-property data, what mixing rules would you propose for the relevant "mixture" pseudo-critical properties?

7.4.C In the previous exercise, a geometric-mean combining rule was proposed as a means to evaluate the energy parameter for a binary mixture of Lennard-Jones particles. From your knowledge of intermolecular interactions, provide a theoretical justification for this combining rule.

7.4.D The so-called Lennard-Jones potential-energy function is given by

$$\Gamma_{ii} = 4\varepsilon_i\left[\left(\frac{\sigma_i}{r}\right)^{12} - \left(\frac{\sigma_i}{r}\right)^{6}\right], \tag{7.70}$$

where σ and ϵ are substance-specific parameters.

(a) From your knowledge of intermolecular interactions, give a theoretical justification for the geometric-mixing rule that is commonly used to evaluate ε_{ij}.

(b) For the Lennard-Jones potential, the second virial coefficient B^* can be evaluated as a function of temperature T^* using the following correlation:

$$B^* = 0.2336 - 0.68\left(\frac{1}{T^*}\right) - 0.406\left(\frac{1}{T^*}\right)^2, \tag{7.71}$$

where the asterisk denotes a quantity reduced with respect to the Lennard-Jones parameters, namely σ and ϵ. For linear alkanes, the following correlation is often used in practice [83]:

$$\frac{B}{v_c} = 0.430 - 0.886\left(\frac{T}{T_c}\right)^{-1} - 0.694\left(\frac{T}{T_c}\right)^{-2} - 0.0375(n-1)\left(\frac{T}{T_c}\right)^{-4.5}, \tag{7.72}$$

where n is the number of carbon atoms in the alkane. Using critical property data for methane and pentane, compare the predictions for B from Eqns. (7.71) and (7.72) at $T^* = 1$ and at $T^* = 2.5$. For which of these two molecules are the two correlations in better agreement? Why?

(c) The second virial coefficient is related to the potential energy of interaction between two molecules through

$$B = 2\pi\tilde{N}_A \int_0^\infty \left(1 - \exp\left[-\frac{\Gamma(r)}{k_B T}\right]\right) r^2\, dr. \tag{7.73}$$

In order to examine the effects of molecular volume on the properties of a gas, you are asked to propose a simple, tractable model in which only repulsive interactions are present. One possibility is to use a "hard-sphere" model, in which the interaction energy between two molecules is of the form

$$\Gamma(r) = \begin{cases} 0, & r \geq d, \\ \infty, & r < d, \end{cases} \tag{7.74}$$

where d is the diameter of the molecules. Please sketch the qualitative shape of $\Gamma(r)$. Derive an expression for the second virial coefficient of a gas of hard spheres. Does it depend on temperature? Why?

7.4.E If we assume that the potential energy of interaction of three molecules is given by the sum of three pairwise-additive terms (corresponding to the three distinct pairs that arise in a group of three molecules), the third virial coefficient can be related to the intermolecular potential-energy function by

$$C = -\frac{8\pi^2 N_A^2}{3} \int_0^\infty \int_0^\infty \int_{|\vec{r}_{12}-\vec{r}_{13}|}^{\vec{r}_{12}+\vec{r}_{13}} f_{12} f_{13} f_{23} r_{12} r_{13} r_{23}\, dr_{12}\, dr_{13}\, dr_{23}, \tag{7.75}$$

where $f_{ij} = \exp[-\Gamma(r_{ij})/(k_B T)] - 1$. Calculate a reduced third virial coefficient and plot it as a function of the reduced temperature T^*. Develop your own personalized correlation for $C^*(T^*)$ by fitting the results of your calculations to a simple polynomial expression.

7.4.F The discussion of second virial coefficients in the text was centered around simple fluids of non-polar, spherical molecules. For engineering work, however, it is important to develop correlations to predict the second virial coefficients of non-spherical molecules.

Several approaches are available. For linear alkanes, for example, it has been proposed [83] that one should estimate second virial coefficients from

$$\frac{B}{v_c} = 0.430 - 0.886\left(\frac{T}{T_c}\right)^{-1} - 0.694\left(\frac{T}{T_c}\right)^{-2} - 0.0375(n-1)\left(\frac{T}{T_c}\right)^{-4.5}, \tag{7.76}$$

where n is the number of carbon atoms in the alkane. Another frequently used approach is to resort to the use of an "acentric factor," usually denoted by ω, which provides a measure of the deviations of the potential-energy function from spherical symmetry. A simple, empirical way of defining ω, due to Pitzer, is given by

$$\omega = -\log_{10}\left(\frac{P^{\text{sat}}}{P_c}\right)_{T/T_c=0.7} - 1, \tag{7.77}$$

where P^{sat} is the saturation pressure evaluated at $T = 0.7T_c$. With this definition, the following correlations can be used to estimate second virial coefficients [100]:

$$\frac{BP_c}{RT_c} = B^{(0)}\left(\frac{T}{T_c}\right) + \omega B^{(1)}\left(\frac{T}{T_c}\right), \tag{7.78}$$

where

$$B^{(0)}\left(\frac{T}{T_c}\right) = 0.1445 - \frac{0.330}{T_r} - \frac{0.1385}{T_r^2} - \frac{0.0121}{T_r^3} - \frac{0.000\,607}{T_r^8}, \tag{7.79}$$

$$B^{(1)}\left(\frac{T}{T_c}\right) = 0.0637 + \frac{0.331}{T_r^2} - \frac{0.423}{T_r^3} - \frac{0.008}{T_r^8}, \tag{7.80}$$

where $T_r = T/T_c$.

(a) Estimate the acentric factors for methane, propane, pentane, octane, and methanol.

(b) Compare the predictions of Eqns. (7.76) and (B.5) (in Appendix B) for methane, propane, pentane, and octane. Are the predictions comparable? Which of these two correlations is more accurate (compare your results with experimental data for alkanes)?

7.4.G Equation (B.5) provides a good representation of the second virial coefficients of non-polar or slightly polar fluids. In order to extend its applicability to polar fluids, it has been proposed [137] that one should add to that equation a third term of the form

$$B^{(2)}\left(\frac{T}{T_c}\right) = \frac{c_1}{T_r^6} - \frac{c_2}{T_r^8}. \tag{7.81}$$

For fluids that do not involve hydrogen bonding, the parameter c_2 can be assumed to be zero. The parameters c_1 and c_2 cannot be generalized for wide classes of fluids. For specific chemical families, however, it is possible to arrive at reasonable correlations that provide c_1 and c_2 as functions of the dipole moment. For ethers and ketones, for example, it is found that

$$c_1 = -2.14 \times 10^{-4}\mu_r^d - 4.308 \times 10^{-21}(\mu_r^d)^8, \tag{7.82}$$

where the reduced dipole moment is defined by

$$\mu_r^d = 0.9869\frac{10^5(\mu^d)^2 P_c}{T_c^2}. \tag{7.83}$$

(a) Propose a correlation for c_1 as a function of μ_r^d for ketones.
(b) What parameters c_1 and c_2 should be used for water?

7.4.H Consider the van der Waals equation of state, given by

$$P = \frac{RT}{v-b} - \frac{a}{v^2}.$$

(a) Show that the parameters a and b of the van der Waals equation are related to the critical constants by

$$a = \frac{27R^2T_c^2}{64P_c}, \tag{7.84}$$

$$b = \frac{RT_c}{8P_c} = \frac{v_c}{3}. \tag{7.85}$$

(b) Show that the parameters a and b appearing in van der Waals' equation can be related to the second virial coefficient by

$$B = b - \frac{a}{RT}. \tag{7.86}$$

(*Hint:* you may use the fact that, for $x \ll 1$, $1/(1-x) \approx 1 + x$)

(c) For a mixture, the parameters a and b depend on composition. This dependence is determined by the so-called mixing rules. For a van der Waals fluid, it is often assumed that

$$a = \sum_i^r \sum_j^r x_i x_j a_{ij} \tag{7.87}$$

and

$$b = \sum_i^r x_i b_i, \tag{7.88}$$

where r is the number of components in the mixture and $a_{ij} = \sqrt{a_i a_j}$. Please provide a clear and concise justification for the functional form of Eqns. (7.87) and (7.88).

(d) The following correlation is available for the second virial coefficient of simple fluids

$$\frac{B}{v_c} = 0.430 - 0.886\left(\frac{T}{T_c}\right)^{-1} - 0.694\left(\frac{T}{T_c}\right)^{-2}. \tag{7.89}$$

Using Eqns. (7.89) and (7.86), derive an expression for a in terms of pure-component critical properties. Under which conditions is your result consistent with Eqn. (7.84)?

7.4.I A storage tank with an internal volume of 1000 l is designed to withstand a total pressure of 100 bar. This tank is used to store 125 kg of pure oxygen at ambient temperature.

(a) On a cold day, the temperature of the atmosphere can be as low as $-10\,^\circ$C. On a warm day, it can be as high as $40\,^\circ$C. What is the maximum amount of excess oxygen (above 125 kg) that you could safely store in that tank?

(b) The supplier of oxygen to your plant has recently had some problems with purity, and is delivering oxygen that can contain as much as 10 mol% nitrogen. What is the maximum amount of oxygen that you can safely store in your tank?

(c) In order to be able to increase the oxygen-storage capacity to 132 kg, it has been suggested that you dilute pure oxygen with a small amount of a second component. You have two additional gases at your disposal: water and carbon dioxide. Which of these two gases should you use and why? How much of the second gas should you add to your mixture in order to be able to increase your storage capacity to 132 kg?

7.4.J The following general correlation has been proposed [139, 138] for the second virial coefficient of a wide variety of fluids:

$$\frac{BP}{RT_c} = B^{(0)} + \omega B^{(1)} + aB^{(3)} + bB^{(4)}, \tag{7.90}$$

where

$$B^{(0)} = 0.1445 - \frac{0.33}{T_r} - \frac{0.1385}{T_r^2} - \frac{0.0121}{T_r^3} - \frac{0.000\,607}{T_r^8},$$

$$B^{(1)} = 0.0637 + \frac{0.331}{T_r^2} - \frac{0.423}{T_r^3} - \frac{0.008}{T_r^8},$$

$$B^{(2)} = \frac{1}{T_r^6},$$

$$B^{(3)} = -\frac{1}{T_r^8}.$$

The values of a and b depend on chemical species and are provided in Table 7.4.

(a) Calculate the second virial coefficient of water at $T = 160\,^\circ\mathrm{C}$ and $P = 15$ bar using the Tsonopoulos *et al.* correlation, Eqn. (7.90). Recalculate this quantity by ignoring the last two terms of the correlation. How important are those two terms?

Table 7.4 Species-dependent parameters for the Tsonopoulos *et al.* [138] second-virial-coefficient correlation. The reduced dipole moment is given by $\mu_r^d = 10^5(\mu^d)^2 P_c/T_c{}^2$, with μ^d in debye, P_c in atm, and T_c in K.

Species	a	b
Simple fluids	0	0
Ketones, aldehydes, alkyl nitriles, ethers, carboxylic acid esters	$-2.14 \times 10^{-4}\mu_r^d - 4.308 \times 10^{-21}(\mu_r^d)^{-8}$	0
Alkyl halides, mercaptans, sulfides, disulfides	$-2.188 \times 10^{-4}(\mu_r^d)^4 - 7.831 \times 10^{-21}(\mu_r^d)^{-8}$	0
Alkanols (except methanol)	0.0878	$0.009\,08 + 0.000\,695\,7\mu_r^d$
Methanol	0.0878	0.0525
Water	−0.0109	0

(b) Calculate the second virial coefficient of acetone at $T = 120\,^{\circ}\mathrm{C}$ and $P = 10\,\mathrm{bar}$ using the Tsonopoulos *et al.* correlation, Eqn. (7.90).

(c) Repeat part (c) of the preceding exercise using the Tsonopoulos *et al.* correlation, Eqn. (7.90), and using the simpler McGlashan correlation [82] of Eqn. (7.69). Are the results very different?

8 Fugacity and vapor–liquid equilibrium

In Chapter 4 we saw how simple stability criteria derived from the postulates have important consequences for the calculation of vapor–liquid equilibria. More specifically, we discussed how the spinodal curve of a pure fluid defines the boundary between locally stable and unstable states, and we also saw how global stability leads to guidelines for the calculation of binodal, or coexistence, curves.

As the number of components of a mixture increases, the stability criteria become increasingly complex and more difficult to apply, although the principles are the same. Calculations of phase equilibria for multicomponent mixtures therefore become more elaborate. These calculations are commonplace in applications, and it is therefore important to have an efficient machinery to address this problem. This chapter presents that machinery.

Previous chapters dealt mostly with pure-component systems. For these we first introduced the idea of fundamental relations for U or for S (Chapter 2), and then, through Legendre transforms, we introduced fundamental relations in the generalized potentials (Chapter 3). In Chapter 4 we showed how these fundamental relations for a one-component system can be derived from two equations of state, typically a mechanical equation of state, and a thermal equation of state. We also showed in Chapter 6 how to derive fundamental relations from simple molecular models.

However, in Chapter 4 we also showed how many things could be accomplished through the mechanical equation of state alone, such as calculations of vapor–liquid equilibrium and residual properties. For multicomponent vapor–liquid equilibrium, we will find in this chapter that mechanical equations of state are sufficient when we use a quantity called **fugacity** in place of the chemical potential. The fugacity has the added advantage of being easier to work with at low pressures than is the chemical potential. Therefore, in reference books, technical articles, and engineering reports, one rarely sees a fundamental relation for multicomponent mixtures.

We begin by recalling the equations for phase equilibria in bulk multicomponent systems. As a specific example, we also revisit a mixture of ideal gases. Before introducing fugacities, we define **partial molar properties** – an important concept in mixtures. This definition, when combined with residual properties, provides a simple definition for fugacities. Unfortunately, the notation in the literature is very cumbersome, and it is necessary to introduce it all here. We recommend that the reader works hard in this chapter to memorize the definitions which have already been mentioned and the ones that follow: partial molar properties, fugacity, the fugacity coefficient, excess properties, activity, the activity coefficient, the Lewis mixing rule (ideal mixing), and the Poynting correction factor. These definitions are the just the preliminary – necessary! – tools.

Chapters 8 and 9 offer an introduction to multi-component phase equilibria. Readers interested in a more in-depth discussion of the subject are referred to the classic treatise by Prausnitz et al. on the subject [160], on which this presentation is based.

8.1 GENERAL EQUATIONS OF PHASE EQUILIBRIA

Engineering problems often require calculation of the distribution of several components through the various phases that constitute a system. Consider, for example, the separation of alcohol and water by distillation. In order to determine how long a water–alcohol mixture must be boiled to achieve a certain degree of separation, it is necessary to know the composition of the vapor and liquid phases in equilibrium with each other at any given temperature and pressure. For the more general case of separation of a multicomponent mixture into its individual constituents (e.g. the separation of crude oil into asphaltenes, gasoline, jet fuel, and light gases), it is important to calculate the concentration of each component throughout a multiphase system as a function of temperature and pressure.

In this section, we summarize general results found earlier in the book that are particularly important for phase-equilibrium calculations and derivations. These relations are valid for all species, all fundamental relations, all equations of state, all mixing models, etc.

Recall that in Sections 2.7 and 2.8 we showed that two subsystems that can exchange energy, volume, and mass must have equal temperature, pressure, and chemical potential. If we have two phases in equilibrium, e.g. two immiscible liquids, or a liquid phase and a gas phase, they exchange heat, mass, and volume, so we can write

$$\begin{aligned} T' &= T'' = \ldots = T''', \\ P' &= P'' = \ldots = P''', \\ \mu_i' &= \mu_i'' = \ldots = \mu_i''', \quad i = 1, \ldots, r, \end{aligned} \tag{8.1}$$

where the single, double, and triple primes denote various phases in equilibrium. Much of the remainder of this book will be concerned with solving Eqns. (8.1) for different situations, depending on the problem at hand.

Because phases in equilibrium are at the same temperature and pressure, the Gibbs potential is the most natural one to work with for these calculations. Recall that the differential for the Gibbs free energy is

$$dG = -S\,dT + V\,dP + \sum_{i=1}^{r} \mu_i\,dN_i. \tag{3.51$'$}$$

If a closed system is held at fixed T and P, then the free energy is minimized.

8.2 MIXTURES OF IDEAL GASES

Ideal gases are made of molecules that rarely interact with one another. In other words, each of the molecules in the volume is not "aware" that there are any other molecules in the box. Hence, if we have two or more species in the same volume, and at the same temperature, then they have the same fundamental relation as if they were pure:

$$F^{\text{ideal}}(T, V, \{N_i\}) = \sum_i F_i^{\text{ideal}}(T, V, N_i), \tag{8.2}$$

where F_i^{ideal} is the Helmholtz potential for pure component i in the ideal state.[1] If we now use the fundamental relation already known for the pure ideal gas, Eqn. (B.2) in Appendix B, we obtain the fundamental Helmholtz relation for a mixture of general ideal gases:

$$f^{\text{ideal}}(T, v, \{x_i\}) = \langle f_0 \rangle - (T - T_0)\langle s_0 \rangle - RT \log\left(\frac{v}{v_0}\right) + \int_{T_0}^{T} \left(\frac{T' - T}{T'}\right) \langle c_v^{\text{ideal}}(T') \rangle dT' + RT \sum_i x_i \log x_i, \tag{8.3}$$

where the angular brackets here indicate a molar average: $\langle m \rangle := \sum_i x_i m_i$, $m_i \equiv s_{0,i}, f_{0,i}, c_{v,i}^{\text{ideal}}$. Remember that the subscript zero refers to the thermodynamic properties at the reference temperature T_0 and specific volume v_0.

Since this is a fundamental relation, we can find all of the equations of state. Of particular interest is to see how the thermodynamic properties at fixed T and P change upon mixing. Using the usual manipulations to obtain equations of state, we find from Eqn. (8.3)

$$\begin{aligned}
u^{\text{ideal}}(T, P, \{x_i\}) &= \sum_i x_i u_i^{\text{pure, ideal}}(T, P), \\
v^{\text{ideal}}(T, P, \{x_i\}) &= \sum_i x_i v_i^{\text{pure, ideal}}(T, P), \\
h^{\text{ideal}}(T, P, \{x_i\}) &= \sum_i x_i h_i^{\text{pure, ideal}}(T, P), \\
s^{\text{ideal}}(T, P, \{x_i\}) &= \sum_i x_i s_i^{\text{pure, ideal}}(T, P) - R \sum_i x_i \log x_i, \\
f^{\text{ideal}}(T, P, \{x_i\}) &= \sum_i x_i f_i^{\text{pure, ideal}}(T, P) + RT \sum_i x_i \log x_i, \\
g^{\text{ideal}}(T, P, \{x_i\}) &= \sum_i x_i g_i^{\text{pure, ideal}}(T, P) + RT \sum_i x_i \log x_i, \\
\mu_j^{\text{ideal}}(T, P, \{x_i\}) &= \mu_j^{\text{pure, ideal}}(T, P) + RT \log x_j.
\end{aligned} \tag{8.4}$$

The first three results are expected, since the molecules do not interact energetically. Physically, these results mean that, if we mix two different ideal gases at the same temperature and pressure, the volume will not change; hence, there is no work exchanged with the environment. Also, the internal energy does not change, so there is no heat exchanged with the environment.

The result for entropy is interesting because of the sign of the mixing term. Since all mole fractions are less than unity, and the logarithm of a fraction is negative, the entropy of mixing is positive; ideal gases mix spontaneously. The results for the free energies follow because of their definitions using entropy. These results are also the basis for the "Lewis mixing" approximation that we encounter below.

[1] Equation (8.2) can also be found from statistical mechanics by taking the logarithm of each side of Eqn. (6.40), and then multiplying by $k_B T$.

8.3 MIXTURES: PARTIAL MOLAR PROPERTIES

8.3.1 Definition of a partial molar property

To introduce the concept of a partial molar property, it is useful to consider the way in which the molar volume of a system changes when two liquids are mixed at constant temperature and pressure. If these two liquids were identical, the molar volume of the resulting binary mixture would be given by

$$v(T,P,x_1) = x_1 v_1(T,P) + x_2 v_2(T,P), \quad \text{identical liquids.} \tag{8.5}$$

Here readers are reminded that $v_i(T,P)$ in Eqn. (8.5) represents the molar volume of pure component i at given T and P. In other words, Eqn. (8.5) states that components 1 and 2 would mix **isometrically**. If the two components were not identical but similar (e.g. pentane and hexane), it would be reasonable to expect Eqn. (8.5) to provide a realistic description of the mixing process. Equation (8.5) is known as **Amagat's "law"**; it states that the volume change associated with mixing is zero. Amagat's law therefore corresponds to some ideal, limiting case and isn't really a law at all. Note that Eqn. (8.5) looks similar to the second line of Eqn. (8.4), but now the volumes are not necessarily those of an ideal gas; in fact, they can be highly non-ideal or liquid-like in Eqn. (8.5). For lack of a better term, mixtures that follow Eqn. (8.5) are typically called "ideal mixtures." Note, however, that an ideal mixture is not necessarily the same as a mixture of ideal gases. In order to avoid confusion, and for reasons that will become apparent later in this chapter, we prefer to use the term **Lewis mixtures** instead of ideal mixtures. Lewis mixtures have a zero volume change of mixing.

Real systems seldom mix isometrically, however. To calculate the correct volume of the mixture, it is useful to introduce the concept of **partial molar volume** for each component of the mixture. To provide an intuitive idea of the meaning of partial molar properties, it is useful to conduct the thought experiment illustrated in Figure 8.1. Consider a binary mixture

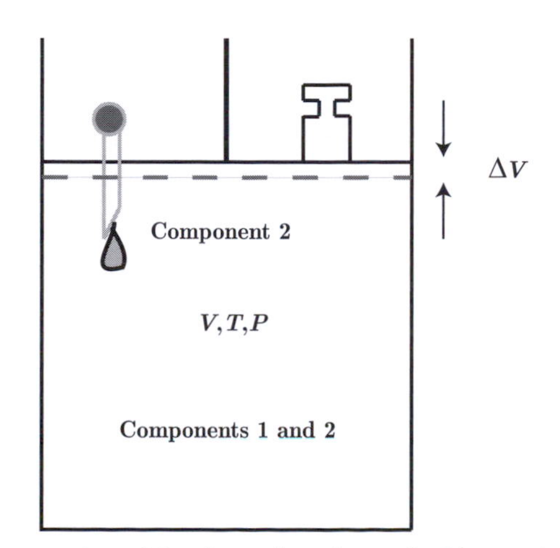

Figure 8.1 A schematic representation of the change in volume of a binary mixture when a droplet of one of the components is added to the mixture at constant temperature and constant pressure.

of components 1 and 2. When a drop of component 2 is added to a mixture, if the system mixed ideally, the total volume would change by an amount $\Delta V^{\text{ideal}} = v_2\,\Delta N_2$, where ΔN_2 is the number of moles of component 2 in the drop. For a real mixture that will no longer be true; the volume change of the mixture will be $\Delta V = v^{\text{eff}}\,\Delta N_2$, where v^{eff} represents the *effective* volume that ΔN_2 moles occupy in the mixture. If an infinitesimal droplet of component 2 is added to the mixture, and we consider the accompanying infinitesimal volume change ΔV, we can write

$$\bar{v}_2 := \frac{\Delta V}{\Delta N_2}, \tag{8.6}$$

where the ratio of these two quantities provides a measure of the effective volume that a mole of component 2 occupies in the real mixture. We use a bar to distinguish it from the molar volume of pure component 2. This thought experiment was carried out at constant temperature, constant pressure, and constant number of moles of component 1. Therefore, a precise definition of $\bar{v}_2$ would be

$$\bar{v}_2 := \left(\frac{\partial V}{\partial N_2}\right)_{T,P,N_1}. \tag{8.7}$$

The volume $\bar{v}_2$ defined by Eqn. (8.7) is called the **partial molar volume** of component 2 in the mixture; it represents the effective molar volume which molecules of component 2 occupy when exposed to the mixture environment. The partial molar volume depends on pressure, temperature, and composition.

EXAMPLE 8.3.1 As we shall discuss later in the text, the addition of a solute to a solvent lowers the freezing point of the solvent. This effect provides the basis for the formulation of antifreeze solutions. Methanol, when added to water, can be used to prepare an antifreeze mixture. In this example we are asked to prepare a liter of an antifreeze solution consisting of 20 mol% methanol in water. Specifically, we must calculate the volumes of pure methanol and of pure water that must be mixed at 50 °C to form 1 l of antifreeze.

The partial and pure-component molar volumes of methanol and water at 50 °C (here (1) denotes methanol and (2) denotes water) are taken from [145]:

$$\bar{v}_1 = 39.34\,\text{cm}^3/\text{mol}, \qquad v_1 = 42.03\,\text{cm}^3/\text{mol},$$
$$\bar{v}_2 = 18.10\,\text{cm}^3/\text{mol}, \qquad v_2 = 18.23\,\text{cm}^3/\text{mol}.$$

Solution. First, we calculate the molar volume of the mixture. As stated in words following Eqn. (8.7), and proven below in Eqn. (8.14), $v = \sum_i x_i \bar{v}_i$:

$$v = x_1\bar{v}_1 + x_2\bar{v}_2 = (0.2)(39.34\,\text{cm}^3/\text{mol}) + (0.8)(18.10\,\text{cm}^3/\text{mol})$$
$$= 22.35\,\text{cm}^3/\text{mol}.$$

Then we use that molar volume to calculate the total number of moles required; the total volume V is 1000 cm^3:

$$\text{Total number of moles required} = n = \frac{1000}{v} = \frac{1000\,\text{cm}^3}{22.35\,\text{cm}^3/\text{mol}} = 44.74\ \text{mol}.$$

Of this, 20% is methanol, 80% is water:

$$n_1 = (0.2)(44.74 \text{ mol}) = 8.95 \text{ mol},$$
$$n_2 = (0.8)(44.74 \text{ mol}) = 35.79 \text{ mol}.$$

The volumes V_i^+ of each component to be added are

$$V_1^+ = (8.95 \text{ mol})(42.03 \text{ cm}^3/\text{mol}) = 376 \text{ cm}^3,$$
$$V_2^+ = (35.79 \text{ mol})(18.23 \text{ cm}^3/\text{mol}) = 652 \text{ cm}^3.$$

This example shows that, due to the volume changes that occur upon mixing, in order to prepare 1 l of mixture you need to mix 1.028 l of the pure components taken separately. □

More generally, an arbitrary **partial molar property** for component i, $\bar{m}_i$, is defined by

$$\bar{m}_i := \left(\frac{\partial Nm}{\partial N_i}\right)_{T,P,N_{j\neq i}} = \left(\frac{\partial M}{\partial N_i}\right)_{T,P,N_{j\neq i}}, \tag{8.8}$$

where N is the total number of moles in the mixture, and M is V, U, S, H, F, or G. It is important to emphasize that partial molar properties are defined as derivatives at constant pressure, temperature, and number of moles of all components other than i. The partial molar entropy of component i in a mixture, for example, is given by

$$\bar{s}_i := \left(\frac{\partial Ns}{\partial N_i}\right)_{T,P,N_{j\neq i}}. \tag{8.9}$$

It is also important to note that the partial molar Gibbs free energy of component i in a mixture is nothing other than its chemical potential:

$$\bar{g}_i := \left(\frac{\partial G}{\partial N_i}\right)_{T,P,N_{j\neq i}} = \left(\frac{\partial Ng}{\partial N_i}\right)_{T,P,N_{j\neq i}} = \mu_i. \tag{8.10}$$

For a single-component, or pure, system, a partial molar quantity is just the specific quantity (e.g. $\bar{v}_i = v$ for a pure system).

8.3.2 General properties of partial molar quantities

To establish some of the useful properties of partial molar quantities, we begin by writing a differential of an arbitrary partial molar property $\bar{m}_i$. From $Nm = M(T, P, \{N_i\})$ we write

$$\begin{aligned} d(Nm) &= N\left(\frac{\partial m}{\partial T}\right)_{P,\{N_j\}} dT + N\left(\frac{\partial m}{\partial P}\right)_{T,\{N_j\}} dP + \sum_{j=1}^{r}\left(\frac{\partial Nm}{\partial N_j}\right)_{T,P,\{N_{i\neq j}\}} dN_j \\ &= N\left(\frac{\partial m}{\partial T}\right)_{P,\{N_j\}} dT + N\left(\frac{\partial m}{\partial P}\right)_{T,\{N_j\}} dP + \sum_{j=1}^{r}\bar{m}_j\, dN_j. \end{aligned} \tag{8.11}$$

The left-hand side of Eqn. (8.11) can be written as $d(Nm) = m\,dN + N\,dm$. The differential of N_j can be written as $dN_j = d(x_jN) = x_j\,dN + N\,dx_j$. After substitution into Eqn. (8.11) and rearrangement, we arrive at

$$0 = \left[dm - \left(\frac{\partial m}{\partial T}\right)_{P,N_j} dT - \left(\frac{\partial m}{\partial P}\right)_{T,N_j} dP - \sum_{j=1}^{r} \bar{m}_j\,dx_j\right] N + \left[m - \sum_{j=1}^{q} \bar{m}_j x_j\right] dN. \quad (8.12)$$

Equation (8.12) is a general thermodynamic expression; it is valid regardless of the particular size N of the system, and is true for arbitrary values of the infinitesimal change dN. These observations and Eqn. (8.12) can hold only if the bits inside the square brackets are each zero:

$$dm = \left(\frac{\partial m}{\partial T}\right)_{P,\{N_j\}} dT + \left(\frac{\partial m}{\partial P}\right)_{T,\{N_j\}} dP + \sum_{j=1}^{r} \bar{m}_j\,dx_j \quad (8.13)$$

and

$$m = \sum_{j=1}^{r} x_j \bar{m}_j. \quad (8.14)$$

Equation (8.14) implies that the mixture value of an arbitrary property m can be expressed as a sum of the corresponding partial molar properties $\bar{m}_i$ weighted over the mole fractions. In a sense, it is a rigorous generalization of Eqn. (8.5).

For example, using Eqns. (8.10) and (8.14), the molar Gibbs free energy of a mixture can be written as

$$g = \sum_{j=1}^{r} \mu_j x_j. \quad (8.15)$$

We can also simplify Eqn. (8.13). On recognizing that $dm = \sum_{j=1}^{q}(\bar{m}_j\,dx_j + x_j\,d\bar{m}_j)$, Eqn. (8.13) can be rewritten as

$$0 = \left(\frac{\partial m}{\partial T}\right)_{P,\{N_j\}} dT + \left(\frac{\partial m}{\partial P}\right)_{T,\{N_j\}} dP - \sum_{j=1}^{r} x_j\,d\bar{m}_j, \quad (8.16)$$

which is another version of the familiar Gibbs–Duhem equation. Note, however, that the expression given here is more general than that derived earlier (i.e. Eqn. (3.22)) because it is valid for an arbitrary property m. For the molar Gibbs free energy we have

$$\left(\frac{\partial g}{\partial T}\right)_{P,N_j} dT + \left(\frac{\partial g}{\partial P}\right)_{T,N_j} dP - \sum_{j=1}^{r} x_j\,d\mu_j = 0, \quad (8.17)$$

which, at constant temperature and pressure, reduces to

$$\sum_{j=1}^{r} x_j\,d\mu_j = 0, \quad \text{constant } T \text{ and } P. \quad (8.18)$$

From Eqn. (8.18) we see that the chemical potentials of different species in a mixture are not independent, but must change in a concerted manner. In general, Eqn. (8.16) indicates that all partial molar properties in a mixture are connected by Gibbs–Duhem-type equations.

EXAMPLE 8.3.2 From the measurement of a thermodynamic property for a binary mixture $m(T, P, x_1)$ at fixed T and P, but varying composition, find a simple graphical means to estimate the partial molar properties m_1 and m_2.

Solution. Since the experiments are carried out at constant temperature and pressure, it is convenient to begin with Eqn. (8.13), which becomes

$$\begin{aligned} dm &= \sum_{i=1}^{2} \bar{m}_i \, dx_i, \quad \text{fixed } T, P, \\ &= \bar{m}_1 \, dx_1 + \bar{m}_2 \, dx_2, \quad \text{fixed } T, P, \\ &= (\bar{m}_1 - \bar{m}_2) dx_1, \quad \text{fixed } T, P, \end{aligned} \tag{8.19}$$

where we used the fact that $x_1 + x_2 = 1$ to obtain $dx_2 = -dx_1$ in the last line. We use Eqn. (8.14) to eliminate $\bar{m}_2$ from this result:

$$dm = \left(\bar{m}_1 - \frac{m - x_1 \bar{m}_1}{1 - x_1} \right) dx_1, \quad \text{fixed } T, P. \tag{8.20}$$

A little algebra yields, then, what we need:

$$\bar{m}_1 = m + (1 - x_1) \left(\frac{\partial m}{\partial x_1} \right)_{T,P}. \tag{8.21}$$

This equation has a straightforward graphical interpretation on a plot of m vs. x_1 at fixed temperature and pressure, as shown in Figure 8.2.

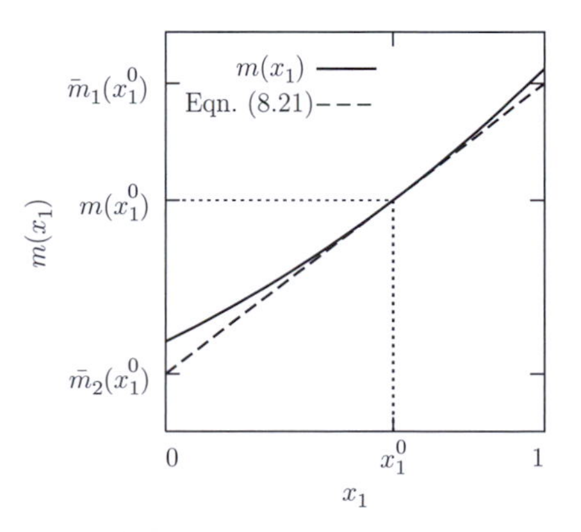

Figure 8.2 A graphical representation of Eqn. (8.21). One first plots the thermodynamic data $m(x_1)$ at fixed temperature and pressure as a function of mole fraction x_1 (solid line). Then, at any given composition x_1^0, one draws a tangent line to the data (dashed line). The x_1 intercept of the dashed line is $\bar{m}_2$ at this composition, and $\bar{m}_1$ is where the straight tangent intercepts the y axis at $x_1 = 1$. □

EXAMPLE 8.3.3 What are the partial molar properties of an ideal-gas mixture?

Solution. One quantity suffices to show the method for finding all the properties. If we pick, say entropy, we begin with its mixing property given in Eqns. (8.4), and multiply by the total number of moles N to make it extensive, to obtain

$$S^{\text{ideal}}(T, P, \{x_i\}) = \sum_i N_i s_i^{\text{ideal}}(T, P) - R \sum_i N_i \log x_i. \tag{8.22}$$

Recall that the left-hand side can be written as the mole-fraction-weighted sum of the partial molar quantities, using Eqn. (8.14),

$$\sum_i N_i \bar{s}_i^{\text{ideal}}(T, P, \{x_i\}) = \sum_i N_i \left[s_i^{\text{ideal}}(T, P) - R \log x_i \right], \tag{8.23}$$

which gives us the expression for the partial molar entropy:

$$\begin{aligned} \bar{s}_i^{\text{ideal}}(T, P, \{x_i\}) &= s_i^{\text{ideal}}(T, P) - R \log x_i \\ &= s_{0,i} + \int_{T_0}^{T} \frac{c_{v,i}^{\text{ideal}}(T')}{T'} dT' + R \log \left(\frac{TP_0}{x_i P T_0} \right). \end{aligned} \tag{8.24}$$

To obtain the second line, we used the expression for the entropy of a generalized, pure ideal gas, Eqn. (3.17). Following the same procedure for the other quantities yields the results

$$\begin{aligned} \bar{v}_i^{\text{ideal}}(T, P, \{x_i\}) &= v_i^{\text{ideal}}(T, P) = \frac{RT}{P}, \\ \bar{u}_i^{\text{ideal}}(T, P, \{x_i\}) &= u_i^{\text{ideal}}(T, P) \\ &= u_{0,i} + \int_{T_0}^{T} c_{v,i}^{\text{ideal}}(T') dT', \\ \bar{h}_i^{\text{ideal}}(T, P, \{x_i\}) &= h_i^{\text{ideal}}(T, P) = u_i^{\text{ideal}}(T, P) + RT, \\ \bar{f}_i^{\text{ideal}}(T, P, \{x_i\}) &= u_i^{\text{ideal}}(T, P) - T\bar{s}_i^{\text{ideal}}(T, P) \\ &= f_{0,i} - s_{0,i}(T - T_0) + \int_{T_0}^{T} \left(\frac{T' - T}{T'} \right) c_{v,i}^{\text{ideal}}(T') dT' \\ &\quad + RT \log \left(\frac{x_i P T_0}{P_0 T} \right), \\ \bar{g}_i^{\text{ideal}}(T, P, \{x_i\}) &= \bar{f}_i^{\text{ideal}}(T, P) + RT. \end{aligned} \tag{8.25}$$

□

EXAMPLE 8.3.4 Find the partial molar volume of an individual component of a gas mixture described by a virial equation of state.

Solution. In Section 7.4 we considered the virial equation of state from a molecular point of view. For the volume-explicit expansion, we discovered that the correct equation of state and mixing rules are

$$z = 1 + \frac{BP}{RT} + \cdots \tag{8.26}$$

and

$$B(T, \{N_i\}) = \sum_{ij} x_i x_j B_{ij}(T). \tag{7.43$'$}$$

By combining these expressions, and using the definition for the compressibility factor $z := PV/(NRT)$, we obtain

$$V = \frac{NRT}{P} + \frac{1}{N}\sum_{ij} N_i N_j B_{ij}(T). \tag{8.27}$$

Following the definition of a partial molar property, we now take the derivative of each side with respect to N_k to obtain

$$\begin{aligned}\bar{v}_k &= \left(\frac{\partial V}{\partial N_k}\right)_{T,P,N_{i\neq k}} && (8.28)\\ &= \frac{RT}{P} + \frac{1}{N}\sum_i N_i B_{ik}(T) + \frac{1}{N}\sum_j N_j B_{kj}(T) - \frac{1}{N^2}\sum_{ij} N_i N_j B_{ik}(T) \\ &= \frac{RT}{P} + 2\sum_i x_i B_{ik}(T) - B. && (8.29)\end{aligned}$$

To obtain the last line, we used the fact that $B_{ij} = B_{ji}$. □

8.3.3 Residual partial molar quantities

The result from Example 8.3.4 shows that the partial molar volume of real gases deviates from that of ideal gases. As it turns out, this deviation from ideality is extremely useful in defining fugacity and performing phase-equilibrium calculations. Hence, we define the **residual partial molar properties**

$$\bar{m}_i^{\mathrm{R}}(T, P, \{x_i\}) := \bar{m}_i(T, P, \{x_i\}) - \bar{m}_i^{\mathrm{ideal}}(T, P, \{x_i\}), \tag{8.30}$$

where $m \equiv v, s, h, u, g, f$ as before. Note that we could equally well have defined this quantity as the "partial molar residual"

$$\bar{m}_i^{\mathrm{R}}(T, P, \{x_i\}) = \left(\frac{\partial M^{\mathrm{R}}}{\partial N_i}\right)_{T,P,\{N_{j\neq i}\}}, \tag{8.31}$$

which is completely equivalent.

From Example 8.3.4 we see that the residual partial molar volume predicted by the virial equation of state is just $\bar{v}_k^{\mathrm{R}} = 2\sum_i x_i B_{ik}(T) - B$.

EXAMPLE 8.3.5 Given a volume-explicit equation of state $z(T, P, \{N_i\})$, find the residual partial molar Gibbs free energy.

Solution. It is easier to use the alternative definition for the residual partial molar Gibbs, Eqn. (8.31), since we already have an expression for the residual Gibbs potential, Eqn. (4.67). We start from the alternative definition

$$\begin{aligned}\frac{\bar{g}_i^{\mathrm{R}}}{RT} &= \left(\frac{\partial}{\partial N_i}\frac{G^R}{RT}\right)_{T,P,\{N_{j\neq i}\}} \\ &= \frac{\partial}{\partial N_i} N \int_0^P \left[z(T, P', \{N_j\}) - 1\right]\frac{dP'}{P'} \\ &= \int_0^P \left[\bar{z}_i(T, P', \{N_j\}) - 1\right]\frac{dP'}{P'},\end{aligned} \tag{8.32}$$

where we have introduced the quantity

$$\bar{z}_i(T, P, \{N_i\}) := \left(\frac{\partial}{\partial N_i} Nz\right)_{T,P,\{N_{j\neq i}\}}. \tag{8.33}$$

Using this expression, the residual partial molar Gibbs free energy can be found for any volume-explicit equation of state when the appropriate mixing rules for the parameters are known. □

The results from the previous example are useful only for a volume-explicit equation of state for mixtures. More common are pressure-explicit equations of state. The analogous result for $\bar{g}_i^{\mathrm{R}}$ given $z(T, V, \{N_i\})$ can be found from Eqn. (8.32) through a change of variable of integration from P' to V'. The result is

$$\frac{\bar{g}_i^{\mathrm{R}}}{RT} = \int_V^\infty \left[\left(\frac{\partial}{\partial N_i} Nz\right)_{T,V,\{N_{j\neq i}\}} - 1\right]\frac{dV'}{V'} - \log z. \tag{8.34}$$

Pressure-explicit equations of state for mixtures are sometimes created from those for pure species, by having mixing rules for the parameters. For example, the van der Waals equation of state for mixtures arises on making the parameters a and b depend upon some weighted sum of the pure-component values, such as given in Eqn. (7.87).

8.4 FUGACITY

We are finally in a position to define the fugacity; there are three main reasons for learning it. First, it is ubiquitous in the technical literature. Secondly, it obviates the need for fundamental relations to predict multicomponent vapor–liquid equilibrium compositions. Thirdly, it avoids difficulties at low pressures. What are these difficulties? Recall that all real gases behave ideally at low densities, and that the chemical potential of an ideal gas can be written in the form (see Eqn. (3.112))

$$\mu_i^{\text{ideal}}(T, P, \{x_j\}) = \mu_i^0(T) + RT\log(x_i P). \tag{8.35}$$

Hence, the chemical potential for gases goes to $-\infty$ as the pressure goes to zero.

8.4.1 Definition of fugacity

Because of the negative infinity in chemical potential at low pressures, it is useful to subtract off the ideal part from the real part of the chemical potential, rather like a residual property. For example, each component of a liquid phase and vapor phase in equilibrium with each other at temperature T and pressure P must have the same chemical potential in both phases:

$$\mu_i^{\text{liquid}}(T, P, \{x_j^{\text{liquid}}\}) = \mu_i^{\text{vapor}}(T, P, \{x_j^{\text{vapor}}\}). \tag{8.36}$$

We would like to subtract off the ideal part – the part that goes to negative infinity – from each side. However, the liquid phase and the vapor phase are likely to have different compositions, so we must be careful. For simplicity, we subtract off the chemical potential of the pure species at the same T and P from each side of this equation:

$$\mu_i^{\text{liquid}}[\{x_j^{\text{liquid}}\}] - \mu_i^{\text{ideal,pure}} = \mu_i^{\text{vapor}}[\{x_j^{\text{vapor}}\}] - \mu_i^{\text{ideal,pure}}. \tag{8.37}$$

All the terms are at the same temperature and pressure, which we stop showing explicitly here. Now we note that we can write the chemical potential for an ideal-gas mixture as $\mu_i^{\text{ideal}}(\{x_i\}) = \mu_i^{\text{ideal,pure}} + RT\log(x_i)$, Eqn. (8.4). Hence, the above becomes

$$\begin{aligned}\mu_i^{\text{liquid}}(\{x_j^{\text{liquid}}\}) - \mu_i^{\text{ideal}}(\{x_j^{\text{liquid}}\}) + RT\log(x_i^{\text{liquid}})\\ = \mu_i^{\text{vapor}}(\{x_j^{\text{vapor}}\}) - \mu_i^{\text{ideal}}(\{x_j^{\text{vapor}}\}) + RT\log(x_i^{\text{vapor}}).\end{aligned} \tag{8.38}$$

We can utilize the fact that the chemical potential is just the partial molar Gibbs free energy, and the definition for a residual property, to write

$$[\bar{g}_i^{\text{R}}(\{x_i^{\text{liquid}}\})]^{\text{liquid}} + RT\log(x_i^{\text{liquid}}) = [\bar{g}_i^{\text{R}}(\{x_j^{\text{vapor}}\})]^{\text{vapor}} + RT\log(x_i^{\text{vapor}}). \tag{8.39}$$

Or, if we divide each side by RT to make it dimensionless and take the exponential, we obtain

$$x_i^{\text{liquid}}\exp\left(\frac{[\bar{g}_i^{\text{R}}]^{\text{liquid}}}{RT}\right) = x_i^{\text{vapor}}\exp\left(\frac{[\bar{g}_i^{\text{R}}]^{\text{vapor}}}{RT}\right). \tag{8.40}$$

This is our desired result, since we have eliminated the ideal contribution to the chemical potential and found a quantity that must be equal in different phases at equilibrium. Historically, there is a slight modification to define fugacity, which we also adopt; both sides of Eqn. (8.40) are multiplied by pressure, thereby giving the following definition for the fugacity f_i of component i in a mixture:

$$f_i := x_i P \exp\left(\frac{\bar{g}_i^{\mathrm{R}}}{RT}\right). \tag{8.41}$$

The definition of fugacity given by Eqn. (8.41) is convenient in that, for ideal gases, it reduces to the partial pressure.

Solving for the compositions of phases in equilibrium can be conveniently accomplished by equating fugacities instead of chemical potentials. As we have just shown, equating fugacities is the same as equating chemical potentials, but the former requires only a PvT equation of state for the mixture.

8.4.2 Properties of fugacity

We can summarize some important properties of fugacity, two of which will require proof.

1. Equal fugacities $\Leftrightarrow$ equal chemical potentials.
2. At low densities, at which real gases behave ideally (and the residual molar Gibbs free energy vanishes), the fugacity becomes the so-called **partial pressure** $x_i P$.
3. Equations of state are sometimes written for the **fugacity coefficient** ϕ_i, which is defined as

$$\phi_i := \frac{f_i}{x_i P}. \tag{8.42}$$

4. The pressure dependence of fugacity is

$$f_i[T, P_2] = f_i[T, P_1] \exp\left[\int_{P_1}^{P_2} \frac{\bar{v}_i[T, P']}{RT} dP'\right]. \tag{8.43}$$

   This form is particularly useful for measuring the fugacity of pure systems.

5. The temperature dependence of fugacity is

$$f_i[T_2, P] = f_i[T_1, P] \exp\left[-\int_{T_1}^{T_2} \frac{\bar{h}_i^{\mathrm{R}}[T', P]}{RT'^2} dT'\right]. \tag{8.44}$$

6. For completeness, we note (but don't prove) that we have complete thermodynamic information about a system if we know the ideal heat capacities and fugacities for all species as functions of temperature, pressure, and composition.

In what follows, we will use the first five properties. We proved number 1 above, and number 2 just follows from the definition of fugacity. Since number 4 is the basis for the Poynting correction factor, we now discuss it in some detail. We begin with the definition of the fugacity, of course, to write

$$\begin{aligned}
\frac{f_i[T,P_2]}{f_i[T,P_1]} &= \frac{P_2}{P_1}\exp\left[\frac{\bar{g}_i^{\mathrm{R}}[T,P_2]-\bar{g}_i^{\mathrm{R}}[T,P_1]}{RT}\right] \\
&= \frac{P_2}{P_1}\exp\left[\frac{1}{RT}\int_{P_1}^{P_2}\left(\frac{\partial \bar{g}_i^{\mathrm{R}}}{\partial P}\right)_{T,\{x_j\}}dP\right] \\
&= \frac{P_2}{P_1}\exp\left[\frac{1}{RT}\int_{P_1}^{P_2}\left\{\left(\frac{\partial \bar{g}_i}{\partial P}\right)_{T,\{x_j\}}-\left(\frac{\partial \bar{g}_i^{\mathrm{ideal}}}{\partial P}\right)_{T,\{x_j\}}\right\}dP\right] \\
&= \frac{P_2}{P_1}\exp\left[\frac{1}{RT}\int_{P_1}^{P_2}\left\{\left(\frac{\partial \mu_i}{\partial P}\right)_{T,\{x_j\}}-\left(\frac{\partial \mu_i^{\mathrm{ideal}}}{\partial P}\right)_{T,\{x_j\}}\right\}dP\right].
\end{aligned} \tag{8.45}$$

We're almost there. The second line uses the fundamental theorem of calculus, Eqn. (A.24) from Appendix A. To obtain the third line, we used the definition of a residual property, Eqn. (4.66). From Eqn. (8.10), we know that the partial molar Gibbs free energy and the chemical potential are equivalent, thereby resulting in the last line. To finish off our derivation, we use the fact that the Gibbs free energy is analytic:

$$\left(\frac{\partial \mu_i}{\partial P}\right)_{T,\{x_j\}} = \left(\frac{\partial^2 G}{\partial P\,\partial N_i}\right)_T = \left(\frac{\partial^2 G}{\partial N_i\,\partial P}\right)_T = \left(\frac{\partial V}{\partial N_i}\right)_{T,P,\{N_{j\neq i}\}} = \bar{v}_i. \tag{8.46}$$

When this result is inserted into Eqn. (8.45), we obtain property number 4. The proof for property 5 is very similar to the one just given.

8.4.3 Estimating the fugacity of a pure vapor or liquid

Having defined fugacity, and its relation to residual properties, we now show how to calculate the fugacity of a fluid numerically. It is precisely in these calculations that the value of the fugacity concept shows up. There are at least three ways to estimate the fugacity of a pure component: (1) using PvT data or a PvT equation of state; (2) for a low pressure gas, we can assume ideality, $f_i^{\mathrm{pure,ideal}} \cong x_i P$; and (3) for liquids, from the saturation pressure and Eqn. (8.43). We examine the third one first.

For a pure, saturated system (i.e. along the binodal curve), the pressure is equal to the vapor pressure at temperature T. When a liquid and a vapor coexist at equilibrium, the chemical potentials of the two phases are equal; from property number 1 of fugacities we have

$$f^{\mathrm{vapor}}[P^{\mathrm{sat}}, v^{\mathrm{vapor}}] = f^{\mathrm{liquid}}[P^{\mathrm{sat}}, v^{\mathrm{liquid}}]. \tag{8.47}$$

The quantities in the square brackets serve to remind us that the fugacity of the vapor is to be evaluated at the vapor pressure and at the saturated vapor molar volume, and that of the liquid is to be evaluated at the saturated liquid molar volume.

Using property number 4, Eqn. (8.43), the fugacity of a pure condensed phase (a liquid or a solid) at temperature T and arbitrary pressure P can therefore be expressed by

$$f_i^{\mathrm{pure}}[P] = f_i^{\mathrm{pure,\ vapor}}[P^{\mathrm{sat}}]\exp\left[\int_{P^{\mathrm{sat}}}^{P}\frac{v[T,P']}{RT}\,dP'\right], \tag{8.48}$$

where $v[T,P']$ is now the molar volume of a condensed phase at pressure P'.

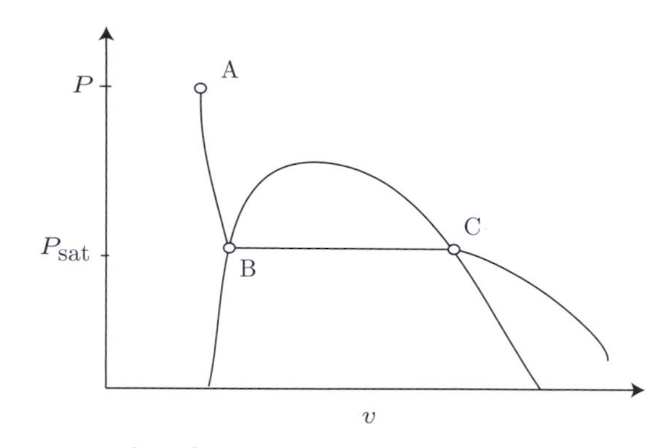

Figure 8.3 A schematic representation of a sub-critical isotherm and the coexistence curve of a pure fluid.

The path outlined above to calculate the fugacity of a condensed fluid is illustrated in Figure 8.3, which provides a schematic representation of a sub-critical isotherm and the coexistence curve of a simple fluid in the $P\!-\!v$ plane. To evaluate the fugacity of the fluid at point A, we follow the isotherm up to point B, and calculate the corresponding change in the liquid-phase fugacity according to

$$f_i^{\text{pure}}[P] = f_i^{\text{pure, liquid}}[P^{\text{sat}}]\exp\left[\int_{P^{\text{sat}}}^{P} \frac{v^{\text{liquid}}[T,P']}{RT}\,dP'\right]. \tag{8.49}$$

Since, according to Eqn. (8.47), the fugacity of a pure fluid at point B is the same as that at point C, we can replace $f_i^{\text{liquid}}[P^{\text{sat}}]$ in Eqn. (8.49) with $f_i^{\text{vapor}}[P^{\text{sat}}]$ to arrive at Eqn. (8.48).

Equation (8.48) is of little use unless we can determine f_i^{vapor} at P^{sat}. We recall that, at the binodal (or coexistence) curve, the fluid can behave rather non-ideally. However, for the moment, it suffices to point out that the vapor pressure of a solid or a liquid is usually relatively small; at such low pressures, within engineering approximation, most vapors are not too far away from ideal-gas behavior. The assumption $f^{\text{vapor}}[P^{\text{sat}}] \approx P^{\text{sat}}$ is therefore justified, and Eqn. (8.48) can be written as

$$f[T,P] \approx P^{\text{sat}}(T) \times \underbrace{\exp\left[\int_{P^{\text{sat}}}^{P} \frac{v[T,P']}{RT}\,dP'\right]}_{\text{Poynting correction factor}}, \quad \text{ideal vapor phase.} \tag{8.50}$$

The physical meaning of a fugacity is now apparent: it is a vapor pressure corrected for the effect of pressure on the condensed phase. By introducing fugacity, we are able to replace an abstract chemical potential with a more intuitive quantity related to the vapor pressure [79].

The exponential factor appearing in Eqn. (8.50) is referred to as the **"Poynting" correction**. In many cases, the molar volume of a condensed phase appearing in the integrand of Eqn. (8.50) does not change appreciably with pressure; if the condensed phase is assumed to be incompressible, then Eqn. (8.50) can be further approximated to give

$$f[P] \approx P^{\text{sat}} \exp\left[\frac{v^{\text{liquid}}(P - P^{\text{sat}})}{RT}\right]. \tag{8.51}$$

Everything on the right-hand side is measurable, so we now have a way of estimating the fugacity of a pure substance. Our only assumptions are incompressibility of the liquid phase and ideal-gas behavior of the vapor phase when the pressure is equal to P^{sat}.

EXAMPLE 8.4.1 Estimate the fugacity of water at 25 °C and 1 bar.

Solution. From Eqn. (8.51) we have

$$f(P) \approx P^{sat} \exp\left[\frac{v(P - P^{sat})}{RT}\right], \tag{8.52}$$

where we have assumed that the molar volume of water at 25 °C and 1 bar is independent of pressure. At 25 °C, the vapor pressure of water is 0.031 43 bar. The molar volume of liquid water is approximately 1.002 g/cm^3. The fugacity at $P = 1$ bar is therefore equal to 0.031 43 bar (i.e. it is essentially identical to the vapor pressure). At $P = 1000$ bar, the fugacity becomes $f = 0.0656$ bar. □

Our second method for estimating the fugacity of a pure system uses PvT data or a PvT equation of state. Note that the definition for fugacity, Eqn. (8.41), simplifies for a pure system, since then $\bar{g}_i^{\mathrm{R}} \equiv g^{\mathrm{R}}$. Hence, we can use the relations for residual properties derived in Section 4.4.2 to find g^{R} from any given PvT relation. This integral has been performed for many of the models given in Appendix B, and expressions for the pure fugacities can be found there. The following example shows the procedure for a volume-explicit virial equation of state.

EXAMPLE 8.4.2 Derive an expression for the fugacity coefficient of a fluid described by a volume-explicit virial equation of state truncated after the second term.

Solution. From Eqn. (8.54) and the compressibility factor of a virial gas, namely $z = 1 + BP/(RT)$, we have

$$\begin{aligned} \log \phi &= \int_0^P \frac{1 + BP/(RT) - 1}{P}\, dP, \quad \text{constant } T \\ &= \frac{BP}{RT}. \end{aligned} \tag{8.53}$$

□

Expressions for models valid over greater ranges of density than the virial expansion give more complex mathematical expressions. The following example tests one of the assumptions in the previous method.

EXAMPLE 8.4.3 Use the Peng–Robinson model to predict the fugacity of methanol at its saturation pressure and 80 °C.

Solution. We can find the critical properties for methanol from the NIST Chemistry Web Book to be $T_c = 513 \pm 1$ K and $P_c = 81 \pm 1$ bar. The reduced temperature is thus $T_r = T/T_c = (273 + 80)/513 = 0.688$. For this reduced temperature, we can estimate the saturation pressure of methanol using the Antoine-like equation found from the Peng–Robinson model, Eqn. (B.55) in Appendix B. The expression requires the acentric factor for methanol, which we find from Table D.3 in Appendix D to be $\omega = 0.272$. Hence, we find $B = -7.134\,65$, $C = -3.073\,99$, and $D = -19.249$. Therefore, from Eqn. (B.55) we find the reduced saturation pressure to be 0.0447, or $P^{\text{sat}} \cong 3.62$ bar.

To find the fugacity, we use Eqn. (B.54). Note that this expression requires the temperature and specific volume, whereas we have the temperature and the pressure. To find the specific volume, we first need the parameters, which we can find from Eqns. (B.49),

$$a_0 \approx 0.457\,236\,7 \frac{(RT_c)^2}{P_c} = 10.2698\ \text{l}^2 \cdot \text{bar/mol}^2,$$

and (B.50),

$$b \approx 0.077\,796 \frac{RT_c}{P_c} = 0.040\,966\,2\ \text{l/mol}.$$

To find the specific volume, we need to find the roots of a cubic equation. Section A.7 of Appendix A can be useful here. We find three roots, of which $v = 7.685\,41$ l/mol is the vapor volume. We can now use the expression for the fugacity for the Peng–Robinson model:

$$f^{\text{pure}}(T, v) = \frac{RT}{v - b} \left[\frac{\sqrt{2}v + \left(2 + \sqrt{2}\right)b}{\sqrt{2}v - \left(2 - \sqrt{2}\right)b} \right]^{a(T)/(2\sqrt{2}RT)} \exp\left[\frac{b}{v - b} + \frac{va(T)}{RT(v - b)^2} \right] \tag{B.54$'$}$$

which yields a fugacity of 3.439 bar. Our previous estimation method (used in Example 8.4.1) would assume that the vapor is ideal, and give $f \cong P^{\text{sat}} = 3.62$ bar, an error of approximately 5%. Hence, the previous method gives a reasonable estimate. □

EXAMPLE 8.4.4 Figure 8.4 shows a P–V isotherm for toluene at 525 K. Calculate the fugacity of the liquid at 4 MPa.

Solution. The saturation pressure P^{sat} is the pressure at which the area of region A equals the area of region B. One can simply estimate this pressure by drawing a line parallel to the volume axis, which is moved up and down until the areas are the same. A better approach would rely on an equation of state to describe the data, which could then be used to calculate the areas numerically and determine P^{sat}:

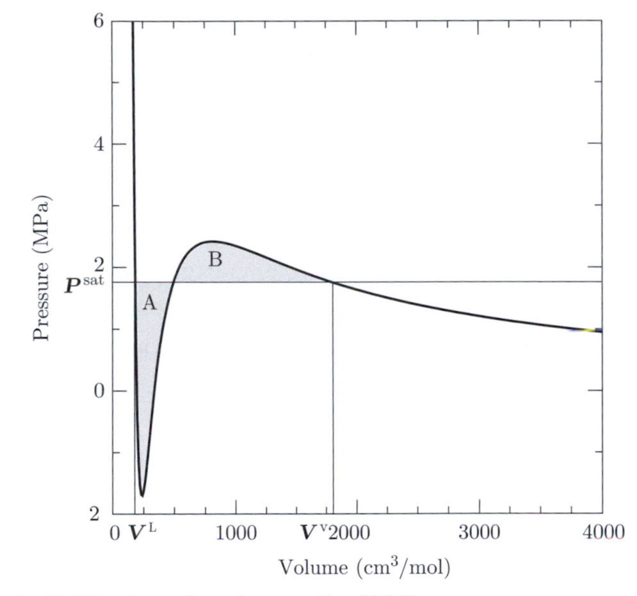

Figure 8.4 Solution: the P–V isotherm for toluene at T = 525 K.

$$P^{sat} \approx 1.8 \text{ MPa},$$
$$v^L = 200 \text{ cm}^3/\text{mol},$$
$$v^V = 1800 \text{ cm}^3/\text{mol}.$$

We assume that the vapor phase in equilibrium with the liquid at 525 K is ideal. At equilibrium,

$$f^L(1.8 \text{ MPa}) = f^V(1.8 \text{ MPa}) = P^{sat} = 1.8 \text{ MPa}.$$

We then calculate the fugacity using the following equation:

$$f_2^L = f_1^L \exp\left(\frac{V^L(P_2 - P_1)}{RT}\right),$$

$$f^L(4 \text{ MPa}) = f^L(1.8 \text{ MPa})\exp\left[\frac{(200 \text{ cm}^3/\text{mol})(4 \text{ MPa} - 1.8 \text{ MPa})}{(8.314 \text{ cm}^3 \text{ MPa/mol} \cdot \text{K})(525 \text{ K})}\right]$$
$$= (1.8 \text{ MPa})(1.2)$$
$$= 2.2 \text{ MPa}.$$

□

8.5 CALCULATION OF FUGACITY COEFFICIENTS OF MIXTURES FROM *PVT* EQUATIONS OF STATE

Having shown how to estimate fugacities for pure systems, we move on to mixtures. From the definition of fugacity, Eqn. (8.41), we need to find the residual Gibbs free energy. However, from Chapter 4, or from Example 8.3.5, we know how to find this residual property from equations of state. On combining these results, we find

$$f_i(T, P, \{N_j\}) = x_i P \exp\left\{\int_0^P \left[\left(\frac{\partial}{\partial N_i} Nz\right)_{T,P,\{N_{j\neq i}\}} - 1\right]\frac{dP}{P}\right\}, \tag{8.54}$$

$$f_i(T, V, \{N_j\}) = \frac{N_i RT}{V} \exp\left\{\int_V^\infty \left[\left(\frac{\partial}{\partial N_i} Nz\right)_{T,V,\{N_{j\neq i}\}} - 1\right]\frac{dV}{V}\right\}. \tag{8.55}$$

The first expression is useful given a volume-explicit *PVT* equation of state, and the second when the equation of state is pressure-explicit. Of course, data can also be used to estimate these derivatives and integrals numerically in lieu of an equation of state.

EXAMPLE 8.5.1 What does the virial equation of state for mixtures predict for the fugacity?

Solution. The (volume-explicit) virial expansion is

$$z = 1 + \frac{BP}{RT}, \tag{8.56}$$

where $B = \sum_{i=1}^{r} \sum_{j=1}^{r} x_i x_j B_{ij}$. Taking the derivative with respect to N_i yields (see Example 8.3.4)

$$\bar{z}_i = 1 + \frac{P}{RT}\left(2\sum_{j=1}^{r} x_j B_{ij} - B\right). \tag{8.57}$$

If we subtract 1 from each side, divide by P, and integrate over pressure from zero to P, we obtain the residual partial molar Gibbs free energy:

$$\frac{\bar{g}_i^{\mathrm{R}}}{RT} = \frac{P}{RT}\left(2\sum_{j=1}^{r} x_j B_{ij} - B\right). \tag{8.58}$$

We can now find the fugacity by exponentiating, and multiplying by the partial pressure,

$$f_i = x_i P \exp\left\{\frac{P}{RT}\left[2\sum_{j=1}^{r} x_j B_{ij} - B\right]\right\}, \tag{8.59}$$

which is the sought-after fugacity as a function of pressure, temperature, and composition. □

EXAMPLE 8.5.2 Consider the van der Waals equation of state, given by

$$P = \frac{RT}{v - b} - \frac{a}{v^2}. \tag{2.35′}$$

Derive an expression for the fugacity coefficient of component i in a mixture.

Solution. The fugacity coefficient of component i can also be written as

$$\begin{aligned}\log \phi_i &= \log\left(\frac{f_i}{x_i P}\right) \\ &= \int_V^{\infty}\left[\left(\frac{\partial}{\partial N_i}Nz\right)_{T,V,\{N_{j\neq i}\}} - 1\right]\frac{dV}{V} - \log z. \end{aligned} \tag{8.60}$$

For a mixture, the parameters a and b depend on composition. This dependence is determined by the so-called mixing rules. For a van der Waals fluid, it is often assumed that

$$a = \sum_i^r \sum_j^r x_i x_j a_{ij} \tag{7.87$'$}$$

and

$$b = \sum_i^r x_i b_i, \tag{7.88$'$}$$

where r is the number of components in the mixture and $a_{ij} = \sqrt{a_i a_j}$. To find the fugacity, we begin by multiplying Eqn. (2.35′) by $v/(RT)$ to obtain the compressibility factor:

$$z = \frac{v}{v-b} - \frac{a}{vRT}.$$

Then

$$\begin{aligned}\left(\frac{\partial}{\partial N_i}Nz\right)_{T,v,\{N_{j\neq i}\}} &= \frac{\partial}{\partial N_i}\left[\frac{VN}{V-bN} - \frac{aN^2}{VRT}\right] \\ &= \frac{V}{V-bN} + \frac{VN}{(V-bN)^2}\frac{\partial}{\partial N_i}\sum_j N_j b_j - \frac{1}{VRT}\frac{\partial}{\partial N_i}\sum_{jk} N_j N_k \sqrt{a_j a_k} \\ &= \frac{v}{v-b} + \frac{v b_i}{(v-b)^2} - \frac{2\sqrt{a_i}}{vRT}\sum_j x_j \sqrt{a_j}. \end{aligned} \tag{8.61}$$

We multiplied both sides of the expression for z by N, and took the derivative with respect to N_i. To obtain the remaining lines, we used the mixing rules for a and b given above. We now subtract 1 from each side of this result, multiply by $1/v$, and integrate over v to find the integral in Eqn. (8.60). Hence, we obtain the following expression for ϕ_i, after some algebra:

$$\log \phi_i = \log\left(\frac{v}{v-b}\right) + \frac{b_i}{v-b} - \frac{2\sqrt{a_i}\sum_{j=1}^r x_j\sqrt{a_j}}{vRT} - \log z. \tag{8.62}$$

Hence, the fugacity is predicted to be

$$f_i = \frac{x_i RT}{v-b}\exp\left[\frac{b_i}{v-b} - \frac{2\sqrt{a_i}\sum_{j=1}^r x_j\sqrt{a_j}}{vRT}\right]. \tag{8.63}$$

This result reduces to the partial pressure, $x_i P$ in the ideal case, which provides a check on our derivation. □

We can use the results of the previous two examples to illustrate the calculation of vapor–liquid equilibrium. Since the van der Waals equation of state is valid both for liquids and for gases, it could, in principle, be used to describe both the vapor and the liquid phases. It is important to note that using a unique set of parameters a and b over the entire pressure range over which the equation must hold would lead to inaccurate predictions. In what follows, however, we set aside the quantitative aspects of a calculation and the van der Waals model is used to illustrate a concept.

Suppose that we had a liquid mixture of r components described by the van der Waals equation. We would have one equation of the sort $P(T, v^{\text{liquid}}, \{x_i^{\text{liquid}}\}) = P(T, v^{\text{vapor}}, \{x_i^{\text{vapor}}\})$. We would have r equations of the sort $f_i(T, v^{\text{liquid}}, \{x_i^{\text{liquid}}\}) = f_i(T, v^{\text{vapor}}, \{x_i^{\text{vapor}}\})$. We would also know that the mole fractions in each of the two phases must sum to unity, giving us two more equations, for a grand total of $r + 3$ equations.

As unknowns, we have T, v^{liquid}, v^{vapor}, and the $2r$ mole fractions in the liquid and vapor phases. Thus, we have $2r + 3$ unknowns, or r degrees of freedom. Therefore, for a given temperature and composition of the liquid phase, these equations could be used, in principle, to find the composition of the vapor phase.

EXAMPLE 8.5.3 Estimate the vapor pressure of an equimolar liquid mixture of water and ethanol at 90 °C using the van der Waals model, with usual mixing rules for the parameters.

Solution. First, we need to estimate the parameters for the model. Calling water species 1 and ethanol species 2, we can find the critical properties from the appendix: $T_{c,1} = 374.2$ °C, $P_{c,1} = 218.3$ atm, $T_{c,2} = 243.1$ °C, and $P_{c,2} = 62.96$ atm. Then, using Eqns. (B.17) and (B.18) in Appendix B, we get $b_1 = 0.030\,432\,6$ l/mol, $b_2 = 0.084\,148\,9$ l/mol, and $a_1 = 5.458\,77$ l^2·atm/mol^2.

We know the temperature and composition of the liquid phase. We cannot find the fugacity of either component in the liquid phase, because we do not know the specific volume in that phase. To find the fugacity in the vapor phase, we need to know its composition and specific volume. It seems that we have three unknowns: x_1^{v}, v^{l}, and v^{v}. Hence, we need three equations. As is typical in such problems, we use the equivalence of pressure,

$$P(T, v^{\text{l}}, x_1^{\text{l}}) = P(T, v^{\text{v}}, x_1^{\text{v}}), \tag{8.64}$$

and the equivalence of fugacities,

$$\begin{aligned} f_1(T, v^{\text{l}}, x_1^{\text{l}}) &= f_1(T, v^{\text{v}}, x_1^{\text{v}}), \\ f_2(T, v^{\text{l}}, x_1^{\text{l}}) &= f_2(T, v^{\text{v}}, x_1^{\text{v}}). \end{aligned} \tag{8.65}$$

We can use the van der Waals equation of state for the first equation, and Eqn. (8.63) for the last two.

However, all of the resulting equations are highly nonlinear in the unknowns. There is no straightforward way to solve them – even numerically. This is an example of a coupled, nonlinear root-finding problem. One numerical method is called the Levenberg–Marquardt method, in which one requires a good initial guess for the unknowns. For our initial estimate, we assume a pressure of 1 atm, and find the resulting specific volumes for species 1 to be approximately 0.08 and 30 l/mol for the liquid and vapor volumes. We use these values for our initial guesses of v_l and v_v. For the vapor composition, we use equimolar, $x_1^v = 0.5$, as an initial guess. The method then makes many iterations in order to find the values for the unknowns that satisfy the equations. We used the software Mathematica to find the roots as $v^l \cong 0.0792$ l/mol, $v^v \cong 1.497$ l/mol, and $x_1^v \cong 0.5568$. One can plug these values into the equations to verify that they do indeed give equal pressures and fugacities. The predicted vapor pressure is rather high, at 17 atm. However, as pointed out earlier, the van der Waals approach is not particularly trustworthy for this mixture. □

8.6 FUGACITY IN IDEAL OR LEWIS MIXTURES

8.6.1 Lewis mixing

In a mixture of ideal gases, the fraction of the total pressure that can be attributed to a particular component i is simply x_iP, the partial pressure of i. Let us therefore examine the consequences of assuming an analogous relationship for the fugacity of component i in a mixture. The only basis for making this assumption is the observation that many binary mixtures exhibit behavior like that shown in Figure 8.5. At high mole fractions, many species have values for the fugacity near the straight line. Note that we have not specified whether it is a mixture of gases, liquids, or solids. In a Lewis mixture, we assume that the fugacity f_i of component i in a mixture is given by

$$f_i(T, P, \{x_i\}) \approx f_i^{\text{Lewis}}(T, P, \{x_i\}) \equiv x_i f_i^{\text{pure}}(T, P), \quad \text{Lewis mixture,} \tag{8.66}$$

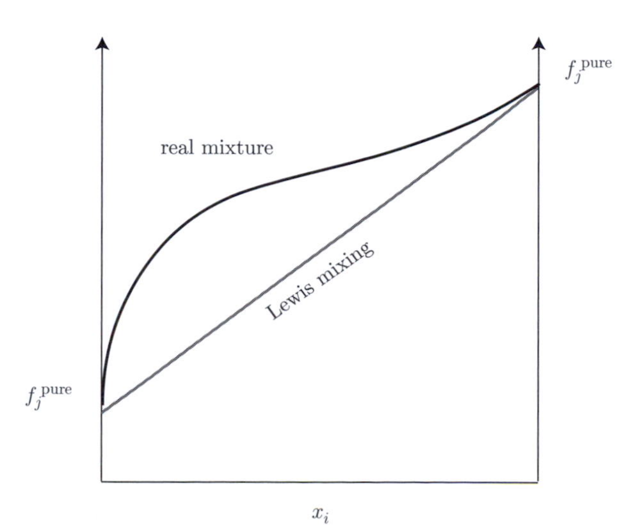

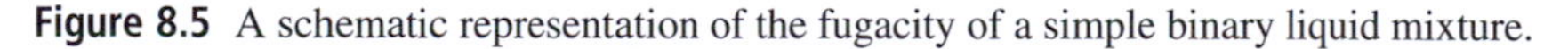
Figure 8.5 A schematic representation of the fugacity of a simple binary liquid mixture.

where f_i^{pure} is used to denote the fugacity of pure component i and x_i is its concentration in the mixture. As we will show, the results of assumption Eqn. (8.66) are *mixing* properties reminiscent of those of ideal gases. The physical implications are quite different. Recall that the molecules of an ideal gas do not interact. A Lewis mixture, on the other hand, has molecular interactions, but here one assumes that the interactions between two different molecules are the same as those between two molecules of the same type.

The relation given by (8.66) is often referred to as Lewis' fugacity rule, and it is used to define what is usually called an *ideal mixture*. Again, it is emphasized that, in order to avoid confusion with ideal-gas mixtures, however, we refer to such a system as a **Lewis mixture**.

8.6.2 Properties of Lewis (ideal) mixtures

Recalling the definition of the fugacity coefficient, we can write

$$\phi_i^{\text{Lewis}} = \frac{f_i^{\text{Lewis}}}{x_i P} = \frac{f_i^{\text{pure}}}{P} = \phi_i^{\text{pure}}. \tag{8.67}$$

We shall see in the next chapter that, since the fugacity coefficient is just that of the pure component, Lewis mixing always predicts miscibility. Given the physical basis for Lewis mixing, this observation is expected.

Lewis' fugacity rule leads to several important consequences for the properties of a mixture. Specifying the fugacity of a component has implications about the chemical potential of that component, as we will now prove. Since the Gibbs free energy is the sum of the chemical potentials, according to Eqn. (8.15), we can write

$$\begin{aligned} g(T, P, \{x_i\}) &= \sum_i x_i \mu_i \\ &= \sum_i x_i \bar{g}_i \\ &= \sum_i x_i \left[\bar{g}_i^{\text{R}} + \bar{g}_i^{\text{ideal}}\right] \\ &= \sum_i x_i \left[RT \log\left(\frac{f_i}{x_i P}\right) + \bar{g}_i^{\text{ideal}}\right]. \end{aligned} \tag{8.68}$$

To obtain the second line we used the equivalence between partial molar Gibbs free energy and chemical potential, Eqn. (8.10). To get to the third line, we used the definition of a residual property. The last line comes from the definition of fugacity, Eqn. (8.41). Hence, for a Lewis mixture we can write

$$\begin{aligned} g^{\text{Lewis}}(T, P, \{x_i\}) &= \sum_i x_i \left[RT \log\left(\frac{f_i^{\text{Lewis}}}{x_i P}\right) + \bar{g}_i^{\text{ideal}}\right] \\ &= \sum_i x_i \left[RT \log\left(\frac{f_i^{\text{pure}}}{P}\right) + \bar{g}_i^{\text{ideal}}\right] \end{aligned}$$

$$
\begin{aligned}
&= \sum_i x_i \left[(g_i^{\mathrm{R}})^{\mathrm{pure}} + \bar{g}_i^{\mathrm{ideal}} \right] \\
&= \sum_i x_i [g_i^{\mathrm{pure}} - g_i^{\mathrm{pure,ideal}} + \bar{g}_i^{\mathrm{ideal}}] \\
&= \sum_i x_i \left[g_i^{\mathrm{pure}} + RT \log x_i \right].
\end{aligned}
\tag{8.69}
$$

This relation is something like a fundamental relation, although it requires that we have fundamental relations for each of the pure species. This lengthy derivation is important only because we wish to derive several more interesting relations. These interesting relations are the characteristic mixing properties of a Lewis mixture, which we summarize here:

$$
v^{\mathrm{Lewis}}(T, P, \{x_i\}) = \sum_{i=1}^{r} x_i v_i^{\mathrm{pure}}(T, P), \tag{8.70}
$$

$$
\frac{u^{\mathrm{Lewis}}(T, P, \{x_i\})}{RT} = \frac{1}{RT} \sum_{i=1}^{r} x_i u_i^{\mathrm{pure}}(T, P), \tag{8.71}
$$

$$
\frac{h^{\mathrm{Lewis}}(T, P, \{x_i\})}{RT} = \frac{1}{RT} \sum_{i=1}^{r} x_i h_i^{\mathrm{pure}}(T, P), \tag{8.72}
$$

$$
\frac{s^{\mathrm{Lewis}}(T, P, \{x_i\})}{R} = \frac{1}{R} \sum_{i=1}^{r} x_i s_i^{\mathrm{pure}}(T, P) - \sum_{i=1}^{r} x_i \log(x_i), \tag{8.73}
$$

$$
\frac{f^{\mathrm{Lewis}}(T, P, \{x_i\})}{RT} = \frac{1}{RT} \sum_{i=1}^{r} x_i f_i^{\mathrm{pure}}(T, P) + \sum_{i=1}^{r} x_i \log(x_i), \tag{8.74}
$$

$$
\frac{g^{\mathrm{Lewis}}(T, P, \{x_i\})}{RT} = \frac{1}{RT} \sum_{i=1}^{r} x_i g_i^{\mathrm{pure}}(T, P) + \sum_{i=1}^{r} x_i \log(x_i), \tag{8.75}
$$

$$
\frac{\mu_i^{\mathrm{Lewis}}(T, P, \{x_i\})}{RT} = \frac{\mu_i^{\mathrm{pure}}(T, P)}{RT} + \log(x_i). \tag{8.76}
$$

It is interesting to note that, while the volume and the enthalpy of a Lewis mixture are given by simple linear combinations of pure-component properties, the molar entropy and Gibbs free energy are not. Earlier in this section, a simple, intuitive explanation was provided as to why Eqn. (8.70) might be valid for mixtures of similar components. The precise forms of the additional "mixing" contributions that arise in the Lewis Gibbs free energy and entropy expressions are perhaps less intuitive, but one can use simple arguments to discuss why they have to be there. If two similar components are mixed, the resulting solution is likely to be homogeneous; the mixing process occurs spontaneously, without a need for external work. The converse is not true; the mixture does not separate into its pure constituents by itself. In order for it to occur spontaneously, the mixing process must lead to an *increase* of the entropy of the system. The $(-\sum_i^r x_i \log(x_i))$ term in Eqn. (8.73) is always positive; it is this term that renders the entropy of the mixture higher than that of the original, "unmixed" system of pure components.

Note also that these results are similar in form to those for mixing general ideal gases, which were shown in Section 8.2. There is an essential difference, however: Lewis mixing rules can be applied to real gases, even liquids. However, these apply only when the species are very similar; if the mixture evolves heat upon mixing, such as happens on mixing sulfuric acid and water, then the mixture is clearly not Lewis. Or, if the mixture changes volume, or is immiscible, then use of the Lewis mixing approximation is a bad idea.

8.6.3 A simple application of Lewis (ideal) mixing: Raoult's law

Consider a vessel containing a vapor phase (′) in equilibrium with a liquid phase (″). We wish to find a relation between the vapor- and liquid-phase compositions, say to predict the bubble point of a mixture. The composition of the vapor phase is given by $\{y_i\}$, and that of the liquid phase is given by $\{x_i\}$. From Eqn. (8.1) we know that, at equilibrium, the chemical potential of each component is the same in both phases; from the definition of fugacity, Eqn. (8.41), such equality can be expressed as

$$f_i' = f_i'', \quad i = 1, \ldots, r. \tag{8.77}$$

If we now assume that the vapor phase (denoted by ′) consists of a mixture of ideal gases, we have

$$f_i' = y_i f_i^{\text{pure ideal gas}}, \quad i = 1, \ldots, r, \quad \text{ideal gas;} \tag{8.78}$$

or, recalling that the fugacity of an ideal gas is the pressure,

$$f_i' = y_i P, \quad i = 1, \ldots, r, \quad \text{ideal gas.} \tag{8.79}$$

If we assume that the liquid phase (denoted by ′) is a Lewis mixture of liquids (i.e. if we invoke Lewis' fugacity rule, Equation (8.66)), we have $f_i'' = x_i f_i^{\text{pure liquid}}$. Equation (8.77) therefore leads to

$$y_i P = x_i f_i^{\text{pure liquid}}, \quad i = 1, \ldots, r. \tag{8.80}$$

We are now faced with the problem of calculating the fugacity of a pure liquid $f_i^{\text{pure liquid}}$. If we also assume that the volume of the liquid v is constant in the range P^{sat} to P, we get Eqn. (8.51); that is,

$$f(P) \approx P^{\text{sat}} \exp\left[\frac{v(P - P^{\text{sat}})}{RT}\right], \quad \text{incompressible liquid, ideal vapor.}$$

Since the molar volume of a liquid is usually much smaller than that of a vapor, at moderate pressures the Poynting correction is negligible and can be omitted from approximate calculations. In that case, $f(P) \approx P^{\text{sat}}$ and Eqn. (8.80) can be written as

$$y_i P \approx x_i P_i^{\text{sat}}, \quad i = 1, \ldots, r. \tag{8.81}$$

- The vapor consists of a mixture of ideal gases.
- The liquid phase is a Lewis (or ideal) mixture; it therefore obeys Lewis' fugacity rule.
- The pressure is moderate, and the Poynting correction can therefore be neglected.

All of the assumptions that have gone into formulating Eqn. (8.81) are listed. The result is known as **Raoult's "law."** It provides a simple means for conducting calculations of vapor–liquid equilibria for mixtures of fluids. Bear in mind, however, that the approximations made in order to derive Raoult's law are severe, and its applicability is therefore highly limited.

Raoult's law can be used to illustrate the calculation of simple vapor–liquid phase diagrams for binary mixtures. Consider a vapor mixture of composition y_i, at a temperature T. If we wish to determine the pressure at which the mixture starts to condense, the so-called dew pressure, we can simply write

$$\begin{aligned} y_1 P &= x_1 P_1^{\text{sat}}, \\ y_2 P &= x_2 P_2^{\text{sat}}. \end{aligned} \tag{8.82}$$

The sum of these two equations leads to the sought-after expression for the pressure:

$$P = x_1 P_1^{\text{sat}} + x_2 P_2^{\text{sat}}. \tag{8.83}$$

The composition of the first drop of liquid that appears can now be determined according to

$$y_i = \frac{x_i P_i^{\text{sat}}}{P}.$$

By repeating this procedure for the whole range of composition extending from $y_1 = 0$ to $y_1 = 1$, the entire dew curve can be generated. The so-called "bubble pressure," the pressure at which a liquid mixture starts to vaporize, can be calculated in an analogous manner. As illustrated in the next few sections, the calculation of vapor–liquid coexistence becomes slightly more cumbersome when the vapor or liquid phases cannot be assumed to behave as ideal gases or Lewis mixtures. In that case, it is necessary to resort to iterative procedures, but the basic structure of the problem remains unchanged.

8.7 SOLUBILITY OF SOLIDS AND LIQUIDS IN COMPRESSED GASES

8.7.1 Phase equilibria between a solid and a compressed gas

The equations for phase equilibria discussed in this chapter are generally applicable, regardless of whether the phases under consideration are solid, liquid, or gaseous. In this section, we briefly discuss their application to solid–fluid equilibria.

In recent years, the extraction of solids using compressed gases has attracted considerable attention. In the food industry, for example, it is now common to extract particular substances (e.g. caffeine from coffee or alcohol from beer) using high-pressure gases such as carbon dioxide. In the environmental industry, there are processes to remove pollutants from soil, also using carbon dioxide. It is therefore useful to discuss briefly the application of thermodynamics for the design of such processes.

For equilibrium between a pure solid and a pure gas phase, Eqn. (8.47) can be written in the following form:

$$f^{\text{solid}}[P^{\text{sat}}] = f^{\text{gas}}[P^{\text{sat}}], \tag{8.84}$$

where the fugacities are evaluated at the vapor pressure of the system. For a mixture, we write

$$f_i^{\text{solid}} = f_i^{\text{gas}}, \qquad i = 1, \ldots, r. \tag{8.85}$$

For simplicity, we assume that the solid phase under consideration is a pure substance; in most cases, the solubility of the other gases in the solid is sparingly small and this is a reasonable assumption. In that event, the fugacity of the solid is given by

$$f_i^{\text{solid}} \approx f_i^{\text{pure solid}} = \phi_i^{\text{sat}} P_i^{\text{sat}} \exp\left(\int_{P^{\text{sat}}}^{P} \frac{v_i^{\text{solid}}}{RT}\, dP\right), \qquad \text{constant } T, \tag{8.86}$$

where ϕ_i^{sat} is the fugacity coefficient of pure i evaluated at pressure P^{sat}, and v_i^{solid} is the molar volume of pure solid i.

The fugacity in the vapor phase is calculated from either Eqn. (8.54) or Eqn. (8.55). Typically, however, one writes this in terms of the fugacity coefficient,

$$f_i = y_i \phi_i P. \tag{8.87}$$

After substitution of Eqns. (8.87) and Eqn. (8.86) into Eqn. (8.85), we arrive at an expression for the solubility of the solid, component i, in the compressed-gas phase:

$$y_i \phi_i(T, P, y_i) P = \phi_i^{\text{sat}} P_i^{\text{sat}} \exp\left(\int_{P^{\text{sat}}}^{P} \frac{v_i^{\text{solid}}}{RT}\, dP\right), \tag{8.88}$$

where it is understood that ϕ_i is found from Eqn. (8.54) or Eqn. (8.55). Note that it is not possible, in general, to find an explicit expression for y_i.

8.7.2 Phase equilibria between a liquid and a compressed gas

The situation for a liquid phase in equilibrium with a compressed gas is slightly more complicated, since the compressed gas is likely to be soluble in the liquid. In this section we discuss only the case in which the solubility of the gas in the liquid is not too high. This is often the case in applications.

Figure 8.5 illustrates schematically the expected behavior of the fugacity of component 1 in a binary mixture. If the two components were miscible in all proportions, the fugacity would follow a behavior similar to that depicted. From the shape of that curve, it is clear that assuming Lewis mixing is reasonable only when the component under consideration is fairly concentrated, almost pure. On the other hand, assuming Lewis mixing for a dilute component results in considerable errors.

To circumvent this problem, it is possible to define a new type of mixing that will be accurate for dilute components; in close analogy to Lewis mixtures, for dilute components we introduce **Henry's law**:

$$f_i \cong f_i^{\text{Henry}} := x_i H_{i,j}(T, P), \tag{8.89}$$

where $H_{i,j}$ is the Henry's-law constant for component i in the solvent, component j, and f_i^{Henry} is a fugacity in the Henry's-law limit of vanishing concentration. The Henry's-law constants depend both on temperature and on pressure. They have been measured for many gases dissolved in a wide variety of liquids. Note, however, that these constants are often measured at low to moderate pressures. In order to extrapolate to high pressures, we can use property 4 of fugacities. By substitution of Eqn. (8.89) into Eqn. (8.43), we get

$$H_{i,j}[T,P_2] = H_{i,j}[T,P_1]\exp\left[\int_{P_1}^{P_2} \frac{\bar{v}_i^\infty[T,P']}{RT}\,dP'\right], \tag{8.90}$$

where the superscript ∞ denotes a quantity evaluated at infinite dilution.

We can now return to the general expressions for phase equilibria between a liquid and a vapor, and write

$$f_i^{\text{liquid}} = f_i^{\text{gas}}, \qquad i = 1,\ldots,r. \tag{8.91}$$

For the liquid phase we use Eqn. (8.90), and for the vapor phase we use Eqn. (8.87). The solubility of the solvent j in the compressed-gas phase is therefore given by

$$y_j\phi_j(T,P,y_j)P = x_i H_{i,j}[T,P_j^{\text{sat}}]\exp\left(\int_{P_j^{\text{sat}}}^{P} \frac{\bar{v}_i^\infty}{RT}\,dP\right), \tag{8.92}$$

where the solubility of component i in the liquid phase is given by

$$x_i = \frac{y_i\phi_i P}{H_{i,j}\exp\left(\int_{P_j^{\text{sat}}}^{P}[\bar{v}_i^\infty/(RT)]dP\right)} \tag{8.93}$$

Summary

In this chapter we introduced the quantity called *fugacity*, which plays a central role in calculating multicomponent vapor–liquid equilibrium. Instead of equating chemical potentials to find compositions of different phases, we equate fugacities. Several ways of measuring and predicting fugacities were spelled out. These methods are briefly summarized here.

- We defined a partial molar quantity by the relation

$$\bar{m}_i := \left(\frac{\partial Nm}{\partial N_i}\right)_{T,P,N_{j\neq i}} = \left(\frac{\partial M}{\partial N_i}\right)_{T,P,N_{j\neq i}}. \tag{8.8}$$

- Fugacity was defined by

$$f_i := x_i P \exp\left(\frac{\bar{g}_i^{\text{R}}}{RT}\right), \tag{8.41}$$

where $\bar{g}_i^{\mathrm{R}}$ is the residual partial molar Gibbs free energy for species i. Five important properties of the fugacity were given in Section 8.4.2.

- When two phases (1) and (2) are in equilibrium, they have the same temperature and pressure, and each component has the same fugacity:

$$T^{(1)} = T^{(2)},$$
$$P^{(1)} = P^{(2)},$$
$$f_i^{(1)} = f_i^{(2)}, \quad i = 1, 2, \ldots, r.$$

- For systems with low vapor pressure, we can estimate the fugacity of a pure (nearly incompressible) liquid from

$$f[T,P] \approx P^{\mathrm{sat}}[T]\exp\left[\frac{v^{\mathrm{liquid}}(P - P^{\mathrm{sat}})}{RT}\right]. \tag{8.51}$$

- We know how to predict fugacities from either pressure-explicit or volume-explicit equations of state for mixtures, from

$$f_i(T,P,\{N_j\}) = x_i P \exp\left\{\int_0^P \left[\left(\frac{\partial}{\partial N_i}Nz\right)_{T,P,\{N_{j\neq i}\}} - 1\right]\frac{dP}{P}\right\} \tag{8.54}$$

and

$$f_i(T,V,\{N_j\}) = \frac{N_i RT}{V}\exp\left\{\int_V^\infty \left[\left(\frac{\partial}{\partial N_i}Nz\right)_{T,V,\{N_{j\neq i}\}} - 1\right]\frac{dV}{V}\right\}. \tag{8.55}$$

- Lewis mixing approximates fugacities of individual components of a mixture in terms of their values in the pure state,

$$f_i \approx f_i^{\mathrm{Lewis}} := x_i f_i^{\mathrm{pure}}. \tag{8.66}$$

This approximation is typically valid for liquid species at high concentration. Raoult's law is a special case, for which Lewis mixing is assumed for the liquid, the Poynting correction is neglected, and the vapor phase is assumed to be an ideal gas.

- For liquid species at low concentration, Henry's law is more useful than Lewis mixing;

$$f_i^{\mathrm{Henry}} := x_i H_{i,j}, \tag{8.89}$$

where $H_{i,j}$ is called the *Henry's-law constant* for dilute species i in concentrated species j.

By the end of this chapter, the student should be able to calculate fugacities for pure components at arbitrary temperature and pressure using any equation of state. Also, the student should be able to devise experiments to estimate fugacities from experiments, and predict temperature and pressure dependences.

Given a fundamental relation for mixtures, the student should be able to estimate fugacities. Typically, these are models for pure substances (e.g. the Peng–Robinson model) with "mixing rules" for the parameters.

Exercises

8.3.A Derive Eqn. (8.34) by changing the variable of integration in Eqn. (8.32). *Hint:* you might find it useful to prove first that

$$\left(\frac{\partial}{\partial N_i} Nz\right)_{T,P,\{N_{j\neq i}\}} = -\frac{zNRT}{V^2}\left(\frac{\partial}{\partial N_i} Nz\right)_{T,V,\{N_{j\neq i}\}}\left(\frac{\partial V}{\partial P}\right)_{T,\{N_i\}}.$$

8.4.A Calculate the fugacity of hexane at 25 °C and at 1 and 50 bar.

8.4.B Calculate the fugacity of mono-ethanol-amine (MEA) at 50 °C and at 1 and 50 bar.

8.4.C Calculate the fugacity of acetone at 50 °C and at 1 and 50 bar.

8.4.D Show that the effects of pressure and temperature on the fugacity coefficient are given by

$$\left(\frac{\partial \log \phi}{\partial P}\right)_{T,N} = \frac{v^{\mathrm{R}}}{RT} \tag{8.94}$$

and

$$\left(\frac{\partial \log \phi}{\partial T}\right)_{P,N} = -\frac{h^{\mathrm{R}}}{RT^2}. \tag{8.95}$$

8.4.E Derive an expression for the temperature dependence of the residual molar volume of a fluid described by a volume-explicit virial equation of state truncated after the second term. You may express your result as a function of dB/dT.

8.4.F What is the residual molar volume of methanol at 50 °C and 1 bar?

8.4.G Show that the effect of temperature on the fugacity of a pure component is given by a special case of Eqn. (8.44), i.e.

$$\left(\frac{\partial \log f_i^{\mathrm{pure}}}{\partial T}\right)_{P,N} = -\frac{h_i^{\mathrm{pure}}}{RT^2}.$$

8.4.H The usual definition of fugacity is

$$RT\, d\log f_i^{\mathrm{pure}} := d\mu_i^{\mathrm{pure}}, \tag{8.96}$$

plus the boundary condition that the fugacity is just the pressure in the ideal state at low pressures. Prove that this definition and Eqn. (8.41) are equivalent.

8.4.I Derive the expression for the fugacity coefficient of a pure fluid described by van der Waals' mechanical equation of state.

8.4.J Derive the expression for the fugacity coefficient of a pure fluid described by the Peng–Robinson mechanical equation of state.

8.4.K Derive the expression for the fugacity coefficient of a pure fluid described by the Redlich–Kwong mechanical equation of state.

8.4.L Derive the expression for the fugacity coefficient of a pure fluid described by the Soave mechanical equation of state.

8.4.M Pitzer has proposed a correlation for the compressibility factor of gases of the form

$$z = \frac{Pv}{RT} = 1 + \left(\frac{BP_c}{RT_c}\right)\frac{P_r}{T_r}, \tag{8.97}$$

where the factor in parentheses is given by

$$\left(\frac{BP_c}{RT_c}\right) = B^{(0)} + \omega B^{(1)}. \tag{8.98}$$

The parameter ω is the so-called Pitzer acentric factor, defined earlier in this text (Eqn. (7.77)). It is meant to provide a measure of the asphericity of simple molecules, and it has been tabulated for a large number of fluids. For many gases, their PVT behavior can be described with the following correlations for $B^{(0)}$ and $B^{(1)}$:

$$B^{(0)} = 0.083 - \frac{0.422}{T_r^{1.6}}, \tag{8.99}$$

$$B^{(1)} = 0.139 - \frac{0.172}{T_r^{4.2}}. \tag{8.100}$$

Derive expressions for the residual molar enthalpy, the residual molar entropy, and the fugacity coefficient of a fluid described by the Pitzer correlation. Please plot your results in the form of generalized or universal charts.

8.4.N Calculate the fugacity coefficient of methanol at 150° C and 20 bar using a volume-explicit virial equation of state truncated after the second term. Repeat your calculation using a van der Waals equation of state with parameters determined from the critical point of methanol. Finally, use a Pitzer correlation for the same calculation (see the previous problem, Exercise 8.4.M, and Exercise 8.5.E). Which of your three estimates is the most trustworthy? Why?

8.4.O The van der Waals mechanical equation of state is given by

$$P = \frac{RT}{v-b} - \frac{a}{v^2}. \tag{8.101}$$

Please show that the second virial coefficient can be related to the van der Waals constants a and b by

$$B = b - \frac{a}{RT}. \tag{8.102}$$

8.4.P Derive the fugacity coefficient of a pure van der Waals fluid.

8.4.Q Carbon dioxide (CO_2) is used extensively in extraction processes, where it is used in the super-critical state as a solvent. Use the van der Waals model to estimate the pressures at which the deviations of CO_2 from ideal-gas behavior are greater than 10% at $T = 50\,^\circ\text{C}$.

8.4.R Perturbation theory can be used to derive free-energy models for strongly interacting or "associating" fluids. In one such model, the so-called "statistical associated-fluid theory," or SAFT [26], the residual Helmholtz free energy of a pure fluid is given by

$$\frac{F^{R}}{RT} = \frac{F^{hs}}{RT} + \frac{F^{assoc}}{RT}, \tag{8.103}$$

where F^{hs} and F^{assoc} are hard-sphere reference and association contributions, respectively, to the free energy. These contributions are given by

$$\frac{F^{hs}}{RT} = \frac{4\eta - 3\eta^2}{(1-\eta)^2}, \tag{8.104}$$

$$\frac{F^{assoc}}{RT} = \sum_S \left(\ln X^{(S)} - \frac{X^{(S)}}{2}\right) + \frac{M}{2}. \tag{8.105}$$

The quantity η in Eqn. (8.104) represents the so-called packing fraction, and it is related to the density through

$$\eta = \frac{\pi \tilde{N}_A}{6} \tilde{\rho} d^3, \tag{8.106}$$

where $\tilde{N}_A$ is Avogadro's number and d is an "effective," temperature-dependent diameter for the molecules, which is given by

$$d = \sigma \left[1 - \exp\left(-\frac{3\epsilon}{kT}\right)\right]. \tag{8.107}$$

The quantity σ represents the actual (temperature-independent) diameter of the molecule, while ϵ represents the interaction energy between different molecules. In Eqn. (8.105), M is the number of "association" sites on a molecule, $X^{(S)}$ is the mole fraction of molecules that are *not* bonded for a given set of conditions, and the summation runs over all association sites of a molecule. The quantity $X^{(S)}$ is given by

$$X^{(S)} = \left[1 + N_A \sum_Y \rho X^{(Y)} \frac{2-\eta}{2(1-\eta)^3} \sigma^3 \kappa^{(SY)} \left(\exp\left(\frac{\epsilon^{(SY)}}{kT}\right) - 1\right)\right]^{-1}. \tag{8.108}$$

The summation in Eqn. (8.108) runs over all types of sites Y in the system. The parameters $\epsilon^{(SY)}$ and $\kappa^{(SY)}$ characterize the interaction energy and interaction range, respectively, for a pair of sites Y–S.

1. Derive an expression for the compressibility factor of a fluid described by Eqn. (8.103).
2. Derive an expression for the fugacity coefficient of a fluid described by Eqn. (8.103).

8.4.S Estimate the fugacity coefficient of pure carbon dioxide as a function of pressure (between 0 and 50 bar) at 320 K. Your plot should look similar to Figure 8.6, which was plotted using the Peng–Robinson model

8.5.A Fill in the details for the derivation of Eqn. (8.60).

8.5.B Derive an expression for the fugacity coefficient of component i in a mixture described by the van der Waals equation. Use the following mixing rules for parameters a and b:

$$a = \sum_i^r \sum_j^r x_i x_j a_{ij}, \quad a_{ij} = \sqrt{a_i a_j}(1 - k_{ij}), \tag{8.109}$$

and

$$b^{1/3} = \sum_i^r x_i b_i^{1/3}, \tag{8.110}$$

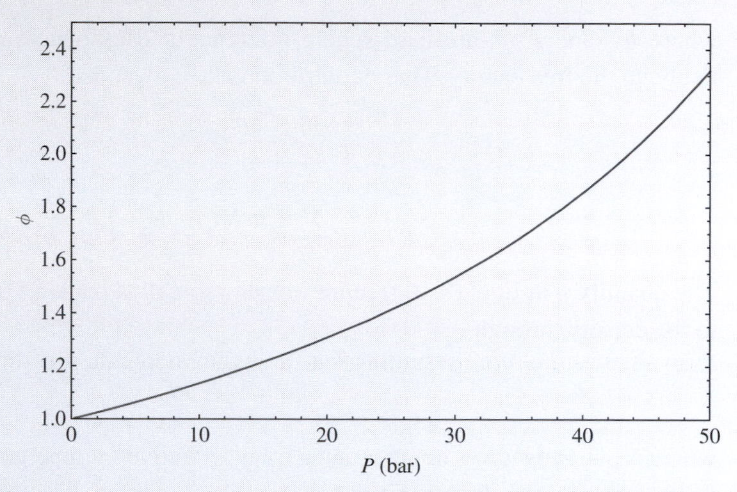

Figure 8.6 The fugacity coefficient as a function of pressure for carbon dioxide at 320 K as estimated by application of the Peng–Robinson model.

where k_{ij} is an adjustable parameter (a constant) that accounts for deviations from a geometric-mean mixing rule.

8.5.C Show that the fugacity coefficient of component i in a mixture described by the Peng–Robinson equation of state is given by

$$\log \phi_i = \frac{b_i}{b}(z-1) - \log\left(z - \frac{bP}{RT}\right) - \frac{a}{2\sqrt{2}bRT}\left(\frac{2\sum_j y_j a_{ij}}{a} - \frac{b_i}{b}\right)\log\left(\frac{v + b(1+\sqrt{2})}{v + b(1-\sqrt{2})}\right). \tag{8.111}$$

Use Eqn. (8.109) as the mixing rule for parameter a, and Eqn. (7.88) for parameter b, where k_{ij} is an adjustable parameter (a constant) that accounts for deviations from a geometric-mean mixing rule.

8.5.D Show that the fugacity coefficient of component i in a mixture described by the Redlich–Kwong equation of state is given by

$$\log \phi_i = \log\left(\frac{v}{v-b}\right) + \frac{b_i}{v-b} - \frac{2\sum_j y_j a_{ji}}{RT^{3/2}b}\log\left(\frac{v+b}{v}\right) + \frac{ab_i}{RT^{3/2}b^2}\left(\log\left(\frac{v+b}{v}\right) - \frac{b}{v+b}\right) - \log z. \tag{8.112}$$

Use the same mixing rules as those suggested for the van der Waals fluid, Eqns. (7.87) and (7.88).

8.5.E Show that, at low to intermediate pressures, the fugacity coefficient of component 1 in a binary mixture described by the van der Waals equation of state can be expressed as

$$\phi_1 = \exp\left(\left[b_1 - \frac{a_1}{RT}\right]\frac{P}{RT} + \frac{[\sqrt{a_1} - \sqrt{a_2}]^2 y_2^2 P}{(RT)^2}\right). \tag{8.113}$$

Use standard van der Waals mixing rules (Eqns. (7.87) and (7.88)) for your derivation.

(*Hint*: as a starting point, write the van der Waals equation in the approximate form $Pv = RT + [b - a/(RT)]P + \cdots$, and find an expression for $\bar{v}_1$.)

8.5.F Show that the fugacity coefficient of component i in a mixture of m components described by a virial equation of state is given by

$$\log \phi_i = \frac{2}{v} \sum_{j=1}^{r} y_j B_{ij} + \frac{3}{2v^2} \sum_{j=1}^{r} \sum_{k=1}^{q} y_j y_k C_{ijk} - \log z. \tag{8.114}$$

8.5.G Consider a binary mixture of two components. In the liquid phase, these two components form a Lewis mixture. Assuming that, at low to moderate pressures, the vapor phase can be described by a virial equation truncated after the second term, could this mixture exhibit an azeotrope? Please explain.

8.5.H Freeze-drying is often used in the food and pharmaceutical industries to remove water from certain products. In this process, a solution of a given product is frozen, thereby yielding a solid material that contains a significant amount of ice crystals. The crystals are subsequently removed by sublimation. In order to design an efficient process to remove these crystals, it is important to determine the concentration of water in a carrier gas as a function of temperature. Ordinarily one would use low pressures to eliminate ice, but in this particular instance you're asked to consider the possibility of employing high-pressure nitrogen in an attempt to arrive at a more efficient drying process. In order to evaluate the feasibility of such a process, please determine the concentration of water in the gas phase at $T = -10\,^{\circ}\mathrm{C}$ and at $P = 0.001,\ 0.1,\ 10,$ and 100 bar. Do you envisage any problems with the operation of the process at 100 bar? Overall, would it be advantageous to conduct the process at 100 bar? You may assume that the vapor pressure of ice at $-10\,^{\circ}\mathrm{C}$ is 1.956 torr, and that the specific volume of ice for the conditions relevant to this problem is approximately 1.09 cm^3/g. You may also assume that the product of interest is completely non-volatile.

8.5.I A stream of moist air at 50 bar is to be cooled to $-2\,^{\circ}\mathrm{C}$. What is the maximum amount of water that the gas phase can contain before water condenses? Please do your calculations using (1) a virial equation truncated after the second term and (2) a Redlich–Kwong equation. You may assume that air consists of a binary mixture of nitrogen (80%) and oxygen (20%). For this problem, you may also assume that the solubility of air in the water is small. How different are your predictions for the virial model and the Redlich–Kwong model?

8.5.J To perform bubble-point and dew-point calculations, it is useful to have a plot of T vs. x_1 at fixed pressure, or P vs. x_1 at fixed temperature for a binary mixture, with two curves.[2] Using the van der Waals model with standard mixing rules, Eqns. (7.87) and (7.88), for benzene and toluene, construct plots of T (at $P = 1$ bar) and P (at $T = 100\,^{\circ}\mathrm{C}$) vs. the mole fraction of benzene.

8.6.A Derive Eqns. (8.73), (8.76), and (8.72).

[2] Examples of such curves are in show in Section 6.4 of [38].

8.6.B Derive Eqn. (8.77).

8.6.C The compressibility factor of a mixture of hard spheres of different diameters can be determined from [88]:

$$z^{\mathrm{hs}} = \frac{6}{\pi N_{\mathrm{A}}\rho}\left(\frac{\zeta_0\zeta_3}{1-\zeta_3} + \frac{3\zeta_1\zeta_2}{(1-\zeta_3)^2} + \frac{(3-\zeta_3){\zeta_2}^3}{(1-\zeta_3)^3}\right), \tag{8.115}$$

where

$$\zeta_k = \frac{\pi N_{\mathrm{A}}\rho}{6}\sum_{i=1}^{r} x_i d_i^k, \tag{8.116}$$

in which ρ is the molar density and d_i is the diameter of the molecules of species i. Use this expression to determine the molar volume (at constant pressure) of hard-sphere binary mixtures of different diameters, ranging from $d_1/d_2 = 1$ to $d_1/d_2 = 5$. Perform your calculations as a function of composition in the range $x_1 = 1$ to $x_1 = 0$. Use a pressure for your calculations such that the packing fraction of the spheres, given by $\eta = (\pi \tilde{N}_{\mathrm{A}}\rho d^3)/6$, is approximately $\eta \approx 0.3$; for a system of hard spheres this packing fraction corresponds to a liquid-like density. For which compositions and diameter ratios does Amagat's law provide a good description of the volume of the mixture?

8.6.D This exercise amounts to filling in the details of Section 8.6.3.

1. Derive Raoult's law for vapor–liquid equilibria in a multicomponent mixture. State clearly all assumptions involved in your derivation.
2. Consider a binary mixture at constant pressure described by Raoult's law. Is the dew-temperature curve a straight line? Is the bubble-temperature curve a straight line? In both cases, please justify your answers.

8.6.E Consider a binary mixture of components 1 and 2. Assume that both the liquid and the vapor form Lewis mixtures. Do not assume, however, that the vapor phase follows ideal-gas behavior.

1. Show that, in that case, Raoult's law is not correct and should be replaced by an expression of the form

$$y_i \phi_i^{\mathrm{pure}} P = x_i P_i^{\mathrm{sat}}, \qquad i = 1, 2, \tag{8.117}$$

where ϕ_i^{pure} is the fugacity coefficient of real gas i in the pure state. State clearly all assumptions involved in your derivation.

2. Show that, if the vapor phase can be described by a pressure-explicit virial equation truncated after the second term, the mole fraction in the vapor phase is given by

$$y_i = \frac{x_i P_i^{\mathrm{sat}} \exp[v_i(P - P_i^{\mathrm{sat}})/(RT)]}{\exp[B_i P/(RT)]}, \tag{8.118}$$

where B_i is the second virial coefficient of component i and v_i is the molar volume of component i in the pure-liquid state.

8.7.A Consider a stream of warm, moist air at 50 bar. This stream is to be used for air-conditioning purposes, and it is to be cooled to 20 °C. What is the maximum amount of water that the air can bear before water starts condensing? You may assume that air is a binary mixture of nitrogen (80%) and oxygen (20%). Please conduct your calculations in the following two ways: (1) assume that the solubility of nitrogen and oxygen in water can be neglected; and (2) take into account the solubility of nitrogen and oxygen in water using Henry's law. The Henry's-law constants for nitrogen and oxygen in water at 20 °C are $\log H_{N_2-H_2O} = 4.8$ bar and $\log H_{O_2-H_2O} = 4.55$ bar. The partial molar volumes of N_2 and O_2 in water at 25 °C and infinite dilution are $\bar{v}_{N_2}^{\infty} = 40$ cm^3/mol and $\bar{v}_{O_2}^{\infty} = 31$ cm^3/mol.

8.7.B Predict the solubility of naphthelene in ethylene using a van der Waals model with the mixing rules used in Example 8.5.2.

The solubility of naphthalene in ethylene at $T = 308.15$ K and $P = 171$ bar is approximately $y_{naph} = 0.01$. The vapor pressure of naphthalene at this temperature is $P^{sat} = 2.9 \times 10^{-4}$ bar. The molar volume of solid naphthalene at this temperature is $v = 125$ cm^3/mol. What is the solubility at $P = 30$ bar and 308.15 K? To simplify your calculations, you may assume that the molar volume of the gas mixture is approximately equal to that of pure ethylene under the conditions of this problem:

Species	T_C (K)	P_C (bar)	M_W (g/mol)
Naphthalene	748.4	40.5	128.2
Ethylene	282.4	50.4	28.05

(*Hint*: you may use critical-property data to estimate the parameters a and b, i.e. $a = 27R^2T_c^2/(64P_c)$ and $b = RT_c/(8P_c)$.)

8.7.C Calculate the solubility of naphthalene in ethylene in the ranges $P = 100$–300 bar and $T = 285$–325 K using the Peng–Robinson equation of state with $k_{12} = 0.02$. Compare your results with the experimental data of Tsekhanskaya *et al.* [136].

9 Activity and equilibrium

In Chapter 8, we discussed how to carry out phase-equilibrium calculations by using an arbitrary equation of state. The formalism presented in that chapter introduced the concept of fugacity as a useful tool to conduct such calculations. Recall that the fugacity coefficient of component i in a mixture can be calculated by combining Eqns. (8.42) and (8.54):

$$\log\phi_i = \int_0^P \frac{\bar{z}_i(T, P, \{N_i\}) - 1}{P}\, dP \tag{9.1}$$

The integral appearing in Eqn. (9.1) covers the range from 0 to pressure P; the model employed for $\bar{z}_i$ must therefore be accurate over that whole range.

In recent years, much progress has been achieved in the development of accurate equation-of-state models for pure fluids and their mixtures. Some of the more sophisticated models are capable of describing the thermodynamic properties of fluids both in low- and in high-density regimes. It is now possible to evaluate integrals such as that appearing in Eqn. (9.1) and arrive at reliable estimates of the fugacity coefficient in mixtures, both vapor and liquid.

For many years, however, such models were not available; a parallel framework was therefore developed for doing phase-equilibrium calculations. In this framework, the fugacities and fugacity coefficients introduced in the previous chapter are used for the vapor phase, and another quantity was introduced for the liquid phase. The liquid- or condensed-phase framework is that of activity coefficients; in this chapter we describe these and present some of the models that are traditionally used in that context.

9.1 EXCESS PROPERTIES AND ACTIVITIES

Recall that fugacities were more useful than chemical potentials for two reasons. First, we did not need a fundamental relation – only a *PVT* equation of state. Secondly, we avoided the problems associated with going to zero pressure for an ideal gas.

For estimating liquid fugacities we use activities, which have a structure analogous to that of fugacities themselves. For fugacities, we defined a residual property as the difference between a real quantity and its ideal counterpart. Now we will introduce an **excess property** as the difference between a real quantity and its Lewis mixing counterpart. Before, we related fugacity to the exponential of the residual partial molar Gibbs free energy. Now, we will relate activity to the exponential of a partial molar excess property. Finally, we introduce the activity coefficient in a way analogous to the fugacity coefficient.

Recall Lewis mixing behavior, presented in Section 8.6.1, which related the properties of mixtures to the properties of pure (real) species. We can then define excess properties as a means to quantify the difference between an actual mixture property and that corresponding to a Lewis mixture. An **excess property** M^{E} is therefore defined by

$$M^{\mathrm{E}}(T, P, \{N_i\}) := M(T, P, \{N_i\}) - M^{\mathrm{Lewis}}(T, P, \{N_i\}). \tag{9.2}$$

Just as we did for fugacity in a mixture, a partial molar excess property for component i is defined by

$$\bar{m}_i^{\mathrm{E}}(T, P, \{x_i\}) := \bar{m}_i(T, P, \{x_i\}) - \bar{m}_i^{\mathrm{Lewis}}(T, P, \{x_i\}); \tag{9.3}$$

and, just as we did for partial molar residual properties, we could also define partial molar excess properties as

$$\bar{m}_i^{\mathrm{E}} := \left(\frac{\partial M^{\mathrm{E}}}{\partial N_i}\right)_{T,P,N_{j\neq i}}. \tag{9.4}$$

From the properties of partial molar quantities, Eqn. (8.14), we can also write

$$m^{\mathrm{E}} = \sum_{i=1}^{r} x_i \bar{m}_i^{\mathrm{E}}. \tag{9.5}$$

Excess properties measure deviations from Lewis mixtures. For most systems, those deviations can be significant, particularly at intermediate to large concentrations. It is instructive to consider experimental data for a few common solvents. Figure 9.1 shows data for the excess volume of mixing of several binary liquid mixtures with n-hexane. Lewis mixtures mix isometrically (recall Amagat's law, Eqn. (8.5)) and, by construction, do not exhibit an excess volume of mixing. In contrast, all of the mixtures of n-hexane shown in Figure 9.1 exhibit large positive or negative excess volumes of mixing. Such data serve to remind us of the importance of deviations from ideal mixing that are encountered in liquid mixtures.

Because of the strong analogy between fugacity and activity, we give the important definitions and relations here without much discussion. The **activity** of component i in a mixture is defined as

$$a_i(T, P, \{x_i\}) := \frac{f_i[T, P, \{x_i\}]}{f_i[T, P^{\mathrm{r}}, \{x_i^{\mathrm{r}}\}]}, \tag{9.6}$$

where $\{x_i\}$ is used collectively to specify the composition of the mixture, and superscript r denotes a fixed, but arbitrarily chosen, reference state. The **activity coefficient** of component i in the mixture is defined by

$$\gamma_i := \frac{a_i}{x_i}. \tag{9.7}$$

The excess partial molar Gibbs free energy of component i in the mixture can be written as

$$\begin{aligned}\bar{g}_i^{\mathrm{E}} &:= \bar{g}_i - \bar{g}_i^{\mathrm{Lewis}} \\ &= RT \log\left(\frac{f_i}{f_i^{\mathrm{Lewis}}}\right)\end{aligned} \tag{9.8}$$

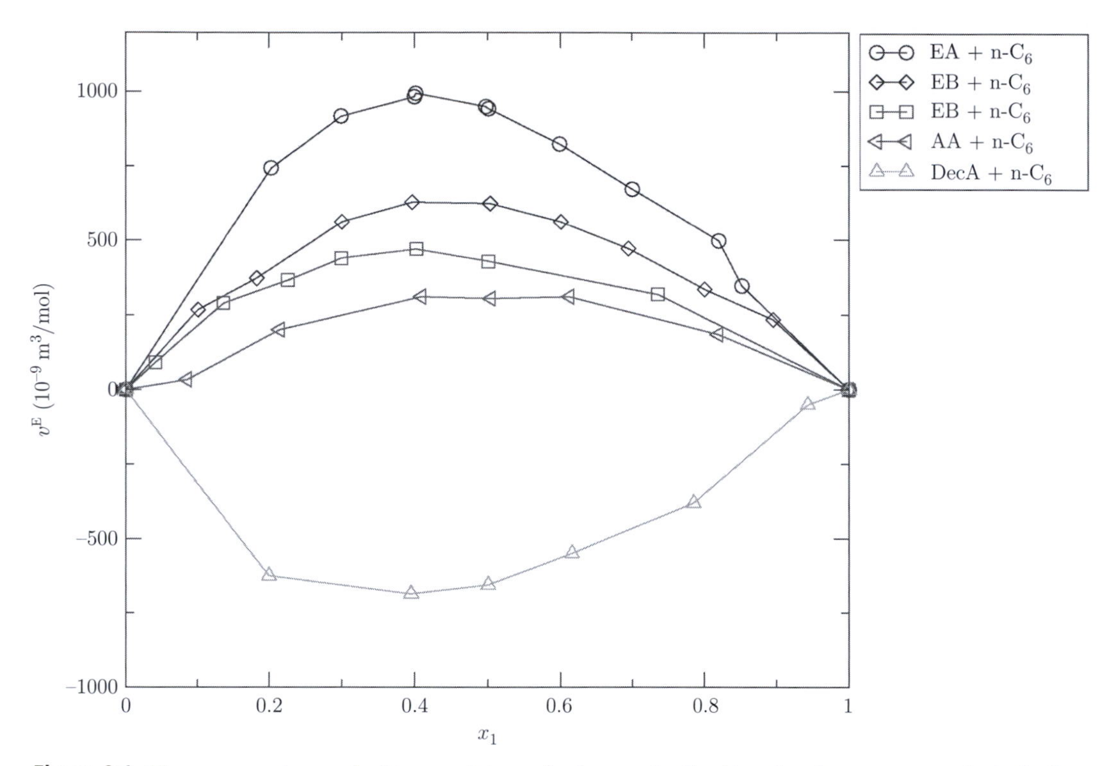

Figure 9.1 The excess volume of a binary mixture of n-hexane (n-C_6, 1) and various components, including ethyl acetate (EA), ethyl propionate (EP), ethyl butyrate (EB), amyl acetate (AA), and decyl acetate (DecA), at $T = 20\,°\mathrm{C}$ and 1 bar. From [93].

by substitution of the fugacity. Since we assume Lewis mixing, we have $f_i^{\text{Lewis}} = x_i f_i^{\text{pure}}$, and Eqn. (9.8) becomes

$$\bar{g}_i^{\mathrm{E}} = RT \log\left(\frac{f_i}{x_i f_i^{\text{pure}}}\right). \tag{9.9}$$

If we now set the fugacity $f_i[T, P^{\mathrm{r}}, \{x_i^{\mathrm{r}}\}]$ in Eqn. (9.6) to be that of the pure component i, Eqn. (9.9) becomes

$$\bar{g}_i^{\mathrm{E}} = RT \log \gamma_i. \tag{9.10}$$

From the general properties of partial molar quantities, by multiplying each side of Eqn. (9.10) by x_i and summing over all i, we also have

$$g^{\mathrm{E}} = RT \sum_i x_i \log \gamma_i. \tag{9.11}$$

Note that, by construction, $\gamma_i \to 1$ when $x_i \to 1$.

It can also be shown that the pressure dependence of an activity coefficient is related to the partial molar volume through an expression of the form

$$\left(\frac{\partial \log \gamma_i}{\partial P}\right)_{T,x} = \frac{\bar{v}_i^{\mathrm{E}}}{RT}. \tag{9.12}$$

The temperature dependence is related to the partial molar enthalpy by an expression of the form

$$\left(\frac{\partial \log \gamma_i}{\partial T}\right)_{P,x} = -\frac{\bar{h}_i^{\mathrm{E}}}{RT^2}. \tag{9.13}$$

9.2 A SUMMARY OF FUGACITY AND ACTIVITY

At this point it is instructive to summarize and compare the approaches presented for the description of condensed-phase non-idealities and those presented earlier for vapor phases. Equilibrium conditions are found by equating temperature, pressure, and fugacities. For vapor–liquid equilibrium, then, we write

$$f_i^{\text{vap}} = f_i^{\text{liq}}, \quad i = 1, \ldots, r. \tag{9.14}$$

The vapor phase is written in terms of fugacity coefficients, and the liquid phase in terms of activity coefficients:

$$\begin{aligned} f_i^{\text{vap}} &= f_i^{\text{liq}}, \quad i = 1, \ldots, r, \\ y_i \phi_i^{\text{vap}} P &= x_i \gamma_i f_i^{\text{pure}}, \quad i = 1, \ldots, r, \end{aligned} \tag{9.15}$$

where y_i denotes the mole fraction of component i in the vapor phase and x_i denotes its mole fraction in the liquid phase. The fugacity coefficient is found by the relation

$$\phi_i = \exp\left[\frac{\bar{g}_i^{\mathrm{R}}}{RT}\right] = \exp\left[\int_0^P \frac{\bar{z}_i(T, P, \{x_j\}) - 1}{P}\, dP\right]. \tag{9.1$'$}$$

Therefore, the fugacity coefficient can be found from a *PVT* equation of state for the mixture in the gas phase.

On the liquid side, the activity coefficient is found from the excess property

$$\gamma_i = \exp\left[\frac{\bar{g}_i^{\mathrm{E}}}{RT}\right], \tag{9.16}$$

which requires an equation of state for the excess property. To find the fugacity in the pure state f_i^{pure}, we require a *PVT* equation of state for the pure component in the vapor phase. The central points are that vapor fugacity is largely determined by pressure, but that for liquids is largely determined by the fugacity of the pure liquid. The coefficients ϕ_i and γ_i are the corrections.

Table 9.1 provides an itemized comparison of the two approaches. As can be seen from this table, the description outlined here in terms of the activity coefficient closely parallels the presentation in terms of the fugacity coefficient; the only substantial difference is that in the latter case the reference fluid was an ideal gas, whereas here the reference fluid is a Lewis, or ideal, mixture.

The remainder of this chapter discusses how to determine the activity coefficient γ_i in the liquid phase.

Table 9.1 A comparison of excess and residual properties.

Condensed phase	Vapor phase
• Measure deviations from Lewis mixing	• Measure deviations from ideal-gas behavior
• Excess properties: $g^{\mathrm{E}} = g - g^{\mathrm{Lewis}}$	• Residual properties: $g^{\mathrm{R}} = g - g^{\text{ideal gas}}$
• Activity coefficient: $\gamma_i = f_i/(x_i f_i^{\text{pure}})$	• Fugacity coefficient: $\phi_i = f_i/(y_i P)$
• $\bar{g}_i^{\mathrm{E}} = RT \log \gamma_i$	• $\bar{g}_i^{\mathrm{R}} = RT \log \phi_i$

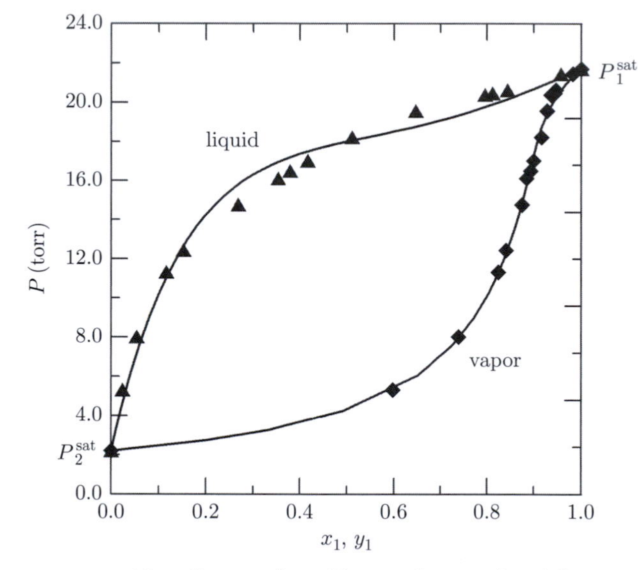

Figure 9.2 The pressure–composition diagram for a binary mixture of cycloheptane (1) and cyclopentanol (2) at 25 °C. The symbols are experimental data from Anand *et al.* [4]. The lines show calculations using a two-suffix Margules equation with $A/(RT) = 1.6385$.

9.3 CORRELATIONS FOR PARTIAL MOLAR EXCESS GIBBS FREE ENERGY

9.3.1 Simple binary systems

For simplicity, we begin by discussing the case of a binary liquid mixture in which the two components are of similar size and chemistry. A general expression for the excess Gibbs free energy of the mixture should be zero for the pure components. In the absence of additional information, one can simply propose the following polynomial to describe excess molar Gibbs free-energy data:

$$g^{\mathrm{E}} = Ax_1x_2, \tag{9.17}$$

where A is a constant, which may depend on temperature and pressure, but not on composition. Equation (9.17) is often referred to as a "two-suffix Margules equation," or also as a

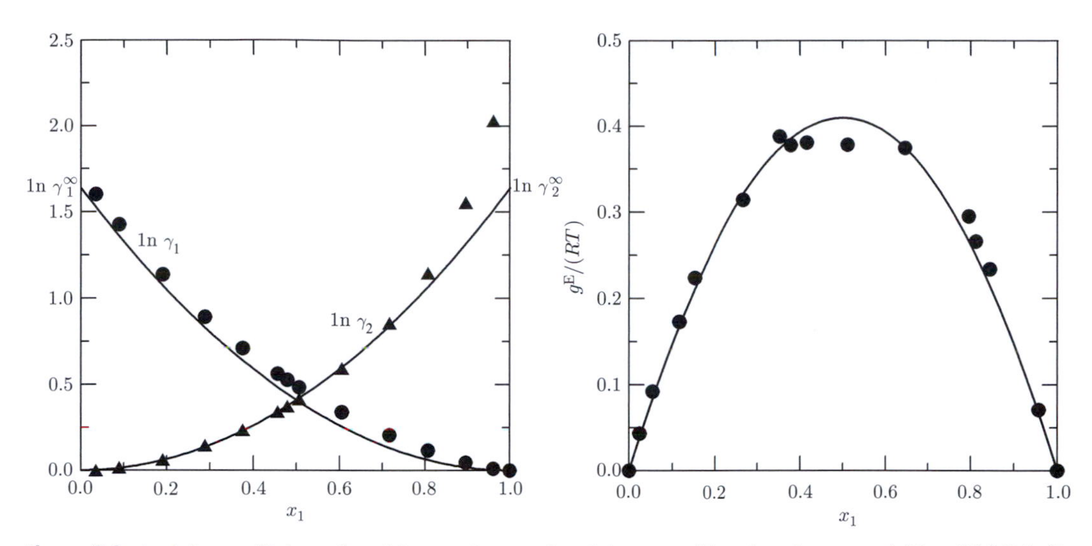

Figure 9.3 Activity coefficients for a binary mixture of cycloheptane (1) and cyclopentanol (2) at 25 °C (left), and the molar excess Gibbs free energy (right). The symbols were determined from the experimental data in Figure 9.2. The lines show calculations using a two-suffix Margules equation with $A/(RT) = 1.6385$.

"one-parameter Margules equation." This expression for g^{E} leads to the following equations for the activity coefficients:

$$\begin{aligned} RT \log \gamma_1 &= A x_2^2, \\ RT \log \gamma_2 &= A x_1^2. \end{aligned} \tag{9.18}$$

Figure 9.2 shows experimental vapor–liquid equilibrium data for a binary mixture of cycloheptane (1) and cyclopentanol (2) at 25 °C. The line is a fit to the data using a two-suffix Margules equation and assuming that the vapor phase is an ideal gas. Figure 9.3 shows the corresponding activity coefficients and excess molar Gibbs free energy for this mixture. The excess Gibbs free-energy data are fairly symmetric and can be described by a two-suffix Margules equation.

EXAMPLE 9.3.1 Use the Margules model to predict the boiling point and vapor composition for a mixture of cycloheptane (species 1) and cyclopentanol (species 2) of arbitrary composition. Assume atmospheric pressure, $P_{\mathrm{atm}} = 1.013\,25$ bar, and the value $A = 4.06$ kJ, as estimated by Anand *et al.* [4].

Solution. The fugacities of each component in the two phases must be the same:

$$\begin{aligned} f_i^{\,\mathrm{l}} &= f_i^{\,\mathrm{v}}, \quad i = 1, 2, \\ x_i \gamma_i^{\mathrm{l}} f_i^{\,\mathrm{pure}} &= y_i \phi_i^{\mathrm{v}} P, \quad i = 1, 2, \end{aligned} \tag{9.19}$$

where y_i is used to denote the composition of the vapor phase and x_i that of the liquid. Superscripts v and l denote a quantity evaluated at the vapor- or liquid-phase composition, respectively. As is customary, we write the liquid phase in terms of activity coefficients and the vapor phase in terms of fugacity coefficients. We do this because the liquid phase is dominated by the energetics of interactions between the molecules, whereas the vapor phase is dominated by the entropy, and, therefore, pressure. To a first approximation, we neglect non-ideality in the vapor phase. We also use the approximation for the pure-liquid fugacity derived earlier, Eqn. (8.50). Therefore, our equal-fugacity equations become

$$x_i \gamma_i^{l} P_i^{\mathrm{sat}} \exp\left[\int_{P_i^{\mathrm{sat}}}^{P_{\mathrm{atm}}} \frac{v_i^{\mathrm{pure}}(T, P')}{RT}\, dP'\right] \cong y_i P_{\mathrm{atm}}, \qquad i = 1, 2. \tag{9.20}$$

At atmospheric pressure, it is safe to neglect the Poynting correction factor for a liquid, so

$$\begin{aligned} x_i \gamma_i^{l} P_i^{\mathrm{sat}} &= y_i P_{\mathrm{atm}}, \qquad i = 1, 2, \\ x_i \exp\left[\frac{A(1 - x_i)^2}{RT}\right] P_i^{\mathrm{sat}} &= y_i P_{\mathrm{atm}}, \qquad i = 1, 2, \end{aligned} \tag{9.21}$$

where we have inserted the Margules estimate for the activity coefficient. We can estimate the saturation pressures for cycloheptane using the Antoine equation, Eqn. (4.49). The parameters are found from Poling, Prausnitz, and O'Connell [102] to be $A_1 = 3.963\,30$, $B_1 = 1322.22$, and $C_1 = -57.853$. The saturation pressure for cyclopentanol uses a generalized form,

$$\log\left(\frac{P_{\mathrm{sat}}}{P_{\mathrm{c}}}\right) = \frac{T_{\mathrm{c}}}{T}\left[a_2\left(1 - \frac{T}{T_{\mathrm{c}}}\right) + b_2\left(1 - \frac{T}{T_{\mathrm{c}}}\right)^{3/2} + c_2\left(1 - \frac{T}{T_{\mathrm{c}}}\right)^{5/2} + d_2\left(1 - \frac{T}{T_{\mathrm{c}}}\right)^{5}\right], \tag{9.22}$$

where $T_{\mathrm{c}} = 619.5\,\mathrm{K}$, $a_2 = -7.409\,84$, $b_2 = 1.718\,52$, $c_2 = -6.8471$, and $d_2 = -4.361\,77$. We can eliminate two of the variables, since $x_2 = 1 - x_1$ and $y_2 = 1 - y_1$. We currently have two unknowns, T and y_1. We can eliminate the latter by summing each side of Eqn. (9.21) for the two components to obtain

$$x_1 \exp\left[\frac{A(1 - x_1)^2}{RT}\right] P_1^{\mathrm{sat}}[T] + (1 - x_1)\exp\left[\frac{A(x_1)^2}{RT}\right] P_2^{\mathrm{sat}}[T] = P_{\mathrm{atm}}. \tag{9.23}$$

This equation has a single unknown, T, if we use the Antoine equation for the saturation pressure of species 1, and the generalized Antoine equation, Eqn. (9.22) for species 2. However, the equation is highly nonlinear, so its solution requires an iterative numerical technique.

For a given liquid composition, we use Eqn. (9.23) to find the boiling point. The vapor composition can then be found from Eqn. (9.21). For example, if we have an equimolar liquid mixture, then Eqn. (9.23) has root $T = 391.8\,\mathrm{K}$, found using the software Mathematica. Many other packages could be used, even a spreadsheet. Then, from Eqn. (9.21) (using $i = 1$), we find the composition of the vapor phase to be $y_1 = 0.679$. We can construct a T–x plot of vapor and liquid compositions and boiling point, as shown in Figure 9.4. For example, we assume a composition, say $x_1^{\mathrm{liq}} = 0.2$, find its boiling point to be approximately 397 K, and add this to the

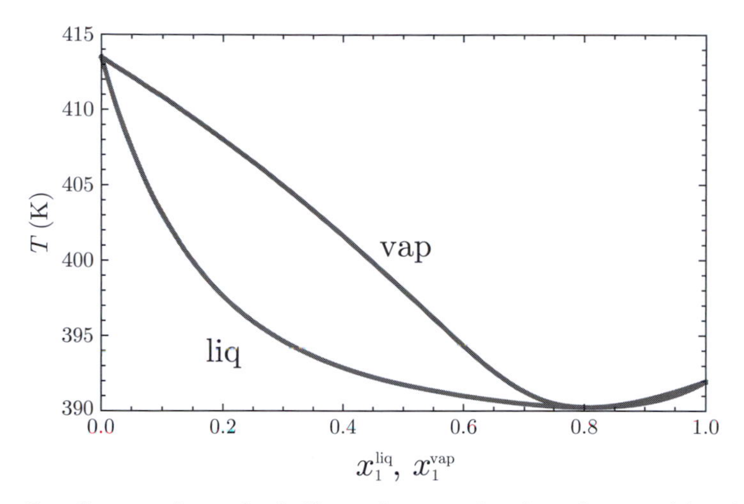

Figure 9.4 This T–x diagram shows the boiling point as a function of composition for a mixture of cycloheptane (1) and cyclopentanol (2) as predicted by the Margules model. The parameter A was found by Anand *et al.*, as shown in Figure 9.2. The "vap" curve shows the composition of the vapor phase in equilibrium with the liquid phase.

liquid curve. The composition of the vapor (which must be at the same temperature) is found to be approximately 0.53, which we add to the vapor curve.

Plots like Figure 9.4 are useful in designing many separation processes, such as distillation. In this case, the Margules model predicts an azeotrope near $x_1^{\text{liq}} \cong 0.8$. At this point, it is no longer possible to separate the components by boiling, since the vapor phase has the same composition as the liquid phase. □

EXAMPLE 9.3.2 Many binary systems are known to exhibit azeotropes. For an azeotropic mixture, the compositions of the vapor and liquid phases at equilibrium are identical, as shown in the previous example. Lewis mixtures cannot give rise to azeotropic behavior; in order to describe azeotropic mixtures, it is necessary to take into account departures from Lewis mixing in the liquid phase.

A classic azeotropic mixture is provided by ethanol (1) and water (2). The azeotropic composition is approximately $x_1 = 0.89$. The boiling temperature of that mixture is approximately 79 °C. The vapor pressures of ethanol and water can be determined from the following Antoine-type equations:[1]

$$\begin{aligned} \log_{10} P_1^{\text{sat}}(kPa) &= 16.6758 - \frac{3674.49}{T\,(^\circ\text{C}) + 226.45}, \\ \log_{10} P_2^{\text{sat}}(kPa) &= 16.262 - \frac{3799.89}{T\,(^\circ\text{C}) + 226.35}. \end{aligned} \tag{9.24}$$

Use this information to estimate the constant A for a one-parameter Margules equation.

[1] See Eqn. (4.49), and Table D.4 in Appendix D for more details.

Solution. For phase equilibrium between a liquid and a vapor, we write

$$\begin{aligned} f_i^{\text{vap}} &= f_i^{\text{liq}}, \quad i = 1, 2, \\ y_i \phi_i^{\text{vap}} P &= x_i \gamma_i f_i^{\text{pure}}, \quad i = 1, 2. \end{aligned} \tag{9.25}$$

For simplicity, we assume that the vapor phase is an ideal gas. For an azeotropic mixture, $x_1 = y_1$; Eqns. (9.25) therefore reduce to

$$P = P_1^{\text{sat}} \gamma_1 = P_2^{\text{sat}} \gamma_2, \tag{9.26}$$

when we neglect the Poynting correction factor. If a one-parameter Margules equation is employed, we have

$$\frac{A}{RT} = \frac{1}{x_2^2 - x_1^2} \log \left(\frac{P_2^{\text{sat}}}{P_1^{\text{sat}}} \right). \tag{9.27}$$

At $T = 79\,^{\circ}\mathrm{C}$, we get

$$\frac{A}{RT} = 1.062. \tag{9.28}$$

It is important to point out that in this example we use a simple one-parameter Margules equation to describe non-ideality in the ethanol–water mixture. This mixture, however, is highly non-ideal, and more elaborate models are necessary in order to describe it over wide ranges of composition, temperature, and pressure. □

In the limit of infinite dilution, i.e. in the limit when the concentration of one of the components approaches zero, the activity coefficients of a one-parameter Margules equation tend to

$$\begin{aligned} \gamma_1^{\infty} := \lim_{x_i \to 0} \gamma_1 &= \exp\left(\frac{A}{RT}\right), \\ \gamma_2^{\infty} &= \exp\left(\frac{A}{RT}\right), \end{aligned} \tag{9.29}$$

which shows one of the shortcomings of the Margules model: the infinite-dilution activity coefficients of components 1 and 2 are identical, regardless of their chemical identity. This is generally not the case, and to avoid this oversimplification it is necessary to resort to a more "flexible" expression containing more adjustable parameters. More generally, the excess molar Gibbs free energy can, for example, be expanded in the order parameter $x_1 - x_2$ to give

$$g^{\mathrm{E}} = x_1 x_2 [A + B(x_1 - x_2) + C(x_1 - x_2)^2 + \cdots], \tag{9.30}$$

where the number of terms in the series depends on the asymmetry and the quality of the data to be correlated with that expression. Equation (9.30) is known as the **Redlich–Kister expansion**.

9.3.2 Thermodynamic consistency

As indicated by Eqn. (9.10), the logarithm of an activity coefficient is a partial molar property. As such, it obeys a relationship of the form

$$\sum_i x_i \, d\log \gamma_i = 0, \qquad \text{constant } T \text{ and } P. \tag{9.31}$$

In other words, the activity coefficients of the various components that constitute a mixture are not independent of each other. In a binary system, Eqn. (9.31) implies that it is possible to determine the activity coefficient of one component from knowledge of the other. Alternatively, Eqn. (9.31) can also be used to test the consistency between experimental phase-equilibrium data for a binary mixture. The following example serves to illustrate these ideas.

EXAMPLE 9.3.3 Let us assume that experimental data for the activity coefficient of component 1 in a binary mixture can be correlated by a simple polynomial of the form

$$\log \gamma_1 = a_1 x_2^2 + a_2 x_2^3 + a_3 x_2^4, \tag{9.32}$$

where the a_i are adjustable parameters. What general form should the correlation for $\log \gamma_2$ follow in order to be consistent with Eqn. (9.32)?

Solution. From Eqn. (9.31), we can write

$$\frac{d\log(\gamma_1/\gamma_2)}{dx_2} = \frac{1}{x_2}\frac{d\log \gamma_1}{dx_2}, \tag{9.33}$$

which, after substitution of Eqn. (9.32), gives

$$\frac{d\log(\gamma_1/\gamma_2)}{dx_2} = 2a_1 + 3a_2 x_2 + 4a_3 x_2^2. \tag{9.34}$$

Integration of that expression results in

$$\log \gamma_2 = \log \gamma_1 - \left(2a_1 x_2 + \frac{3}{2}a_2 x_2^2 + \frac{4}{3}a_3 x_2^3\right) + C, \tag{9.35}$$

where C is a constant of integration. To evaluate it, we recall that for $x_2 = 1$ we should recover $\gamma_2 = 1$; the result for C is

$$C = a_1 + \frac{1}{2}a_2 + \frac{1}{3}a_3. \tag{9.36}$$

The final expression for $\log \gamma_2$ is therefore given by

$$\log \gamma_2 = \left(a_1 + \frac{3}{2}a_2 + 2a_3\right)x_1^2 - \left(a_2 + \frac{8}{3}a_3\right)x_1^3 + a_3 x_1^4. \tag{9.37}$$

□

EXAMPLE 9.3.4 Let us assume that experimental data are available for the activity coefficients of both components of a binary mixture. Show that

$$\int_0^1 \log\left(\frac{\gamma_1}{\gamma_2}\right) dx_1 = 0. \tag{9.38}$$

Solution. From the general properties of partial molar quantities we can write

$$\frac{g^{\mathrm{E}}}{RT} = x_1 \log \gamma_1 + x_2 \log \gamma_2. \tag{9.39}$$

After differentiation with respect to x_1 we have

$$\frac{dg^{\mathrm{E}}/(RT)}{dx_1} = x_1 \frac{\partial \log \gamma_1}{\partial x_1} + \log \gamma_1 + x_2 \frac{\partial \log \gamma_2}{\partial x_1} + \log \gamma_2 \frac{dx_2}{dx_1}, \tag{9.40}$$

which, given that $dx_1 = -dx_2$, can also be written as

$$\frac{dg^{\mathrm{E}}/(RT)}{dx_1} = \log\left(\frac{\gamma_1}{\gamma_2}\right). \tag{9.41}$$

Integrating with respect to x_1 gives

$$\int_0^1 \log\left(\frac{\gamma_1}{\gamma_2}\right) dx_1 = 0, \tag{9.42}$$

where we have used the fact that $g^{\mathrm{E}} = 0$ for the pure components. Equation (9.38) serves as the basis for the so-called *area test* of phase-equilibrium data. □

The molar excess Gibbs free energy is related to the molar excess enthalpy of mixing by

$$\left(\frac{\partial g^{\mathrm{E}}/(RT)}{\partial T}\right)_{P,x} = -\frac{h^{\mathrm{E}}}{RT^2}. \tag{9.43}$$

Similarly, it is related to the molar excess volume of mixing by

$$\left(\frac{\partial g^{\mathrm{E}}/(RT)}{\partial P}\right)_{T,x} = \frac{v^{\mathrm{E}}}{RT}. \tag{9.44}$$

These two relations allow us to establish thermodynamic consistency among enthalpy-of-mixing, volume-of-mixing, and vapor–liquid-equilibrium data.

9.4 SEMI-THEORETICAL EXPRESSIONS FOR ACTIVITY COEFFICIENTS

In the previous section we provided simple polynomial expressions to correlate activity coefficient data. Over the last three decades a considerable effort has been directed towards development of semi-theoretical models for the molar excess Gibbs free energy of mixtures. For an extensive discussion of excess Gibbs free-energy models readers are referred to [102].

These models have been shown to be extremely useful for applications; furthermore, when coupled to a group-contribution representation of a system, they can be used as a predictive tool. This section describes some of the more widely used excess Gibbs free-energy models.

9.4.1 The van Laar equation

The **model of van Laar** provides a two-parameter expression for the molar excess Gibbs free energy of the form:

$$\frac{g^{\mathrm{E}}}{RT} = \frac{ABx_1x_2}{Ax_1 + Bx_2}, \tag{9.45}$$

where A and B are two adjustable parameters that may depend on T and P, but not on composition. The corresponding activity coefficients are given by

$$\begin{aligned} \log \gamma_1 &= \frac{A}{[1 + (A/B)x_1/x_2]^2}, \\ \log \gamma_2 &= \frac{B}{[1 + (B/A)x_2/x_1]^2}. \end{aligned} \tag{9.46}$$

In principle, the model of van Laar should be used to describe mixtures of relatively simple components. In practice, however, it is found that van Laar's equation is often able to correlate data for more complex fluids. Figure 9.5 shows experimental data for a binary mixture of tetrachloromethane and acetonitrile. This mixture exhibits an azeotrope, which the van Laar equation is able to capture with reasonable accuracy. The corresponding molar excess Gibbs free energy is also shown in Figure 9.5; as can be seen from that figure, the excess Gibbs free energy is slightly asymmetric and a two-parameter expression is required in order to fit the data.

9.4.2 Wilson's equation

In Wilson's equation, the molar excess Gibbs free energy is given by

$$\frac{g^{\mathrm{E}}}{RT} = -x_1 \log(x_1 + \Lambda_{12}x_2) - x_2 \log(x_2 + \Lambda_{21}x_1), \tag{9.47}$$

where Λ_{21} and Λ_{12} are two adjustable parameters. These parameters are assumed to depend on temperature according to

$$\Lambda_{12} = \frac{v_2}{v_1} \exp\left(-\frac{\lambda_{12} - \lambda_{11}}{RT}\right), \tag{9.48}$$

$$\Lambda_{21} = \frac{v_1}{v_2} \exp\left(-\frac{\lambda_{12} - \lambda_{22}}{RT}\right), \tag{9.49}$$

where v_i represents the molar volume of pure component i, and $\lambda_{ij} - \lambda_{ii}$ represents the difference between the interaction energy of ij pairs and that of ii pairs. These differences are not too sensitive to temperature, and are often assumed to be constant. Note, however, that in practice, parameters Λ_{21} and Λ_{12} are determined directly by regression of experimental data, and little regard is given to the functional form suggested by Eqns. (9.48) and (9.49).

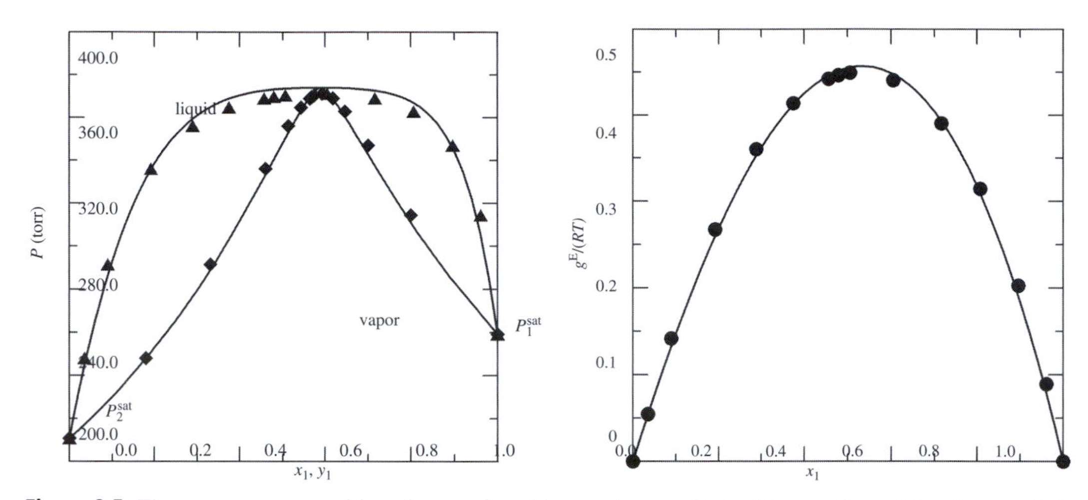

Figure 9.5 The pressure–composition diagram for a binary mixture of tetrachloromethane (1) and acetonitrile (2) at 45 °C (left), and the molar excess Gibbs free energy (right). The symbols are experimental data from Brown and Smith [15]. The lines show calculations using a van Laar equation with parameters $A = 1.6$ and $B = 2.106$.

It can be shown that the activity coefficients derived from Wilson's excess Gibbs free-energy model are given by

$$\log \gamma_1 = -\log(x_1 + \Lambda_{12}x_2) + x_2\left(\frac{\Lambda_{12}}{x_1 + \Lambda_{12}x_2} - \frac{\Lambda_{21}}{x_2 + \Lambda_{21}x_1}\right), \tag{9.50}$$

$$\log \gamma_2 = -\log(x_2 + \Lambda_{21}x_1) - x_1\left(\frac{\Lambda_{12}}{x_1 + \Lambda_{12}x_2} - \frac{\Lambda_{21}}{x_2 + \Lambda_{21}x_1}\right). \tag{9.51}$$

9.4.3 The NRTL equation

The non-random two-liquid (NRTL) model equation is often used to describe the activity coefficients of highly non-ideal mixtures. In this model, the excess Gibbs free energy is given by

$$\frac{g^{\mathrm{E}}}{RT} = x_1x_2\left(\frac{\tau_{21}G_{21}}{x_1 + x_2G_{21}} + \frac{\tau_{12}G_{12}}{x_2 + x_1G_{12}}\right), \tag{9.52}$$

where τ_{21} and τ_{12} are two adjustable parameters whose physical significance is loosely related to energy parameters characteristic of the 1–1, 2–2, and 1–2 interactions. The G_{ij} parameters are related to the τ_{ij} through

$$G_{12} = \exp\left(-\alpha\frac{\tau_{12}}{RT}\right), \quad G_{21} = \exp\left(-\alpha\frac{\tau_{21}}{RT}\right), \tag{9.53}$$

where α is also an adjustable parameter, usually set to a value of 0.2. The activity coefficients corresponding to Eqn. (9.52) are

$$
\begin{aligned}
\log \gamma_1 &= x_2^2 \left[\tau_{21} \left(\frac{G_{21}}{x_1 + x_2 G_{21}} \right)^2 + \frac{\tau_{12} G_{12}}{(x_2 + x_1 G_{12})^2} \right], \\
\log \gamma_2 &= x_1^2 \left[\tau_{12} \left(\frac{G_{12}}{x_2 + x_1 G_{12}} \right)^2 + \frac{\tau_{21} G_{21}}{(x_1 + x_2 G_{21})^2} \right].
\end{aligned}
\tag{9.54}
$$

9.4.4 The UNIQUAC model

In the universal quasi-chemical (UNIQUAC) model, the molar excess Gibbs free energy is written as the sum of *combinatorial* contributions, arising from entropic contributions, and *residual* contributions, mostly due to intermolecular forces. Since the combinatorial part of the free energy is entropic in nature, it depends only on composition and on pure-component data. The residual part, however, includes two adjustable parameters which are meant to consider intermolecular interactions.

The molar excess Gibbs free energy is given by

$$
\frac{g^{\mathrm{E}}}{RT} = \left(\frac{g^{\mathrm{E}}}{RT} \right)_{\text{combinatorial}} + \left(\frac{g^{\mathrm{E}}}{RT} \right)_{\text{residual}}, \tag{9.55}
$$

where, for a binary mixture,

$$
\begin{aligned}
\left(\frac{g^{\mathrm{E}}}{RT} \right)_{\text{combinatorial}} &= x_1 \log \left(\frac{\Phi_1}{x_1} \right) + x_2 \log \left(\frac{\Phi_2}{x_2} \right) + \frac{z_1}{2} \left(x_1 q_1 \log \left(\frac{\theta_1}{\Phi_1} \right) + x_2 q_2 \log \left(\frac{\theta_2}{\Phi_2} \right) \right), \\
\left(\frac{g^{\mathrm{E}}}{RT} \right)_{\text{residual}} &= -x_1 q_1' \log(\theta_1' + \theta_2' \tau_{21}) - x_2 q_2' \log(\theta_2' + \theta_1' \tau_{12}),
\end{aligned}
\tag{9.56}
$$

where z_1 has the meaning of a coordination number, and is usually set to 10. The parameter Φ has the meaning of a segment fraction, and θ and θ' have the meaning of area fractions. They are given by

$$
\begin{aligned}
\Phi_1 &= \frac{x_1 r_1}{x_1 r_1 + x_2 r_2}, \\
\Phi_2 &= \frac{x_2 r_2}{x_1 r_1 + x_2 r_2}, \\
\theta_1 &= \frac{x_1 q_1}{x_1 q_1 + x_2 q_2}, \\
\theta_2 &= \frac{x_2 q_2}{x_1 q_1 + x_2 q_2}, \\
\theta_1' &= \frac{x_1 q_1'}{x_1 q_1' + x_2 q_2'}, \\
\theta_2' &= \frac{x_2 q_2'}{x_1 q_1' + x_2 q_2'},
\end{aligned}
\tag{9.57}
$$

where r, q, and q' are pure-component, molecular-structure parameters. The temperature dependence of g^{E} enters the UNIQUAC model through the τ_{ij}, which are expressed in terms of adjustable binary parameters a_{ij} through

$$
\tau_{12} = \exp\left(-\frac{a_{12}}{T} \right), \quad \tau_{21} = \exp\left(-\frac{a_{21}}{T} \right). \tag{9.58}
$$

The corresponding activity coefficients are given by

$$\begin{aligned}
\log \gamma_1 &= \log\left(\frac{\Phi_1}{x_1}\right) + \frac{z}{2} q_1 \log\left(\frac{\theta_1}{\Phi_1}\right) + \Phi_2\left(l_1 - \frac{r_1}{r_2} l_2\right) \\
&\quad - q_1' \log(\theta_1' + \theta_2' \tau_{21}) + \theta_2' q_1' \left(\frac{\tau_{21}}{\theta_1' + \theta_2' \tau_{21}} - \frac{\tau_{12}}{\theta_2' + \theta_1' \tau_{12}}\right), \\
\log \gamma_2 &= \log\left(\frac{\Phi_2}{x_2}\right) + \frac{z}{2} q_2 \log\left(\frac{\theta_2}{\Phi_2}\right) + \Phi_2\left(l_2 - \frac{r_1}{r_2} l_1\right) \\
&\quad - q_2' \log(\theta_2' + \theta_1' \tau_{12}) + \theta_1' q_2' \left(\frac{\tau_{12}}{\theta_2' + \theta_1' \tau_{12}} - \frac{\tau_{21}}{\theta_1' + \theta_2' \tau_{21}}\right),
\end{aligned} \tag{9.59}$$

where

$$\begin{aligned}
l_1 &= \frac{z_1}{2}(r_1 - q_1) - (r_1 - 1), \\
l_2 &= \frac{z_1}{2}(r_2 - q_2) - (r_2 - 1).
\end{aligned}$$

9.5 DILUTE MIXTURES: HENRY'S CONSTANTS

Our previous discussion of activity coefficients has implicitly assumed that the proposed models are valid over the full range of composition. As discussed in Chapter 8, however, there are systems for which it is not necessary to consider the full spectrum of composition. In solutions of sparingly soluble gases in liquids, for example, the concentration of the gas in the liquid phase seldom exceeds a few weight per cent. As illustrated in Example 9.5.2, the solubility of nitrogen in water at ambient conditions is on the order of 5×10^{-4} mol/l. That number is small but, as illustrated in Exercise 9.5.A, its precise calculation is important for applications, including the study of certain diseases.

In those cases, i.e. for dilute solutions of one component in the other, we have seen that it is convenient to introduce an alternative reference state to that proposed in Eqns. (9.9) and (9.10), where we set $f_i[T, P^{\mathrm{r}}, \{x_j^{\mathrm{r}}\}] = f_i^{\,\mathrm{pure}}(P^{\mathrm{r}} = 1\ \mathrm{bar})$ and defined γ_i according to

$$\gamma_i := \frac{f_i}{x_i f_i^{\,\mathrm{pure}}}. \tag{9.60}$$

For simplicity, we consider again the case of a binary mixture. As discussed in the previous chapter, Lewis' fugacity rule for component 1 is an excellent approximation when component 1 is in excess, or when the mixture consists of almost pure 1. It is, however, a poor approximation when a component is dilute. If component 1 is almost pure, then component 2 must necessarily be highly dilute; Lewis's fugacity rule is therefore a poor approximation for that component. The activity coefficients defined by Eqn. (9.60) serve to measure deviations from ideal mixing in the sense of Lewis' fugacity rule; for a binary mixture, it is given by $f_1 = x_1 f_1^{\mathrm{pure}}$. We can introduce an alternative definition of activity coefficients to measure deviations from $f_2 = x_2 H_{2,1}$, where $H_{2,1}$ would be chosen in such a way as to provide an accurate representation of f_2 when component 2 is highly dilute, i.e. in the limit $x_2 \to 0$. That would be the case if $H_{2,1}$ were chosen as the hypothetical value that f_2 would have for the case of pure component 2; in that event, the relationship $f_2 = x_2 H_{2,1}$, i.e. Henry's law, would be appropriate, where $H_{2,1}$ is the Henry's-law constant for component 2 dissolved in the solvent, component 1.

Figure 8.5 illustrates schematically the behavior of fugacity for a generic binary mixture. Lewis' fugacity rule is shown to be accurate for component 1 when $x_1 \to 1$. For component 2, in the limit $x_2 \to 0$, the fugacity is described reasonably well by $f_2 = x_2 H_2$. For this mixture, the activity coefficients of the two components are defined according to

$$\begin{aligned} \gamma_1 &:= \frac{f_1}{x_1 f_1^{\text{pure}}}, \quad x_1 \to 1, \\ \gamma_2^* &:= \frac{f_2}{x_2 H_{2,1}}, \quad x_2 \to 0, \end{aligned} \tag{9.61}$$

where an asterisk is used to denote an activity coefficient referred to a Henry's-law type of behavior. Note that, in contrast to activity coefficients referred to a Lewis mixture, for Henry's-law mixtures we have $\gamma_2^* \to 1$ as $x_2 \to 0$.

Activity coefficients that are normalized with respect to a Lewis mixture can be related to activity coefficients in which one component is defined with respect to a Henry's-law constant. In a binary mixture where component 2 is dilute, for example, we have the following two definitions:

$$\begin{aligned} \gamma_2 &:= \frac{f_2}{x_2 f_2^{\text{pure}}}, \\ \gamma_2^* &:= \frac{f_2}{x_2 H_{2,1}}. \end{aligned}$$

We therefore have

$$\frac{\gamma_2}{\gamma_2^*} = \frac{H_{2,1}}{f_2^{\text{pure}}}. \tag{9.62}$$

Since

$$\lim_{x_2 \to 0} \gamma_2^* = 1, \tag{9.63}$$

we can write

$$\lim_{x_2 \to 0} \gamma_2 = \frac{H_{2,1}}{f_2^{\text{pure}}}, \tag{9.64}$$

and, following Eqn. (9.62), we get

$$\frac{\gamma_2}{\gamma_2^*} = \lim_{x_2 \to 0} \gamma_2. \tag{9.65}$$

As discussed previously, the Henry's-law constants depend not only on the nature of the solute–solvent pair, but also on temperature and pressure. The following example illustrates these effects.

EXAMPLE 9.5.1 Derive expressions for the temperature and pressure dependence of the Henry's-law constant for a binary mixture of solute 2 in solvent 1.

Solution. It was shown in Chapter 8 that the fugacity of component i in a liquid mixture changes with pressure according to (see Eqn. (8.43))

$$\left(\frac{\partial \log f_i}{\partial P}\right)_{T,x} = \frac{\bar{v}_i}{RT}, \tag{9.66}$$

where $\bar{v}_i$ is the partial molar volume of i in a liquid phase. The Henry's-law constant can be formally written as

$$H_{2,1} = \lim_{x_2 \to 0} \frac{f_2}{x_2}. \tag{9.67}$$

Substitution into Eqn. (9.66) leads to

$$\left(\frac{\partial \log H_{2,1}}{\partial P}\right)_T = \frac{\bar{v}_2^\infty}{RT}, \tag{9.68}$$

where superscript ∞ indicates that the partial molar volume of the solute (component 2) is evaluated at infinite dilution in the solvent.

For the temperature dependence, we also found

$$\left(\frac{\partial \log f_i}{\partial T}\right)_{P,x} = -\frac{\bar{h}_i^{\mathrm{R}}}{RT^2}, \tag{9.69}$$

leading to

$$\left(\frac{\partial \log H_{2,1}}{\partial T}\right)_{P,x} = -\frac{\bar{h}_2^\infty}{RT^2}. \tag{9.70}$$

□

EXAMPLE 9.5.2 Estimate the solubility of nitrogen in water at 25 °C and at 200, 500, and 1000 bar. The Henry's-law constant for nitrogen in water, evaluated at the vapor pressure of water (i.e. $P^{\mathrm{sat}} = 0.032$ bar at 25 °C), is $H_{2,1} = 86\,000$ bar. The partial molar volume of nitrogen in water at infinite dilution is $\bar{v}_2^\infty = 32.8\ \mathrm{cm^3\,mol^{-1}}$. The virial coefficient of nitrogen at 25 °C is $B_{N_2} = 43.3\ \mathrm{cm^3\,mol^{-1}}$.

Solution. The solubility of nitrogen in water can be estimated from the relationship

$$\begin{aligned} f_{N_2}^{\mathrm{vapor}} &= f_{N_2}^{\mathrm{liquid}}, \\ \phi_{N_2} y_{N_2} P &= H_{N_2,H_2O}[P,T] x_{N_2}, \end{aligned} \tag{9.71}$$

where we have assumed that the fugacity of the liquid phase can be described by an ideal solution in the sense of Henry's law, and where the notation $H_{N_2,H_2O}[P,T]$ emphasizes the fact that the Henry's-law constant must be evaluated at the temperature and pressure for which a solubility is to be determined. For simplicity, we assume that the vapor phase consists of pure nitrogen. We estimate the fugacity coefficient ϕ_{N_2} from a virial equation truncated after the second term. In that event, the solubility x_{N_2} of nitrogen in water is given by

$$x_{N_2} \approx \frac{P \exp[B_{N_2} P/(RT)]}{H_{N_2,H_2O}[P,T]}. \tag{9.72}$$

Assuming that $H_{N_2,H_2O}[P^r, T]$ is available only at a reference pressure P^r, we integrate Eqn. (9.68) with respect to pressure and substitute the result into Eqn. (9.72) to get

$$x_{N_2} \approx \frac{P\exp[B_{N_2}P/(RT)]}{H_{N_2,H_2O}[P^r, T]\exp[\bar{v}_{N_2}^{\infty}(P - P^r)/(RT)]}, \tag{9.73}$$

where we have approximated the partial molar volume of nitrogen at infinite dilution $\bar{v}_{N_2}^{\infty}$ to be independent of P. Since B_{N_2}, $H_{N_2,H_2O}[P^{sat}, T]$, and $\bar{v}_{N_2}^{\infty}$ are all available at 25 °C, this equation can now be used to estimate solubilities of nitrogen in water over a range of pressures (at that temperature). The corresponding results at 200, 500 and 1000 kbar are $x_{N_2} = 0.0025$, 0.0072 and 0.0178, respectively.

Note that by assuming that the liquid phase is an ideal solution in the sense of Henry's law, we implicitly assumed that the activity coefficient of the nitrogen in the liquid phase is unity. At high enough pressures, however, the concentration of nitrogen in the liquid can become sufficiently high to introduce appreciable deviations from ideal mixing. In that event, liquid-phase non-idealities should be taken into account, i.e.

$$f_{N_2}^{liq} = x_{N_2}\gamma_{N_2}^{*}H_{N_2,H_2O}, \tag{9.74}$$

where, as mentioned above, the asterisk serves to remind us that the activity coefficient for nitrogen is normalized with respect to a Henry's-law frame of reference. For the purposes of this example, a simple activity-coefficient model suffices; we use a one-suffix Margules equation,

$$\log \gamma_{H_2O} = \frac{A}{RT}x_{N_2}^2. \tag{9.75}$$

For nitrogen, following Eqn. (9.65), we have

$$\log \gamma_{N_2}^{*} = \frac{A}{RT}(x_{H_2O}^2 - 1). \tag{9.76}$$

The fugacity of nitrogen in the liquid can therefore be written as

$$\log\left(\frac{f_{N_2}^{liq}}{x_{N_2}}\right) = \log H_{N_2,H_2O}[P_{H_2O}^{sat}, T] + \frac{A}{RT}(x_{H_2O}^2 - 1) + \frac{\bar{v}_{N_2}^{\infty}(P - P_{H_2O}^{sat})}{RT}. \tag{9.77}$$

Equation (9.77) can now be used to estimate the fugacity of nitrogen in water, and the compositions of the vapor and liquid phases can be estimated from

$$\begin{aligned} \phi_{N_2}y_{N_2}P &= x_{N_2}H_{N_2,H_2O}[P_{H_2O}^{sat}, T]\exp\left(\frac{A}{RT}x_{N_2}^2 + \frac{\bar{v}_{N_2}^{\infty}(P - P_{H_2O}^{sat})}{RT}\right), \\ \phi_{H_2O}y_{H_2O}P &= x_{H_2O}\gamma_{H_2O}P_{H_2O}^{sat}. \end{aligned} \tag{9.78}$$

□

9.5.1 Measurement of activity coefficients

Freezing-point depression

Upon addition of a solute to a liquid, the freezing point of the newly formed solution decreases. This observation can be explained using stability arguments. It is possible to show that this

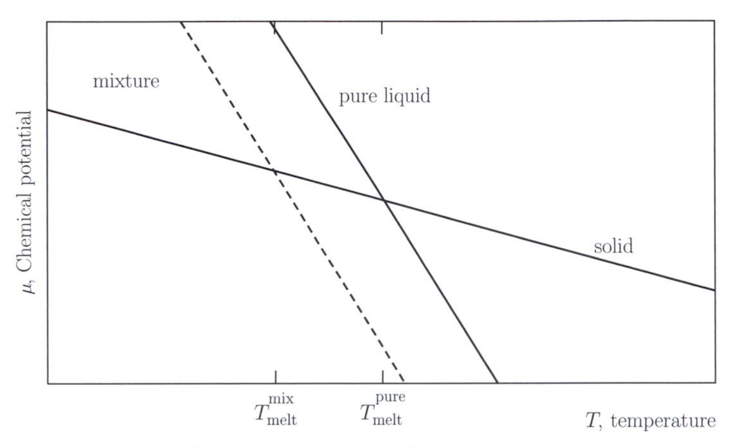

Figure 9.6 The chemical potential of a solvent plotted against temperature at fixed pressure for a pure liquid solvent, a pure solid solute, and a mixture of solvent and solute (dashed line). The text proves that these curves must have this general shape and relation to one another (though they are not generally straight lines).

feature is general, and that it works for any solute in the liquid. Our proof begins with a figure similar to Figure 4.7. In that figure, the saturation pressure is where the chemical potentials of the two phases intersect. To look at freezing-point depression now we plot the chemical potential vs. *temperature*, for a *solid* and liquid. Such a sketch is shown in Figure 9.6. First, we can show that the liquid line is always steeper than the solid line. The slope of each line on this plot is the derivative

$$\left(\frac{\partial \mu}{\partial T}\right)_P = \left(\frac{\partial g}{\partial T}\right)_P = -s. \tag{9.79}$$

Since the entropy of the liquid is always greater than that of the solid, it must have a steeper slope on this plot, and the sketch must have this general form. Note that the melting temperature of the solvent is the intersection of these two curves, $T_{\text{melt}}^{\text{pure}}$, as shown in Figure 9.6. Now we add solute to the solvent. The chemical potential of the solid remains unchanged, since the solute enters only the liquid. We can find the direction that the liquid line moves on this plot. Actually, it is easier to consider how the liquid line moves as we add more solvent (species 1), diluting the solute (species 2),

$$\left(\frac{\partial \mu_1}{\partial N_1}\right)_{T,P,N_2} = \left(\frac{\partial^2 G}{\partial N_1^2}\right)_{T,P,N_2} > 0, \tag{9.80}$$

which follows from the definition of the chemical potential, and from the stability arguments of Section 4.1. Recall from that section that stability requires all second-order derivatives with respect to extensive independent variables of any potential to be positive. Since G is a potential and N_1 is extensive, the derivative must be positive as shown. In words, this relation shows that the chemical potential of the solvent must increase with increasing solvent, which means that it must decrease with increasing solute. Therefore, the chemical potential of the solute–solvent mixture must lie below the line for the pure solvent, as shown in Figure 9.6. The new melting point is where the chemical potential of the solvent intersects with the chemical potential of the pure solid. As shown, this new melting point, $T_{\text{melt}}^{\text{mix}}$ must be lower than the pure melting point. Note that we made no assumptions to derive this result, which must be general.

The so-called freezing-point depression can be used to measure the activity coefficient of the solvent in such a dilute mixture. If subscript 1 is used to denote the solvent, the chemical potential of the solvent in the mixture can be written as

$$\mu_1[T,P] = \mu_1^{\text{pure liquid}}[T,P] + RT\log a_1, \tag{9.81}$$

where a_1 is the activity. If the solid phase in equilibrium with the solution is assumed to be pure, at the melting point of the solution, T', we have

$$\mu_1[T',P] = \mu_1^{\text{pure solid}}[T',P]. \tag{9.82}$$

The activity of component 1 in solution is therefore given by

$$\begin{aligned}\log a_1 &= \frac{\mu_1^{\text{pure solid}}[T',P] - \mu_1^{\text{pure liquid}}[T',P]}{RT'}\\ &= -\frac{\Delta g^{\text{melting}}[T',P]}{RT'}.\end{aligned} \tag{9.83}$$

Equation (9.83) can be differentiated with respect to temperature to give

$$\begin{aligned}\frac{\partial \log a_1}{\partial T'} &= \frac{\Delta s^{\text{melting}}[T',P]}{RT'} + \frac{\Delta g^{\text{melting}}[T',P]}{RT'^2}\\ &= \frac{T'\Delta s^{\text{melting}}[T',P] + \Delta h^{\text{melting}}[T',P] - T'\Delta s^{\text{melting}}[T',P]}{RT'^2}\\ &= \frac{\Delta h^{\text{melting}}[T',P]}{RT'^2}.\end{aligned} \tag{9.84}$$

Equation (9.84) can now be integrated with respect to temperature from the pure state (pure solvent, $a_1 = 1$) to the composition of the solution:

$$\begin{aligned}\log a_1 &= \log(\gamma_1 x_1)\\ &= \int_{T^{\text{melting,pure}}}^{T'} \frac{\Delta h^{\text{melting}}[T,P]}{RT^2}\,dT.\end{aligned} \tag{9.85}$$

For small concentrations of the solute (component 2) the logarithm in Eqn. (9.85) can be expanded in a series. If a two-suffix Margules equation (9.18) is used for the activity coefficient, we get

$$\begin{aligned}\log(\gamma_1 x_1) &= \log\gamma_1 - x_2 - \frac{1}{2}x_2^2 + \cdots\\ &= \left(A - \frac{1}{2}\right)x_2^2 - x_2.\end{aligned} \tag{9.86}$$

Furthermore, for small concentrations of the solute the freezing-point depression is likely to be small; in that case we can expect $\Delta h^{\text{melting,pure}}$ to be approximately constant in the interval $T^{\text{melting,pure}} - T'$, and Eqns. (9.85) and 9.86 to be approximately equivalent to

$$A \cong \frac{1}{2} + \frac{1}{x_2^2}\left[x_2 + \frac{\Delta h_1^{\text{melting,pure}}}{R}\left(\frac{1}{T^{\text{melting,pure}}} - \frac{1}{T'}\right)\right], \tag{9.87}$$

which provides a means for estimating the constant A from knowledge of the enthalpy change of melting, the concentration, and the freezing-point depression experienced for that concentration.

Osmotic pressure

For solutions of macromolecules (e.g. polymers or proteins), osmotic-pressure measurements provide a convenient means for estimating activity coefficients. In these measurements, a macromolecule solution is separated from a solvent reservoir by a semi-permeable membrane. See Figure 9.7. The pressure on the macromolecule-solution side rises above that on the solvent-reservoir side by an amount $\Pi[x_2, T]$, i.e. the **osmotic** pressure of the solution at concentration x_2 of the solute.

At equilibrium, the chemical potential of the solvent must be the same on both sides of the membrane. We therefore can write for the solvent

$$f_1^{\text{pure}}[P,T] = f_1[x_1, P+\Pi, T], \tag{9.88}$$

or

$$f_1^{\text{pure}}[P,T] = x_1 \gamma_1[x_1, P+\Pi, T] f_1^{\text{pure}}[P+\Pi, T]. \tag{9.89}$$

Since, by virtue of (8.43),

$$\log\left(\frac{f_1^{\text{pure}}[P+\Pi, T]}{f_1^{\text{pure}}[P,T]}\right) = \int_P^{P+\Pi} \frac{v_1}{RT}\, dP, \tag{9.90}$$

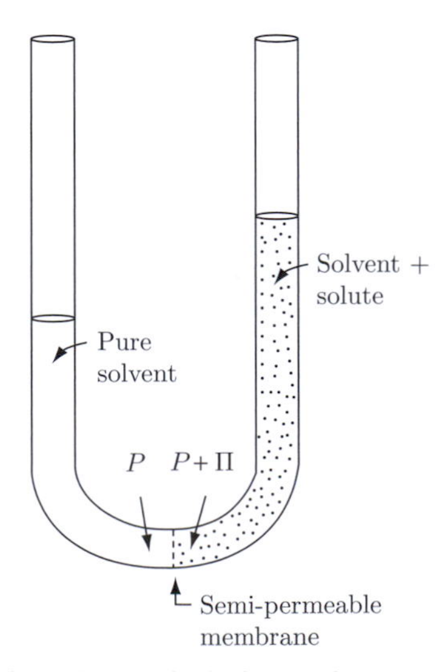

Figure 9.7 A schematic representation of a simple device used to measure osmotic pressure. The membrane at the bottom of the U-shaped tube allows solvent to pass, but not solute. The membrane is rigid, so it can support a pressure drop. The difference in height of the liquid on the two sides creates a pressure drop, whose magnitude can be calculated from the difference in height.

we can write our relation as

$$\log(\gamma_1 x_1) = -\int_P^{P+\Pi} \frac{v_1[T,P]}{RT}\, dP. \tag{9.91}$$

If we assume that the specific volume of the pure solvent, v_1, doesn't change appreciably in the range $[P; P+\Pi]$, Eqn. (9.91) can be simplified to give

$$\Pi = -RT\frac{\log(\gamma_1 x_1)}{v_1}. \tag{9.92}$$

Equation (9.92) can now be used to measure γ_1 from knowledge of Π.

9.6 THE BLOOD–BRAIN BARRIER

Human and animal cells are enveloped by membranes. These membranes are rather complicated, but are composed primarily of a bilayer of phospholipids. Figure 9.8 shows a sketch of a small section of cell membrane. One sees that the bulk of the membrane is composed of *amphiphilic* molecules called *phospholipids*, whose "heads" are hydrophilic (water-loving), and whose two "tails" are hydrophobic (water-avoidant). In an aqueous solution, such molecules will naturally, then, form a bilayer to keep the tails in a hydrophilic environment and the heads exposed to the water. By tuning the degree of hydrophilicity in the head, and the length and structure of the tails, many structures can be produced, such as bilayers, cylinders, bicontinuous phases, etc. Predicting these structures is a very interesting thermodynamics problem and an area of active research. However, we leave aside these interesting questions for the moment. Instead we focus on phospholipids, which are composed of two hydrophobic fatty-acid tails attached to an alcohol, such as glycerol. This

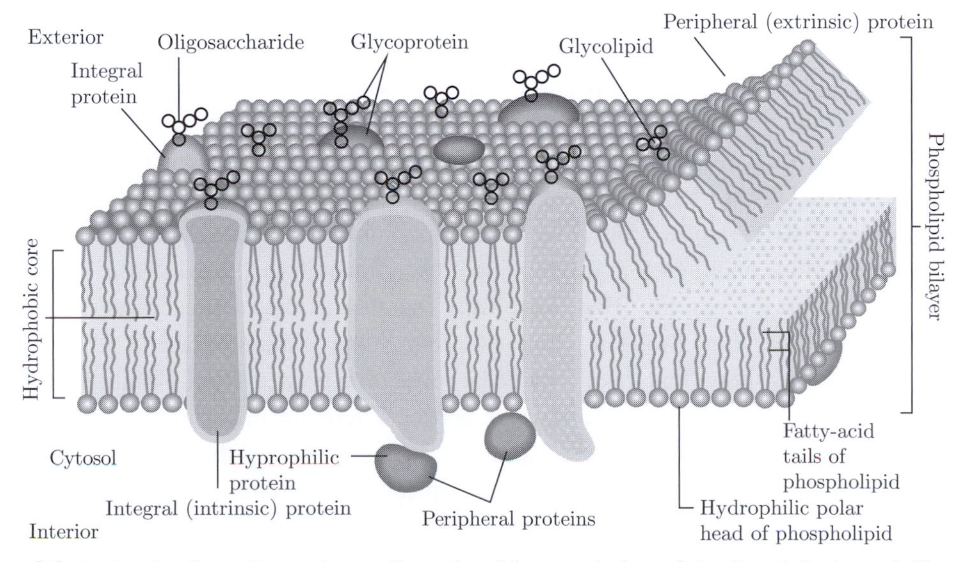

Figure 9.8 A sketch of a cell membrane. Reproduced by permission of the Royal Society of Chemistry from [156].

alcohol also contains a phosphate group, making this head hydrophilic. There are many sorts of phospholipids, depending on the chemical details, but they all have this basic structure.

A given membrane, like that shown in Figure 9.8, made up of phospholipids, forms a barrier to molecules that are hydrophilic, preventing them from passing into the cell easily. In fact, the other proteins embedded in the membrane walls are believed to have several functions, including guiding desirable ions through the membrane. On the other hand, certain hydrophobic molecules pass favorably into the cell walls. Therefore, we expect molecules that are soluble in the phospholipid tails to have potential biological impact on animals (e.g. humans). We are interested in predicting this impact for a given molecule, either because it is a drug or because it may be toxic.

A good thermodynamic mimic for the phospholipid tails, but which does not spontaneously form vesicles, is n-octanol, $CH_3C_7H_2OH$, since it has approximately the same length as the hydrophobic tails of many phospholipids. For this reason, a strong correlation has been observed between solubility with octanol and ability to stay in the body. A useful predictor of potential biological impact is then the **partition coefficient** between octanol and water. Since octanol is rather hydrophobic, an equimolar mixture of water and octanol will phase separate, although the octanol-rich phase will contain some water. The partition coefficient for species i is defined as

$$K_i^{\mathrm{ow}} = \frac{c_i^{\mathrm{o}}}{c_i^{\mathrm{w}}}, \tag{9.93}$$

where c_i^{w} is the concentration (moles/volume) of i in the water-rich phase, and c_i^{o} is the concentration in the octanol-rich phase. Hence, we expect that a substance with a large partition coefficient might have high "solubility" in the body. Then, one could use the partition coefficient to estimate the concentration of a species in the body, if the concentration in water were known. In fact, such calculations are often done to estimate the environmental impact of pollution.

Here we consider a more positive aspect of water/octanol partitioning, namely as an estimate for how efficiently drugs might be absorbed in the body, particularly the brain. For protection, there exists a barrier between the blood system, on the one hand, and the brain or spinal cord on the other. This barrier is composed of endothelial lipid membranes, and called the blood–brain barrier (BBB). Hence, drugs with a larger partition coefficient are expected to cross the BBB more rapidly, and therefore avoid being broken down in the blood stream. The rate of delivery to the brain through the BBB is typically reported as the permeability–surface-area product, *PS*, which has dimensions of (volume of drug)/(time $\times$ mass of brain).

Figure 9.9 shows that there is indeed a correlation between the solubility of a drug in octanol and its ability to pass the blood–brain barrier. There are important exceptions, however. The drugs that lie above the correlation utilize carrier molecules that mediate passage through the BBB. The drugs that lie below the line typically have higher molecular weights, which may make them too large to pass through the membranes. In fact, finding a method to mediate passage through the phospholipid membranes that make up the barrier is an important area of research for drug delivery to the brain [98].

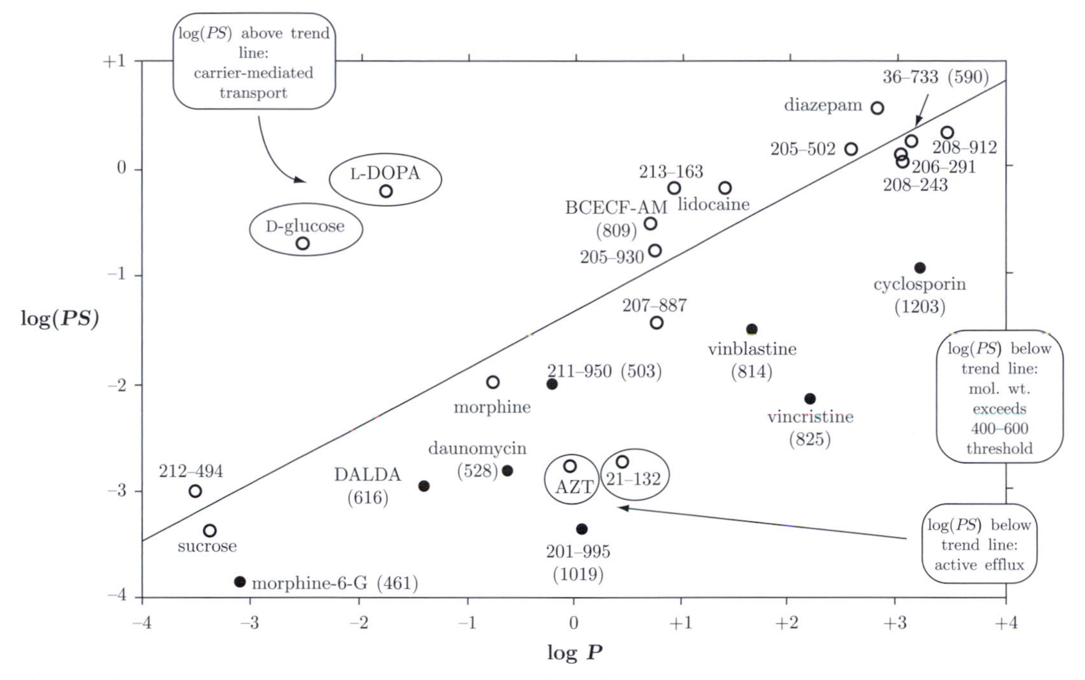

Figure 9.9 The permeability–surface-area product [μl/g · min] versus the water/octanol partition coefficient (on a log–log scale). Reproduced from [97], with permission.

9.7 PARTIAL MISCIBILITY

9.7.1 Thermodynamic stability

Excess properties provide a useful framework within which to describe partially miscible liquid mixtures. Partial miscibility serves as the basis for many industrial separation processes; it is therefore important to have a methodology to describe such mixtures quantitatively. This includes calculation of the phase diagrams of binary, ternary, and multicomponent systems.

As was the case for pure substances, partial miscibility and liquid–liquid phase transitions in binary mixtures arise as a consequence of thermodynamic instability. To illustrate this point, it is instructive to recall our previous discussion of thermodynamic stability for a pure component, in Section 4.1. There we saw that second derivatives of potentials with extensive quantities were required to be positive. Therefore, linear stability also requires

$$\left(\frac{\partial^2 G}{\partial N_1^2}\right)_{P,T,N_2} > 0. \tag{9.94}$$

For binary mixtures it is often more convenient to work with x_1 and N as independent variables rather than N_1 and N_2. Thus, we wish to cast Eqn. (9.94) in terms of mole fractions instead of mole numbers. In order to do this, we use the chain rule of partial differentiation, Eqn. (A.8) in Appendix A, which in this case becomes

$$\left(\frac{\partial}{\partial N_1}\right)_{T,P,N_2} = \left(\frac{\partial x_1}{\partial N_1}\right)_{T,P,N_2}\left(\frac{\partial}{\partial x_1}\right)_{T,P,N} + \left(\frac{\partial N}{\partial N_1}\right)_{T,P,N_2}\left(\frac{\partial}{\partial N}\right)_{T,P,x_1}$$
$$= \frac{1-x_1}{N}\left(\frac{\partial}{\partial x_1}\right)_{T,P,N} + \left(\frac{\partial}{\partial N}\right)_{T,P,x_1}. \tag{9.95}$$

Hence, we can write

$$\left(\frac{\partial G}{\partial N_1}\right)_{T,P,N_2} = \frac{1-x_1}{N}\left(\frac{\partial G}{\partial x_1}\right)_{T,P,N} + \left(\frac{\partial G}{\partial N}\right)_{T,P,x_1}$$
$$= (1-x_1)\left(\frac{\partial g}{\partial x_1}\right)_{T,P,N} + g. \tag{9.96}$$

To obtain the second line we have exploited the extensivity of the free energy: $G(T,P,x_1,N) = Ng(T,P,x_1)$. Using the chain rule once more on this result gives

$$\left(\frac{\partial^2 G}{\partial N_1^2}\right)_{T,P,N_2} = \frac{(1-x_1)^2}{N}\left(\frac{\partial^2 g}{\partial x_1^2}\right)_{T,P} - \frac{1-x_1}{N}\left(\frac{\partial g}{\partial x_1}\right)_{T,P} + \frac{1-x_1}{N}\left(\frac{\partial g}{\partial x_1}\right)_{T,P}$$
$$= \frac{(1-x_1)^2}{N}\left(\frac{\partial^2 g}{\partial x_1^2}\right)_{T,P}. \tag{9.97}$$

A very similar result can be derived for species 2. Hence, we arrive at the simple result

$$\left(\frac{\partial^2 g}{\partial x_i^2}\right)_{P,T} > 0, \quad i = 1,2, \quad \text{locally stable,} \tag{9.98}$$

where the derivative is now taken with respect to the mole fraction (as opposed to the number of moles).

It is important to emphasize that g in Eqn. (9.98) refers to the free energy of the mixture. The total molar Gibbs free energy of a binary mixture can be written as the sum of pure-component contributions and a free-energy change of mixing

$$g^{\text{mixture}}(T,P,\{x_i\}) = x_1 g_1^{\text{pure}}(T,P) + x_2 g_2^{\text{pure}}(T,P) + \Delta g^{\text{mixing}}(T,P,\{x_i\}). \tag{9.99}$$

Note that this equation defines the free energy of mixing, Δg^{mixing}. Putting this expression into Eqn. (9.98) says that thermodynamic stability requires

$$\left(\frac{\partial^2 \Delta g^{\text{mixing}}}{\partial x_i^2}\right)_{P,T} > 0, \quad i = 1,2, \quad \text{locally stable.} \tag{9.100}$$

Just as for the pure-component case, for given values of temperature and pressure, the point at which Eqn. (9.100) is violated defines the spinodal composition:

$$\left(\frac{\partial^2 \Delta g^{\text{mixing}}}{\partial x_i^2}\right)_{P,T} = 0 \quad \text{at spinodal composition.} \tag{9.101}$$

In terms of the excess molar Gibbs free energy, thermodynamic stability is satisfied when

$$\left(\frac{\partial^2 g^{\text{E}}}{\partial x_1^2}\right)_{P,T} + RT\left(\frac{1}{x_1} + \frac{1}{x_2}\right) > 0. \tag{9.102}$$

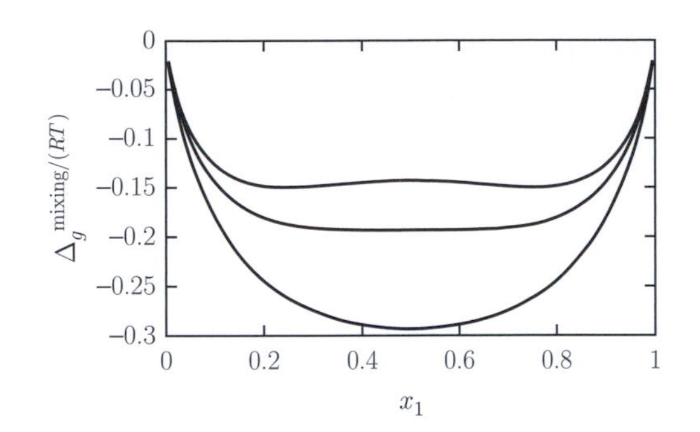

Figure 9.10 Change of Gibbs potential upon mixing as predicted by the Margules model. Shown are curves for three values of the parameter A. The lower curve is always stable, the upper curve predicts phase splitting, and the middle curve is the cross-over.

Figure 9.10 shows the molar Gibbs free energy of mixing predicted by a simple two-suffix Margules equation as a function of composition. Three curves are shown: one for which the system is stable at all compositions; one for which the system starts to develop an instability, precisely at $x = 0.5$; and one for which the system is unstable over a range of compositions. The following example discusses in more detail the temperature at which the instability first appears.

EXAMPLE 9.7.1 Consider a binary liquid mixture whose molar excess Gibbs free energy can be described by an equation of the form

$$g^{\mathrm{E}} = Ax_1x_2.$$

For what values of the constant A will the two liquids be miscible?

Solution. The second derivative of the molar excess Gibbs free energy with respect to mole fraction is given by

$$\left(\frac{\partial^2 g^{\mathrm{E}}}{\partial x_1^2}\right)_{P,T} = -2A. \tag{9.103}$$

After substitution into Eqn. (9.102), we find that the mixture is stable (miscible) whenever

$$A < \frac{RT}{2x_1x_2}. \tag{9.104}$$

Furthermore, the largest value of A that satisfies Eqn. (9.104) is

$$A = 2RT, \tag{9.105}$$

and the two liquids phase separate whenever

$$\frac{A}{RT} > 2. \tag{9.106}$$

□

As mentioned earlier, liquid–liquid equilibria provide the basis for numerous industrial separation processes, including liquid extraction. To design such processes, one requires knowledge of the coexistence curve. The following example describes the calculation of the liquid–liquid coexistence curve for a binary mixture.

EXAMPLE 9.7.2 Consider a binary liquid mixture whose excess Gibbs free energy is given by a two-suffix Margules model

$$g^{\mathrm{E}} = A(T,P)x_1x_2. \tag{9.107}$$

For a given value of the parameter A, find the spinodal compositions. By varying A, construct a phase diagram.

Solution. The spinodal point is the boundary between local stability, given by Eqn. (9.102), and local instability, where the greater-than symbol changes direction. Namely, the spinodal point is

$$\left(\frac{\partial^2}{\partial x_1^2}\frac{g^{\mathrm{E}}}{RT}\right)_{P,T} + \frac{1}{x_1} + \frac{1}{x_2} = 0. \tag{9.108}$$

If we take the derivative (twice) of the excess Gibbs free energy given in Eqn. (9.107) and insert the result above, we find

$$-\frac{2A}{RT} + \frac{1}{x_1^{\mathrm{s}}} + \frac{1}{1-x_1^{\mathrm{s}}} = 0, \tag{9.109}$$

where x_1^{s} is the composition of the spinodal point for a given value of $A/(RT)$. This curve is shown in Figure 9.11.

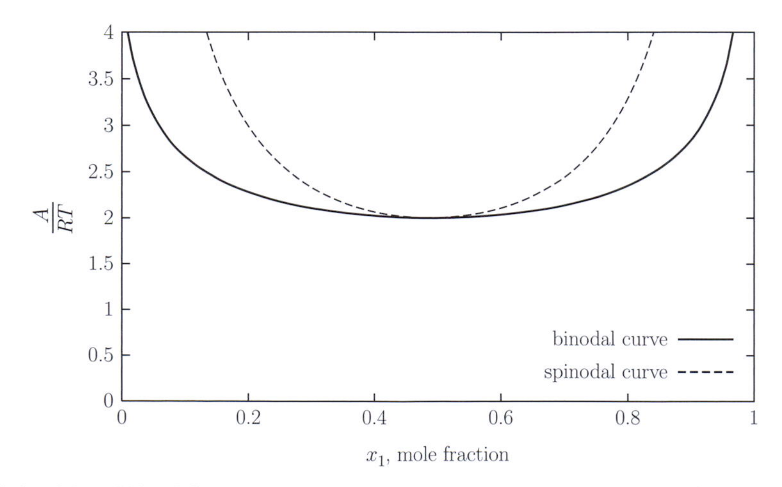

Figure 9.11 Spinodal and binodal curves predicted for the Margules model, Eqn. (9.107).

□

EXAMPLE 9.7.3 What is the critical point of the "two-suffix Margules model"?

Solution. The solution follows from the previous example. For a given value of the constant A, we find that the critical solution temperature T_c is given by

$$T_c = \frac{A(T_c, P)}{2R}, \tag{9.110}$$

and the critical composition is given by

$$x_c = \frac{1}{2}. \tag{9.111}$$

□

EXAMPLE 9.7.4 For a binary liquid mixture whose excess Gibbs free energy is described by a two-suffix Margules equation, propose a procedure to calculate the coexistence curve.

Solution. At equilibrium, the fugacity of both components should be the same in both phases. We can therefore write

$$\begin{aligned} f_1' &= f_1'', \\ f_2' &= f_2'', \end{aligned} \tag{9.112}$$

where $'$ and $''$ serve to denote the two coexisting phases. Equation (9.112) can be written as

$$\begin{aligned} x_1'\gamma_1' &= x_1''\gamma_1'', \\ x_2'\gamma_2' &= x_2''\gamma_2''. \end{aligned} \tag{9.113}$$

After substituting expressions for the activity coefficient of a two-suffix Margules equation, we have

$$\begin{aligned} x_1' \exp(A(1-x_1')^2) &= x_1'' \exp\left(\frac{A(1-x_1'')^2}{RT}\right), \\ (1-x_1')\exp\left(\frac{A(x_1')^2}{RT}\right) &= (1-x_1'')\exp\left(\frac{Ax_1''^2}{RT}\right). \end{aligned} \tag{9.114}$$

Equations (9.114) are equivalent to each other. They can be simplified by recognizing that the two-suffix Margules model is symmetric in composition, and therefore $x_1' = 1 - x_1''$. With that simplification, we have

$$x_1' \exp\left(\frac{A(1-x_1')^2}{RT}\right) = (1-x_1')\exp\left(\frac{Ax_1'^2}{RT}\right), \tag{9.115}$$

which can now be solved for arbitrary values of A. To construct the binodal curve in Figure 9.11, we can solve this equation numerically for $A/(RT) = \log[(1-x_1')/x_1']/(1-2x_1')$.

□

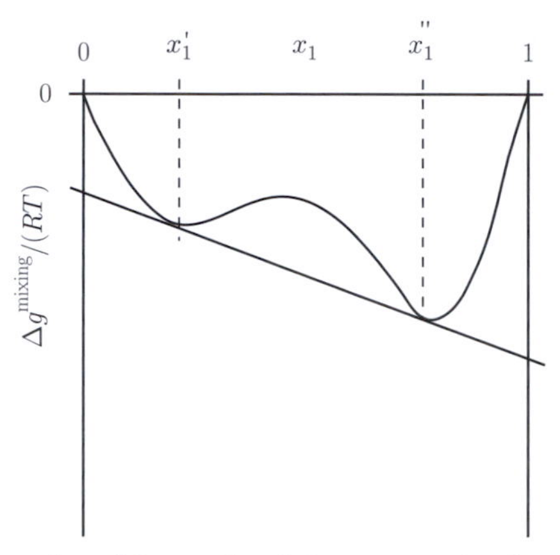

Figure 9.12 A graphical representation of the predicted phase separation for a model of the excess Gibbs free energy g^{E} of a binary mixture. The specific Gibbs free energy of mixing $\Delta g = g^{\mathrm{E}} + x_1 \log x_1 + x_2 \log x_2$ has the general curved shape shown in the figure, when the mixture is unstable. The straight line touches both lobes tangentially, determining the composition of the "1-rich" phase, x_1'', and the "2-rich" phase x_1'.

In the previous example, we saw how the composition of two coexisting liquid phases can be calculated by equating the fugacities of each component in each phase. For a binary mixture, it can also be shown that, at coexistence, the following relations are satisfied:

$$\left(\frac{\partial \Delta g^{\text{mixing}}}{\partial x_1}\right)' = \left(\frac{\partial \Delta g^{\text{mixing}}}{\partial x_1}\right)'', \tag{9.116}$$

$$\Delta g^{\text{mixing}'} - x_1' \left(\frac{\partial \Delta g^{\text{mixing}}}{\partial x_1}\right)' = \Delta g^{\text{mixing}''} - x_1'' \left(\frac{\partial \Delta g^{\text{mixing}}}{\partial x_1}\right)''. \tag{9.117}$$

In other words, at coexistence, the molar Gibbs free energy of mixing has a common tangent line, as sketched in Figure 9.12. The points at which this common tangent line intersects the Gibbs-free-energy-of-mixing curve correspond to the compositions of the two coexisting phases. Note that, for a simple two-suffix Margules equation, these two points correspond to the minima of the molar Gibbs free energy of mixing. For more realistic, asymmetric models, the compositions at coexistence and those at the relative minima in the free energy of mixing are not symmetric.

9.7.2 Liquid–liquid equilibria in ternary mixtures

Liquid–liquid extraction processes usually involve three or more components. The familiar triangular liquid–liquid diagrams are simply phase diagrams for liquid mixtures calculated at constant temperature and pressure and plotted in "triangular" coordinates.

The calculation of such phase diagrams is slightly more involved than that of their binary counterparts. For a three-component system, the compositions of the two coexisting phases must be calculated, i.e. there are four unknowns (x_1', x_2', x_1'', and x_2''). The equality of the fugacity of each component throughout the system provides just three equations. In order to arrive at a solution, it is therefore necessary to specify one of the unknown compositions, say x_1'. In that event, the value specified for x_1' should be such that a solution exists for the set of equations to be solved. In other words, a solution will not be found unless the specified compositions are in the two-phase region.

Alternatively, it is possible to perform a "flash" calculation to calculate a ternary phase diagram. In that case, in addition to the three equations for equality of fugacity, a mass balance can be written for each component of the mixture:

$$x_1'L' + x_1''L'' = z_1, \tag{9.118}$$

$$x_2'L' + x_2''L'' = z_2, \tag{9.119}$$

$$L' + L'' = 1, \tag{9.120}$$

where z_i denotes the overall (feed) composition of component i, L' and L'' are the total numbers of moles in phases $'$ and $''$, respectively, and the basis for calculations is one mole of feed mixture. There are now six equations and six unknowns, namely x_1', x_2', x_1'', x_2'', L', and L''.

It is interesting to point out that the coexistence curve for a ternary mixture can be determined through a geometrical construct analogous to that mentioned above for binary systems. It can be shown that, at coexistence, the following relations are true:

$$\left(\frac{\partial \Delta g^{\text{mixing}}}{\partial x_1}\right)' = \left(\frac{\partial \Delta g^{\text{mixing}}}{\partial x_1}\right)'', \tag{9.121}$$

$$\left(\frac{\partial \Delta g^{\text{mixing}}}{\partial x_2}\right)' = \left(\frac{\partial \Delta g^{\text{mixing}}}{\partial x_2}\right)'', \tag{9.122}$$

and

$$\begin{aligned} &\Delta g^{\text{mixing}'} - x_1'\left(\frac{\partial \Delta g^{\text{mixing}}}{\partial x_1}\right)' - x_2'\left(\frac{\partial \Delta g^{\text{mixing}}}{\partial x_2}\right)' \\ &= \Delta g^{\text{mixing}''} - x_1''\left(\frac{\partial \Delta g^{\text{mixing}}}{\partial x_1}\right)'' - x_2''\left(\frac{\partial \Delta g^{\text{mixing}}}{\partial x_2}\right)''. \end{aligned} \tag{9.123}$$

Equations (9.121)–(9.123) state that at coexistence the molar Gibbs free energy of mixing surface has a common tangent plane. The coexistence curve can be obtained by rolling a plane over that surface and projecting the intersection points onto the composition plane.

It is convenient to represent the phases in ternary mixtures using triangular diagrams of the sort shown in Figure 9.13. The mole fraction of the mixture determines the location inside the triangle. The pure components are located at the vertices of the triangle. Note the straight line that runs parallel to the side of the triangle opposite the vertex. The fractional distance from this opposite side to the vertex indicates the mole fraction of that species. The thick curved line is the binodal curve. Inside this region the mixture separates into two phases – a water-rich phase and a benzene-rich phase. These two phases are connected by a straight tie line.

9.7.3 Critical points

Just as for the case of thermodynamic stability, there exists an exact correspondence between the critical point observed in a pure fluid and the critical points that occur in binary liquid mixtures. For the pure-component case we showed that, at the critical point,

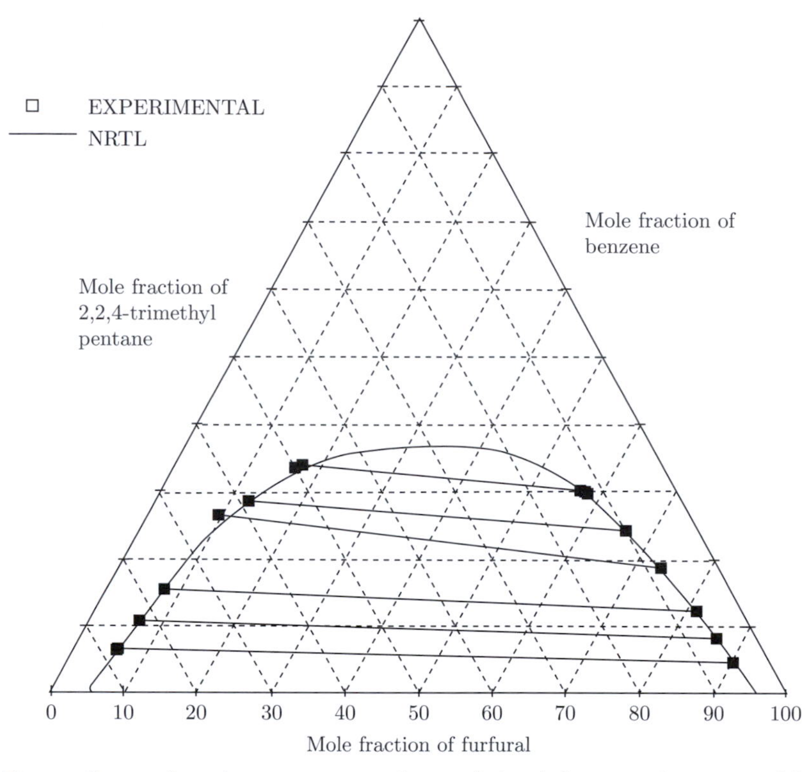

Figure 9.13 Ternary diagram for a three component mixture of trimethyl pentane, benzene, and furfural at $T = 25\,°\mathrm{C}$. All fractions are expressed as percentages. The location of a point inside the triangle indicates the mole fraction of each component. The thick line is the binodal curve, which delineates the region of composition where the mixture separates into two phases. The solid squares show experimental data. The straight lines connecting the data points on the two sides of the binodal (which are in equilibrium with each other) are called tie-lines. The binodal curve and the tie lines shown in the figure were calculated using the NRTL model. Reproduced from [56], with permission.

$$\left(\frac{\partial P}{\partial v}\right)_T = 0,$$

$$\left(\frac{\partial^2 P}{\partial v^2}\right)_T = 0.$$

For a binary liquid mixture, at the **critical solution point**, we have

$$\left(\frac{\partial^2 g^{\text{mixture}}}{\partial x_i^2}\right)_{T,P} = 0, \tag{9.124}$$

$$\left(\frac{\partial^3 g^{\text{mixture}}}{\partial x_i^3}\right)_{T,P} = 0. \tag{9.125}$$

For any given excess Gibbs free-energy model, Eqns. (9.124) and (9.125) can therefore be used to determine the position of the critical point.

9.8 SIMPLE FREE-ENERGY MODELS FROM STATISTICAL MECHANICS

Some of the models employed in this chapter can be derived using the methods of statistical mechanics. In this section we discuss how to derive Lewis mixing rules and the Margules equation of state using lattice models. The assumptions necessary for these derivations indicate when we can expect these equations to be applicable.

Lattice models are commonly used to describe liquids and solids. In these models, molecules are typically assumed to occupy distinct lattice sites, and the system of interest is generally assumed to be incompressible (i.e. the volume is constant). This assumption of constant volume makes it easier to work in the canonical ensemble $(T, V, \{N_i\})$, instead of the usual grand canonical ensemble $(T, P, \{N_i\})$ of the excess Gibbs free-energy models.

In the simplest case, the lattice can be assumed to be cubic and fully occupied. The coordination number of the lattice, denoted by z_l, specifies the number of nearest neighbors for a lattice site. In two dimensions, each molecule of the system has four nearest neighbors; in three dimensions each molecule has six nearest neighbors. For simplicity, only interactions between nearest neighbors are considered in the derivations that follow. With these assumptions in mind, we can write the canonical partition function as

$$Q \sim q_{\mathrm{AA}}(T, v)^{\tilde{N}_{\mathrm{AA}}} q_{\mathrm{BB}}(T, v)^{\tilde{N}_{\mathrm{BB}}} q_{\mathrm{AB}}(T, v)^{\tilde{N}_{\mathrm{AB}}}, \tag{9.126}$$

where $\tilde{N}_{\mathrm{AA}}$, $\tilde{N}_{\mathrm{BB}}$, and $\tilde{N}_{\mathrm{AB}}$ denote the numbers of nearest-neighbor AA, BB, or AB pairs. We have also used the pairwise partition functions $q_{ij}(T, v)$, for a lattice of fixed specific volume.

At this point it is instructive to note that, for a given composition and a given size of the system, several distinct configurations of the system can have exactly the same energy. Rather than summing over all distinct configurations of the system in Eqn. (9.126), we can therefore sum over configurations having a certain number of AB pairs. The quantity $\omega_{\mathrm{d}}(\tilde{N}_{\mathrm{AB}}, \tilde{N}_{\mathrm{A}}, \tilde{N}_{\mathrm{B}})$ is called the **degeneracy of the system** and, for any given composition, it corresponds to the number of distinct arrangements or configurations of the system that have exactly $\tilde{N}_{\mathrm{AB}}$ pairs.

For a given composition of the system, the number of AA (or BB) pairs on the lattice is related to the coordination number through

$$z_l \tilde{N}_{\mathrm{A}} = 2\tilde{N}_{\mathrm{AA}} + \tilde{N}_{\mathrm{AB}}, \tag{9.127}$$

$$z_l \tilde{N}_{\mathrm{B}} = 2\tilde{N}_{\mathrm{BB}} + \tilde{N}_{\mathrm{AB}}, \tag{9.128}$$

where $\tilde{N}_{\mathrm{A}}$ and $\tilde{N}_{\mathrm{B}}$ are the numbers of molecules of type A and type B on the lattice, respectively. Hence, using the degeneracy, and these relations for $\tilde{N}_{\mathrm{AA}}$ and $\tilde{N}_{\mathrm{BB}}$, we can rewrite Eqn. (9.126) as

$$Q = q_{\mathrm{AA}}(T, v)^{z_l \tilde{N}_{\mathrm{A}}/2} q_{\mathrm{BB}}(T, v)^{z_l \tilde{N}_{\mathrm{B}}/2} \sum_{\tilde{N}_{\mathrm{AB}}} \omega_{\mathrm{d}}(\tilde{N}_{\mathrm{AB}}, \tilde{N}_{\mathrm{A}}, \tilde{N}_{\mathrm{B}}) \lambda(T, v)^{\tilde{N}_{\mathrm{AB}}}, \tag{9.129}$$

where

$$\lambda(T, v) := q_{\mathrm{AB}}(T, v) / \sqrt{q_{\mathrm{AA}}(T, v) q_{\mathrm{BB}}(T, v)}. \tag{9.130}$$

If we take the logarithm of each side of Eqn. (9.129), and multiply by $-k_B T$, we obtain

$$F = -\frac{1}{2} z_1 N_A RT \log q_{AA}(T, v) - \frac{1}{2} z_1 N_B RT \log q_{BB}(T, v) - k_B T \log \left[\sum_{\tilde{N}_{AB}} \omega_d(\tilde{N}_{AB}, \tilde{N}_A, \tilde{N}_B) \lambda(T, v)^{\tilde{N}_{AB}} \right], \tag{9.131}$$

where we used the relationship between the free energy and the partition function, Eqn. (6.12). The first two quantities on the right are the free energies for the pure species (imagine filling a lattice of the same specific volume with only A and finding the partition function). Hence, we can rewrite this expression as

$$F = F^{\text{pure A}}(T, v, N_A) + F^{\text{pure B}}(T, v, N_B) + \Delta F^{\text{mixing}}(T, v, \tilde{N}_A, \tilde{N}_B), \tag{9.132}$$

where we have found an expression for the free energy of mixing:

$$\Delta F^{\text{mixing}}(T, v, \tilde{N}_A, \tilde{N}_B) = -k_B T \log \left[\sum_{\tilde{N}_{AB}} \omega_d(\tilde{N}_{AB}, \tilde{N}_A, \tilde{N}_B) \lambda(T, v)^{\tilde{N}_{AB}} \right]. \tag{9.133}$$

Note that this is the change of mixing at constant *volume* instead of at constant *pressure*, so the equivalence with our earlier models is valid only for

$$\left(\frac{\partial \log \lambda}{\partial \log v} \right)_T \cong 0.$$

At any rate, we have a starting point to find a fundamental F relation for the mixture.

The problem of estimating the free energy of the system has now been reduced to that of estimating the degeneracy of the system and carrying out the summation appearing in Eqn. (9.133). At this point, it is necessary to introduce approximations that will allow us to produce such estimates.

9.8.1 Lewis mixing

As a starting point, we consider the case for which $q_{AB}(T, v) = \sqrt{q_{AA}(T, v) q_{BB}(T, v)}$, so that $\lambda(T, v) = 1$. In this case, the interactions of an AB pair are the same, on average, as those of an AA pair and a BB pair. For such a mixture we have

$$\Delta F^{\text{mixing}}(T, v, \tilde{N}_A, \tilde{N}_B) = -k_B T \log \left[\sum_{\tilde{N}_{AB}} \omega_d(\tilde{N}_{AB}, \tilde{N}_A, \tilde{N}_B) \right]. \tag{9.134}$$

The summation in Eqn. (9.134) represents the total number of ways of arranging $\tilde{N}_A$ and $\tilde{N}_B$ objects randomly on precisely $\tilde{N} = \tilde{N}_A + \tilde{N}_B$ lattice sites; as such, it is given by (see Section A.8 of Appendix A)

$$\sum_{\tilde{N}_{AB}} \omega_d(\tilde{N}_{AB}, \tilde{N}_A, \tilde{N}_B, \tilde{N}_{AB}) = \frac{\tilde{N}!}{\tilde{N}_A! \tilde{N}_B!}. \tag{9.135}$$

The Helmholtz free energy of the random mixture is therefore given by

$$\Delta F^{\text{mixing}}(T, v, \tilde{N}_{\text{A}}, \tilde{N}_{\text{B}}) = -k_{\text{B}} T \log\left[\frac{(\tilde{N}_{\text{A}} + \tilde{N}_{\text{B}})!}{\tilde{N}_{\text{A}}!\tilde{N}_{\text{B}}!}\right]. \tag{9.136}$$

This equation can be simplified by invoking Stirling's approximation, which states that, for sufficiently large values of N, the natural logarithm of a factorial number can be approximated according to

$$\log N! \approx N \log N - N. \tag{9.137}$$

We can therefore rewrite Eqn. (9.136) as

$$\frac{\Delta F^{\text{mixing}}(T, v, x_{\text{A}}, x_{\text{B}})}{NRT} \cong x_{\text{A}} \log x_{\text{A}} + x_{\text{B}} \log x_{\text{B}}. \tag{9.138}$$

This result shows that the free energy of a mixture with no change in the energies of interactions upon mixing corresponds to a Lewis mixture. The difference is that the definition of Lewis mixing in Eqn. (8.66) in purely macroscopic terms resorted to Lewis' fugacity rule. The random mixture considered here therefore provides a molecular foundation for some of the concepts introduced earlier using phenomenological arguments.

9.8.2 The Margules model

We can now go beyond the random (or Lewis) mixture by introducing the idea of a "mean field." To first order, we could assume that the number of AB pairs appearing in the exponential of Eqn. (9.133) can be replaced by its average or mean value for a random mixture. If we use $\tilde{N}^*_{\text{AB}}$ to denote this random mean value, we have

$$\begin{aligned}
\Delta F^{\text{mixing}}(T, v, \tilde{N}_{\text{A}}, \tilde{N}_{\text{B}}) &\cong -k_{\text{B}} T \log\left[\sum_{\tilde{N}_{\text{AB}}} \omega_{\text{d}}(\tilde{N}_{\text{AB}}, \tilde{N}_{\text{A}}, \tilde{N}_{\text{B}}) \lambda(T, v)^{\tilde{N}^*_{\text{AB}}}\right] \\
&= -k_{\text{B}} T \log\left[\lambda(T, v)^{\tilde{N}^*_{\text{AB}}} \sum_{\tilde{N}_{\text{AB}}} \omega_{\text{d}}(\tilde{N}_{\text{AB}}, \tilde{N}_{\text{A}}, \tilde{N}_{\text{B}})\right] \\
&= -k_{\text{B}} T \log\left[\lambda(T, v)^{\tilde{N}^*_{\text{AB}}} \frac{(\tilde{N}_{\text{A}} + \tilde{N}_{\text{B}})!}{\tilde{N}_{\text{A}}!\tilde{N}_{\text{B}}!}\right] \\
&= -\tilde{N}^*_{\text{AB}} k_{\text{B}} T \log \lambda(T, v) + k_{\text{B}} T \left(\tilde{N}_{\text{A}} \log x_{\text{A}} + \tilde{N}_{\text{B}} \log x_{\text{B}}\right).
\end{aligned} \tag{9.139}$$

The second term on the right-hand side of Eqn. (9.139) is given by Eqn. (9.135). The first term can be evaluated using probabilistic arguments. We have $\tilde{N}_{\text{A}}$ A molecules on the lattice. Each one of these has z_1 adjacent sites. The probability that one of these sites has a B molecule is $\tilde{N}_{\text{B}}/\tilde{N}$, if the molecules are distributed randomly. Therefore, the number of AB pairs is estimated to be

$$\tilde{N}^*_{\text{AB}} \approx \tilde{N}_{\text{A}} z_1 \frac{\tilde{N}_{\text{B}}}{\tilde{N}}. \tag{9.140}$$

This simplification is called the **random-phase approximation**. The average number of AB pairs for a random mixture is therefore given by

$$\tilde{N}^*_{\text{AB}} = z_l \frac{\tilde{N}_{\text{A}}\tilde{N}_{\text{B}}}{\tilde{N}_{\text{A}} + \tilde{N}_{\text{B}}}. \tag{9.141}$$

After a few algebraic manipulations, Eqn. (9.139) can be rewritten in terms of mole fractions as

$$\frac{\Delta F^{\text{mixing}}}{NRT} = x_{\text{A}} \log x_{\text{A}} + x_{\text{B}} \log x_{\text{B}} - z_l \log\left(\frac{q_{\text{AB}}(T,v)}{\sqrt{q_{\text{AA}}(T,v)q_{\text{BB}}(T,v)}}\right) x_{\text{A}}x_{\text{B}}, \tag{9.142}$$

which is the expression introduced earlier (within our constant-volume assumption) as the "two-suffix Margules" equation using purely phenomenological arguments, with $A/(RT) = -z_l \log\left[q_{\text{AB}}/\sqrt{q_{\text{AA}}q_{\text{BB}}}\right]$.

9.8.3 Exact solution of the lattice model

The simple theory outlined above can be refined systematically by eliminating or improving some of the approximations involved in the derivation of Eqn. (9.142). In the exercises we present another analytic approach called the "quasi-chemical approximation." Alternatively, one can solve the equation numerically using a Monte Carlo simulation. Those refinements, however, are beyond the scope of this text, and interested readers are referred to the original literature on the subject.

Summary

This chapter introduces the student to concepts necessary for predicting the thermodynamic properties, particularly equilibrium compositions, of multiphase miscible and partially miscible liquid mixtures. A student should now be able to perform vapor–liquid, liquid–liquid, and solid–liquid phase-equilibrium calculations using the concepts put forth in this chapter.

- We introduced the concept of excess properties, M^{E}, Eqn. (9.2), to quantify deviations from Lewis (or ideal) mixing behavior:

$$M^{\text{E}}(T,P,\{N_i\}) = M(T,P,\{N_i\}) - M^{\text{Lewis}}(T,P,\{N_i\}).$$

- The molar excess Gibbs free energy g^{E} was introduced as a useful means to define activity coefficients:

$$\log \gamma_i := \frac{\bar{g}_i^{\text{E}}}{RT}.$$

- The general phase-equilibrium problem for an r-component system can now be stated in terms of fugacity coefficients (for a vapor, fluid phase, or liquid phase) and activity coefficients (for a condensed phase) according to

$$x_i \gamma_i f_i^{\text{pure}} = y_i \phi_i P, \quad i = 1, \ldots, r.$$

- Several models of $G^{\mathrm{E}}(T, P, \{N_i\})$ can be used to predict activities in mixtures. In particular, we considered the Margules model in some detail. However, other models, such as van Laar's model, Wilson's model, the NRTL model, and the UNIQUAC model, were introduced in Section 9.4. Using these models, the student should be able to predict compositions of liquid and vapor phases in equilibrium. We also discussed how several experiments, including the measurement of freezing points or osmotic pressure, could be used to determine activity coefficients.
- In Section 9.5 we introduced Henry's law for phase-equilibrium problems involving dilute components (e.g. gases) in liquid mixtures. The temperature and pressure dependences of the Henry's-law constant, $H_{i,j}$ were also found. Hence, thermodynamic data are used to predict changes in solubility with temperature and pressure. As an example, the solubility of nitrogen in water was predicted as a function of pressure.
- We introduced the concept of the water/octanol partition coefficient, and how it could be used to predict the ability of specific chemical compounds to cross the blood–brain barrier. Drugs for treating the brain must be able to pass through this barrier, whereas organic toxins are potentially more hazardous when they have a large water/octanol partition coefficient.
- Some liquids are not completely miscible. Section 9.7 illustrates how to use excess molar Gibbs free-energy models to predict the miscibility of binary and ternary mixtures. Given such a model, the student should be able to predict the coexistence curve and spinodal curve for a binary, ternary, or multicomponent mixture of species. The resulting binary and ternary diagrams are essential for the design of liquid–liquid extraction processes.
- The final section of the chapter shows how some of the simpler excess Gibbs free-energy models can be derived using a lattice model from statistical mechanics. That section is important for understanding the molecular origins (and the assumptions) of specific models for G^{E}, which in Section 9.4 were introduced in purely empirical terms.

Exercises

9.3.A Show that the activity coefficient of component 1 of a binary mixture can be obtained from the following relation:

$$RT \log \gamma_1 = g^{\mathrm{E}} + x_2 \left(\frac{\partial g^{\mathrm{E}}}{\partial x_1} \right)_{T,P}. \tag{9.143}$$

9.3.B Derive Eqn. (9.18).

9.3.C Derive an expression for the molar excess enthalpy of mixing of a mixture described by a molar excess Gibbs free energy of the form

$$g^{\mathrm{E}} = x_1 x_2 (A_{21} x_1 + A_{12} x_2), \tag{9.144}$$

where the constants A_{21} and A_{12} are independent of temperature.

9.3.D For a ternary mixture, the simplest version of the Margules equation takes the form

$$\frac{g^{\mathrm{E}}}{RT} = A_{12}x_1x_2 + A_{13}x_1x_3 + A_{23}x_2x_3, \tag{9.145}$$

where the A_{ij} are adjustable parameters determined from binary data for the ij pairs. Show that the activity coefficients of components 1, 2, and 3 are given by

$$\log \gamma_1 = A_{12}x_2^2 + A_{13}x_3^2 + (A_{12} + A_{13} - A_{23})x_2x_3, \tag{9.146}$$

$$\log \gamma_2 = A_{12}x_1^2 + A_{23}x_3^2 + (A_{12} + A_{23} - A_{13})x_1x_3, \tag{9.147}$$

$$\log \gamma_3 = A_{13}x_1^2 + A_{23}x_2^2 + (A_{13} + A_{23} - A_{12})x_1x_2. \tag{9.148}$$

9.3.E Show that the activity coefficients corresponding to Eqn. (9.30) are of the form

$$RT \log \gamma_1 = a'x_2^2 + b'x_2^3 + c'x_2^4 + d'x_2^5 + \cdots,$$

$$RT \log \gamma_2 = a''x_1^2 + b''x_1^3 + c''x_1^4 + d''x_1^5 + \cdots, \tag{9.149}$$

where the lower-case constants are related to those appearing in Eqn. (9.30) through

$$a' = A + 3B + 5C + 7D,$$
$$a'' = A - 3B - 5C - 7D,$$
$$b' = -4B - 16C - 36D,$$
$$b'' = 4B - 16C + 36D,$$
$$c' = 12C + 60D,$$
$$c'' = 12C - 60D,$$
$$d' = -32D,$$
$$d'' = 32D.$$

9.3.F Show that, for a binary azeotropic mixture at constant pressure, the temperature at the azeotrope corresponds to the maximum or the minimum of the temperature–composition curve.

9.3.G Derive Eqns. (9.12) and (9.13).

9.3.H[2] The molar excess Gibbs free energy for the Flory–Huggins model Eqn. (11.1) can be written as

$$\frac{g^{\mathrm{E}}}{RT} = x_{\mathrm{s}} \log\left(\frac{\phi_{\mathrm{s}}}{x_{\mathrm{s}}}\right) + x_{\mathrm{p}} \log\left(\frac{\phi_{\mathrm{p}}}{x_{\mathrm{p}}}\right) + \chi(x_{\mathrm{s}} + r_{\mathrm{p}}x_{\mathrm{p}})\phi_{\mathrm{s}}\phi_{\mathrm{p}}, \tag{9.150}$$

where the volume fractions of the solvent and polymer are given by

$$\phi_{\mathrm{s}} = \frac{x_{\mathrm{s}}}{x_{\mathrm{s}} + r_{\mathrm{p}}x_{\mathrm{p}}}, \quad \phi_{\mathrm{p}} = \frac{r_{\mathrm{p}}x_{\mathrm{p}}}{x_{\mathrm{s}} + r_{\mathrm{p}}x_{\mathrm{p}}}, \tag{9.151}$$

where $r_{\mathrm{p}} = v_{\mathrm{p}}/v_{\mathrm{s}}$ provides a measure of the asymmetry between the volume of the polymer and solvent molecules.

1. Experimental data for the solubility of polystyrene in methylcyclohexane indicate that this mixture exhibits partial miscibility. Show that the critical volume fraction and the critical χ parameter are given by

[2] Based on a problem in [163].

$$\phi_c = \frac{1}{1+\sqrt{r_p}}, \tag{9.152}$$

$$\chi_c = \frac{1}{2}\left(1+\frac{1}{\sqrt{r_p}}\right)^2. \tag{9.153}$$

2. Show that, for the Flory–Huggins model, the activity coefficient of the solvent can be written as

$$\log \gamma_s = \log\left(\frac{\phi_s}{x_s}\right) + \left(1-\frac{1}{r_p}\right)\phi_p + \chi\phi_p^2, \tag{9.154}$$

where ϕ_s is the volume fraction of the solvent.

3. For a molecular weight of $M_w = 4.64 \times 10^4$, this mixture first becomes partially immiscible at a temperature of $T = 40\,°\mathrm{C}$ and a volume fraction of the polymer of $\phi = 0.13$. Calculate the solvent partial pressure above a 60 wt% polystyrene solution in methylcyclohexane at 320 K.

   Use the following data. The molar volume of methylcyclohexane is approximately 88 cm^3/mol. Its molecular weight is 99 g/mol, and its vapor pressure is $P^{\mathrm{sat}} = 0.13$ bar at $T = 298$ K and $P^{\mathrm{sat}} = 0.24$ bar at $T = 312$ K. The molecular weight of a polymer segment in polystyrene is 104 g/mol, and the volume of the monomer is approximately 150 cm^3/mol. If you cannot calculate χ or are uncertain about your result, perform your calculations assuming $\chi = 1$.

9.3.1 Biological molecules (such as amino acids and proteins) and heat-sensitive hydrocarbons are often purified by crystallization processes. The design of such processes requires precise knowledge of the activity of the liquid in equilibrium with the solid. This problem considers the thermodynamics of solid–liquid equilibria.

1. Show that the "fugacity ratio" of a pure substance (the ratio of the fugacities of the solid and the supercooled liquid) is given by an expression of the form

$$\log\left(\frac{f_i^{\text{ solid}}}{f_i^{\text{ liquid}}}\right) = \frac{\Delta h_{i,\text{melting}}}{RT_{t,i}}\left(\frac{T_{t,i}}{T}-1\right) - \frac{\Delta C_{p,i}}{R}\left(\frac{T_{t,i}}{T}-1\right) + \frac{\Delta C_{p,i}}{R}\log\left(\frac{T_{t,i}}{T}\right), \tag{9.155}$$

where $\Delta h_{i,\text{melting}}$ is the enthalpy change of melting of component i, and $\Delta C_{p,i}$ is the difference between the molar heat capacity of supercooled liquid i and that of solid i at the same temperature, i.e. $\Delta C_p = C_p^{\text{liquid}} - C_p^{\text{solid}}$. In Eqn. (9.155) T denotes the temperature at which the fugacity ratio is being evaluated, and $T_{t,i}$ denotes the triple-point temperature of i.

(*Hint*: Eqn. (9.155) can be derived by constructing a thermodynamic cycle consisting of the following three steps.

- Cool the solid from the temperature of interest (T) to the triple-point temperature (T_t).
- Melt the solid into a liquid at the triple-point temperature.
- Heat the supercooled liquid from the triple-point temperature to the temperature of interest.

For each of these processes, one can evaluate the change in free energy, noting that $g = h - Ts$.)

2. Derive an expression for the ideal solubility of a solid solute in a liquid solvent (an expression that assumes that the liquid phase is ideal).
3. Experimental data (see below) are available for the solubility of glucose and sucrose in water at $T = 343$ K. For simplicity, the excess molar Gibbs free energy of this ternary mixture in the liquid phase can be described by an expression of the form

$$\frac{g^{\mathrm{E}}}{RT} = A_{12}x_1x_2 + A_{13}x_1x_3 + A_{23}x_2x_3. \tag{9.156}$$

Please determine appropriate values for the parameters A_{ij}. You may assume that sucrose and glucose form pure solids. You may also assume that the terms involving the heat capacity in Eqn. (9.155) can be neglected. To a first approximation, you can also replace the triple-point temperature in Eqn. (9.155) with the melting temperature T_{m} to get

$$\log\left(\frac{f_i^{\text{ solid}}}{f_i^{\text{ liquid}}}\right) \approx \frac{\Delta h_{i,\text{melting}}}{RT_{i,\mathrm{m}}}\left(\frac{T_{i,\mathrm{m}}}{T} - 1\right). \tag{9.157}$$

Use the following data. The enthalpy changes of melting for glucose (1) and sucrose (2) are $\Delta h_{1,\text{melting}} = 32.4$ kJ/mol and $\Delta h_{2,\text{melting}} = 46.2$ kJ/mol. The melting temperatures are $T_{1,\mathrm{m}} = 423$ K and $T_{2,\mathrm{m}} = 460$ K. At $T = 343$ K the solubility of glucose in pure water is $x_1 = 0.259$ and the solubility of sucrose in pure water is $x_2 = 0.146$. When $x_1 = 0.123$, the solubility of sucrose is $x_2 = 0.118$.

9.3.J The so-called van 't Hoff equation stipulates a linear dependence of the osmotic pressure on concentration and temperature:

$$\Pi = CRT. \tag{9.158}$$

Please provide a derivation of this equation and clearly indicate under which assumptions or in which limit it is justified.

9.4.A Show that for a two-parameter Margules equation for the molar excess Gibbs free energy,

$$g^{\mathrm{E}} = x_1x_2(A_{21}x_1 + A_{12}x_2), \tag{9.159}$$

the activity coefficients of components 1 and 2 are given by

$$\log\gamma_1 = \frac{1}{RT}\left[x_2^2A_{12} + 2x_1(A_{21} - A_{12})\right], \tag{9.160}$$

$$\log\gamma_2 = \frac{1}{RT}\left[x_1^2A_{21} + 2x_2(A_{12} - A_{21})\right]. \tag{9.161}$$

9.4.B Show that, at infinite dilution, the activity coefficients of a binary mixture described by Wilson's equation are given by

$$\log\gamma_1^{\infty} = 1 - \log\Lambda_{12} - \Lambda_{21}, \tag{9.162}$$

$$\log\gamma_2^{\infty} = 1 - \log\Lambda_{21} - \Lambda_{12}. \tag{9.163}$$

9.4.C The molar excess Gibbs free energy of a binary mixture of ethanol (1) and water (2) can be described reasonably well by Wilson's equation. Regression of vapor–liquid equilibrium data at a total pressure of 0.133 bar yields the following values for the Wilson parameters:

$$\Lambda_{12} = 100.7919,$$
$$\Lambda_{21} = 872.6804.$$

At this pressure, does the mixture exhibit an azeotrope? If so, at what temperature?

9.4.D Consider a binary mixture of ethanol (1) and water (2) described by Wilson's equation. Experimental data are available at two temperatures, namely 74.79 and 39.76 °C. Regression of experimental data yields the following values for the Wilson parameters: at $T = 74.79$ °C

$$\Lambda_{12} = 422.3398,$$
$$\Lambda_{21} = 931.8602,$$

and at $T = 39.76$ °C

$$\Lambda_{12} = 312.2493,$$
$$\Lambda_{21} = 855.3572.$$

The molar volume of pure water at room temperature is $v_2 = 1.007$ cm^3/g, and that of ethanol is $v_1 = 1.267$ cm^3/g. Assuming that these molar volumes are not too sensitive to temperature, and assuming that the interaction energies (λ_{ij}) are independent of temperature, are these parameters consistent with the functional form of the Wilson equation?

9.4.E The experimental data in Table 9.2 are available for the infinite-dilution activity coefficient of dichloroethane (2) in water (1). Can these data be used to determine parameters for a Wilson equation capable of describing the excess Gibbs free energy of dichloroethane–water mixtures?

Table 9.2 The infinite-dilution activity coefficient of dichloroethane (2) in water (1).

T (°C)	γ_2^∞
12.5	200
20.0	209
35.0	242

9.4.F Consider a binary liquid mixture consisting of components 1 and 2. This mixture can be assumed to follow an equation of state of the form

$$P = \frac{RT}{v - b} - \frac{a}{v^2} \tag{9.164}$$

with mixing rules given by

$$a = a_1 x_1^2 + a_2 x_2^2 + 2x_1 x_2 \sqrt{a_1 a_2}(1 - k_{12}), \tag{9.165}$$

$$b = x_1 b_1 + x_2 b_2. \tag{9.166}$$

The parameter k_{12} in Eqn. (9.165) depends on the nature of the mixture; it is intended to account for deviations from a geometric-mean combining rule. Assuming that both the molar excess volume of mixing and the molar excess entropy of mixing are negligible, derive an expression for the molar excess free energy of such a mixture. (*Hint*: describe the mixing process by devising a three-step thermodynamic cycle that comprises expansion of the two pure liquids, mixing of the resulting gases, and compression of the gas mixture into a liquid state. Evaluate the internal energy corresponding to each of the three steps by using the equation of state above.)

9.4.G Table 9.3 provides VLE data for the methanol (1)–2-methyl-1-propanol (2) system at 60 °C (from [139]). The Antoine equation, Eqn. (4.49), can be used to correlate vapor pressures. The constants for each of the two components in this problem are provided in Table 9.4, for P_i^{sat} the vapor pressure of pure component i (in mm Hg) and T the temperature (in °C).

1. Find parameters for the van Laar equation that provide a best fit to the experimental vapor-phase composition. Prepare a P_{xy} diagram that compares the experimental data with the curves determined from the calculations.

Table 9.3 Experimental data for the methanol (1)–2-methyl-1-propanol (2) system at 60 °C.

Pressure (mm Hg)	x_1	y_1
96.00	0.0	0.000
163.90	0.1	0.471
227.40	0.2	0.660
287.10	0.3	0.761
342.50	0.4	0.827
394.80	0.5	0.873
443.70	0.6	0.908
490.40	0.7	0.936
535.70	0.8	0.960
579.60	0.9	0.981
620.00	1.0	1.000

Table 9.4 Constants for Eqn. (4.49), when the pressure is in mm Hg and the temperature is in Celsius.

Component	A	B	C
1	8.080 97	1582.271	239.726
2	8.535 16	1950.940	237.147

2. Estimate the excess Gibbs free energy, $g^{\mathrm{E}}/(RT)$, directly from the experimental data given in Table 9.3 and compare it with the $g^{\mathrm{E}}/(RT)$ estimate that you obtained in part 1 using the van Laar equation. Discuss the agreement (or the lack thereof) between the two resulting estimates of g^{E} and explain your findings.
3. Perform a consistency check on the two sets of activity coefficients (from parts 1 and 2) by evaluating the following integral:

$$\int_0^1 \log\left(\frac{\gamma_1}{\gamma_2}\right) dx. \tag{9.167}$$

With reference to the values that you get for this integral, please comment on the accuracy of the activity coefficients determined from the van Laar equation in part 1 and those determined directly from experimental data in part 2?

9.4.H (An exercise based on a problem from Modell and Reid [88]). Solutions of monosaccharides or disaccharides provide good model systems for numerous solutions encountered in the food and pharmaceutical industries. Understanding the properties of sugars in water is therefore of interest in a wide variety of applications. Table 9.5 gives experimental data for the vapor pressure of water over sucrose solutions of different concentrations. The molality m is defined as the number of moles of sucrose dissolved in 1 kg of water. The chemical formula of sucrose is $C_{10}H_{22}O_{11}$; its molecular weight is 342.3 g/mol.

Table 9.5 Experimental data for the vapor pressure of water over sucrose solutions.

$T = 273$ K		$T = 298$ K		$T = 373$ K	
m	p_{H_2O}	m	p_{H_2O}	m	p_{H_2O}
0	0.611	0	3.166	0	101.33
0.2	0.609	0.1	3.160	0.3	100.78
0.5	0.605	0.4	3.14	0.6	100.21
1.0	0.599	1.0	3.11	0.8	99.81
3.5	0.560	2.0	3.03	1.2	99.01
4.5	0.542	3.0	2.95	1.5	98.35
5.0	0.533	4.0	2.87		
6.0	0.516	5.0	2.78		
7.0	0.499				

1. Propose an expression for the activity coefficient of sucrose in aqueous solutions. Use the data in Table 9.5 to estimate any necessary parameters in your expression.
2. Propose a model to predict the enthalpy of mixing ΔH_{mix} and the partial molar enthalpies of water and sucrose in the range 273–373 K as functions of sucrose concentration.

Table 9.6 gives the osmotic pressure of sucrose solutions at $T = 273$ K as a function of concentration. Use these data to parameterize the same activity coefficient that you

Table 9.6 Osmotic pressure for a sucrose solution as a function of concentration, from [109, 125].

Π (bar)	Concentration (mol/l)
0.02	0.1
0.1	1.45
0.3	6.5
0.5	12.1
0.8	23.5

Table 9.7 Vapor–liquid equilibrium data for 1-propanol (species 1) and water. The liquid and vapor mole fractions in equilibrium with one another are x_1^{liq} and x_1^{vap}, respectively. The pressure is fixed at 1.01 bar.

x_1^{liq}	x_1^{vap}	T (°C)
0.075	0.375	89.05
0.179	0.388	87.95
0.482	0.438	87.80
0.712	0.560	89.20
0.850	0.685	91.70

used for part 1 of this problem. Are the two sets of data (vapor pressure and osmotic pressure) consistent with each other? Discuss your answer.

9.4.I Table 9.7 contains vapor–liquid equilibrium data for a mixture of 1-propanol (1) and water (2) at $P = 1.01$ bar, taken from [89].[3] Estimate the excess Gibbs free energy for the liquid at each composition given. Estimate the parameters in the van Laar equation of state for this system.

9.5.A Scuba divers must take care to avoid the "bends" (also known as "Caisson's disease"), a disease that is often fatal, but which is preventable once you perform a few thermodynamic calculations. As a scuba diver descends below sea level, the pressure increases. The pressure of the air that she or he breathes must match the pressure of the environment (the surrounding water). The bends is a problem of gas solubility in the blood. The solubility of air in the blood of a diver at 20 m below sea level is considerably higher than that at ambient pressure. If that diver rises rapidly to sea level (i.e. ambient pressure), the increased amount of air in the blood stream will form bubbles that can get trapped in the blood and tissues in a process called "aeroembolism." Such bubbles can lead to serious complications and death.

Estimate the solubility of air in blood at 25 °C and 1 bar. Please construct a graph of air concentration in the blood as a function of depth. Estimate the solubility of air in

[3] This problem has been adapted from Poling, Prausnitz, and O'Connell, [102, 8.25–29].

blood at 20 m below sea level. What is the volume of air (per liter of blood) that would be released in the form of bubbles if that diver came up to ambient pressure "instantaneously," i.e. without letting the air levels in his or her blood gradually decrease? For simplicity you may assume that blood is pure water, and that air is a binary mixture of 80% N_2 and 20% O_2. The values of the Henry's-law constant for nitrogen and oxygen in water at 25 °C and ambient pressure are 1600 and 760 atm · l/mol, respectively. The density of water is 1000 kg/l. The partial molar volume of nitrogen in water at infinite dilution is $\bar{v}_2^{\infty} = 32.8\ \text{cm}^3/\text{mol}$. Would there be an advantage to filling the scuba diver's tank with a greater proportion of oxygen (i.e. more than 20%)? Please explain. Are divers in cold water (e.g. 4 °C) at greater risk of the bends than those in warm water (25 °C)? It is of interest to note that, at high concentrations, nitrogen can be intoxicating and may lead to impaired judgment (this is called nitrogen narcosis). Partly for that reason, 30 m has been set as the limit (or maximum depth) for sport diving.

9.5.B Five grams of a protein are dissolved in 10 l of water at room temperature. The osmotic pressure Π is measured to be 103 Pa for this mixture. What is the molecular weight of the protein?

9.7.A Show that, for a binary mixture at constant pressure, the critical solution composition corresponds to the maximum temperature of the liquid–liquid coexistence curve.

9.7.B Show that, for a binary mixture whose excess Gibbs free energy is described by the model of van Laar in the form

$$g^{\mathrm{E}} = \frac{A' x_1 x_2}{(A'/B')x_1 + x_2} \tag{9.168}$$

the coordinates of the critical solution temperature are given by

$$T_{\mathrm{c}} = \frac{2 x_1 x_2 A'^2/B'}{R((A'/B')x_1 + x_2)^3} \tag{9.169}$$

and

$$x_{1,\mathrm{c}} = \frac{[(A'/B')^2 + 1 - (A'/B')]^{1/2} - A'/B'}{1 - A'/B'}. \tag{9.170}$$

9.7.C 1. For a ternary mixture, the simple two-suffix Margules equation takes the form

$$\frac{g^{\mathrm{E}}}{RT} = A_{12}x_1x_2 + A_{13}x_1x_3 + A_{23}x_2x_3, \tag{9.171}$$

where A_{ij} is an adjustable parameter that characterizes interactions between components i and j of the mixture. The activity coefficients of components 1, 2, and 3 are given by

$$\log \gamma_1 = A_{12}x_2^2 + A_{13}x_3^2 + (A_{12} + A_{13} - A_{23})x_2x_3, \tag{9.172}$$

$$\log \gamma_2 = A_{12}x_1^2 + A_{23}x_3^2 + (A_{12} + A_{23} - A_{13})x_1x_3, \tag{9.173}$$

$$\log \gamma_3 = A_{13}x_1^2 + A_{23}x_2^2 + (A_{13} + A_{23} - A_{12})x_1x_2. \tag{9.174}$$

Please derive one of these expressions (e.g. that for $\log \gamma_1$).

2. The following information is available.

- Components 2 and 3 are partially miscible; the compositions of the two liquid phases denoted by ′ and ″ in a binary 2–3 mixture are $x_2' = 0.05$ and $x_2'' = 0.95$. Use this information to determine the parameter A_{23}.
- Components 1 and 2 form an azeotrope of composition $x_2 = 0.845$ at $P = 1$ bar; the vapor pressure of component 1 at this azeotrope is $P_1^{\text{sat}} = 0.7$ bar. Use this information to determine A_{12}.
- At $P = 1$ bar, the ternary mixture exhibits a liquid phase in equilibrium with a vapor phase; the composition of the liquid phase is $x_1 = x_2 = x_3 = 1/3$. At the temperature of interest, the vapor pressures of components 1, 2, and 3 are $P_1^{\text{sat}} = 0.7$ bar, $P_2^{\text{sat}} = 0.988$ bar and $P_3^{\text{sat}} = 0.1$ bar, respectively. Note that components 1 and 3 are fully miscible at the temperature of interest. The excess Gibbs free energy of the ternary mixture is $g^{\text{E}}/(RT) = 0.452$. Use these data to determine A_{13}.

What is the composition of the vapor phase in equilibrium with this equimolar liquid phase?

9.7.D 1. Show that for a ternary liquid mixture at equilibrium the condition for phase coexistence between two liquid phases (denoted by ′ and ″, respectively) can be written as

$$\left(\frac{\partial \Delta g^{\text{mixing}}}{\partial x_1}\right)' = \left(\frac{\partial \Delta g^{\text{mixing}}}{\partial x_1}\right)'', \tag{9.121′}$$

$$\left(\frac{\partial \Delta g^{\text{mixing}}}{\partial x_2}\right)' = \left(\frac{\partial \Delta g^{\text{mixing}}}{\partial x_2}\right)'', \tag{9.122′}$$

$$\begin{aligned}&\left(\Delta g^{\text{mixing}} - x_1\left(\frac{\partial \Delta g^{\text{mixing}}}{\partial x_1}\right) - x_2\left(\frac{\partial \Delta g^{\text{mixing}}}{\partial x_2}\right)\right)' \\ &= \left(\Delta g^{\text{mixing}} - x_1\left(\frac{\partial \Delta g^{\text{mixing}}}{\partial x_1}\right) - x_2\left(\frac{\partial \Delta g^{\text{mixing}}}{\partial x_2}\right)\right)''.\end{aligned} \tag{9.123′}$$

Equations (9.121)–(9.123) state that at coexistence the molar Gibbs free energy of mixing surface has a common tangent plane.

2. Outline clearly and concisely an algorithm to calculate the liquid–liquid coexistence curve of a ternary mixture.

9.7.E Consider a dilute aqueous solution of a high-molecular-weight polymer (a very large molecule). The osmotic pressure π of this solution can be measured in the device depicted in Figure 9.7. The container is divided into two distinct sections by a membrane. On one side of the membrane there is pure water. The polymer solution is on the other side of the membrane. The membrane is permeable only to water; the polymer is too large to diffuse through it. At any given temperature T and pressure P, water flows out of the polymer-rich solution until the pressure on the pure-water side becomes equal to $P + \pi$; at that pressure, the system reaches equilibrium and the net flow of water across the membrane ceases.

Show that if the solution is described by a two-suffix Margules equation the constant A appearing in the model is approximately given by

$$A = \frac{1}{2} + \frac{1}{x_{\mathrm{p}}^2}\left(x_{\mathrm{p}} - \frac{\pi\, v_{\mathrm{H_2O}}}{RT}\right), \tag{9.175}$$

where x_{p} is the mole fraction of the polymer in the polymer-rich solution at equilibrium, and $v_{\mathrm{H_2O}}$ is the molar volume of pure water at temperature T and pressure P.

(*Hint:* remember that $(\partial g/\partial P)_T = v$.)

9.8.A Another approximation to the lattice model considered in Section 9.8 is called the quasi-chemical approximation. Rather than assume that $\tilde{N}^*_{\mathrm{AB}}$ is well estimated by the random-phase approximation, we assume that it can be estimated from something like a chemical equilibrium equation,

$$\frac{\tilde{N}_{\mathrm{AA}}\tilde{N}_{\mathrm{BB}}}{\left(\tilde{N}^*_{\mathrm{AB}}\right)^2} = \frac{1}{4}\left(\frac{q_{\mathrm{AB}}^2}{q_{\mathrm{AA}}q_{\mathrm{BB}}}\right). \tag{9.176}$$

This expression incorporates the idea that unfavorable energetics for AB interactions (compared with AA and BB energetics) leads to a decrease in the number of AB pairs.

Using the quasi-chemical approximation, find the predicted excess Gibbs potential. From this potential construct a plot of the spinodal and binodal curves as functions of $\chi(T) := \frac{1}{4}\log[q_{\mathrm{AB}}^2/(q_{\mathrm{AA}}q_{\mathrm{BB}})]$.

10 Reaction equilibrium

Reactions are an essential component of chemical engineering, and reaction engineering itself is a very broad discipline [43, 107]. However, before we study the *time evolution* of chemical reactions, it is important to know the final *equilibrium* state that a system can reach. The equilibrium state of a chemical reaction is determined by thermodynamic principles.

In this chapter we examine the thermodynamic principles that govern chemical reactions, and find methods for calculating the final composition of a mixture that results from chemical reactions. The methods consist of two general steps: determination of the *equilibrium constant(s)* from thermodynamic properties of the constituent chemical compounds; and determination of the chemical composition of the equilibrium mixture from the initial composition and the equilibrium constant(s). In general, finding this composition requires the use of models with good estimates for the activities (or fugacities) of the constituent species, as shown in the previous two chapters. The equilibrium constant depends only on the temperature and the species present, not on the composition.

We also consider how thermodynamic driving forces can influence reaction rates. However, few details are given in this book, since that is a topic best left for other courses and textbooks. Denaturation of DNA strands is discussed in Section 10.9, where denaturation is applied to polymerase chain reactions (PCR). PCR is a technique that can be used to amplify small amounts of genetic material. Finally, in the last section, Section 10.10, connections are made to statistical mechanics.

10.1 A SIMPLE PICTURE: THE REACTION COORDINATE

Before constructing the rigorous mathematical machinery necessary to solve problems, we introduce a simplified picture to aid our thinking about kinetics and the equilibrium composition of reacting systems. We start with the simple, first-order reaction

$$A \rightleftharpoons B, \tag{10.1}$$

but the picture applies to second-order reactions as well.

We imagine that the molecule A must go through some changes in order for it to become molecule B. For example, bonds might be strained and broken, or bond angles might be modified. These intermediate states must have a higher energy than the state A, otherwise the reaction would be instantaneous. These changes might be very complicated, but we characterize the path by a single number called the *reaction coordinate*. Therefore, we draw the simple picture shown in Figure 10.1. The reaction coordinate is typically the path of minimum free energy between the reactants and the products. However, real reactions may fluctuate around this path, or there may even be multiple paths.

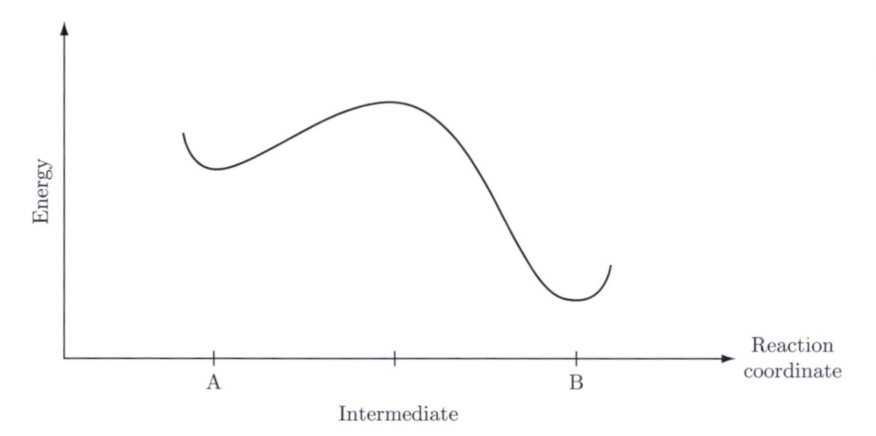

Figure 10.1 A simplified picture for thinking about chemical reactions.

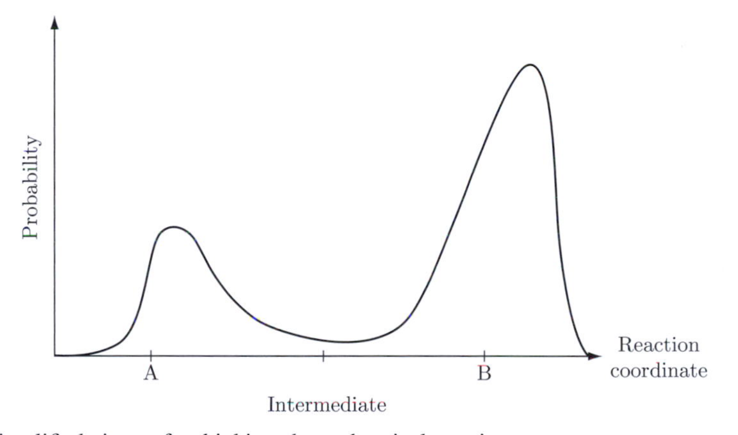

Figure 10.2 A simplified picture for thinking about chemical reactions.

As you can see from this sketch, the same arguments apply for the reverse reaction from B to A. This simple picture leads to three simple predictions, all of which are qualitatively correct.

Our first observation is that we expect to have both A and B molecules present at equilibrium. If the energy barrier is not too high, we expect thermal fluctuations experienced by the molecules to cause molecules to constantly go back and forth across the barrier. From statistical mechanics, we know that the probability of a molecule's being in a given energetic state E (see Section 6.2) is proportional to $\exp[-E/(k_B T)]$. Therefore, from our sketch, Figure 10.1, we can estimate the probability of a single molecule's having a given reaction coordinate. We sketch this idea in Figure 10.2. Any molecule with reaction coordinate to the left of the barrier is called molecule A. From statistical mechanics, we have then

$$\frac{\tilde{N}_A}{\tilde{N}_B} = \exp\left[\frac{E^B - E^A}{k_B T}\right], \tag{10.2}$$

where E^A is the energy of one molecule of A.

Our second prediction comes from the idea that the rate of the reaction A $\to$ B depends upon the height of the energy barrier to be overcome. Some simple calculations outside the scope of this text [143] suggest that the rate for a single molecule to overcome the barrier is also proportional to the exponential of the height:

$$\text{“rate of one molecule A} \to \text{B”} \sim \exp -\left[\frac{E^{\text{barrier}} - E^{\text{A}}}{k_{\text{B}}T}\right]. \tag{10.3}$$

This rate for a single molecule can be multiplied by the number of molecules $\tilde{N}_{\text{A}}$ to get the total rate of molecules A going to B per unit time. Of course, the same argument applies for the reverse reaction.

At equilibrium, the forward and reverse rates must be equal. Hence, Eqn. (10.3) leads to the correct equilibrium composition, Eqn. (10.2).

Our third, and final, prediction from the simple picture deals with catalysts and enzymes. In synthetic chemistry, one often uses a metallic surface, such as platinum or nickel, to speed up a reaction. In biological systems, reactions are aided by enzymes. By definition, these quantities are neither produced nor consumed by the reaction. Therefore, catalysts cannot modify the difference in energy between the reactants and products, $E_{\text{A}} - E_{\text{B}}$, and the equilibrium composition is unaffected. What catalysts and enzymes *do* change is the height of the activation-energy barrier. In the presence of catalysts, the reaction proceeds more quickly in both directions.

In fact, sometimes a reaction will not proceed at all under certain conditions without the presence of a catalyst, even though the energy of the product is much lower than that of the reactant – the barrier is much too high. A catalyst, or enzyme, can sometimes lower the barrier sufficiently to allow a reaction that otherwise would not occur.

10.2 EXTENT OF REACTION

If we begin with a certain number of initial moles in a reacting system, then not all compositions are available, since the number of moles of atoms must be conserved. Hence, the number of moles of each of the species $\{N_i\}$ has too many degrees of freedom. For example, if we have one mole of hydrogen and one mole of oxygen, then we can make at most one mole of water, leaving half a mole of oxygen unreacted. If we make less water, we have more oxygen left over – the numbers of moles of oxygen and water are not independent. Therefore, we cast the criterion for chemical equilibrium using a new independent variable called **extent of reaction**. In general, we can write a chemical reaction in the form

$$0 \rightleftharpoons \sum_i^r \nu_i B_i, \tag{10.4}$$

where a species is represented by B_i, and its **stoichiometric coefficient** is ν_i. If the stoichiometric coefficient of a species is negative, it would normally show up on the left-hand side of the equation as a reactant. For example, the combustion of ethene can be written as

$$C_2H_4 + 3O_2 \rightleftharpoons 2CO_2 + 2H_2O. \tag{10.5}$$

Here, the stoichiometric coefficients are $\nu_{C_2H_4} = -1$ and $\nu_{CO_2} = 2$ for ethene and carbon dioxide, respectively. Note that there is a restriction on the stoichiometric coefficients, since the numbers of atoms must be the same on each side of the equation; for example, there are two carbon atoms, four hydrogen atoms, and six oxygen atoms on each side of the ethene reaction. However, in principle, all coefficients could be multiplied by an integer to yield a different set of equally valid values. Hence, it is important to be certain that all thermodynamic quantities considered below are associated with the *same set of values* for the stoichiometric coefficients.

For a closed system, the only way in which the number of moles of a chemical species can change is by reaction. Hence, for a single chemical reaction, we can write

$$\frac{dN_1}{\nu_1} = \frac{dN_2}{\nu_2} = \ldots = \frac{dN_m}{\nu_m} =: d\varepsilon, \quad \text{single reaction,} \tag{10.6}$$

which defines the **extent of reaction** ε, for a single reaction. The extent of reaction is typically (though arbitrarily) taken to be zero when the components are mixed. Hence, if we integrate just one of the equations above, $dN_i = \nu_i\varepsilon$, from the point of mixing, to some other time, we obtain

$$N_i = N_{i,0} + \nu_i\varepsilon, \tag{10.7}$$

where $N_{i,0}$ is the initial, non-equilibrium number of moles of species i.

If the reaction is taking place in a single phase, we can sum Eqn. (10.7) over all species to find the total number of moles,

$$N_{\mathrm{T}} = N_{\mathrm{T},0} + \varepsilon\nu, \quad \text{single phase,} \tag{10.8}$$

where we define ν as

$$\nu := \sum_{i=1}^{r} \nu_i. \tag{10.9}$$

Again, if we are working in a single phase, we can divide Eqn. (10.7) by Eqn. (10.8) to find the mole fraction of species i:

$$x_i = \frac{N_{i,0} + \nu_i\varepsilon}{N_{\mathrm{T},0} + \nu\varepsilon}, \quad \text{single phase, single reaction.} \tag{10.10}$$

If there are q simultaneous reactions taking place in the phase, then we can write q reactions:

$$0 \rightleftharpoons \sum_{i}^{r} \nu_i^1 B_i, \quad 0 \rightleftharpoons \sum_{i}^{r} \nu_i^2 B_i, \quad \ldots, \quad 0 \rightleftharpoons \sum_{i}^{r} \nu_i^q B_i. \tag{10.11}$$

The change in mole number for species i is now the result of all the reactions, so we can write

$$dN_i = \nu_i^1 d\varepsilon_1 + \nu_i^2 d\varepsilon_2 + \cdots + \nu_i^q\, d\varepsilon_q = \sum_{j=1}^{q} \nu_i^j\, d\varepsilon_j,$$

$$\rightarrow N_i = N_{i,0} + \sum_{j=1}^{q} \nu_i^j \varepsilon_j, \tag{10.12}$$

where ε_j is the extent of reaction j. The last line, as above, is found by integration from the initial state. Now, instead of working with the number of moles of the species, we can work with the extents of reaction to ensure that atoms are conserved and we are working with the minimum number of degrees of freedom necessary.

EXAMPLE 10.2.1 Assuming that the reaction given in Eqn. (10.5) occurs in the gas phase, find the mole fractions of each component as a function of the extent of reaction.

Solution. For convenience, we rename the components as $B_1 \equiv C_2H_4, B_2 \equiv O_2, B_3 \equiv CO_2$, and $B_4 \equiv H_2O$. Then, the mole-balance equation (10.7) becomes

$$\begin{aligned} N_1 &= N_{1,0} - \varepsilon, \\ N_2 &= N_{2,0} - 3\varepsilon, \\ N_3 &= N_{3,0} + 2\varepsilon, \\ N_4 &= N_{4,0} + 2\varepsilon. \end{aligned} \tag{10.13}$$

On summing these four equations, we obtain the total number of moles in the gas phase:

$$N_T = N_{1,0} + N_{2,0} + N_{3,0} + N_{4,0} \equiv N_{T,0}. \tag{10.14}$$

For this example, the total number of moles in the gas phase does not change. Therefore, if we divide each side of Eqn. (10.13) by each side of Eqn. (10.14), we find the composition of the reacting phase as a function of the extent of reaction

$$\begin{aligned} x_1 &= \frac{N_{1,0} - \varepsilon}{N_{1,0} + N_{2,0} + N_{3,0} + N_{4,0}}, \\ x_2 &= \frac{N_{2,0} - 3\varepsilon}{N_{1,0} + N_{2,0} + N_{3,0} + N_{4,0}}, \\ x_3 &= \frac{N_{3,0} + 2\varepsilon}{N_{1,0} + N_{2,0} + N_{3,0} + N_{4,0}}, \\ x_4 &= \frac{N_{4,0} + 2\varepsilon}{N_{1,0} + N_{2,0} + N_{3,0} + N_{4,0}}. \end{aligned} \tag{10.15}$$

□

10.3 THE EQUILIBRIUM CRITERION

Consider a reaction taking place in a system that is in contact with a thermal and pressure reservoir. For constrained temperature and pressure, we have already seen in Section 3.2 that the Gibbs potential of the system is minimized. Hence, for constant T and P, Eqn. (3.51) becomes

$$dG = \sum_{i=1}^{r} \mu_i \, dN_i$$
$$= \left(\sum_{i=1}^{r} \mu_i \nu_i \right) d\varepsilon, \quad \text{single reaction,} \tag{10.16}$$

where we have used Eqn. (10.6) to obtain the second line. The minimization requirement at equilibrium for the Gibbs potential leads to

$$\left(\frac{\partial G}{\partial \varepsilon} \right)_{T,P} = \sum_{i=1}^{r} \mu_i \nu_i = 0. \tag{10.17}$$

Equation (10.17) states that the reaction proceeds until the Gibbs potential is minimized; we can predict satisfaction of the criterion by studying the chemical potentials of all the species.

If there are multiple reactions, G must be minimized with respect to all the extents of reaction:

$$\left(\frac{\partial G}{\partial \varepsilon_i} \right)_{T,P,\{\varepsilon_{j \neq i}\}} = 0, \quad i = 1, 2, \ldots, q. \tag{10.18}$$

Note that the last two results apply only for an isothermal and isobaric system. An isolated system would maximize the entropy instead of minimizing G, for example.

10.4 THE REACTION EQUILIBRIUM CONSTANT

From Sections 8.4–8.6, we know that we can also find chemical potentials from estimation of fugacities, fugacity coefficients, and activity coefficients. For reactions, it is customary and convenient to use the *activity* $a_i(T, P)$ of species i through the following. First, we define the standard reference state to be the pure substance at $P^{\text{ref}} = 1$ bar, and, if the species is a gas at this pressure, as an ideal gas. Note that this reference state, which we shall denote by the superscript $^\circ$ is at arbitrarily specified temperature. Then, from the definition of fugacity, Eqn. (8.41), we can write the chemical potential relative to this reference state as

$$a_i := \frac{f_i(T, P, \{x_i\})}{f_i^{\text{pure}}(T, P^{\text{ref}})}$$
$$= \left(\frac{x_i P}{P^{\text{ref}}} \right) \exp\left[\frac{\mu_i(T, P, \{x_i\}) - \mu_i^{\text{ideal}}(T, P, \{x_i\})}{RT} \right]$$
$$\times \exp\left[-\frac{\mu_i^{\text{pure}}(T, P^{\text{ref}}) - \mu_i^{\text{pure,ideal}}(T, P^{\text{ref}})}{RT} \right]$$
$$= \left(\frac{x_i P}{P^{\text{ref}}} \right) \exp\left[\frac{\mu_i(T, P, \{x_i\}) - \mu_i^{\text{pure}}(T, P^{\text{ref}}) - RT \log(x_i P / P^{\text{ref}})}{RT} \right]$$
$$= \exp\left[\frac{\mu_i(T, P, \{x_i\}) - \mu_i^{\text{pure}}(T, P^{\text{ref}})}{RT} \right], \tag{10.19}$$

taking the logarithm of each side. To obtain the second line, we used the definition of fugacities. We used the expression for the chemical potential of a mixture of ideal gases to obtain the third line, and the properties of logarithms to obtain the fourth.

The equilibrium criterion can be written in terms of activity by substituting Eqn. (10.19) into Eqn. (10.17). After a bit of algebra, we arrive at the equilibrium criterion in the form

$$\prod_{i=1}^{r} a_i^{\nu_i} = K(T), \quad \text{equilibrium criterion, single reaction,} \tag{10.20}$$

where we define the **equilibrium constant** K as

$$K(T) := \exp\left(-\frac{\Delta g^\circ}{RT}\right), \tag{10.21}$$

and the **standard Gibbs free-energy change of reaction** is given by

$$\Delta g^\circ(T) := \sum_{i=1}^{r} \nu_i \mu_i^\circ, \tag{10.22}$$

where $\mu_i^\circ := \mu_i^{\text{pure}}[P^{\text{ref}}]$. These are now the rigorous results for a system held at fixed temperature and pressure, to replace the qualitative results from Section 10.1. Instead of mole number, the activity is important; and instead of energy differences, free-energy differences determine the equilibrium composition.

For solving problems, our task is now split into two parts: first we find the equilibrium constant K from Eqns. (10.21) and (10.22), which is a function solely of temperature and which species are present; once we know K, we determine the compositions that give the activities satisfying Eqn. (10.20). It is important to use the same reference state for finding Δg° or K as is used for the activities.

Methods for finding K are discussed in Section 10.6, and methods of finding compositions based on K are discussed in Section 10.7.

10.5 STANDARD PROPERTY CHANGES

In addition to the standard Gibbs free-energy change, it is convenient to define a general standard property change for property M:

$$\Delta M^\circ := \sum_i \nu_i M_i^\circ. \tag{10.23}$$

Of particular importance in reactions are the standard enthalpy changes, ΔH°, and the standard entropy changes, ΔS°. Note that the standard enthalpy change is equivalent to the standard heat of reaction used in reaction energy balances. For example, from the definition of the standard enthalpy change and the Gibbs–Duhem relation (8.17) we can write

$$
\begin{aligned}
\frac{\Delta H^\circ}{RT^2} &= \sum_i \frac{\nu_i H_i^\circ}{RT^2} \\
&= -\sum_i \nu_i \frac{d}{dT}\left(\frac{G_i^\circ}{RT}\right) \\
&= -\frac{d}{dT}\left(\sum_i \nu_i \frac{G_i^\circ}{RT}\right) \\
&= -\frac{d}{dT}\left(\frac{\Delta G^\circ}{RT}\right) \\
&= \frac{d}{dT}\log K.
\end{aligned}
\tag{10.24}
$$

The fourth line follows from the definition of ΔG°, and the last line follows from the definition of K.

We can make several useful observations from Eqn. (10.24).

- Equation (10.24) provides a relatively easy means to calculate the standard heat of reaction by measuring the equilibrium constant at several temperatures.
- If $\Delta H^\circ < 0$, then the enthalpy of the products is smaller than the enthalpy of the reactants. Hence, energy is released and the reaction is called **exothermic**. From Eqn. (10.24) we see that the equilibrium constant for exothermic reactions decreases when the temperature is raised. The opposite is true for **endothermic reactions**: raising the temperature increases the equilibrium constant. Hence, one should lower the temperature of exothermic reactions and raise the temperature of endothermic ones to increase yield.[1]
- If the heat of reaction is already known, Eqn. (10.24) can be used to estimate the temperature dependence of the equilibrium constant, by approximating ΔH° as constant.

10.6 ESTIMATING THE EQUILIBRIUM CONSTANT

To estimate the equilibrium constant, Eqn. (10.21), it is necessary to find the standard Gibbs free-energy change of reaction at the temperature of interest. From the definition of ΔG°, we need the chemical potentials for each component at the standard pressure, and the reaction temperature. If we choose pure elements in their natural state at a standard pressure of $P^{\text{ref}} = 1$ bar as a reference state, we can find ΔG° from the Gibbs free energy of formation of each compound.

[1] This observation is sometimes called **Le Châtelier's principle**, after the French chemist Henry-Louis Le Châtelier. However, no results from stability analysis are necessary in order to arrive at the conclusion.

EXAMPLE 10.6.1 Consider the ethene reaction given in Eqn. (10.5). Find the corresponding standard change in Gibbs free energy from the energy of formation of each compound.

Solution. We can write this reaction as the sum of the four reactions of formation of each of the species present:

$$\begin{array}{lrl} -1\times(& 2C(s)+2H_2(g)\rightleftharpoons C_2H_4 &) \\ +2\times(& C(s)+O_2(g)\rightleftharpoons CO_2 &) \\ +2\times(& H_2(g)+\tfrac{1}{2}O_2(g)\rightleftharpoons H_2O &) \\ -3\times(& O_2(g)\rightleftharpoons O_2(g) &) \\ \hline & C_2H_4+3O_2\rightleftharpoons 2CO_2+2H_2O & \end{array} \tag{10.25}$$

Note that we simply multiply the formation reaction by the stoichiometric coefficient of the compound, and add the result to the sum. In an analogous way, we can add up the standard property changes for each of the formation reactions

$$\begin{array}{lrl} -1\times(& \Delta g_f^\circ(C_2H_4)=\mu^\circ_{C_2H_4}-2\mu^\circ_C-2\mu^\circ_{H_2} &) \\ +2\times(& \Delta g_f^\circ(CO_2)=\mu^\circ_{CO_2}-\mu^\circ_C-\mu^\circ_{O_2} &) \\ +2\times(& \Delta g_f^\circ(H_2O)=\mu^\circ_{H_2O}-\mu^\circ_{H_2}-\tfrac{1}{2}\mu^\circ_{O_2} &) \\ -3\times(& \Delta g_f^\circ(O_2)=\mu^\circ_{O_2}-\mu^\circ_{O_2}=0 &) \\ \hline \end{array} \tag{10.26}$$

$$\Delta g^\circ = 2\,\Delta g_f^\circ(CO_2)+2\,\Delta g_f^\circ(H_2O)-\Delta g_f^\circ(C_2H_4)-3\,\Delta g_f^\circ(O_2)$$

It is important to point out that the values of the free energy of formation which are usually tabulated (such as in Section D.2 of Appendix D) are also at a reference temperature. Hence, the standard free-energy change found above would be valid at that temperature only. □

The general construction is a straightforward generalization of Eqn. (10.26):

$$\Delta M^\circ = \sum_i \nu_i\,\Delta M^\circ_{f,i}. \tag{10.27}$$

If values for the Gibbs free energy of formation are tabulated, we can use Eqn. (10.27) with $M \equiv G$ directly. Or, if values for H and S are tabulated at some standard temperature T_0, we can use the definition of G:

$$\begin{aligned} \Delta G^\circ(T) &= \sum_i \nu_i\,\Delta G^\circ_{f,i}(T) \\ &= \sum_i \nu_i(\Delta H^\circ_{f,i}(T)-T\,\Delta S^\circ_{f,i}(T)) \\ &= \Delta H^\circ(T)-T\,\Delta S^\circ(T). \end{aligned} \tag{10.28}$$

Usually, standard property changes of formation are tabulated not only at a standard pressure, but at a standard temperature T^{ref} as well (such as in Section D.2 of Appendix D), whereas we may need the equilibrium constant at arbitrary temperature T. In order to find the standard change of Gibbs free energy at arbitrary temperature T instead of the standard tabulated temperature T^{ref}, we may use the heat capacity of the constituents and Eqn. (10.28):

$$\begin{aligned}
\Delta G^{\circ}[T] &= \sum_i \nu_i \left[\Delta H^{\circ}_{\mathrm{f},i}[T] - T\,\Delta S^{\circ}_{\mathrm{f},i}[T]\right] \\
&= \sum_i \nu_i \left[\Delta H^{\circ}_{\mathrm{f},i}[T^{\text{ref}}] + \int_{T^{\text{ref}}}^{T} C_{P,i}[T']dT' - TS^{\circ}_i[T^{\text{ref}}] - T\int_{T^{\text{ref}}}^{T} \frac{C_{P,i}[T']}{T'}\,dT'\right] \\
&= \Delta G^{\circ}[T^{\text{ref}}] - (T - T^{\text{ref}})\Delta S^{\circ}[T^{\text{ref}}] + \int_{T^{\text{ref}}}^{T} \left(\frac{T'-T}{T'}\right)\Delta C^{\circ}_P[T']dT',
\end{aligned} \tag{10.29}$$

where $\Delta C^{\circ}_P[T'] := \sum_i \nu_i C_{P,i}[T', P^{\text{ref}}]$. For the reference pressure used here ($P^{\text{ref}} = 1$ bar), ideal heat capacities can be used for gases.

EXAMPLE 10.6.2 For the chemical vapor deposition of solid silicon in semiconductor manufacturing, the decomposition of silane at elevated temperatures is used:

$$SiH_4 \rightleftharpoons Si(s) + 2H_2, \tag{10.30}$$

where, when not indicated otherwise, components are gases. However, there are also two undesirable side reactions that can occur:

$$\begin{aligned}
SiH_4 &\rightleftharpoons SiH_2 + H_2, \\
SiH_4 + SiH_2 &\rightleftharpoons Si_2H_6.
\end{aligned} \tag{10.31}$$

We would like to minimize the extent of the two side reactions to increase the efficiency of the process. Before studying the dynamics of the reactions, we first want to check to see which reactions are favored at equilibrium. Find the equilibrium constant for the three reactions shown at 1 bar and 600 K.

Solution. For simplicity, we will call the species ($B_1 \equiv SiH_4, B_2 \equiv Si_2H_6, B_3 \equiv SiH_2, B_4 \equiv H_2, B_5 \equiv Si(s)$). Then, the reactions are written

$$\begin{aligned}
B_1 &\rightleftharpoons B_5 + 2B_4, \text{ reaction 1,} \\
B_1 &\rightleftharpoons B_3 + B_4, \text{ reaction 2,} \\
B_1 + B_3 &\rightleftharpoons B_2, \text{ reaction 3.}
\end{aligned} \tag{10.32}$$

The stoichiometric coefficients for each of the reactions are shown in the following table:

	Stoichiometric coefficients, ν_i^j				
	B_1	B_2	B_3	B_4	B_5
Reaction 1	−1	0	0	2	1
Reaction 2	−1	0	1	1	0
Reaction 3	−1	1	−1	0	0

We show the explicit calculation for the first reaction only. In the following table are the standard heats of formation at $T^{\text{ref}} = 298.15\,\text{K}$ given in Section D.2 of Appendix D, taken originally from [25]:

	$\Delta h^\circ_{\text{f},i}$ (kJ/mol)	s_i° (J/mol · K)
B_1	34.309	204.653
B_4	0	130.680
B_5	0	18.82

From these values, we can find the standard changes in enthalpy and entropy for reaction 1:

$$\begin{aligned}\Delta h^\circ[T^{\text{ref}}] &= (0) + 2(0) - 34.309\ \text{kJ/mol}\\ &= -34.309\ \text{kJ/mol},\\ \Delta s^\circ[T^{\text{ref}}] &= 18.82\ \text{J/mol}\cdot\text{K} + 2(130.680\ \text{J/mol}\cdot\text{K}) - 204.653\ \text{J/mol}\cdot\text{K}\\ &= 75.527\ \text{J/mol}\cdot\text{K}.\end{aligned} \tag{10.33}$$

With these values we can calculate ΔG° of Eqn. (10.29):

$$\begin{aligned}\Delta g^\circ[T^{\text{ref}}] &= \Delta h^\circ[T^{\text{ref}}] - T^{\text{ref}}\,\Delta s^\circ[T^{\text{ref}}],\\ &= -56.827\ \text{kJ/mol}.\end{aligned} \tag{10.34}$$

Note that this is the same as the value for ΔG_f° for silane given in Table D.2 in Appendix D, as it should be. However, we still need to account for the difference between the reaction temperature and the tabulated standard temperature through the integrals of the heat capacity. For the components in the reaction, the heat capacities are approximated by the expressions for the ideal heat capacities given in Table D.2:

$$c_P^{\text{ideal}} = A_0 + A_1 T + A_2 T^2 + A_3 T^3 + A_4 T^4 + A_5 T^5 + A_6/T + A_7/T^2. \tag{10.35}$$

This table contains values for hydrogen, but not the other two compounds. However, these may be found in the NIST Chemistry Web Book (see Section D.1). The values for all species are shown in Table 10.1.

Table 10.1 Heat-capacity coefficients for the compounds present in reaction 1. The units are such that the heat capacity is J/mol when the temperature is degrees Kelvin and Eqn. (10.35) is used. The bottom line of each half of this table contains the ΔA_i used in Eqn. (10.36). The values for silicon and silane are taken from the NIST Chemistry Web Book (see Section D.1 of Appendix D).

	$A_0 \times 10^1$	$A_1 \times 10^3$	$A_2 \times 10^6$	$A_3 \times 10^9$
B_1	0.606 018 9	139.9632	−77.884 74	16.240 95
B_4	1.7835	9.0511	−0.434 44	−0.488 36
B_5	22.817 19	3.899 510	−0.082 885	0.042 111
$\sum_i \nu_i^1(\ldots)$	25.7782	−117.9615	76.9330	−17.1756
	$A_4 \times 10^{13}$	$A_5 \times 10^{17}$	$A_6 \times 10^{-3}$	$A_7 \times 10^{-5}$
B_1	0	0	0	1.355 09
B_4	1.1977	−0.837 80	4.9289	−7.2026
B_5	0	0	0	−3.540 63
$\sum_i \nu_i^1(\ldots)$	2.3954	−1.6756	9.8578	−19.3001

The values for the coefficients given in Table D.2 yield an expression for $\Delta C_P(T)$:

$$\Delta c_P^{\text{ideal}} = \Delta A_0 + \Delta A_1 T + \Delta A_2 T^2 + \Delta A_3 T^3 + \Delta A_4 T^4 + \Delta A_5 T^5 + \Delta A_6/T + \Delta A_7/T^2, \tag{10.36}$$

where the values are also given in Table 10.1.

We can now perform the integrations on the right-hand side of Eqn. (10.29). Upon inserting Eqn. (10.36) into the integral in Eqn. (10.29), we obtain

$$\begin{aligned}\int_{T^{\text{ref}}}^{T} \left(\frac{T'-T}{T'}\right) \Delta c_P^{\text{ideal}}(T')dT' = {} & (\Delta A_0 - T\,\Delta A_1)\left(T - T^{\text{ref}}\right) \\ & + (\Delta A_1 - T\,\Delta A_2)\frac{1}{2}\left(T^2 - (T^{\text{ref}})^2\right) \\ & + (\Delta A_2 - T\,\Delta A_3)\frac{1}{3}\left(T^3 - (T^{\text{ref}})^3\right) \\ & + (\Delta A_3 - T\,\Delta A_4)\frac{1}{4}\left(T^4 - (T^{\text{ref}})^4\right) \\ & + (\Delta A_4 - T\,\Delta A_5)\frac{1}{5}\left(T^5 - (T^{\text{ref}})^5\right) \\ & + \Delta A_5 \frac{1}{6}\left(T^6 - (T^{\text{ref}})^6\right)\end{aligned}$$

$$\begin{aligned}
&+ (\Delta A_6 - T\,\Delta A_0)\log(T/T^{\rm ref}) \\
&+ \left(\Delta A_6 - \frac{\Delta A_7}{T}\right)\left(1 - \frac{T}{T^{\rm ref}}\right) \\
&+ \frac{1}{2}T\,\Delta A_7\left(\frac{1}{T^2} - \frac{1}{(T^{\rm ref})^2}\right) \\
&= -925.88\ \text{J/mol}.
\end{aligned} \tag{10.37}$$

Upon inserting this result and the results shown in Eqn. (10.33) into Eqn. (10.29), we find that

$$\Delta g^\circ(T = 600\,\text{K}) = -80.863\,\text{kJ/mol}. \tag{10.38}$$

From the definition of the equilibrium constant, Eqn. (10.21), we can then find

$$K(T = 600\,\text{K}) = 10.96 \times 10^6, \tag{10.39}$$

which indicates that the reaction is extremely favorable. Under most reactor conditions, the reaction does not proceed to anywhere close to its equilibrium composition. □

10.7 DETERMINATION OF EQUILIBRIUM COMPOSITIONS

Once we have determined the equilibrium constants for the possible reactions in a system, we use Eqn. (10.20) to determine the compositions. Therefore, we require useful expressions for the activities. Fortunately, these have already been found in the previous two chapters, and we summarize them here. For gases,

$$\begin{aligned}
a_i &:= f_i/f_i^\circ \\
&= \frac{x_i\phi_i P}{P^{\rm ref}}, \quad \text{moderate reference pressure} && (10.40)\\
&\approx \frac{x_i\phi_i^{\rm pure} P}{P^{\rm ref}}, \quad \text{Lewis mixtures} && (10.41)\\
&\approx \frac{x_i P}{P^{\rm ref}}, \quad \text{ideal gas.} && (10.42)
\end{aligned}$$

For the second line, one then needs a *PVT* equation of state for mixtures in order to estimate the fugacity coefficient. For the third line, a *PVT* equation of state for the pure vapor is sufficient. For liquids,

$$\begin{aligned}
a_i &= x_i\gamma_i \exp\left[\int_{P^{\rm ref}}^{P} \frac{v_i^{\rm pure}(T,P)}{RT}\,dP\right] \\
&\approx x_i\gamma_i \exp\left[\frac{v_i^{\rm pure}(P - P^{\rm ref})}{RT}\,dP\right], \quad \text{incompressible} \\
&\approx x_i\gamma_i, \quad \text{moderate pressure} \\
&\approx x_i, \quad \text{Lewis mixture.}
\end{aligned} \tag{10.43}$$

For dilute liquid components, we can also assume Henry's law:

$$a_i \approx \frac{x_i H_{i,j}}{f_i^\circ}, \quad \text{Henry's law.} \tag{10.44}$$

To estimate the activity coefficient γ_i, an equation of state for mixtures (e.g. Margules) is needed.

EXAMPLE 10.7.1 Find the equilibrium composition for the decomposition of silane at 600 K and 1 atm.

Solution. From the last example we know how to find the values for the equilibrium constant K_i of each reaction i. For each reaction we can write

$$K_i = \prod_{j=1}^{5} (a_j)^{(\nu_j^i)}, \quad i = 1, 2, 3. \tag{10.45}$$

Since one of the species (A_5) is solid, P is near P^{ref}, and $x_5 \approx 1$, so $\hat{a}_5 \approx 1$. At a pressure of 1 atm it is safe to assume that the other species are ideal gases. Therefore, we can write Eqn. (10.45) as

$$K_i = \left(\frac{P}{P^{\text{ref}}}\right)^{\left(\sum_{j=1}^{4} \nu_j^i\right)} \prod_{j=1}^{4} (x_j^{\text{vap}})^{(\nu_j^i)}, \quad i = 1, 2, 3. \tag{10.46}$$

It is important to note that the terms in the product (and in the sum occurring in the exponent) go to 4 instead of to 5. The above equation represents only three equations, whereas there are four unknowns, y_1, y_2, y_3, and y_4. We could eliminate one of the unknowns by the constraint that the mole fractions must sum to one. However, we will use here the more general approach of introducing the extent of reaction. The mole balance for each species can be written by integrating Eqn. (10.12) to find

$$N_i = N_{i,0} + \sum_{j=1}^{3} \nu_i^j \varepsilon_i^j, \quad i = 1, 2, 3, 4. \tag{10.47}$$

If we sum each side of Eqn. (10.47) over i from 1 to 4, and divide the same equation by the result, we obtain an expression for each of the mole fractions. Note that we sum only to $i = 4$ because the fifth species does not enter the gas phase:

$$x_i^{\text{vap}} = \frac{N_{i,0} + \sum_{j=1}^{3} \nu_i^j \varepsilon_i^j}{N_{\text{T},0} + \varepsilon_1 + \varepsilon_2 - \varepsilon_3}. \tag{10.48}$$

When we insert Eqn. (10.47) into Eqn. (10.45), we obtain the three equations

$$\begin{aligned}
K_1 &= \left(\frac{P}{P^{\text{ref}}}\right) \frac{(N_{2,0} + 2\varepsilon_1 + \varepsilon_2)^2}{(N_{\text{T},0} + \varepsilon_1 + \varepsilon_2 - \varepsilon_3)(N_{1,0} - \varepsilon_1 - \varepsilon_2 - \varepsilon_3)}, \\
K_2 &= \left(\frac{P}{P^{\text{ref}}}\right) \frac{(N_{2,0} + \varepsilon_2 - \varepsilon_3)(N_{3,0} + 2\varepsilon_1 + \varepsilon_2)}{(N_{\text{T},0} + \varepsilon_1 + \varepsilon_2 - \varepsilon_3)(N_{1,0} - \varepsilon_1 - \varepsilon_2 - \varepsilon_3)}, \\
K_3 &= \left(\frac{P}{P^{\text{ref}}}\right)^{-1} \frac{(N_{\text{T},0} + \varepsilon_1 + \varepsilon_2 - \varepsilon_3)(N_{2,0} - \varepsilon_3)}{(N_{3,0} + 2\varepsilon_1 + \varepsilon_2)(N_{1,0} - \varepsilon_1 - \varepsilon_2 - \varepsilon_3)}.
\end{aligned} \tag{10.49}$$

In the reaction we introduce only species 1 initially. Thus, we can assume a basis of one mole for species 1, and zero for the rest: $N_{T,0} = N_{1,0} = 1, N_{2,0} = N_{3,0} = N_{4,0} = N_{5,0} = 0$. Therefore, since $P = P^{\text{ref}}$, Eqn. (10.49) has the three unknowns $\varepsilon_1, \varepsilon_2$, and ε_3, which may be found numerically with a root-finding routine, using a spreadsheet, MATLAB, Mathematica, Maxima, MathCAD, or programming [103, Chapter 9].

If we assume, however, that K_1 is much larger than the others, we can write $\varepsilon_1 \gg \varepsilon_2, \varepsilon_3$, the other extents are small, and

$$K_1 \approx \frac{(2\varepsilon_1)^2}{(1+\varepsilon_1)(1-\varepsilon_1)}. \tag{10.50}$$

On solving for ε_1, we find

$$\varepsilon_1 \approx \sqrt{\frac{K_1}{4+K_1}} \approx 1. \tag{10.51}$$

□

10.8 ENZYMATIC CATALYSIS: THE MICHAELIS–MENTEN MODEL

Many enzymatic reactions (reactions catalyzed by enzymes), can be modeled as a two-step process. The first step binds the enzyme En to the reactant, or substrate, Su, and the second step makes the product Pr:

$$\text{En} + \text{Su} \underset{k_2}{\overset{k_1}{\rightleftharpoons}} \text{EnSu} \overset{k_3}{\longrightarrow} \text{En} + \text{Pr}, \tag{10.52}$$

where EnSu indicates the bound enzyme–substrate complex. The products are written here as Pr, although this might indicate more than one species. The second step is written as irreversible here, which is the case when one of the products is driven out of the system, insoluble, or further consumed, or when the concentration is low and the reverse rate is negligible.

The Michaelis–Menten model describes the situation when $k_1, k_2 \gg k_3$, so that the concentration of the EnSu complex is in steady state with the enzyme and substrate concentrations, and the second step is rate limiting. Such is often the case, for example, in the microbial breakdown of contaminants in soil – as used in cleaning up contaminated land, i.e. site remediation.

We neglect non-idealities in mixing, and write the reaction rates in the simple forms

$$\begin{aligned}
\frac{dx_{\text{Su}}}{dt} &= k_2 x_{\text{EnSu}} - k_1 x_{\text{En}} x_{\text{Su}}, \\
\frac{dx_{\text{EnSu}}}{dt} &= k_1 x_{\text{En}} x_{\text{Su}} - k_2 x_{\text{EnSu}} - k_3 x_{\text{EnSu}}, \\
\frac{dx_{\text{Pr}}}{dt} &= k_3 x_{\text{EnSu}}.
\end{aligned} \tag{10.53}$$

The EnSu concentration is difficult to measure. However, we expect that the enzyme and substrate are rapidly binding and unbinding, and only occasionally does the bound complex react to form products. Mathematically, this means that rate 1 and rate 2 are much faster than rate 3, and, hence, the EnSu concentration is at steady state. Setting $dx_{\mathrm{EnSu}}/dt = 0$ above allows us to solve for the concentration of EnSu as

$$x_{\mathrm{EnSu}} \cong \frac{k_1}{k_2 + k_3} x_{\mathrm{En}} x_{\mathrm{Su}} \\ = K_{\mathrm{m}}^{-1} x_{\mathrm{En}} x_{\mathrm{Su}}, \tag{10.54}$$

where K_{m} is the **Michaelis dissociation constant**.

The next important idea is that we would like to write the rate in terms of the (constant) total enzyme concentration $x_{\mathrm{En}}^{\mathrm{T}}$, which includes both bound and unbound enzyme molecules. Remember, enzymes act as catalysts, so they are not consumed in the reaction. Therefore, we write

$$x_{\mathrm{En}}^{\mathrm{T}} = x_{\mathrm{En}} + x_{\mathrm{EnSu}} \\ = x_{\mathrm{En}}(1 + K_{\mathrm{m}}^{-1} x_{\mathrm{Su}}). \tag{10.55}$$

The rate of production of Pr is

$$\frac{dx_{\mathrm{Pr}}}{dt} = k_3 x_{\mathrm{EnSu}} \\ = \frac{k_3 x_{\mathrm{En}}^{\mathrm{T}} x_{\mathrm{Su}}}{K_{\mathrm{m}} + x_{\mathrm{Su}}}. \tag{10.56}$$

This is our desired result, which is called the **Michaelis–Menten equation**. There are several interesting things to note about this result. First, unlike first-order reaction kinetics, as we increase the concentration of substrate, the rate of production approaches a maximum: $(dx_{\mathrm{Pr}}/dt)_{\max} = k_3 x_{\mathrm{En}}^{\mathrm{T}}$.

Secondly, the rate has the same form as the Langmuir adsorption isotherm, Section 3.7.3, where K_{m}^{-1} plays the role of χ, and concentration plays the role of pressure. This similarity is not surprising, since adsorption plays a key role in both cases.

In fact, plots like the Langmuir adsorption isotherm can be constructed in order to find the rate constants k_3 and K_m. However, it is more convenient to find these constants using a so-called **Lineweaver–Burke plot** of rate^{-1} vs. x_{Su}^{-1}. Such a plot yields a straight line with slope $K_{\mathrm{m}}/(k_3 x_{\mathrm{En}}^{\mathrm{T}})$ and intercept $1/(k_3 x_{\mathrm{En}}^{\mathrm{T}})$.

10.9 DENATURATION OF DNA AND POLYMERASE CHAIN REACTIONS

Deoxyribonucleic acid (DNA) contains the genetic information that makes up living creatures like us. The chemical structure of DNA mutates. It also combines with other DNA during reproduction to make offspring. In fact, one might say that the point of a person is to make another DNA.[2] DNA is made up of two very long polymers in a double-helix formation. The monomers of the polymer molecules are nucleotides with side groups called *residues* that stick into the

[2] Stephen Jay Gould wrote that one could say that the point of a chicken is to make another egg.

center of the helix, and hydrogen-bond with their partners on the other chain. There are four different kinds of residues, C, T, A, and G (cytosine, thymine, adenine, and guanine), and the particular order constitutes the genetic information carried by the chain. Each type of residue pairs with a particular type: C always pairs with G, and A pairs with T. Each individual of a given species has a unique code given by the sequence of these CG and AT pairs. Depending on its environment, DNA can take several types of helical forms, but we will ignore that detail here.

Given a small sample of DNA, say from a single cell, for example in a drop of saliva, it is extremely useful to be able to reproduce the chain to produce many identical copies. These copies can be used for identification purposes in legal questions, for cloning individuals, or for genetic modification of plants and animals. The method used for replicating copies of small samples of DNA is called **polymerase chain reaction**, or PCR. The first step of PCR is separating the two chains in the coil, which we now discuss in some detail. In a rough sense, denaturation can be seen as a chemical reaction, whereby an attached nucleotide can become detached.

10.9.1 Denaturation

One can imagine that the entropy of two free, coiled polymer chains is much higher than that of the hydrogen-bonded double helix. Clearly, many more conformations are possible when the helix separates, or **denatures**. However, at some temperatures and in some solvent environments, the double helix is the thermodynamically favorable state. Since this formation is *entropically unfavorable*, it must be *energetically favorable*.

A simple statistical-mechanical model exists to describe this behavior qualitatively. The model ignores the differences between residues, and assumes that all pairs are the same. However, it does make a distinction between the first bound pair and all subsequent pairs when considering the coil $\rightarrow$ helix transition [151]. The model also assumes that, after the first residue pair has bound, an adjacent pair can then bind, and then the next adjacent pair, and so on. Hence, it is referred to as the **zipper model** [32, p. 31].

The model ignores pressure dependence of the energies, and uses the variables $(T, \tilde{M})$, where $\tilde{M}$ is the number of total pairs, bound or unbound. The fundamental relation derived from statistical mechanics is

$$\Delta G^{\text{helix}} = -k_{\text{B}} T \log \left\{ 1 + \frac{\sigma q_{\text{base}} \left[\tilde{M}(1 - q_{\text{base}}) - q_{\text{base}}(1 - q_{\text{base}}^{\tilde{M}}) \right]}{(1 - q_{\text{base}})^2} \right\}, \tag{10.57}$$

where ΔG^{helix} is the change in Gibbs potential for two coils to bind at least one pair of residues. This expression contains the nucleation probability $\sigma \ll 1$ of binding the first pair, and the change in free energy associated with binding subsequent pairs Δg_{base} through the simple function

$$q_{\text{base}}(T) := \exp\left[-\frac{\Delta g_{\text{base}}(T)}{k_{\text{B}} T} \right]. \tag{10.58}$$

The remaining parameter σ is sometimes called the **nucleation parameter**, because it is related to the likelihood that the first base pair attaches, making it easier for all subsequent pairs to attach. Note that the model does not include the solvent explicitly. However, Δg_{base} is the change on free energy on going from the solvated base pair to attachment with its complement on the other chain. Hence, the value depends on the solvent present. It should also depend on the type of pair (CG or AT), but to a first approximation, the model replaces these values with some average of the two, which is then independent of the type of nucleotide. In reality, we also know that an important energy of interaction is that from the "stacking" of adjacent bases on the same strand. This detail is ignored, and we assume that q_{base} contains this information.

At low temperatures all pairs will be bound, and the DNA will form a helix. At high temperatures, the two chains will completely separate, and form coils. This process is called **denaturation**. At intermediate temperatures, some number of pairs $\tilde{N} \leq \tilde{M}$ will be bound. We can find the fraction of bound pairs in the following way.

We call the change in Helmholtz potential of the chain on going from the coil to the helix ΔH^{helix}. This energy must be proportional to the number of bound pairs by $\Delta H^{\text{helix}} \cong \tilde{N}\,\Delta h_{\text{base}}$, according to the zipper model. We can find the change in enthalpy as a function of temperature from our fundamental relation

$$\begin{aligned}\frac{\Delta H^{\text{helix}}}{k_B T^2} &= -\left(\frac{\partial}{\partial T}\frac{\Delta G^{\text{helix}}}{k_B T}\right)_{\tilde{M}} \\ &= -\left(\frac{dq_{\text{base}}}{dT}\right)\left(\frac{\partial}{\partial q_{\text{base}}}\frac{\Delta G^{\text{helix}}}{k_B T}\right)_{\tilde{M},\sigma} \\ &= -\left(\frac{\Delta h_{\text{base}}}{k_B T^2}\right) q_{\text{base}}\left(\frac{\partial}{\partial q_{\text{base}}}\frac{\Delta G^{\text{helix}}}{k_B T}\right)_{\tilde{M},\sigma}.\end{aligned} \tag{10.59}$$

We used the chain rule to obtain the second line, and the definition for q_{base}, Eqn. (10.58), to obtain the third. Therefore, since $\Delta H^{\text{helix}} = \tilde{N}\,\Delta h_{\text{base}}$ we find that

$$\begin{aligned}\tilde{N} &= -q_{\text{base}}(T)\left(\frac{\partial}{\partial q_{\text{base}}}\frac{\Delta G^{\text{helix}}}{k_B T}\right)_{\tilde{M},\sigma} \\ &= \frac{q_{\text{base}}\sigma\left[2q_{\text{base}}\left(1-q_{\text{base}}^{\tilde{M}}\right)-\tilde{M}\left(1-q_{\text{base}}\right)\left(1+q_{\text{base}}^{\tilde{M}+1}\right)\right]}{\left(q_{\text{base}}-1\right)\left\{1+q_{\text{base}}\left[q_{\text{base}}-2+\tilde{M}\sigma-\left(\tilde{M}+1\right)q_{\text{base}}\sigma+q_{\text{base}}^{\tilde{M}+1}\sigma\right]\right\}}.\end{aligned} \tag{10.60}$$

As the temperature is increased, we expect Δg_{base} to decrease. At low temperatures, the energy of binding dominates the entropy of binding, and $\Delta g_{\text{base}} < 0$. At high temperatures, entropy dominates, and $\Delta g_{\text{base}} > 0$. At some intermediate temperature, $\Delta g_{\text{base}} = 0$, and $q_{\text{base}} = 1$. Figure 10.3 shows the fraction of attached base pairs as a function of $q_{\text{base}}(T)$ for the zipper model. From this figure, we see how the denaturation depends on the length of the strand, $\tilde{M}$. As the chain gets longer, the strand tends to become either completely separated or completely attached. The sharpness of this transition is called **cooperativity**.

The fraction of attached base pairs can be measured experimentally by observing how polarized light is rotated through the solution. The helix is always formed with a right-handed twist, thereby rotating circularly polarized light. As the chains increase their helicity, they increasingly rotate the light.

There exist more complicated denaturation models for DNA, which is an active field of research. An improved model (called the "matrix model") for longer chains is considered in the exercises.

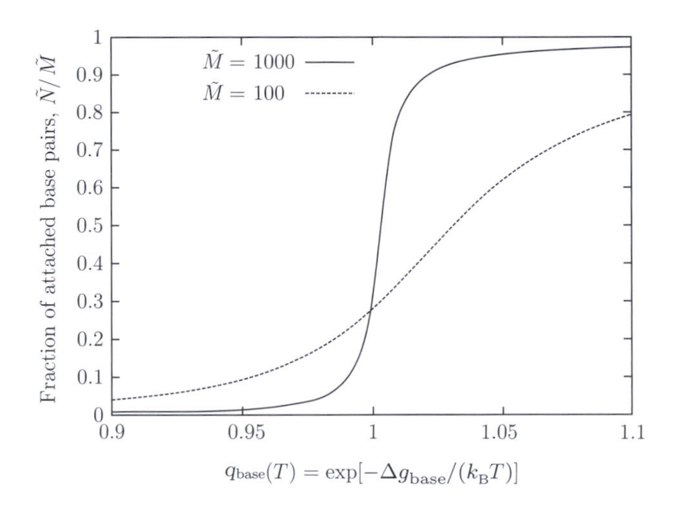

Figure 10.3 The fraction of attached base pairs in DNA strands of lengths 100 and 1000 pairs as predicted by the zipper model. Here the *nucleation parameter* σ is set to 0.001. In aqueous solutions, we expect q_{base} to decrease with temperature, leading to denaturation of the strand at high temperatures.

10.9.2 Polymerase chain reaction

Polymerase chain reaction (PCR) exploits denaturation in the following way. We put a single two-stranded DNA helix in a solution of the four nucleotides, CGAT. Also dissolved in the solution is a rather large concentration of primer, and an enzyme called polymerase. Primer is a short section of one of the chains whose sequence of base pairs matches that of one end of the single chain. The temperature of the solution is raised to approximately 95 °C, denaturing the DNA in a couple of minutes. As the temperature is lowered to near 50 °C, the primer attaches first to the single chains, before the helices can re-form. This primer attachment (over helix formation) happens for two reasons. First, because primer is present in excess concentration, and secondly, because, as we see from Figure 10.3, short chains attach at higher temperatures than do long chains. The attachment of the primer also takes a couple of minutes.

The temperature is raised to 72 °C, and the polymerase now catalyzes the reaction to form the complementary chain to which the primer is attached. In a few minutes, the two denatured strands are now two pairs of attached helices. The temperature is raised again, causing denaturation, and the process is repeated many times. A sketch of the temperature protocol for this procedure is given in Figure 10.4. If the cycle happens n times, we end up with 2^n times more helices than we started with, and our goal is accomplished. In two hours, $n \cong 20$, and we have $2^n \cong 1$ million times as many DNA strands as we started with. We have produced millions of replicas of the chains without even knowing what their sequences are. The idea earned windsurfer Kari Mullis a Nobel Prize in 1993.

There is an optimal range for choosing the length of the primer. If the primer is too short, it might attach at several places along the chain, instead of just at the end. If the primer is too long, it might take too long to attach.

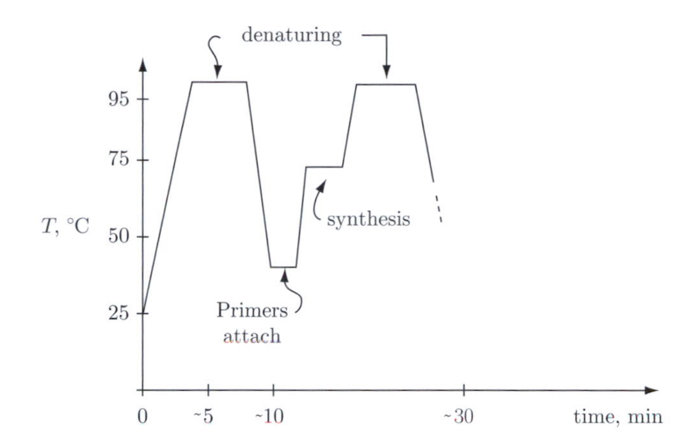

Figure 10.4 A sketch of the temperature protocol for performing PCR.

10.10 STATISTICAL MECHANICS OF REACTIONS AND DENATURATION

10.10.1 Stochastic fluctuations in reactions

Engineers are becoming increasingly interested in reactions in small volumes. For example, reactions inside cells may involve only a few molecules of a large species; some reactors use small particles inside of which a few molecules are polymerizing; and, some companies are forming "labs on a chip," where small amounts of quantities, like blood, are tested in tiny reactors created on silicon chips. In these situations, the traditional assumptions about large reactors no longer apply. Such is the case in small systems, similarly to what we saw in Section 6.7.

On the one hand, the usual reaction-rate equations as ordinary differential equations (ODEs), such as those learned in basic chemistry or reaction engineering, can be derived from molecular-level stochastic dynamical equations [143, Chapter IX]. Such a derivation is beyond the scope of this text, so we will not consider it in detail. On the other hand, the same starting point applies to small systems; we consider reaction dynamics in general, large or small. The importance of fluctuations in reactions has recently been considered in an undergraduate chemical-reactor textbook by Rawlings and Eckerdt [107, Section 4.8].

Stochastic dynamics in a linear, reversible reaction

We consider the simplest possible example to elucidate the technique, namely a reversible, first-order reaction

$$\mathrm{A} \underset{k_2}{\overset{k_1}{\rightleftharpoons}} \mathrm{B}. \tag{10.61}$$

In this context, k_1 is the probability that a single molecule of A reacts to form B, *per unit time.*

Imagine a box of volume V at temperature T, into which you place $\tilde{N}_{\mathrm{A}}^{\circ}$ molecules of A. After some period of time, there would be both A and B molecules such that $\tilde{N}_{\mathrm{A}} + \tilde{N}_{\mathrm{B}} = \tilde{N}_{\mathrm{A}}^{\circ}$. The number of such molecules $\tilde{N}_A$ would fluctuate with time; hence, there would be a probability $p(\tilde{N}_A;t)$ of having $\tilde{N}_A$ at time t. We can find this probability through the idea of an *ensemble* of identical such boxes. If we had, say, millions of identical boxes, each in the state $(T, V, \tilde{N}_A^\circ)$, then, after some time, there would be a distribution of mole numbers – each box would have its own $\tilde{N}_A$. The physical definition of the probability, is that $p(\tilde{N}_A;t)$ is the fraction of boxes at time t with $\tilde{N}_A$ molecules of A, in the limit of infinitely many boxes.

The trick now is to focus on $p(\tilde{N}_A;t)$, instead of on an individual box. The fraction of boxes with a given number of molecules changes because of the reactions. When a box with $\tilde{N}_A + 1$ molecules of A forms a new molecule of B, then $p(\tilde{N}_A;t)$ must increase. The probability that one molecule reacts to form B is k_1. There are $\tilde{N}_A + 1$ such molecules, and fraction $p(\tilde{N}_A + 1;t)$ such boxes. Hence, we can write

$$\frac{\partial p(\tilde{N}_A;t)}{\partial t} \sim k_1(\tilde{N}_A + 1)p(\tilde{N}_A + 1;t). \tag{10.62}$$

There are three other ways in which the fraction of boxes with $\tilde{N}_A$ molecules can change: when a molecule of B reacts to form A, and the reverse of these two cases:

$$\begin{aligned}\frac{\partial p(\tilde{N}_A;t)}{\partial t} &= k_1\left(\tilde{N}_A + 1\right)p(\tilde{N}_A + 1;t) + k_2\left(\tilde{N}_A^\circ - \tilde{N}_A + 1\right)p(\tilde{N}_A - 1;t) \\ &\quad - k_1\tilde{N}_A p(\tilde{N}_A;t) - k_2\left(\tilde{N}_A^\circ - \tilde{N}_A\right)p(\tilde{N}_A;t).\end{aligned} \tag{10.63}$$

Note that we used the conservation of molecules to eliminate $\tilde{N}_B = \tilde{N}_A^\circ - \tilde{N}_A$ from this expression. All the physical ideas are in this equation, and the rest is just mathematics, so make sure that you understand the origin of each term in the equation.

We know that large systems make only very small fluctuations around the average number of molecules, $\langle\tilde{N}_A\rangle := \sum_{\tilde{N}_A}\tilde{N}_A p(\tilde{N}_A;t)$ (see Section 6.7). From Eqn. (10.63), it is possible to construct an evolution equation for the average. This is accomplished by multiplying with $\tilde{N}_A$, and summing over all possible values for $\tilde{N}_A$:

$$\begin{aligned}\sum_{\tilde{N}_A=0}^{\infty}\tilde{N}_A\frac{\partial p(\tilde{N}_A;t)}{\partial t} &= k_1\sum_{\tilde{N}_A=0}^{\infty}\tilde{N}_A\left(\tilde{N}_A + 1\right)p(\tilde{N}_A + 1;t) \\ &\quad + k_2\sum_{\tilde{N}_A=0}^{\infty}\tilde{N}_A\left(\tilde{N}_A^\circ - \tilde{N}_A + 1\right)p(\tilde{N}_A - 1;t) \\ &\quad - k_1\sum_{\tilde{N}_A=0}^{\infty}\tilde{N}_A^2 p(\tilde{N}_A;t) - k_2\sum_{\tilde{N}_A=0}^{\infty}\tilde{N}_A\left(\tilde{N}_A^\circ - \tilde{N}_A\right)p(\tilde{N}_A;t).\end{aligned} \tag{10.64}$$

To simplify this equation, we can pull the partial derivative out of the term on the left-hand side, and change the dummy summation variable in the first ($\tilde{N}_A \to n - 1$) and second ($\tilde{N}_A \to n + 1$) terms on the right-hand side. The third term on the right-hand side involves the average of the second moment. Hence, we obtain

$$\frac{\partial}{\partial t}\sum_{\tilde{N}_A=0}^{\infty}\tilde{N}_A p(\tilde{N}_A;t) = k_1\sum_{n=1}^{\infty}(n-1)np(n;t) + k_2\sum_{n=0}^{\infty}(n+1)(\tilde{N}_A^\circ - n)p(n;t)$$
$$- k_1\langle\tilde{N}_A^2\rangle - k_2\left(\tilde{N}_A^\circ\langle\tilde{N}_A\rangle - \langle\tilde{N}_A^2\rangle\right),$$
$$\frac{d}{dt}\langle\tilde{N}_A\rangle = k_1\left[\langle\tilde{N}_A^2\rangle - \langle\tilde{N}_A\rangle\right] + k_2\left[\tilde{N}_A^\circ\left(\langle\tilde{N}_A\rangle + 1\right) - \langle\tilde{N}_A^2\rangle - \langle\tilde{N}_A\rangle\right]$$
$$- k_1\langle\tilde{N}_A^2\rangle - k_2\left(\tilde{N}_A^\circ\langle\tilde{N}_A\rangle - \langle\tilde{N}_A^2\rangle\right),$$
$$\frac{d}{dt}\langle\tilde{N}_A\rangle = k_2\left(\tilde{N}_A^\circ - \langle\tilde{N}_A\rangle\right) - k_1\langle\tilde{N}_A\rangle. \tag{10.65}$$

Note the simple final result. If we divide each side by the volume, we obtain an equation familiar from basic chemistry, or from reaction engineering

$$\frac{d}{dt}c_A = -k_1 c_A + k_2(c_A^\circ - c_A), \tag{10.66}$$

where $c_A = \langle N_A\rangle/V$ is the concentration of A (and the concentration of B is, from conservation again, $c_B = c_A^\circ - c_A$). In other words, the probability evolution equation, Eqn. (10.63), contains the simple reaction evolution equation with which we are familiar. However, it also contains more information, since we can also derive the second moment evolution equation

$$\frac{d}{dt}\langle\tilde{N}_A^2\rangle = k_1\left[-2\langle\tilde{N}_A^2\rangle + \langle\tilde{N}_A\rangle\right] + k_2\left[\tilde{N}_A^\circ\left(2\langle\tilde{N}_A\rangle + 1\right) - 2\langle\tilde{N}_A^2\rangle - \langle\tilde{N}_A\rangle\right]. \tag{10.67}$$

This equation is derived in a fashion very similar to that used to derive the evolution equation for the average, Eqn. (10.65). Perhaps more useful is the evolution equation for the variance, $\sigma_N^2 := \langle\tilde{N}_A^2\rangle - \langle\tilde{N}_A\rangle^2$

$$\frac{d}{dt}\sigma_N^2 = -2(k_1 + k_2)\sigma_N^2 + k_1\langle\tilde{N}_A\rangle + k_2\left[\tilde{N}_A^\circ - \langle\tilde{N}_A\rangle\right]. \tag{10.68}$$

The ODEs describing $\langle\tilde{N}_A\rangle$ and σ_N^2, Eqns. (10.65) and (10.68), can be solved analytically, since they are just linear ODEs:

$$\langle\tilde{N}_A\rangle = \tilde{N}_A^\circ \exp[-(k_1 + k_2)t] + \frac{k_2\tilde{N}_A^\circ}{k_1 + k_2}\{1 - \exp[-(k_1 + k_2)t]\},$$
$$\sigma_N^2 = \frac{k_1\tilde{N}_A^\circ}{(k_1 + k_2)^2}\{k_2 + (k_1 - k_2)\exp[-(k_1 + k_2)t] - k_1\exp[-2(k_1 + k_2)t]\}. \tag{10.69}$$

Figure 10.5 shows the normalized mole number $\langle\tilde{N}_A\rangle/\tilde{N}_A^\circ$ and the normalized variance $\sigma_N^2(k_1 + k_2)^2/(k_1 k_2 \tilde{N}_A^\circ)$ as functions of dimensionless time $(k_1 + k_2)t$. There are a few interesting things to note from this plot.

First, the variance starts off at zero, since all of the molecules are A initially, and $\sigma_N^2(t = 0) = (\tilde{N}_A^\circ)^2 - (\tilde{N}_A^\circ)^2 = 0$. Secondly, the final variance is $k_1 k_2\tilde{N}_A^\circ/(k_1 + k_2)^2$, which means that the deviation, σ_N, rises with the square root of mole number. Hence, for large systems, $\sigma_N^2/\langle\tilde{N}_A\rangle^2$ goes to zero, and fluctuations are unimportant. Thirdly, we see that the variance can rise non-monotonically with time when $k_1 > k_2$. Finally, we see that the usual expression (10.66) applies even to small systems.

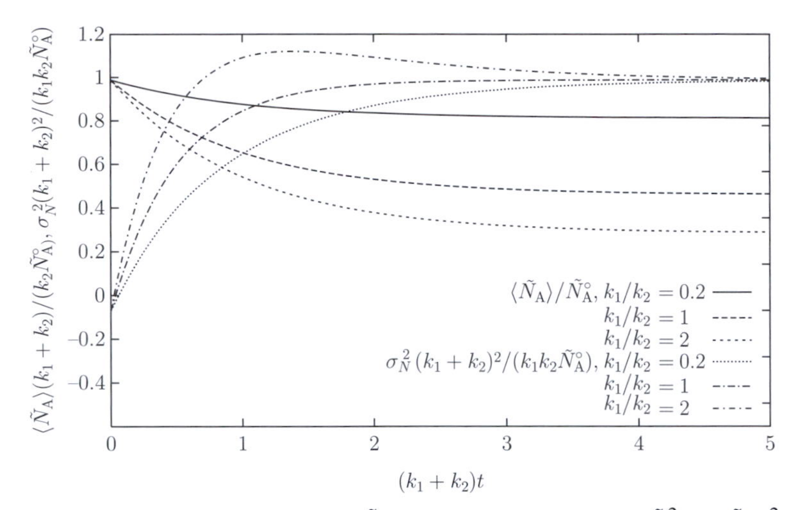

Figure 10.5 The average number of A molecules $\langle\tilde{N}_A\rangle$ and the variance $\sigma_N := \langle\tilde{N}_A^2\rangle - \langle\tilde{N}_A\rangle^2$, normalized as functions of dimensionless time for the reversible, first-order reaction given in Eqn. (10.1.)

Stochastic simulation of reactions

It is illustrative to look at what can happen in a single such system, when the mole number is small. The random dynamics of a single box in our ensemble is called a **realization** of a **stochastic process**. A stochastic process is what a time-dependent random variable is called. We can construct a realization using a computer program and a pseudo-random-number generator. Most programs have available a subroutine that returns a random number between 0 and 1, with all numbers equally likely. Technically, this is called a uniform random number on the domain zero to one. From our probability evolution equation, Eqn. (10.63), it is possible to construct a computer algorithm.

The computer algorithm that we will construct here is not necessarily the most efficient, but it is the simplest to understand, and sufficiently fast on even cheap and old computers for our purposes. There is a second reason to construct such an algorithm. Only for first-order reactions is it possible to find estimates for the averages and variances analytically, in general. Van Kampen [143] has devised a way to find the analytic evolution equations for the average and variance, but only in the limit of very large systems, and we are interested here in the behavior of small systems. More sophisticated algorithms may be found, for example, in [65].

To construct our algorithm, we integrate Eqn. (10.63) over a small time step Δt. The term on the left-hand side can be integrated analytically, using the fundamental theorem of calculus, Section A.4 of Appendix A. The terms on the right-hand side, however, cannot be integrated analytically, since the probability function p depends on time. However, if we make our time-step size sufficiently small, we can approximate the integrand to be nearly constant over the integral. Hence, we write

$$\begin{aligned} p(\tilde{N}_A; t+\Delta t) \cong\ & p(\tilde{N}_A; t) + k_1\left(\tilde{N}_A + 1\right)p(\tilde{N}_A + 1; t)\Delta t \\ & + k_2\left(\tilde{N}_A^\circ - \tilde{N}_A + 1\right)p(\tilde{N}_A - 1; t)\Delta t \\ & - k_1\tilde{N}_A p(\tilde{N}_A; t)\Delta t - k_2\left(\tilde{N}_A^\circ - \tilde{N}_A\right)p(\tilde{N}_A; t)\Delta t. \end{aligned} \tag{10.70}$$

We use the algorithm in the following way. At the initial time $t = 0$, we begin with $\tilde{N}_A = \tilde{N}_A^\circ$ and $\tilde{N}_B = 0$. Hence, we know the distribution function p: $p(\tilde{N}_A^\circ; t = 0) = 1$, and all other p are zero. Then, Eqn. (10.70) becomes

$$\begin{aligned} p(\tilde{N}_A^\circ; t = \Delta t) &= 1 - k_1 \tilde{N}_A^\circ \, \Delta t, \\ p(\tilde{N}_A^\circ - 1; t = \Delta t) &= k_1 \tilde{N}_A^\circ \, \Delta t, \end{aligned} \tag{10.71}$$

and $p(\tilde{N}_A; t = \Delta t)$ for all other values of $\tilde{N}_A$ are zero.

For this first time step, we use our pseudo-random-number generator to give us $X_0 \in (0, 1)$. If $X_0 < k_1 \tilde{N}_A^\circ \, \Delta t$, we set $\tilde{N}_A(t = \Delta t) = \tilde{N}_A^\circ - 1$; otherwise, we leave the number of moles of A unchanged.

For the second and subsequent time steps we use again Eqn. (10.70). At each step, we know the probability distribution from the results for our previous time step. Say that our previous time step n gave us the number of moles of A, $\tilde{N}_A(t = n \, \Delta t)$. Then, the algorithm gives us three possibilities for the next time step $\tilde{N}_A(t = (n + 1)\Delta t)$:

$$\begin{aligned} p(\tilde{N}_A(n \, \Delta t); t = (n+1)\Delta t) &= 1 - k_1 \tilde{N}_A(n \, \Delta t)\Delta t - k_2 \left[\tilde{N}_A^\circ - \tilde{N}_A(n \, \Delta t)\right] \Delta t, \\ p(\tilde{N}_A(n \, \Delta t) - 1; t = (n+1)\Delta t) &= k_1 \tilde{N}_A(n \, \Delta t)\Delta t \equiv p_1, \\ p(\tilde{N}_A(n \, \Delta t) + 1; t = (n+1)\Delta t) &= k_2 \left[\tilde{N}_A^\circ - \tilde{N}_A(n \, \Delta t)\right] \Delta t \equiv p_2. \end{aligned} \tag{10.72}$$

So, at each time step n, we draw a random number X_n between zero and one. If $X_n < p_1$, we decrease $\tilde{N}_A$ by one; if $p_1 < X_n < (p_1 + p_2)$, we increase $\tilde{N}_A$ by one; otherwise, we leave the number of moles fixed. Because of the approximation used in the construction of our algorithm, one must pick a time-step size sufficiently small that the results are independent of Δt.

Figure 10.6 shows the results of two such simulations compared with our analytic result above. In both simulations we consider the same reaction rate constants $k_1 = 2k_2$. However, on the left-hand side we have a box with 100 molecules of A initially, and on the right a box with 5000.

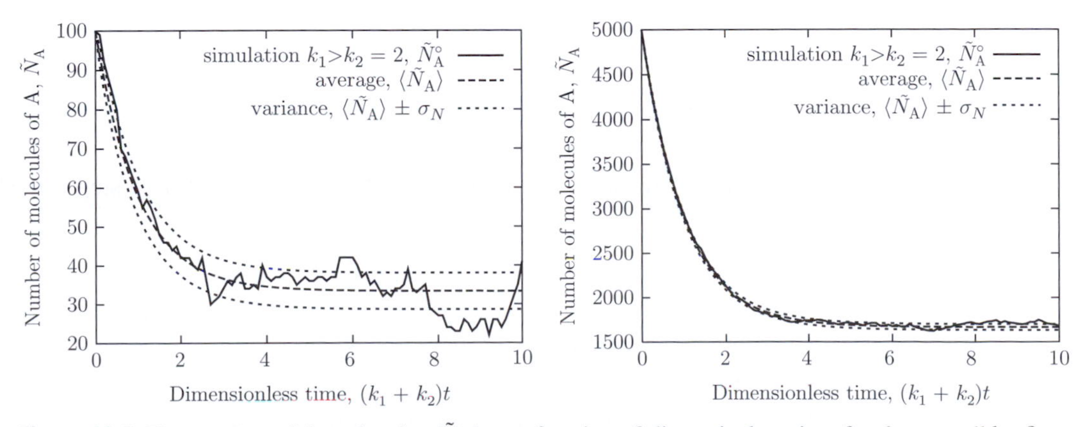

Figure 10.6 The number of A molecules $\langle \tilde{N}_A \rangle$ as a function of dimensionless time for the reversible, first-order reaction given in Eqn. (10.1.) Also shown are the analytic results for the average and the magnitude of the expected variation. Note that the plot on the left, which begins with fewer molecules, has larger relative fluctuations.

We end this section by considering generalization of this simple example to higher-order, or multiple, reactions.

Generalizations to multi-step, higher-order reactions

There is no possible way to derive evolution equations for the averages or variances σ_N, when the reactions are of second or higher order, except in the limit $V \to \infty$. On the other hand, changing the computer simulation from the previous subsection is trivial. Hence, it is important to consider what form the evolution equation for the probability distribution takes for higher-order reactions. Here we discover that an important subtlety exists for higher-order reactions.

The subtlety is perhaps easiest to see with a simple example. For a second-order reaction of the type

$$2\mathrm{A} \underset{k_2}{\overset{k_1}{\rightleftharpoons}} \mathrm{B}, \tag{10.73}$$

the evolution equation for the probability distribution becomes

$$\begin{aligned}\frac{\partial p(\tilde{N}_\mathrm{A};t)}{\partial t} &= k_1\left(\tilde{N}_\mathrm{A}+2\right)\left(\tilde{N}_\mathrm{A}+1\right)p(\tilde{N}_\mathrm{A}+2;t) + k_2\left(\frac{\tilde{N}^\circ_\mathrm{A}-\tilde{N}_\mathrm{A}+2}{2}\right)p(\tilde{N}_\mathrm{A}-2;t)\\ &- k_1\tilde{N}_\mathrm{A}(\tilde{N}_\mathrm{A}-1)p(\tilde{N}_\mathrm{A};t) - k_2\left(\frac{\tilde{N}^\circ_\mathrm{A}-\tilde{N}_\mathrm{A}}{2}\right)p(\tilde{N}_\mathrm{A};t).\end{aligned} \tag{10.74}$$

Note the subtle difference from the evolution equation that arises for the average in the thermodynamic limit. Namely, the forward reaction is proportional to $\tilde{N}_\mathrm{A}(\tilde{N}_\mathrm{A}-1)$ instead of $\tilde{N}_\mathrm{A}^2$. Physically, this makes sense, since, if one molecule is already involved in the reaction, there remain only $\tilde{N}_\mathrm{A}-1$ molecules. For small systems, this difference is very important.

Secondly, we note that the evolution equations for the averages – the thermodynamic limit of large systems – considered in this way result in reactions for Lewis mixtures, or ideal gases. For more general systems, one requires that the rate coefficients depend on concentrations (or, equivalently, mole numbers).

Thirdly, for multiple reactions, we should recall from Section 10.2 that there are typically fewer degrees of freedom than there are species. In the last example, we considered two species A and B, but the state of the system could be characterized by knowing $\tilde{N}_\mathrm{A}$ only. For multiple reactions, it is better to use the extents of reaction (Eqn. (10.6) on a molecular, instead of molar, basis). Therefore, one writes down the evolution equation for the probability distribution $p(\varepsilon_1, \varepsilon_2; t)$, for a two-step reaction, where the molecule numbers are replaced with expressions involving the initial molecule number and extent of reaction.

Finally, we note that irreversible, multiple reactions can be fundamentally different, since extinction can occur. If one of your species in the reaction is a bacterium, for example, it is possible that all of the bacteria die. At that point, the reactions cease. If one is considering the traditional average concentration, the possibility for extinction is ignored, and qualitatively different results for small and large systems can be observed.

10.10.2 DNA denaturation

We derive here the zipper model introduced in Section 10.9.1. We begin with a simplified picture of the coil → helix transition for DNA. We assume the following.

- All base pairs have the same free energy of binding Δg_{base}.
- All bound pairs occupy adjacent sites on the chain – in other words, there is only one continuous section of attached nucleotides on the chain, and there are no "bubbles" (see Figure 10.7).
- The free-energy change associated with the first bound pair is much greater than that for the other pairs; this increase is primarily entropic, and therefore temperature-independent.
- The volume of the system is constant.

We picture two complementary chains of $\tilde{M}$ nucleotides each. Adjacent bases do not interact energetically to any significant extent. Since the adjacent base pairs are non-interacting, we can write the partition function as a product of individual base-pair partitions. We also need to sum up over the number of all possible attached pairs. Therefore, we write the chain partition function as

$$Q = q_{\text{u}}^{\tilde{M}} + q_{\text{u}}^{\tilde{M}-1} q_{\text{fa}} \omega_{\text{d}}(1) + q_{\text{u}}^{\tilde{M}-2} q_{\text{fa}} q_{\text{a}} \omega_{\text{d}}(2) + \cdots, \tag{10.75}$$

where $q_{\text{u}}(T)$ is the partition function for an unattached pair, $q_{\text{fa}}(T)$ is the partition function for the first attached pair, $q_{\text{a}}(T)$ is the partition function for each subsequently attached pair, and $\omega_{\text{d}}(\tilde{N})$ is the degeneracy, or number of ways in which a chain can have $\tilde{N}$ attached base pairs. The first term on the right-hand side is the contribution from two chains that are completely separated. The second term is for two chains attached by a single base pair, and so on.

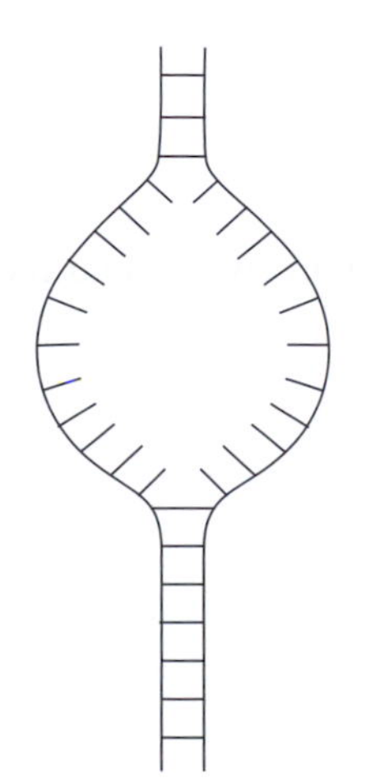

Figure 10.7 When an interior section of DNA has unbound base pairs, the structure is called a "bubble."

We assume that $\sigma := q_{\rm fa}/q_{\rm a}$, which is the extra entropic penalty for bringing the two chains together, is temperature-independent. This ratio is sometimes called the **nucleation parameter**. We expect the penalty to be large, so $\sigma \ll 1$. Putting this expression for σ into Eqn. (10.75) gives us

$$Q = q_{\rm u}(T)^{\tilde{M}} + \sigma \sum_{\tilde{N}=1}^{\tilde{M}} \omega_{\rm d}(\tilde{N}) q_{\rm u}(T)^{\tilde{M}-\tilde{N}} q_{\rm a}(T)^{\tilde{N}}. \tag{10.76}$$

The degeneracy $\omega_{\rm d}(\tilde{N})$ keeps track of the number of ways in which a chain can have $\tilde{N}$ base pairs attached. Recall that these bound pairs are assumed to be adjacent. Hence, $\omega_{\rm d} = \tilde{M} - \tilde{N} + 1$. We also define

$$q_{\rm base}(T) := \frac{q_{\rm a}(T)}{q_{\rm u}(T)}, \tag{10.77}$$

which, by virtue of its definition, is $\exp[-\Delta g_{\rm base}/(k_{\rm B}T)]$, where $\Delta g_{\rm base}$ is the change in free energy of a base pair upon attachment. Therefore, our partition function becomes

$$\begin{aligned} Q &= q_{\rm u}^{\tilde{M}} \left\{ 1 + \sigma \sum_{\tilde{N}=1}^{\tilde{M}} (\tilde{M} - \tilde{N} + 1) q_{\rm base}(T)^{\tilde{N}} \right\} \\ &= q_{\rm u}^{\tilde{M}} \left\{ 1 + \sigma \left[(\tilde{M}+1) \sum_{\tilde{N}=1}^{\tilde{M}} q_{\rm base}^{\tilde{N}} - \sum_{\tilde{N}=1}^{\tilde{M}} \tilde{N} q_{\rm base}^{\tilde{N}} \right] \right\} \\ &= q_{\rm u}^{\tilde{M}} \left\{ 1 + \sigma \left[(\tilde{M}+1) - q_{\rm base} \frac{\partial}{\partial q_{\rm base}} \right] \sum_{\tilde{N}=1}^{\tilde{M}} q_{\rm base}^{\tilde{N}} \right\}. \end{aligned} \tag{10.78}$$

It is possible to find an algebraic expression for the sum in the following way:

$$\begin{aligned} \sum_{i=1}^{n} \lambda^i &= \sum_{i=0}^{\infty} \lambda^i - 1 - \sum_{i=n+1}^{\infty} \lambda^i \\ &= \frac{1}{1-\lambda} - 1 - \lambda^{n+1} \sum_{i=0}^{\infty} \lambda^i \\ &= \frac{\lambda}{1-\lambda} - \frac{\lambda^{n+1}}{1-\lambda} \\ &= \frac{\lambda(1-\lambda^n)}{1-\lambda}. \end{aligned} \tag{10.79}$$

Hence, with a little more algebra, we obtain

$$Q = q_{\rm u}^{\tilde{M}} \left\{ 1 + \frac{\sigma q_{\rm base} \left[\tilde{M}(1 - q_{\rm base}) - q_{\rm base}(1 - q_{\rm b}^{\tilde{M}}) \right]}{(1 - q_{\rm base})^2} \right\}. \tag{10.80}$$

Recall that the free energy is $-k_{\rm B}T$ times the logarithm of the partition function. We split the free energy into that of the detached chains and that of the attached chains of various amounts of helicity

$$F = F_{\rm unattached} + \Delta F^{\rm helix}. \tag{10.81}$$

From our partition function, then, we have found the free energy for the unattached chains,

$$\frac{F_{\text{unattached}}}{k_{\text{B}}T} = -\tilde{M}\log q_{\text{u}}(T), \tag{10.82}$$

and the bound chains,

$$\frac{\Delta F^{\text{helix}}}{k_{\text{B}}T} = -\log\left\{1 + \frac{\sigma q_{\text{base}}\left[\tilde{M}(1-q_{\text{base}}) - q_{\text{base}}(1-q_{\text{base}}^{\tilde{M}})\right]}{(1-q_{\text{base}})^2}\right\}. \tag{10.83}$$

The expression used in Section 10.9.1, namely Eqn. (10.57), replaces the Helmholtz potential with the Gibbs free energy as an approximation neglecting compressibility. From the derivation, one can see the assumptions necessary to derive the expression.

From the first line of Eqn. (10.78), we see that the probability of having $\tilde{N}$ attached base pairs is

$$p_{\tilde{N}} = \frac{\sigma(\tilde{M} - \tilde{N} + 1)q_{\text{u}}^{\tilde{M}} q_{\text{base}}(T)^{\tilde{N}}}{Q(T,\tilde{M})}. \tag{10.84}$$

The average number of attached pairs can then be found using this probability as another way to obtain Eqn. (10.60).

A somewhat more realistic model allows multiple sections of attached nucleotides on the chains. In other words, there can be a string of attached base pairs, then a string of detached base pairs, and then another string of attached pairs, as shown in Figure 10.7. The resulting fundamental relation of the **Zimm–bragg matrix model** is

$$\frac{\Delta F(T,\tilde{M})}{k_{\text{B}}T} = \log\left[\frac{q_{\text{base}}\sigma\left(1 + q_{\text{base}} + \sqrt{(q_{\text{base}}-1)^2 + 4q_{\text{base}}\sigma}\right)^{\tilde{M}+1}}{2^{\tilde{M}}\sqrt{(q_{\text{base}}-1)^2 + 4q_{\text{base}}\sigma}\left(q_{\text{base}} - 1 + \sqrt{(q_{\text{base}}-1)^2 + 4q_{\text{base}}\sigma}\right)}\right]. \tag{10.85}$$

The derivation of this model is considered as an exercise. The functions showing up here have the same meaning as those in the zipper model. We can expect these two models to give the same results when chains are not expected to have more than one portion of the chain attached.

Summary

Upon completion of this chapter, the student should be able to do the following.

- Find the equilibrium composition for a reacting system, which we may write as

$$0 \rightleftharpoons \sum_i^r \nu_i^j B_i, \quad j = 1, 2, \ldots, q, \tag{10.11$'$}$$

  for r species and q reactions. In order to solve such problems, we found it useful to introduce two new entities: the extent of reaction ε and the equilibrium constant K. Since the total number of atoms is conserved in a reaction, the number of degrees of freedom is typically smaller than the number of moles of all species present. Introducing the extent of reaction allows us to replace the number of moles N_i with ε and the initial number of moles $N_{i,0}$:

$$N_i = N_{i,0} + \sum_{j=1}^{q} \nu_i^j \varepsilon_j, \quad i = 1, 2, \ldots, r. \tag{10.12$'$}$$

- Estimate the equilibrium constant K defined as

$$K(T) := \exp\left(-\frac{\Delta g^\circ}{RT}\right), \tag{10.21}$$

where the standard Gibbs free-energy change of reaction is defined as

$$\Delta g^\circ(T) := \sum_{i=1}^{r} \nu_i \mu_i^\circ. \tag{10.22}$$

Here we use the chemical potential of the pure component at arbitrary T, but reference pressure P^{ref}: $\mu_i^\circ := \mu_i^{\text{pure}}[P^{\text{ref}}]$.

- Find the equilibrium compositions from the expression

$$\prod_{i=1}^{r} a_i^{\nu_i} = K(T), \quad \text{equilibrium criterion, single reaction,} \tag{10.20$'$}$$

where we use the activity defined by

$$a_i := \frac{f_i(T, P, \{x_i\})}{f_i^{\text{pure}}(T, P^{\text{ref}})}, \tag{10.19$'$}$$

when the equilibrium constant has already been found. The activities are found as functions of temperature, pressure, and composition using the techniques introduced in Chapters 8 and 9.

As examples, we considered the decomposition of silane used in chemical vapor deposition for producing silicon wafers (Example 10.6.2), production of hydrogen from biomass, and DNA denaturation using a simple zipper model. Michaelis–Menten (or Monod) kinetics were also considered for two-step reactions.

- Estimate fluctuations in small reacting systems. We found a way to estimate the size of the fluctuations analytically for first-order reactions, and showed how to construct a simple computer algorithm for higher-order reactions.
- Derive simple statistical-mechanical models for predicting the denaturation of DNA.

Exercises

10.4.A As part of the design of a smokestack scrubber, your engineering team will need to estimate the importance of various reactions at many different temperatures. To aid in the design, make a plot of the equilibrium constant vs. temperature for the reaction

$$SO_2 + \frac{1}{2}O_2 \rightleftharpoons SO_3, \tag{10.86}$$

in which all of the species participating are gases. Make sure that your data are valid in the range of your plot, and use 1 bar as your reference pressure.

10.4.B Make a plot of the equilibrium constant vs. temperature for the reaction

$$CO + \frac{1}{2}O_2 \rightleftharpoons CO_2, \tag{10.87}$$

in which all of the species participating are gases. Your temperature range should be 2000–4000 K, and the reference pressure is 1 bar. In what temperature range is carbon dioxide favored over carbon monoxide? What does this say about the catalytic converter in your car?

10.6.A[3] It has been suggested that a new catalyst will be appropriate for use in the manufacture of methanol and ethanol from synthesis gas, a mixture of carbon monoxide and hydrogen. Consider the following reactions:

$$\begin{aligned} CO + 2H_2 &\rightleftharpoons CH_3OH, \\ 2CO + 4H_2 &\rightleftharpoons C_2H_5OH + H_2O. \end{aligned} \tag{10.88}$$

Determine the equilibrium composition which will be achieved at 400 bar and 650 K when the initial mole ratio of hydrogen to carbon monoxide is 2. Use the enthalpy and Gibbs free energy of formation data from the NIST Chemistry Web Book. You may neglect the variation of the standard heat of reaction with temperature. Assume Lewis mixing, and use the Peng–Robinson model to estimate the pure fugacity for each species.

10.6.B One of your chemist colleagues at eRenewableEnergyTechLabs claims to have developed a new catalyst to convert 95% pure ethane to ethylene and hydrogen,

$$C_2H_6 \rightleftharpoons C_2H_4 + H_2, \tag{10.89}$$

at 1000 K and 1 atm. Estimate the necessary pressure to obtain 95% conversion at 1000 K, and the necessary temperature to obtain the same conversion at 1 atm. Is your colleague's claim plausible?

10.6.C A mixture of carbon dioxide and hydrogen is flowing over a catalyst at 1000 K, at a pressure yet to be determined. Assuming that equilibrium is attained, you need to determine for what range of pressures carbon will be deposited on the catalyst. You have good reason to believe that the following reactions are the important ones:

$$\begin{aligned} C + H_2O &\rightleftharpoons CO + H_2, \\ C + 2H_2O &\rightleftharpoons CO_2 + 2H_2, \\ CO_2 + C &\rightleftharpoons 2CO, \\ CO + H_2O &\rightleftharpoons CO_2 + H_2. \end{aligned} \tag{10.90}$$

[3] Exercises 10.6.A–10.6.E are based on material in [157]. © 2014 John Wiley & Sons. Reproduced with permission.

At the given temperature, the reactions have equilibrium constants of 3.16, 5.01, 2.00, and 1.58, respectively. The reference pressure is 1 bar. Our inlet flow stream has twice as many moles of hydrogen as carbon dioxide.

1. First, convince yourself that there are only two independent reactions.
2. Since we do not know the pressure yet, make the simplest possible assumption for all species: ideal gas. Now calculate the extent of reaction when no carbon is deposited (e.g. there is only one reaction). Find the exiting, equilibrium composition. Did we need to make any assumptions about the value for the pressure?
3. Now we consider what happens when carbon is deposited and one of the other reactions comes into play. We wish to find the pressure at which carbon is just starting to be deposited. Find this by taking the composition you just found and find the pressure at which one of the other equilibrium constants is satisfied. Assume unit activity for the solid carbon. This is the pressure at which the first tiny bit of carbon is deposited.
4. Now check your assumption for ideality at this pressure.
5. Finally, find the pressure necessary to convert half the incoming moles of carbon atoms to solid carbon, assuming ideality. Is this a good assumption?

10.6.D Ethylbenzene is being synthesized from an equimolar mixture of ethylene and benzene fed to a reactor at 400 °C and 2 bar. All necessary thermodynamic data are given in Table 10.2. You might find Eqn. (10.2) useful.

Assuming that the reaction is isothermal and reaches equilibrium, find the exiting composition and the amount of heat that must be removed per mole of feed.

Note that at this pressure ideality is probably a safe assumption.

Table 10.2 Standard Gibbs free energy of formation data, standard heat of formation data and average constant-pressure ideal heat capacity data for ethylene, benzene, and ethylbenzene. The standard quantities are at 25 °C and 1 bar. The heat capacity is in cal/gmol · °C. In other words, you can assume that the heat capacity is constant over the range of temperature of interest.

Quantity	Ethylene	Benzene	Ethylbenzene
$\Delta G_f^\circ(25\,°C)$ (kcal/gmol)	16.282	40.989	31.208
$\Delta H_f^\circ(25\,°C)$ (kJ/gmol)	52.5	49	29.8
$\langle c_P^{ideal} \rangle$ (cal/gmol · °C)	20.5	45.9	68.3

10.6.E An equimolar mixture of ethylene and water is mixed at 254 °C and 100 atm, where they undergo a reaction to make ethanol:

$$C_2H_4 + H_2O \rightleftharpoons C_2H_5OH. \tag{10.91}$$

Assuming that both vapor and liquid exist at equilibrium, find the compositions. You may assume Lewis mixing, but not that the system can be modeled as an ideal gas.

10.6.F Naghibi *et al.* [90] measured the equilibrium constant for the binding of cytidine 2′-monophosphate (2′-CMP, L) to ribonuclease A (S) in 0.2 M potassium acetate buffer,

$$S + L \rightleftharpoons SL, \tag{10.92}$$

using a technique called calorimetric titration. The results are shown in Table 10.3. The estimates for the equilibrium constant were made assuming Lewis mixing. First, use these data to estimate the enthalpy of binding, assuming that it is constant. Then, assume that the enthalpy of binding depends linearly on temperature, $\Delta h(T) \cong \Delta h^{\circ} + A(T - T_0)$. What are Δh° and A? Using this assumption, estimate the enthalpy of mixing at $T_0 = 25\,^{\circ}\mathrm{C}$ and the difference in heat capacity, Δc_P. Which assumption is better?

Table 10.3 Equilibrium-constant data for the binding of 2′-CMP to ribonuclease A, from [90].

T (°C)	K (10^5 M^{-1})
15	1.410
15	1.240
16	1.340
20	1.110
20	1.040
20	1.090
25	0.886
25	0.941
30	0.661
30	0.661
30	0.641
35	0.447
35	0.488
35	0.501
40	0.357
40	0.353

10.6.G Gaseous nitrogen peroxide is kept in a container at 1 atm. The nitrogen peroxide can associate to form gaseous dinitrogen tetroxide:

$$2NO_2 \rightleftharpoons N_2O_4. \tag{10.93}$$

At 350 K, the gas is found to be 83% nitrogen peroxide. Estimate the equilibrium constant. If the pressure is doubled, what is the predicted composition of the gas? How does your estimate of the equilibrium constant compare with that predicted from tabulated properties of nitrogen peroxide and nitrogen tetroxide?

10.6.H You were asked in Exercise 10.4.B to find the equilibrium constant for the combustion of carbon monoxide to form carbon dioxide at various temperatures. There you should have found that $K(T = 2000\,\text{K}) = 756.2$ and $K(T = 3000\,\text{K}) = 3.016\,68$. Here we wish to estimate the maximum effectiveness we can expect from a catalytic converter operating at these temperatures and a pressure of 2 bar.

To estimate the composition of the exhaust coming from our engine, we use a basis of one mole of air. Of that one mole, approximately 0.21 moles are oxygen (A = CO, B = O_2, C = CO_2). We assume that half of that is consumed in the engine, leaving an initial mole number of $N_{B,0} = 0.11$ for oxygen. We assume that half of the oxygen atoms are tied up in carbon monoxide and half in carbon dioxide. A real combustion would produce also NO_x, but we ignore that for the sake of simplification. This means that the initial number of moles of carbon dioxide is $N_{C,0} = 0.10$, and that for carbon monoxide is $N_{A,0} = 0.20$. The nitrogen and other components of air are assumed inert, contributing 0.79 moles to inerts, all of which we call D. However, there is also water produced from the hydrocarbon combustion. For simplicity, we assume that there are 2.2 atoms of hydrogen for each carbon atom consumed, or a total of $2.2(0.10 + 0.20) = 0.66$. Hence, $N_{D,0} = 0.66 + 0.79 = 1.45$. Find the final composition of the gas exiting the catalytic converter at these two temperatures assuming that equilibrium is reached.

10.6.I Assuming the same initial compositions as in Exercise 10.6.H, find the final composition of a mixture at 2000 K and 1000 bar. Note that it is unclear whether we can assume an ideal gas at this pressure. However, we can assume Lewis mixing and use the Peng–Robinson model to estimate the pure fugacities.

10.8.A An enzyme is rarely alone inside a cell. There also exist other species that can inhibit or activate a reaction catalyzed by an enzyme. Consider an enzyme En in the presence of a single inhibitor I and a single activator A. The enzyme binds to the substrate Su to produce product Pr, but can also form complexes with the inhibitor and activator:

$$\begin{aligned}
\text{En} + \text{Su} &\underset{k^{\text{r}}_{\text{Su}}}{\overset{k^{\text{f}}_{\text{Su}}}{\rightleftharpoons}} \text{EnSu} \overset{k_{\text{p}}}{\longrightarrow} \text{En} + \text{Pr}, \\
\text{En} + \text{A} &\underset{k^{\text{r}}_{\text{A}}}{\overset{k^{\text{f}}_{\text{A}}}{\rightleftharpoons}} \text{EnA}, \\
\text{EnA} + \text{Su} &\underset{k^{\text{r}}_{\text{ASu}}}{\overset{k^{\text{f}}_{\text{ASu}}}{\rightleftharpoons}} \text{EnSuA} \overset{bk_{\text{p}}}{\longrightarrow} \text{EnA} + \text{Pr}, \\
\text{En} + \text{I} &\underset{k^{\text{r}}_{\text{I}}}{\overset{k^{\text{f}}_{\text{I}}}{\rightleftharpoons}} \text{EnI}, \\
\text{EnI} + \text{Su} &\underset{k^{\text{r}}_{\text{ISu}}}{\overset{k^{\text{f}}_{\text{ISu}}}{\rightleftharpoons}} \text{EnSuI} \overset{ak_{\text{p}}}{\longrightarrow} \text{EnI} + \text{Pr},
\end{aligned} \tag{10.94}$$

$$\mathrm{EnSu} + \mathrm{I} \underset{k^{\mathrm{r}}_{\mathrm{cI}}}{\overset{k^{\mathrm{f}}_{\mathrm{cI}}}{\rightleftharpoons}} \mathrm{EnSuI},$$

$$\mathrm{EnSu} + \mathrm{A} \underset{k^{\mathrm{r}}_{\mathrm{cA}}}{\overset{k^{\mathrm{f}}_{\mathrm{cA}}}{\rightleftharpoons}} \mathrm{EnSuA}.$$

Just as we did for Michaelis–Menten kinetics, assume here that all of the enzyme complexes, EnSu, EnSuA, and EnSuI, reach steady-state concentrations. Further, ignore non-idealities so that concentration can be used. Find an expression for the rate of production of dx_{Pr}/dt divided by the total enzyme concentration, $x^{\mathrm{T}}_{\mathrm{E}}$, analogous to expression (10.56). In other words, your answer should be given solely in terms of the rate constants (or equilibrium constants: $K_{\mathrm{Su}} := k^{\mathrm{r}}_{\mathrm{Su}}/k^{\mathrm{f}}_{\mathrm{Su}}$, $\beta K_{\mathrm{Su}} := k^{\mathrm{r}}_{\mathrm{ASu}}/k^{\mathrm{f}}_{\mathrm{ASu}}$, $\alpha K_{\mathrm{Su}} := k^{\mathrm{r}}_{\mathrm{ISu}}/k^{\mathrm{f}}_{\mathrm{ISu}}$, $K_{\mathrm{A}} := k^{\mathrm{r}}_{\mathrm{A}}/k^{\mathrm{f}}_{\mathrm{A}}$, $\beta K_{\mathrm{A}} := k^{\mathrm{r}}_{\mathrm{cA}}/k^{\mathrm{f}}_{\mathrm{cA}}$, $K_{\mathrm{I}} := k^{\mathrm{r}}_{\mathrm{I}}/k^{\mathrm{f}}_{\mathrm{I}}$, and $\alpha K_{\mathrm{I}} := k^{\mathrm{r}}_{\mathrm{cI}}/k^{\mathrm{f}}_{\mathrm{cI}}$), and the concentrations of substrate, inhibitor, and activator.

Can you put your answer into the following form:

$$\frac{1}{x^{\mathrm{T}}_{\mathrm{E}}}\frac{dx_{\mathrm{Pr}}}{dt} = \frac{x_{\mathrm{Su}}}{m(x_{\mathrm{A}}, x_{\mathrm{I}})K_{\mathrm{Su}} + b(x_{\mathrm{A}}, x_{\mathrm{I}})x_{\mathrm{Su}}}? \tag{10.95}$$

What are the expressions for m and b? Why is this form useful?

10.9.A Another model for predicting the denaturation of DNA is called the Zimm–Bragg matrix model [147, 148]. Its primary difference from the zipper model is that it allows multiple nucleation events on the chain, instead of just a single nucleation as in the zipper model. It contains the same components, σ and $q_{\mathrm{base}}(T)$, as the zipper model, with the same interpretation. It has the fundamental relation

$$\frac{\Delta F(T,\tilde{M})}{k_{\mathrm{B}}T} = \log\left[\frac{q_{\mathrm{base}}\sigma\left(1 + q_{\mathrm{base}} + \sqrt{(q_{\mathrm{base}}-1)^2 + 4q_{\mathrm{base}}\sigma}\right)^{\tilde{M}+1}}{2^{\tilde{M}}\sqrt{(q_{\mathrm{base}}-1)^2 + 4q_{\mathrm{base}}\sigma}\left(q_{\mathrm{base}} - 1 + \sqrt{(q_{\mathrm{base}}-1)^2 + 4q_{\mathrm{base}}\sigma}\right)}\right]. \tag{10.85}$$

Find the fraction θ of attached base pairs predicted by the model as a function of temperature and length $\tilde{M}$. Compare the result with that given by the zipper model for several values of length and σ by constructing plots like those in Figure 10.3. Do your results make sense? When do you expect these two models to give similar results? Is this what you observe? Is the dependence on length reasonable? Please explain your answers.

10.9.B The zipper model may also be applied to simpler polymers that form helices, such as poly-γ-benzyl-L-glutamate. Zimm *et al.* estimate that this polymer has $\Delta h_{\mathrm{base}} \cong +890\,\mathrm{cal/mol}$, and a melt temperature $T_{\mathrm{m}} = 11.8\,^{\circ}\mathrm{C}$ for a solution that is 70% dicholoracetic acid and 30% ethylene dichloride. The melt temperature is that at which $\theta \cong 0.5$ at high molecular weights. In practice, we can find this temperature from $q_{\text{base}}(T_{\text{m}}) = 1$. Plot the fractional helicity θ as a function of $T - T_{\text{m}}$ as predicted by the zipper model with $\sigma = 0.0002$, and for $\tilde{M} = 26, 46$, and 1500. Does your plot show the trends that you expected? What is different from the description in the text, and why is that?

10.10.A Consider the second-order reaction shown in Eqn. (10.73). Find the equations for the evolution of the average $\langle \tilde{N}_{\text{A}} \rangle$ and variance. Can you solve these equations? Explain clearly why not.

10.10.B Write the evolution equation for the probability distribution function $p(\varepsilon_1, \varepsilon_2; t)$ for the two-step reaction

$$\begin{aligned} \text{A} + \text{B} &\underset{k_2}{\overset{k_1}{\rightleftharpoons}} \text{D}, \\ \text{C} + \text{B} &\overset{k_3}{\rightarrow} \text{D}. \end{aligned} \tag{10.96}$$

Assume that you have arbitrary initial numbers of molecules.

10.10.C Consider the second-order reaction shown in Eqn. (10.73). Since it is not possible to find the variance and average analytically, it is necessary to perform a stochastic simulation to estimate the size of the fluctuations in a micro-reactor. Using MATLAB, Mathematica, Basic, FORTRAN, or some other tool, write a short program to simulate the average of several realizations of the reaction. Assume that $k_1 = 4k_2$, and that the initial molecule numbers are $\tilde{N}_{\text{A}} = \tilde{N}_{\text{A}}^{\circ}$, and $\tilde{N}_{\text{B}} = 0$. Compare your results for the average with those that arise in the thermodynamic limit (i.e. the usual $dc_{\text{A}}/dt \sim -k_1 c_{\text{A}}^2 + \cdots$).

You should first write the code to repeat the example in Section 10.10. When you are able to reproduce the results there, it is a small change to simulate the second-order reaction.

10.10.D The Zimm–Bragg model relaxes one of the assumptions used to derive the zipper model. Namely, the helix is now allowed to have multiple nucleation sites, or beginning points, for base-pair association.

One constructs the partition function consecutively. The first pair on the end of the chain is either unattached, with partition function $q_{\text{u}}(T)$, or attached with partition $\sigma q_{\text{a}}(T)$. Hence, the partition function of a chain with one base pair is $q_{\text{a}} + \sigma q_{\text{u}}$. It is convenient to write this as the sum of the elements of a column vector:

$$[Q_1] := \begin{pmatrix} q_{\text{u}} \\ \sigma q_{\text{a}} \end{pmatrix}.$$

The second base pair is again either attached or unattached. However, the partition function of the second pair depends on whether or not the first pair is attached. If the second pair is unattached, it has partition function q_{u}. If the second pair is attached and the first pair is attached, the partition function of the second pair is q_{a}. However, if the first pair is unattached, and the second is attached, the partition function of the second pair is σq_a. It is convenient, again, to write this as the sum of the column vector Q_2,

$$[Q_2] = \begin{pmatrix} q_u(q_u + \sigma q_a) \\ \sigma q_a(q_u + q_a) \end{pmatrix},$$

which we can write as a matrix equation,

$$\begin{aligned}[Q_2] &= \begin{pmatrix} q_u & q_u \\ \sigma q_a & q_a \end{pmatrix} \begin{pmatrix} q_u \\ \sigma q_a \end{pmatrix} \\ &= [M][Q_1], \end{aligned} \tag{10.97}$$

where we have introduced the matrix $[M]$.

1. Convince yourself that this works for $\tilde{M}$ base pairs to arrive at

$$[Q_{\tilde{M}}] = [M]^{\tilde{M}-1}[Q_1]. \tag{10.98}$$

2. To solve this equation, it is useful to use the eigenvalues and eigenvectors for the matrix $[M]$. Verify that the eigenvectors for the matrix are

$$[x_1] = \begin{pmatrix} \dfrac{q_u - q_a - \sqrt{(q_a - q_u)^2 + 4\sigma q_a q_u}}{2\sigma q_a} \\ 1 \end{pmatrix},$$

$$[x_2] = \begin{pmatrix} \dfrac{q_u - q_a + \sqrt{(q_a - q_u)^2 + 4\sigma q_a q_u}}{2\sigma q_a} \\ 1 \end{pmatrix},$$

and that the eigenvalues are

$$\begin{aligned} \lambda_1 &= \frac{1}{2}\left(q_a + q_u - \sqrt{(q_a - q_u)^2 + 4\sigma q_a q_u}\right) \\ \lambda_2 &= \frac{1}{2}\left(q_a + q_u + \sqrt{(q_a - q_u)^2 + 4\sigma q_a q_u}\right) \end{aligned} \tag{10.99}$$

3. Decompose the $[Q_1]$ vector into a sum of the two eigenvectors,

$$[Q_1] = a_1[x_1] + a_2[x_2].$$

Find the constants a_1 and a_2.

4. Put this expression for $[Q_1]$ into the expression for $[Q_{\tilde{M}}]$ we found above, Eqn. (10.98). Using the properties of eigenvectors, you should arrive at two terms for the partition function. One of these is much larger than the other. Which one?
5. Now show how to arrive at the fundamental relation given in Eqn. (10.85).

10.10.E Using the expression for the probability of attached base pairs, Eqn. (10.84), show that the average is given by Eqn. (10.60).

10.10.F Here we consider a simple model for reproduction of a bacterium B, which consumes nutrients A to reproduce and create waste W. The nutrients are fed to the system at a constant rate $k_1 c_F$:

$$\mathrm{F} \xrightarrow{k_1} \mathrm{A},$$

$$\mathrm{A} + \mathrm{B} \xrightarrow{k_2} 2\mathrm{B} + \mathrm{W}, \tag{10.100}$$

$$\mathrm{B} \xrightarrow{k_3} \mathrm{D},$$

where D is a dead bacterium. Hence, the first equation is the nutrient feed, the second is reproduction, and the third is death, reducing the number of bacteria by one.

We first consider a thermodynamically large system so that we can write down the usual ODEs for the concentrations using simple mass-action kinetics. Then, we consider the stochastic modeling of the system, since it might be small.

1. Write down these ODEs for the rates of production of c_{A} and c_{B}. Do you need any other rate equations to make this set well defined? In other words, could you find the concentration of B with just these two equations, or do you need more?
2. Find the steady-state concentration of bacterium c_{B} for this model.
3. Now write a code to simulate the number of bacteria $\tilde{N}_{\mathrm{B}}$ in a volume V using this model. Take special note of the fact that, if this number ever reaches zero, all of the bacteria are dead and no further reproduction can happen.
4. Do your simulation results agree with your analytic steady-state result for the large system when your initial number of bacteria is large (say $\tilde{N}_{\mathrm{B}} = 10\,000$)? Why, or why not? What about when the system is much smaller? Try several runs where you change the seed to your random-number generator, and begin with only 100 bacteria.

11 Thermodynamics of polymers

The word *polymer* comes from the Greek *polymeres*, meaning many units. It is used to designate molecules with large molecular weights. There is no exact cutoff to determine when a large molecule is actually polymeric, but chained molecules of molecular weight greater than a few thousand are typically considered to be polymers.

Scientists and engineers are concerned both with naturally occurring and with synthetic polymers. Naturally occurring polymers are important in biotechnological and biomedical fields, for example as DNA, RNA, microtubules, or actin filament, and in other fields, such as in the form of cellulosics in high-strength fibers.

Synthetic polymers play important roles in advanced materials such as plastics, liquid crystals, and fiber-reinforced composites. The manufacture of these materials, as well as numerous applications, can often require that the polymer be dissolved in solvents, for example in spin-coating processes. Alternatively, material scientists often wish to combine the desirable properties of two different polymers in a plastic alloy by blending different mixtures of the two.

In this chapter, we consider the phase behavior of polymer solutions and the miscibility of polymer blends. The starting point for describing these properties is the Flory–Huggins equation for the Gibbs free energy of mixing. In Section 11.1 we compare the experimental data for the solubility of various polymer–solvent mixtures with the Flory–Huggins equation of state. We see that the theory is insufficient to describe all systems. Hence, in Section 11.2 we also consider generalizations to the Flory–Huggins approach.

Finally, in Section 11.4 we derive the Flory–Huggins equation of state using the lattice theory introduced in Section 9.8.

11.1 SOLUBILITY AND MISCIBILITY OF POLYMER SOLUTIONS

In Section 8.6 we saw that assuming Lewis mixing rules leads to zero energy of mixing, and zero change in volume upon mixing. In Section 9.8.1 we saw how a statistical-mechanical model for binary mixtures of simple molecules that assumes constant volume and no heat of mixing leads to Lewis mixing. Hence, at first glance, we might be tempted to assume that a polymer solution that has no heat of mixing and constant volume would also exhibit ideal mixing. However, the real behavior of polymer solutions is much more complex. Much of the discrepancy between Lewis mixing and the behavior of polymers can be attributed to the fact that polymers can stretch out and take many different conformations, and, hence, occupy much greater volume than the solvent molecules.

Therefore, the important parameters of description for a polymer solution are the volume fraction of polymer and the size of the polymer chain. The Flory–Huggins model for binary mixtures of polymers predicts that the free energy of mixing is given by

$$\frac{\Delta g_{\rm mix}}{RT} = \frac{\phi_{\rm A}}{r_{\rm A}} \log \phi_{\rm A} + \frac{1-\phi_{\rm A}}{r_{\rm B}} \log(1-\phi_{\rm A}) + \chi(T)\phi_{\rm A}(1-\phi_{\rm A}), \tag{11.1}$$

where $\phi_{\rm A} = N_{\rm A}r_{\rm A}/(N_{\rm A}r_{\rm A} + N_{\rm B}r_{\rm B})$ is the volume fraction of species A, and $r_{\rm A}$ is the number of **monomers** (single units) that make up the polymer chain, called **the degree of polymerization**. We use the subscripts A and B to denote the type of polymer when we are considering blends of polymers. If we are considering a polymer solution, then B is a solvent, so $r_{\rm B} \equiv r_{\rm s} = 1$. The free energy of mixing as given in Eqn. (11.1) is actually *per mole of monomer*, not per mole of polymer. Hence, we multiply Eqn. (11.1) by $(N_{\rm A}r_{\rm A} + N_{\rm B}r_{\rm B})$ to obtain

$$\frac{\Delta G_{\rm mix}}{RT} = N_{\rm A} \log \phi_{\rm A} + N_{\rm B} \log(1-\phi_{\rm A}) + \chi(T)N_{\rm B}\phi_{\rm A}r_{\rm B}. \tag{11.2}$$

The parameter $\chi(T)$ is meant to describe the energy of interaction between the solvent and the monomers. According to the theory of Flory and of Huggins, it may be a function of temperature, but not of composition, and decreases with increasing temperature.

If we plot the free energy of mixing for a polymer solution, with $r_{\rm A} = 50$, $r_{\rm B} = 1$, and χ 50% larger than its critical value, as a function of the mole fraction of polymer, Figure 11.1 results. We see that the mixture is predicted to be globally unstable over a range of volume fractions. Hence, the mixture should separate into a polymer-rich phase $x_{\rm A}''''$ and a polymer-poor phase $x_{\rm A}'$. These compositions are called *binodal points*, and the polymer mixture is always stable outside these points. Here we have used a relatively small value of $r_{\rm A}$ to make the plot clearer; however, the results are qualitatively similar for larger values.

We also know from Section 4.1 that there is a condition of *local* instability $\phi_{\rm A}'' \le \phi_{\rm A} \le \phi_{\rm A}'''$ between the two binodal points $\phi_{\rm A}'$ and $\phi_{\rm A}''''$. We can find the region of local instability using the criterion of Section 4.2.3. These points are found from the second-order derivative of the Gibbs free energy with respect to mole number. In the following example, we find expressions for the

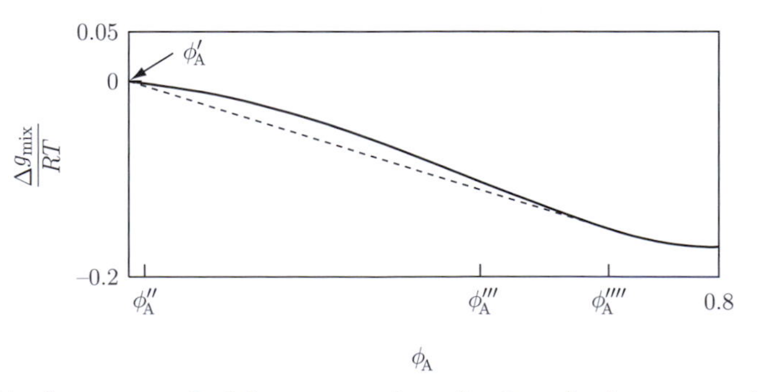

Figure 11.1 The Gibbs free energy of mixing versus volume fraction of polymer as predicted by the Flory–Huggins theory, Eqn. (11.1), for a polymer–solvent mixture ($r_{\rm B} = 1$, $r_{\rm A} = 50$, and $\chi = 1.5\chi_{\rm c}$). The compositions $\phi_{\rm A}''$ and $\phi_{\rm A}'''$ represent the spinodal points for the limit of local stability, and $\phi_{\rm A}''''$ is the upper binodal point. The lower binodal point, $\phi_{\rm A}' = 1.1 \times 10^{-5}$, is too small to label clearly on this plot, but the dashed line indicates the region of global instability.

spinodal points and the critical value for the Flory–Huggins interaction parameter. In the next section, we also show a method of finding the binodal points.

EXAMPLE 11.1.1 Find the spinodal curve and critical points for a polymer solution as predicted by the Flory–Huggins model.

Solution. For a polymer solution, the component B is solvent, so we set $r_{\mathrm{B}} \equiv r_{\mathrm{s}} = 1$, and species A is the polymer "p." The spinodal criterion can be written from Eqn. (9.94) as

$$\left.\left(\frac{\partial}{\partial N_{\mathrm{s}}}\frac{\mu_{\mathrm{s}}-\mu_{\mathrm{s}}^{\mathrm{pure}}}{RT}\right)_{T,P,N_{\mathrm{p}}}\right|_{\phi_{\mathrm{p}}=\phi_{\mathrm{p}}^{\mathrm{spinodal}}} = 0, \tag{11.3}$$

where $\mu_{\mathrm{s}}^{\mathrm{pure}}(T,P)$ is the chemical potential of the pure solvent at the same temperature and pressure as the mixture. The definition of the chemical potential is

$$\begin{aligned}
\frac{\mu_{\mathrm{s}}}{RT} &:= \frac{1}{RT}\left(\frac{\partial G}{\partial N_{\mathrm{s}}}\right)_{T,P,N_{\mathrm{p}}} \\
&= \left(\frac{\partial}{\partial N_{\mathrm{s}}}\frac{\Delta G_{\mathrm{mix}} + N_{\mathrm{s}}\mu_{\mathrm{s}}^{\mathrm{pure}} + N_{\mathrm{p}}\mu_{\mathrm{p}}^{\mathrm{pure}}}{RT}\right)_{T,P,N_{\mathrm{p}}} \\
&= \left(\frac{\partial}{\partial N_{\mathrm{s}}}\frac{\Delta G_{\mathrm{mix}}}{RT}\right)_{T,P,N_{\mathrm{p}}} + \frac{\mu_{\mathrm{s}}^{\mathrm{pure}}}{RT},
\end{aligned} \tag{11.4}$$

where the second line follows from the definition of a mixing property, Eqn. (9.99). Using the extensive form of the Flory–Huggins mixing law Eqn. (11.2), we find

$$\frac{\mu_{\mathrm{s}}-\mu_{\mathrm{s}}^{\mathrm{pure}}}{RT} = \log(1-\phi_{\mathrm{p}}) + \phi_{\mathrm{p}}\left(1-\frac{1}{r_{\mathrm{p}}}\right) + \chi\phi_{\mathrm{p}}^2, \tag{11.5}$$

or, similarly,

$$\frac{\mu_{\mathrm{p}}-\mu_{\mathrm{p}}^{\mathrm{pure}}}{RT} = \log\phi_{\mathrm{p}} - (r_{\mathrm{p}}-1)(1-\phi_{\mathrm{p}}) + \chi r_{\mathrm{p}}(1-\phi_{\mathrm{p}})^2. \tag{11.6}$$

We can now use this expression for the solvent chemical potential Eqn. (11.5) to find the spinodal points. However, since the expression is in terms of volume fraction, it is convenient to write

$$\begin{aligned}
\left(\frac{\partial}{\partial N_{\mathrm{s}}}\frac{\mu_{\mathrm{s}}-\mu_{\mathrm{s}}^{\mathrm{pure}}}{RT}\right)_{T,P,N_{\mathrm{p}}} &= \left(\frac{\partial\phi_{\mathrm{p}}}{\partial N_{\mathrm{s}}}\right)_{N_{\mathrm{p}}}\left(\frac{\partial}{\partial\phi_{\mathrm{p}}}\frac{\mu_{\mathrm{s}}-\mu_{\mathrm{s}}^{\mathrm{pure}}}{RT}\right)_{T,P,N_{\mathrm{p}}} \\
&= -\frac{\phi_{\mathrm{p}}(1-\phi_{\mathrm{p}})}{N_{\mathrm{s}}}\left(\frac{\partial}{\partial\phi_{\mathrm{p}}}\frac{\mu_{\mathrm{s}}-\mu_{\mathrm{s}}^{\mathrm{pure}}}{RT}\right)_{T,P,N_{\mathrm{p}}}.
\end{aligned}$$

Putting this result into Eqn. (11.3) implies the simpler (and general) spinodal criterion

$$\left.\left(\frac{\partial}{\partial\phi_{\mathrm{p}}}\frac{\mu_{\mathrm{s}}-\mu_{\mathrm{s}}^{\mathrm{pure}}}{RT}\right)_{T,P,N_{\mathrm{p}}}\right|_{\phi_{\mathrm{p}}=\phi_{\mathrm{p}}^{\mathrm{spinodal}}} = 0. \tag{11.7}$$

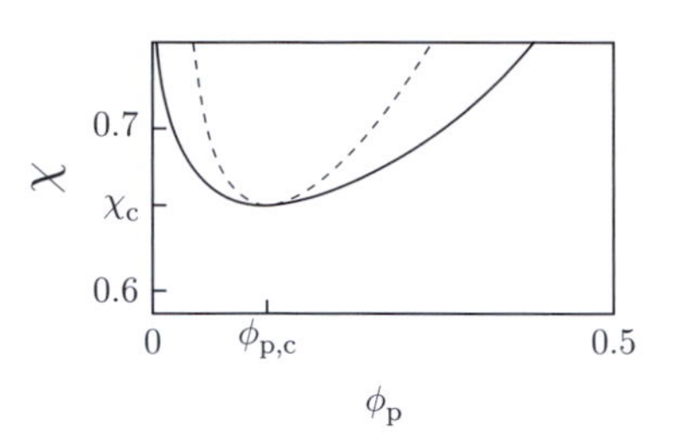

Figure 11.2 The phase diagram for a polymer–solvent mixture as predicted by the Flory–Huggins theory, $r_p = 50$. $\chi(T)$ is the only temperature-dependent parameter in the model, so it plays the role of inverse temperature for the classical Flory–Huggins theory. The ordinate is the volume fraction of polymer. The dashed line represents the spinodal curve, and the solid line is the coexistence curve.

Using our expression for the solvent chemical potential, Eqn. (11.5), we find

$$\left(\frac{\partial}{\partial \phi_p} \frac{\mu_s - \mu_s^{pure}}{RT}\right)_{T,P,N_p}\Bigg|_{\phi_p = \phi_p^{spinodal}} = -\frac{1}{1-\phi_p^{spinodal}} + 1 - \frac{1}{r_p} + 2\chi\phi_p^{spinodal}$$

and hence

$$0 = -\frac{1}{1-\phi_p^{spinodal}} + 1 - \frac{1}{r_p} + 2\chi\phi_p^{spinodal}. \tag{11.8}$$

This quadratic equation can be solved for $\phi_p^{spinodal}$ to find

$$\phi_p^{spinodal} = \frac{2\chi - 1 + 1/r_p \pm \sqrt{(2\chi - 1 + 1/r_p)^2 - 8\chi/r_p}}{4\chi}. \tag{11.9}$$

From this expression it is possible to find the critical point by inspection. Namely, the two spinodal curves meet at the critical point, which occurs when the radical is zero. Setting the radical to zero, and solving for χ yields the critical interaction parameter χ_c,

$$\chi_c = \frac{\left(\sqrt{r_p} + 1\right)^2}{2r_p}. \tag{11.10}$$

Setting $\chi = \chi_c$ in the expression for the spinodal curve yields the critical volume fraction of polymer:

$$\phi_{p,c} = \phi_p^{spinodal}(\chi = \chi_c) = \frac{1}{\sqrt{r_p} + 1}. \tag{11.11}$$

The spinodal curve is shown in Figure 11.2, for a solution with $r_p = 100$. Note how the critical point and spinodal curve are largely skewed towards low polymer concentrations, and the curve is not symmetric. As r_p is increased, the skewing becomes even more pronounced. □

Spinodal compositions are thermodynamically unstable, but may be long-lived. Since polymer molecules are very large, they diffuse very slowly in mixtures. Hence, if the system is globally unstable, but locally stable, the driving force for phase separation is very small, and the mixture may appear to be stable for very long periods of time. Hence, it is of great practical interest to find not just the binodal curve of a polymer liquid, but also the spinodal curve, where many polymer solutions remain.

If we find the spinodal and binodal points for all values of χ, then, similarly to a T–x_A phase diagram, we can construct a χ–ϕ_A phase diagram. Figure 11.2 shows both the spinodal (dashed line) curve and the binodal (solid line) curve as predicted by the Flory–Huggins theory. We see that, for sufficiently small, positive values of χ, or all negative values of χ, the mixture is stable and no phase separation occurs. At a critical value χ_c, the solution begins to phase separate.

In Section 11.4 we will see that the simple statistical-mechanical theory used to derive the Flory–Huggins mixing expression allows arbitrary temperature dependence on χ. Therefore, the T–ϕ_A diagram is able to predict many possible shapes for the miscibility curve. In fact, real systems exhibit a variety of such shapes on such diagrams.

The theory also predicts that the longer the polymer chains become, the lower the value for χ_c becomes, reaching a lower limit of 1/2, according to Eqn. (11.10). This trend is observed experimentally. Experiments also reveal that the polymer volume fraction, rather than the mole fraction, is important. However, the shape of the coexistence curve is typically different from what Flory–Huggins theory predicts. In other words, Figure 11.2 is qualitatively correct for many polymer systems, such as the system shown in Figure 11.3(a). The Flory–Huggins expression is superior to the Lewis mixing rule, even when the heat of mixing and volume changes are negligible (e.g. Figure 11.3(a)).

Consider, for example, a mixture of deuterated and hydrogenated polybutadiene. Replacing the hydrogens on the chain with deuterium does not change the electronic structure of the system, but rather, because of the increased mass, changes the lengths of the bonds, albeit only slightly. Hence, from a chemical point of view, the difference between the deuterated and undeuterated chains is nearly indistinguishable. So, at first it might seem surprising that such a system would phase separate at all.

However, the result can be predicted from Eqn. (11.1). Note that the first two terms on the right-hand side represent the entropy of mixing for the system. These terms are always negative, and hence are always stabilizing. The only source of instability is the energy of mixing, the third term on the right-hand side of Eqn. (11.1). However, unlike the case for Lewis mixing, or the Margules equation, the stabilizing terms are weighted by the inverse of the degree of

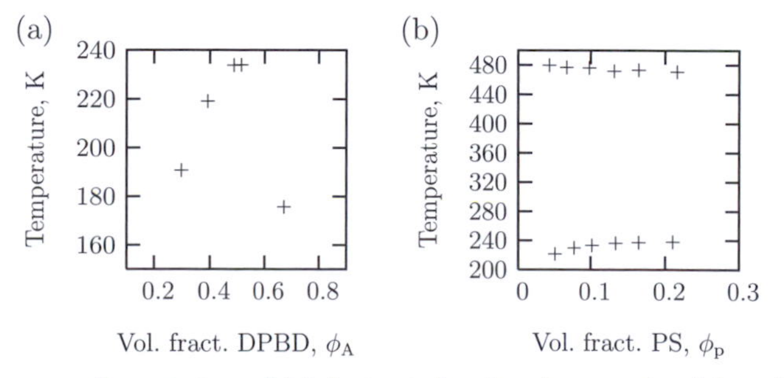

Figure 11.3 Phase diagrams for a system of (a) deuterated and undeuterated polybutadiene, as a function of volume fraction of the deuterated chains, where an upper-critical solution temperature (UCST) is observed; and (b) polystyrene of molecular weight 4800 in acetone, which exhibits a combined UCST and LCST (lower-critical solution temperature).

polymerization. Since the degree of polymerization is very large, these stabilizing terms are greatly diminished. Hence, very small energies of mixing can cause the system to phase separate. Consequently, mixing different chemical species can be extremely difficult for polymer blending.

On the other hand, many polymer systems exhibit much more complex phase behavior. Figure 11.3(b) shows the phase behavior of polystyrene in acetone, which exhibits two critical points: both an upper-critical solution temperature (UCST, the curve on the bottom) and a lower-critical solution temperature (LCST, the curve on the top). Systems that exhibit both a UCST and an LCST, at sufficiently high molecular weights, can exhibit an hourglass shape. The hourglass shape results from the convergence of the UCST and LCST curves, resulting in there being (observably) no critical point.

Finally, Figure 11.5 on p. 373 shows a solution that exhibits a closed-loop phase diagram. For this last example, there is an intermediate region of temperature where the polymer solution phase separates. Both at high and at low temperatures, the polymer is fully soluble.

The Flory–Huggins theory as given by Eqn. (11.1) can explain all of this behavior *qualitatively* if the interaction parameter is allowed to have a non-monotonic temperature dependence. However, this generalization is not sufficient to describe experimental data *quantitatively*. Thus, in the following section, we consider other models that are able to describe these more complicated systems by allowing the interaction parameter to depend on both temperature and composition.

11.2 GENERALIZATIONS OF THE FLORY–HUGGINS THEORY

As we saw in the last section, although it provides great insight into the thermodynamics of polymer solutions and blends, the Flory–Huggins theory has a limited capability of describing experimental data quantitatively. Also, the model ignores the compressibility of polymer solutions. Here we consider three generalizations to the Flory–Huggins theory. The first equation allows more quantitative predictions of liquid–liquid equilibrium data, and the second allows compressibility. We consider also a model that uses the quasi-chemical approximation as a modification to the Flory–Huggins theory. The fourth modification, a model not considered in detail here, but widely used industrially, is the PC-SAFT model derived from statistical mechanics [50].

11.2.1 The generalization of Qian *et al.*

Qian *et al.* [104] considered an *ad hoc* modification to Eqn. (11.1) that is able to predict all five types of phase behavior described in the last section: UCST, LCST, combined UCST and LCST, hourglass, and closed loop. With the addition of adjustable parameters, the model is then able to describe more data quantitatively.

The free energy of mixing studied by Qian *et al.* is

$$\frac{\Delta G_{\mathrm{mix}}}{RT} = N_{\mathrm{A}} \log \phi_{\mathrm{A}} + N_{\mathrm{B}} \log(1-\phi_{\mathrm{A}}) + N_{\mathrm{A}} r_{\mathrm{A}} D(T) \int_{\phi_{\mathrm{A}}}^{1} B(\phi) d\phi. \tag{11.12}$$

Note that, if we set $B = 1$, then we recover the Flory–Huggins expression with the function $D(T)$ playing the role of χ. We wish to have Flory–Huggins as a special case, so we consider the dependence

$$B(\phi) = 1 + b_1\phi + b_2\phi^2. \tag{11.13}$$

Then, the Flory–Huggins expression is recovered when $b_1 = b_2 = 0$.

For the sake of simplicity, we focus on the shape of the spinodal curve predicted by Eqn. (11.12), and recognize that the binodal curve must have a similar shape. Using the same procedure as that used in Example 11.1.1, we find the following prediction for the solvent chemical potential:

$$\frac{\mu_{\mathrm{B}} - \mu_{\mathrm{B}}^{\mathrm{pure}}}{RT} = \log(1 - \phi_{\mathrm{A}}) + \left(\frac{r_{\mathrm{A}} - r_{\mathrm{B}}}{r_{\mathrm{A}}}\right)\phi_{\mathrm{A}} + D(T)B(\phi_{\mathrm{A}})r_{\mathrm{B}}\phi_{\mathrm{A}}^2; \tag{11.14}$$

and for the other species

$$\frac{\mu_{\mathrm{A}} - \mu_{\mathrm{A}}^{\mathrm{pure}}}{RT} = \log\phi_{\mathrm{A}} - \left(\frac{r_{\mathrm{A}} - r_{\mathrm{B}}}{r_{\mathrm{B}}}\right)(1 - \phi_{\mathrm{A}}) + r_{\mathrm{A}}D\left[\int_{\phi_{\mathrm{A}}}^{1} B(\phi)d\phi - B(\phi_{\mathrm{A}})\phi_{\mathrm{A}}(1 - \phi_{\mathrm{A}})\right]. \tag{11.15}$$

The criterion for a spinodal point, Eqn. (11.7), becomes

$$0 = \frac{1}{r_{\mathrm{A}}} - \frac{1}{r_{\mathrm{B}}} + \frac{1}{r_{\mathrm{B}}(1 - \phi_{\mathrm{A}}^{\mathrm{spinodal}})} - D(T)C(\phi_{\mathrm{A}}^{\mathrm{spinodal}}), \tag{11.16}$$

where

$$C(\phi) := 2B(\phi) + \phi\frac{dB}{d\phi} = 2 + 3b_1\phi + 4b_2\phi^2. \tag{11.17}$$

We require that C always be positive. Since the first two terms on the right-hand side of Eqn. (11.16) are necessarily positive, we find that the solution is predicted to be homogeneous (locally stable) whenever D is negative. However, when $D(T)$ is positive, then we have the possibility of phase separation. In order to observe all of the types of phase behavior mentioned at the beginning of the section, we require a functional form for D that can be of all four forms shown in Figure 11.4. Wherever one of these curves lies sufficiently above the x axis, phase separation can occur. For example, Figure 11.4(a) may exhibit closed-loop phase behavior: at low and high temperatures, the system is homogeneous (stable), but at intermediate temperatures, it phase separates.

One possible expression for $D(T)$ that exhibits this behavior is [104, Eqn. (23)]

$$D(T) = d_0 + \frac{d_1}{T} + d_2\log T. \tag{11.18}$$

If $d_2 = 0$, then either of Figure 11.4(c) or Figure 11.4(d) can result, depending on whether d_1 is negative or positive, respectively. Hence, either an LCST or a UCST is predicted. If $d_2 \neq 0$, then either Figure 11.4(a) or Figure 11.4(b) results, depending on whether d_2 is negative or

Table 11.1 Predicted phase diagrams for the equation of state proposed by Qian *et al.*

d_0	d_1	d_2	Figure	Type of phase diagram
>0	<0	<0	11.4(a)	Closed loop
<0	>0	>0	11.4(b)	UCST + LCST, or hourglass
>0	<0	=0	11.4(c)	LCST only
<0	>0	=0	11.4(d)	UCST only

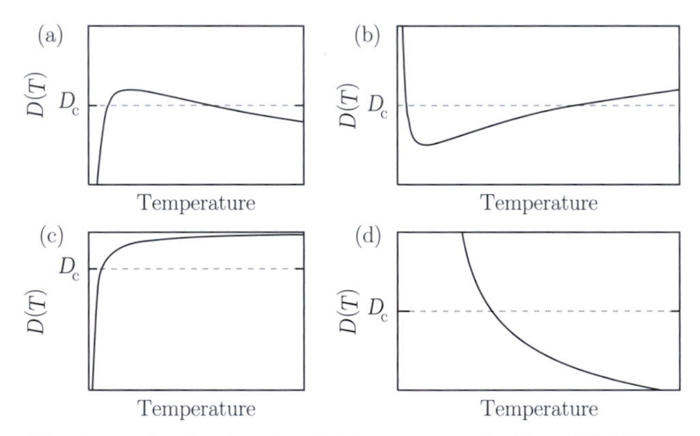

Figure 11.4 Four possible shapes for the function $D(T)$ necessary for Eqn. (11.12) to exhibit (a) closed-loop, (b) both UCST and LCST, or hourglass, (c) LCST, and (d) UCST phase behavior. The equation can predict hourglass behavior in (b) only if the minimum lies above the critical value, $D_c > 0$.

positive, respectively. This behavior in D can be seen by studying the extremum in D. First we find the temperature T_* at which D is an extremum,

$$\begin{aligned} 0 &= \left.\frac{dD}{dT}\right|_{T=T_*} \\ &= -\frac{d_1}{T_*^2} + \frac{d_2}{T_*} \rightarrow T_* = \frac{d_1}{d_2}, \end{aligned} \tag{11.19}$$

then we evaluate the second derivative at T_*:

$$\left.\frac{d^2D}{dT^2}\right|_{T=T_*} = \frac{2d_1}{T_*^3} - \frac{d_2}{T_*^2} = \frac{d_2^3}{d_1^2}. \tag{11.20}$$

Therefore, if d_2 is less (greater) than zero, then D is a maximum (minimum), Figure 11.4(a) (Figure 11.4(b)) results, and a closed-loop (combined UCST and LCST, or hourglass) diagram can result. These results are summarized in Table 11.1.

EXAMPLE 11.2.1 For an aqueous solution of a copolymer, Qian et al. recommend the following parameters: $r_A = 5000, r_B = 1, b_1 = 0.650, b_2 = 0, d_0 = 2.582, d_1 = -111.9\,\text{K}$, and $d_2 = -0.30$. Plot the spinodal and binodal curves predicted for these values of the parameters. Note that the first two are determined *a priori* from the molecular weight of the polymer chains divided by the molecular weight of a monomer.

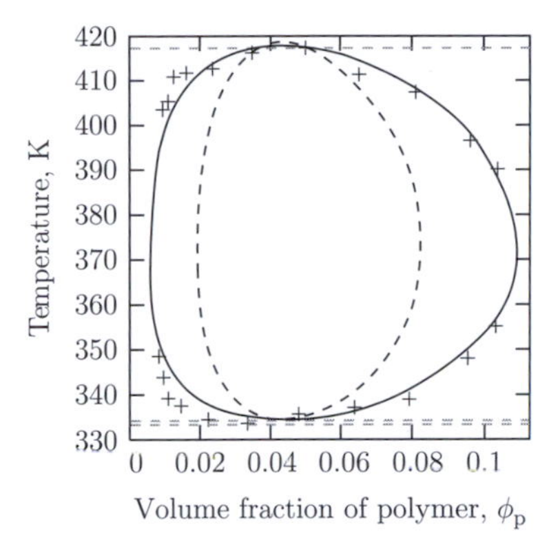

Figure 11.5 Spinodal (dashed line) and binodal (solid line) curves predicted for an aqueous solution of the copolymer poly(vinyl alcohol)$_{93}$-co-(vinyl acetate)$_7$ using the generalized Flory–Huggins expression. The symbols (+) are data taken from [104].

Solution. Since d_1 and d_2 are both less than zero, we see from Table 11.1 that a closed-loop phase diagram should be predicted. The spinodal criterion is given by Eqn. (11.16). When we insert the definition for C, Eqn. (11.17), into this equation, we obtain

$$0 = r_A\phi_A + r_B(1 - \phi_A) - D(T)r_A r_B\phi_A(1 - \phi_A)[2 + 3b_1\phi_A + 4b_2\phi_A^2], \tag{11.21}$$

which is a quartic equation in ϕ_A. To find the spinodal points for a given temperature, we must find the roots of this equation, and throw out those that do not lie between 0 and 1. For this particular example $b_2 = 0$, however, so the equation is cubic, and the analytic expression for cubic roots given in Section A.7 of Appendix A may be used. The coefficients are

$$a_0 = \frac{1}{3Dr_A b_1}, \quad a_1 = \frac{r_A - r_B}{3Dr_A r_B b_1} - \frac{2}{3b_1}, \quad a_2 = \frac{2}{3b_1} - 1. \tag{11.22}$$

Using the expressions for the roots, Eqn. (A.33), it is then straightforward to construct the spinodal curves shown in Fig. 11.5. We find that a closed-loop phase diagram is predicted by the model, which is indeed consistent with experimental results.

To find the binodal curve we must satisfy two nonlinear equations simultaneously. Namely, we need to find the two compositions ϕ_A' and ϕ_A'' that satisfy

$$\mu_A(\phi_A') = \mu_A(\phi_A''), \quad \mu_B(\phi_A') = \mu_B(\phi_A''). \tag{11.23}$$

Perhaps surprisingly, finding the zeros of coupled, nonlinear algebraic equations is not a numerically simple task [103, Section 9.6]. However, for the two-component case considered here, there is a guaranteed way that is not computationally expensive. For a given temperature, we use three-dimensional plotting software to plot $\mu_A(\phi_A') - \mu_A(\phi_A'')$ as a function of ϕ_A' and ϕ_A''. We also use the software to find the contour lines of the surface, in particular the contour line where this surface is zero. Figure 11.6 shows where $\mu_A(\phi_A') - \mu_A(\phi_A'') = 0$ for a temperature of 410 K. The straight line is the uninteresting set where $\phi_A' = \phi_A''$. The binodal points must lie somewhere on the oval-shaped contour line. Exactly which point is the correct one is determined by the intersection of this contour line with the corresponding contour line for $\mu_B(\phi_A') - \mu_B(\phi_A'')$, which is also shown in Figure 11.6. Therefore, from this figure we find that,

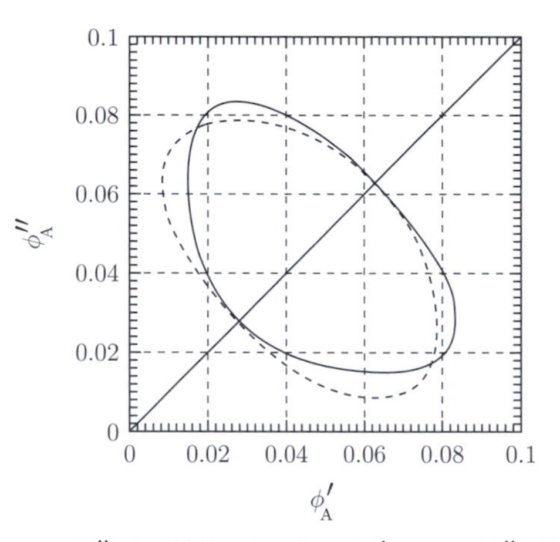

Figure 11.6 Zeros of $\mu_A(\phi'_A) - \mu_A(\phi''_A)$ (solid lines) and $\mu_B(\phi'_A) - \mu_B(\phi''_A)$ (dashed lines) for a temperature of 410 K for the generalized Flory–Huggins model of Eqn. (11.12). The parameter values are given in Example 11.2.1. The binodal compositions are seen from this plot to be $\phi'_A = 0.0174$ and $\phi''_A = 0.077$.

at 410 K, the binodal points are $\phi'_A = 0.0174$ and $\phi''_A = 0.077$. If we go through this procedure for several values of the temperature, we can construct the binodal curves shown in Figure 11.5. The model shows good quantitative agreement with the data. The same curve can be found more quickly using MATLAB, Mathematica, MathCAD, Octave, or similar software. □

11.2.2 The Sanchez–Lacombe equation of state

The two polymer models considered so far for polymer liquids assume that they are incompressible. However, there are important industrial processes that exploit the compressibility of polymers. Polymer foaming is an example.

Microcellular polymer foams (MPFs) are porous polymers that are generally characterized by pore densities greater than 10^9 cells/cm^3 and cell sizes smaller than 10 microns (see Figure 11.7). These novel polymeric materials have high strength-to-weight ratios and low densities, and are widely used in sports equipment, adsorbents, filters, and catalyst supports. Microcellular foams are generated by initially saturating the polymer with a super-critical fluid (SCF), usually carbon dioxide or nitrogen. Thermodynamic instability is induced by imposing a rapid pressure drop so that the solubility of the super-critical fluid in the mixture decreases, thereby generating bubbles in the polymer. The nucleation of bubble formation is also an interesting thermodynamic problem, but here we are concerned with calculating the initial condition of the foaming process: the saturation concentration and density of super-critical fluid in the polymer.

For foams to be generated in a continuous process, it is necessary that the solubility of the SCF in the polymer as well as the density of the polymer–SCF mixture be known in order to fix the other operating parameters. Typically, isothermal curves of density versus pressure are generated for different solubilities of fluid in the polymer by using an equation of state to predict the density of the mixture.

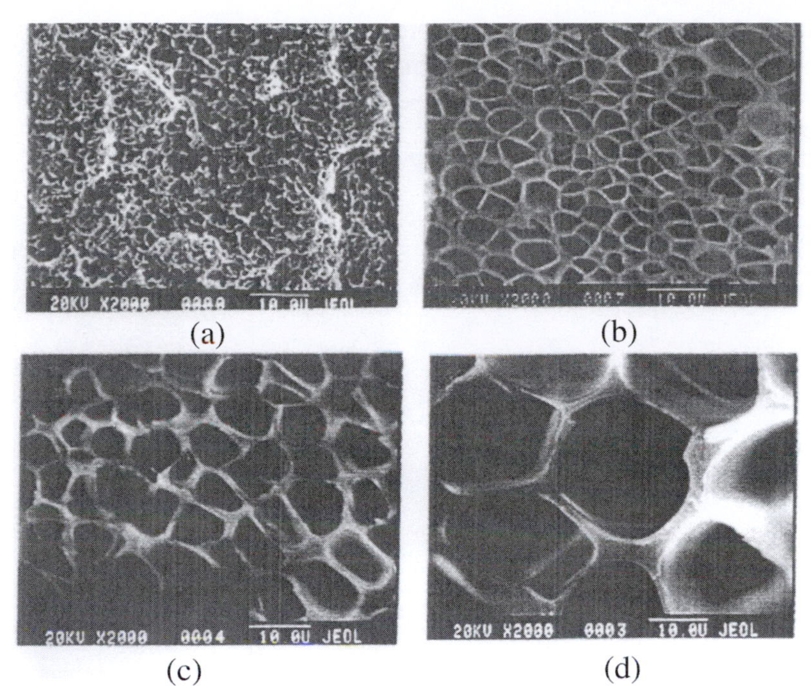

Figure 11.7 Scanning electron micrographs for polystyrene foams generated with CO_2 at different temperatures: (a) 40 °C, (b) 80 °C , (c) 100 °C , and (d) 120 °C. All foams were prepared at 3530 psi. The scale in the lower right-hand corner of each micrograph indicates 10 microns. Reprinted with permission from [7].

To model such a system one requires a fundamental relation that is applicable over a large range of densities. The Sanchez–Lacombe fundamental relation is one such PvT equation of state for polymers [77]. The model here is also lattice-based, and uses a generalization to the Flory–Huggins theory [113], where

$$\frac{G}{rN} = RT^* \left(\frac{P\bar{v}}{P^*} - \frac{1}{\bar{v}} \right) + RT \left[(\bar{v} - 1) \log \left(1 - \frac{1}{\bar{v}} \right) - \frac{1}{r} \log \bar{v} \right]. \tag{11.24}$$

The reduced specific volume $\bar{v}$ is defined as

$$\bar{v} := \frac{P^* v}{rRT^*}. \tag{11.25}$$

We have introduced the characteristic pressure P^* and temperature T^*. These are material parameters for either gas or polymer, and have been tabulated for many substances in Table 11.2. We have also introduced a dimensionless size parameter r, which is similar to that used in the Flory–Huggins theory. The parameter usually tabulated is ρ^*, which is related to r by

$$r = \frac{M_w P^*}{RT^* \rho^*}. \tag{11.26}$$

Recall that a fundamental relation for G requires independent parameters (T, P, N). However, Eqn. (11.24) also has the specific volume on the right-hand side. In other words, the relation as it stands is over-specified. Sanchez and Lacombe complete the fundamental relation by finding

Table 11.2 Characteristic parameters for the Sanchez–Lacombe equation of state. These values are from [115, 110].

Substance	P^* (MPa)	T^* (K)	ρ^* (kg/m^3)
CO_2	720.3	262	1580
N_2	103.6	159.0	803.4
Polypropylene	297.5	692.0	882.8
HDPE	288.7	736.0	867.0
Polystyrene	387.0	739.9	1108
LDPE	349.4	679	886.1
PMMA	488.3	742	1249.8
PDMS	277.4	501	1085.7

an algebraic equation relating v, T, P, and N, which fixes v. This equation is found by finding the value of $\bar{v}$ which minimizes the free energy. By setting the derivative of Eqn. (11.24) with respect to $\bar{v}$ equal to zero, we arrive at the additional equation of state

$$\frac{P}{P^*} = -\frac{1}{\bar{v}^2} - \frac{T}{T^*}\left[\log\left(1 - \frac{1}{\bar{v}}\right) + \left(1 - \frac{1}{r}\right)\frac{1}{\bar{v}}\right], \tag{11.27}$$

which determines $\bar{v}$ in Eqn. (11.24).

Equation (11.24) is for pure-component species. The fundamental relation for a gas–polymer mixture also requires entropic mixing terms of the Flory–Huggins type, leading to the fundamental relation

$$\frac{G}{rN} = RT^*\left(\frac{P\bar{v}}{P^*} - \frac{1}{\bar{v}}\right) + RT\left[(\bar{v} - 1)\log\left(1 - \frac{1}{\bar{v}}\right) - \frac{1}{r}\log\bar{v} + \frac{\phi_1}{r_1}\log\phi_1 + \frac{\phi_2}{r_2}\log\phi_2\right], \tag{11.28}$$

where ϕ_i is the volume fraction of species i, and r_i is its dimensionless volume parameter. We also need a suitable mixing rule for the parameter values, which will be analogous to what has been done for simple fluid mixtures in Section 8.5 and Problem 7.4.H. A commonly used set of mixing rules for the Sanchez–Lacombe equation of state is

$$\begin{aligned} P^* &= \phi_1 P_1^* + \phi_2 P_2^* - \phi_1\phi_2\,\Delta P^*, \\ \Delta P^* &:= P_1^* + P_2^* - 2(1 - k_{12})\sqrt{P_1^* P_2^*}, \\ r &= \frac{N_1 r_1 + r_2 N_2}{N}, \end{aligned} \tag{11.29}$$

Table 11.3 The binary interaction parameter as a function of temperature for several solvent–polymer mixtures.

Polymer	Solvent	T (°C)	k_{12}
PDMS[a]	CO_2	50	0.075
		80	0.122
		100	0.128
Polystyrene[a]	1,1-Difluoroethane	135	0.034
		160	0.064
Polystyrene[b]	CO_2	100.2	-0.088
		140.2	-0.117
		180.2	-0.132
Polystyrene[b]	N_2	100.2	0.223
		140.2	0.213
		180.2	0.199
Polyethylene[c]	N_2	100.2	0.250
		140.2	0.251
		180.2	0.237
Polypropylene[c]	N_2	180.2	0.245
		200.2	0.212

[a]Data taken from [47].
[b]Data taken from [116].
[c]Data taken from [115].

where k_{12} is a binary interaction parameter, which can be a function of temperature, but not composition. The characteristic temperature is found from the equation

$$\frac{P^*}{T^*} = \frac{\phi_1 P_1^*}{T_1^*} + \frac{\phi_2 P_2^*}{T_2^*}. \tag{11.30}$$

EXAMPLE 11.2.2 Find the density of a mixture of polypropylene ($M_W = 150\,000$) and super-critical CO_2 at 15 MPa and 453 K, which are the typical values at the start of the foaming process for this system. Take the solubility of CO_2 to be 10 wt%, and the binary interaction parameter as $k_{12} = -0.2555$ under these conditions.

Solution. We can use the equation of state, Eqn. (11.27), to find the density

$$\frac{P}{P^*} = -\frac{1}{\bar{v}^2} - \frac{T}{T^*}\left[\log\left(1 - \frac{1}{\bar{v}}\right) + \left(1 - \frac{1}{r}\right)\frac{1}{\bar{v}}\right]. \tag{11.31}$$

We then find the necessary parameters, labeling CO_2 as component 1, and PP as component 2. From Table 11.2, we can find $P_1^* = 720.3\,\text{MPa}$, $P_2^* = 297.5\,\text{MPa}$, $T_1^* = 262\,\text{K}$, $T_2^* = 692\,\text{K}$, $\rho_1^* = 1580\,\text{kg/m}^3$, and $\rho_2^* = 882.8\,\text{kg/m}^3$.

We convert from the given weight fraction to the volume fraction,

$$\begin{aligned}\phi_1 &= \frac{w_1/\rho_1^*}{w_1/\rho_1^* + w_2/\rho_2^*} \\ &= \frac{0.10/1580\,\text{kg/m}^3}{0.10/1580\,\text{kg/m}^3 + 0.90/882.8\,\text{kg/m}^3} \\ &= 0.0585,\end{aligned} \tag{11.32}$$

and, hence, $\phi_2 = 0.9415$. The dimensionless size parameters for each species are found using Eqn. (11.26):

$$\begin{aligned}r_1 &= \frac{M_i P_i^*}{RT_i^* \rho_i^*} \\ &= \frac{(44\,\text{g/mol})(720.3\,\text{MPa})}{(8.314\,\text{J/mol}\cdot\text{K})(1580\,\text{kg/m}^3)(262\,\text{K})} \\ &= 9.21,\end{aligned} \tag{11.33}$$

and $r_2 = 8786.17$ by similar means. Hence, the dimensionless volume for the mixture is

$$\begin{aligned}r &= \frac{N_1 r_1 + N_2 r_2}{N_1 + N_2} \\ &= \frac{w_1 r_1/M_1 + w_2 r_2/M_2}{w_1/M_1 + w_2/M_2} \\ &= \frac{(0.10)(9.21)/(44\,\text{g/mol}) + (0.90)(8786.17)/(150\,000\,\text{g/mol})}{(0.10)/(44\,\text{g/mol}) + (0.90)/(150\,000\,\text{g/mol})} \\ &= 32.32.\end{aligned} \tag{11.34}$$

The contribution of mixing to P^* is

$$\begin{aligned}\Delta P^* &:= P_1^* + P_2^* - 2(1 - k_{12})\sqrt{P_1^* P_2^*} \\ &= 720.3\,\text{MPa} + 297.5\,\text{MPa} - 2(1 + 0.255)\sqrt{(720.3\,\text{MPa})(297.5\,\text{MPa})} \\ &= -144.6\,\text{MPa}.\end{aligned} \tag{11.35}$$

Therefore, the characteristic pressure for the mixture is

$$\begin{aligned}P^* &= \phi_1 P_1^* + \phi_2 P_2^* - \phi_1\phi_2\,\Delta P^* \\ &= (0.0585)(720.3\,\text{MPa}) + (0.9415)(297.5\,\text{MPa}) + (0.0585)(0.9415)(144.6\,\text{MPa}) \\ &= 330.2\,\text{MPa}.\end{aligned} \tag{11.36}$$

Using the mixing rule for the characteristic temperature yields

$$\begin{aligned} T^* &= \frac{P^*}{\phi_1 P_1^*/T_1^* + \phi_2 P_2^*/T_2^*} \\ &= \frac{330.2\,\text{MPa}}{(0.0585)(720.3\,\text{MPa})/262\,\text{K} + (0.9415)(297.5\,\text{MPa})/692\,\text{K}} \\ &= 584\,\text{K}. \end{aligned} \tag{11.37}$$

To obtain the last required parameter, we first need the molecular weight of the mixture, which is

$$\begin{aligned} M_\text{w} &= x_1 M_1 + x_2 M_2 \\ &= \frac{w_1}{w_1/M_1 + w_2/M_2} + \frac{w_2}{w_1/M_1 + w_2/M_2} \\ &= \frac{0.10}{0.10/(44\,\text{g/mol}) + 0.90/(150\,000\,\text{g/mol})} \\ &\quad + \frac{0.90}{0.10/(44\,\text{g/mol}) + 0.90/(150\,000\,\text{g/mol})} \\ &= 394.96\,\text{g/mol}. \end{aligned} \tag{11.38}$$

Then, we may finally find the characteristic density

$$\begin{aligned} \rho^* &= \frac{M_\text{w} P^*}{RT^* r} \\ &= \frac{(394.96\,\text{g/mol})(330.2\,\text{MPa})}{(8.314\,\text{MPa}\cdot\text{cm}^3/\text{mol}\cdot\text{K})(584\,\text{K})(32.32)} \\ &= 0.8311\,\text{g/cm}^3. \end{aligned} \tag{11.39}$$

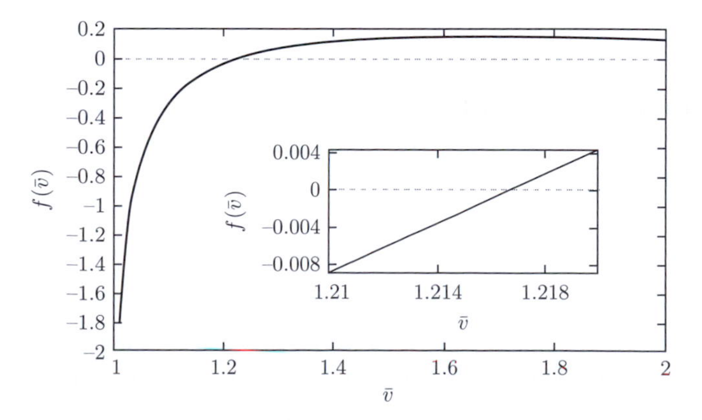

Figure 11.8 A plot for finding the dimensionless density in the Sanchez–Lacombe equation of state in Example 11.2.2. We are seeking the root to f, defined by Eqn. (11.41). The inset shows an enlarged view near the root.

We may now find the necessary dimensionless thermodynamic variables:

$$\frac{P}{P^*} = \frac{15\,\text{MPa}}{330.2\,\text{MPa}} = 0.045\,43 \tag{11.40}$$
$$\frac{T}{T^*} = \frac{453\,\text{K}}{584\,\text{K}} = 0.775\,95.$$

The Sanchez–Lacombe equation of state, Eqn. (11.27), may be written in the form $f(\bar{v}) = 0$, where

$$f(\bar{v}) := \bar{P} + \frac{1}{\bar{v}^2} + \bar{T}\left[\log\left(1 - \frac{1}{\bar{v}}\right) + \left(1 - \frac{1}{r}\right)\frac{1}{\bar{v}}\right], \tag{11.41}$$
$$= 0.045\,43 + \frac{1}{\bar{v}^2} + (0.775\,95)\left[\log\left(1 - \frac{1}{\bar{v}}\right) + \left(1 - \frac{1}{32.32}\right)\frac{1}{\bar{v}}\right].$$

In other words, the density is the root of the equation $f(\bar{v}) = 0$. The root may be found using Excel, MATLAB, Mathematica, Maxima, FORTRAN, or other such programs. If we zoom in on where f crosses the axis (see Figure 11.8), we can read off the zero to be 1.2167. Thus, the dimensional density of the CO_2–polypropylene mixture is

$$v = \frac{M_w \bar{v}}{\rho^*} = \frac{(438.8\,\text{g/mol})(1.2167)}{(0.8311\,\text{g/cm}^3)} = 642.4\,\text{cm}^3/\text{mol}. \tag{11.42}$$

□

The Sanchez–Lacombe model can also be used to predict the solubility of gas in the polymer, by equating the chemical potential of the solvent in the gas phase and that of the solvent dissolved in the polymer. The chemical potential of the solvent can be found by taking the derivative of the fundamental relation with respect to N_1, keeping T, P, and N_2 constant, giving

$$\frac{\mu_1}{RT} = \log\phi_1 + \left(1 - \frac{r_1}{r_2}\right)\log\phi_2 - \log\bar{v} + \frac{r_1 T^*}{TP^*}\left(P\bar{v} - \frac{P_1^*}{\bar{v}} + \frac{\Delta P_1^* \phi_2^2}{\bar{v}}\right)$$
$$+ \frac{r_1 T^* P_1^*}{T_1^* P}(\bar{v} - 1)\log\left(\frac{\bar{v} - 1}{\bar{v}}\right). \tag{11.43}$$

Again, this expression contains the dimensionless volume $\bar{v}$, which is determined implicitly by Eqn. (11.27). Hence, two coupled, nonlinear equations must be solved simultaneously in order to obtain the solubility.

11.2.3 The BGY model

Another compressible model derived from statistical mechanics has been suggested by Lipson and co-workers [131], and has the fundamental Helmholtz relation

$$\frac{F}{RTN_1} = \sum_{i=0}^{r}\phi_i\left\{\log\phi_i + \frac{z_1}{2}\log\left[\frac{\gamma_i}{\sum_{j=0}^{q}\gamma_j\phi_j q_{ij}(T)}\right]\right\}. \tag{11.44}$$

As before, the volume fraction of species i is denoted by $\phi_i := r_i N_i \bar{v}/V$. This species also has a weighting parameter associated with it, namely

$$\gamma_i := z_l - 2 + \frac{2}{r_i}, \tag{11.45}$$

which is the average coordination number for each monomer of a chain with r_i monomers. The monomer–monomer energetics give rise to a partition function $q_{ij}(T)$ for interactions between a monomer of type i and one of type j. Typically, one assumes a single energy of interaction so that $q_{ij}(T) = \exp[-\epsilon_{ij}/(RT)]$.

The remaining parameter N_l is the number of lattice sites, which can be related to the volume, V, and to the volume $\bar{v}$ of a lattice site,

$$N_l := \frac{V}{\bar{v}}. \tag{11.46}$$

Then we have the parameters $\bar{v}$, z_l, and r_i, and the $\{\epsilon_{ij}\}$. To date z_l has always been taken to be 6 in the model, and r_i is the number of monomers in the chain.

What might seem strange about this expression is that there exists a species $i = 0$ in Eqn. (11.44). These terms represent empty sites on the lattice, allowing for compressibility in the model. Therefore, $\phi_0 = 1 - \sum_{i=1}^{q} \phi_i$.

What is particularly useful about this approach is that most of the parameters can be found strictly from compressibility data of the pure species. For example, the *PVT* equation of state for a single-component system can be found from the model by setting $q = 1$, multiplying each side of Eqn. (11.44) by N_l, and taking the derivative with respect to volume to obtain

$$\begin{aligned}\frac{P\bar{v}}{RT} &= -\log(1 - \phi_1) + \frac{z_l \gamma_1^2 [1 - q_{11}(T)]\phi_1^2}{2[(\gamma_1 - 1)\phi_1 + 1][1 - \phi_1 + \gamma_1 q(T)\phi_1]} \\ &\quad + \frac{z_l}{2} \log[(\gamma_1 - 1)\phi_1 + 1].\end{aligned} \tag{11.47}$$

It is important to note that there is only one species here. From this expression, we can then find expressions for either the isothermal compressibility, κ_T, or the isothermal compressibility, α. These measurable quantities can be fit to the resulting expressions to estimate the parameters. Since the resulting equations are rather nonlinear, the fits require good initial guesses. Typically, $z_l = 6$ is used, though values up to 12 are physically meaningful. The number of monomers in the chain is a good estimate for r_i, which is sometimes also allowed to be fit to experiment. The size of a monomer is a good estimate for $\bar{v}$, though fit values are typically larger, in the neighborhood of $\bar{v} \cong 0.01$ l/mol. The energetics of interaction should be negative, and of magnitude similar to RT.

EXAMPLE 11.2.3 Estimate the parameters for a polystyrene melt of molecular weight 279 000 Da.

Solution. Tambasco and Lipson [131] measured the coefficient of thermal expansion for polystyrene at four temperatures, and the results are given in Table 11.4.

Table 11.4 Data of Tambasco and Lipson [131] for polystyrene.

T (K)	$T\alpha$
333	0.23
353	0.245
373	0.262
393	0.278

We need to find an expression for the coefficient of thermal expansion using a pressure-explicit equation of state. Using the definition for the coefficient of thermal expansion,

$$\alpha := -\frac{1}{V}\left(\frac{\partial V}{\partial P}\right)_T = -\frac{1}{V(\partial P/\partial V)_T}. \tag{11.48}$$

All volume dependence is in the volume fraction, so we use the chain rule

$$\left(\frac{\partial P}{\partial V}\right)_T = \left(\frac{\partial \phi}{\partial V}\right)\left(\frac{\partial P}{\partial \phi}\right)_T = -\frac{\phi}{V}\left(\frac{\partial P}{\partial \phi}\right)_T. \tag{11.49}$$

Placing the last expression into Eqn. (11.48) yields

$$\alpha := \left(\frac{1}{V}\frac{\partial V}{\partial P}\right)_T = \frac{1}{\phi(\partial P/\partial \phi)_T}. \tag{11.50}$$

Now using the PVT equation of state for the pure system, Eqn. (11.47), yields

$$\alpha := \frac{2\bar{v}}{RT\phi_{\mathrm{PS}}\left(\dfrac{z_1\gamma_{\mathrm{PS}}}{(\gamma_{\mathrm{PS}}q(T)\phi_{\mathrm{PS}} - \phi_{\mathrm{PS}} + 1)^2} + \dfrac{z_1\left((\gamma_{\mathrm{PS}} - 1)^2\phi_{\mathrm{PS}} - 1\right)}{((\gamma_{\mathrm{PS}} - 1)\phi_{\mathrm{PS}} + 1)^2} - \dfrac{2}{\phi_{\mathrm{PS}} - 1}\right)}. \tag{11.51}$$

We can replace ϕ in this expression with $rN\bar{v}/V = r\rho\bar{v}/M_{\mathrm{w}}$, where $\rho \cong 1000$ g/l is the density and $M_{\mathrm{w}} = 279\,000$ g/mol is the molecular weight. We also replace the monomer partition function with $q(T) = \exp[-\epsilon_{\mathrm{PS}}/(RT)]$.

We then have the parameters $\bar{v}$, ϵ_{PS}, and r that can be fit to the data. However, the molecular weight of a polystyrene monomer is a little over 100, so $r \cong 2500$. So, we fix its value at that recommended by Tambasco and Lipson, namely 2200. We use the values $\epsilon_{\mathrm{PS}} \cong -2413$ J/mol

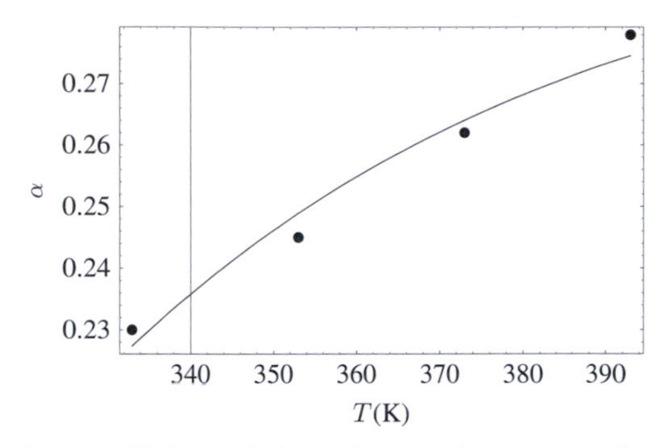

Figure 11.9 The dimensionless coefficient of thermal expansion α as a function of temperature for a polystyrene liquid. The symbols are data taken from Tambasco and Lipson [131], and the curve is the fit to Eqn. (11.51). The parameter values are given in the text.

and $\bar{v} \cong 0.01$l/mol for our initial guesses, and perform a nonlinear least-squares fit (Levenberg–Marquardt) to the data in Table 11.4. The optimal values are found to be $\epsilon_{\text{PS}} \cong -2188.69$ J/mol and $\bar{v} \cong 0.01153$ l/mol. The fit is shown in Figure 11.9. □

For mixtures, the cross energies ϵ_{12} must also be found. These are typically assumed to be very near the usual mixing-rule estimate $g_{12}\sqrt{\epsilon_{11}\epsilon_{22}}$, where $g_{12} \cong 1$. The exact value for g_{12} can be determined from experiments on mixtures.

11.3 BLOCK COPOLYMERS

The term copolymer is used for chain molecules comprising two or more types of monomers. An example might look like

$$—\text{A}—\text{A}—\text{A}—\text{A}—\text{B}—\text{B}—\text{B}—\text{B}—\text{A}—\text{A}—\text{A}—\text{A}—\text{B}—\text{B}—\text{B}—\text{B}—.$$

Polyethylene, which accounts for a significant fraction of the world's polymer production, is actually produced by the co-polymerization of ethylene and an alkene (e.g. hexene); it is therefore a **statistical copolymer**. In **block copolymers**, distinct monomers of the polymer are arranged in block sequences. Block copolymers are a particularly interesting class of materials in that they can exhibit a number of unusual properties, such as the ability to exhibit microphase separation.

For concreteness, in this section we discuss only the case of diblock copolymers. More extensive discussions of multiblock copolymers and of random copolymers can be found in the specialized literature. Consider a diblock copolymer of monomers A and B. The block copolymer shown above provides a schematic representation of a **symmetric diblock copolymer**, namely one in which the degree of polymerization is the same for both blocks. If $\bar{f}_{\text{A}}$ is used to denote the fraction of A monomers in a molecule, then for a symmetric diblock copolymer $\bar{f}_{\text{A}} = \bar{f}_{\text{B}} = \frac{1}{2}$. If the two monomers were similar, one would expect the behavior of the diblock

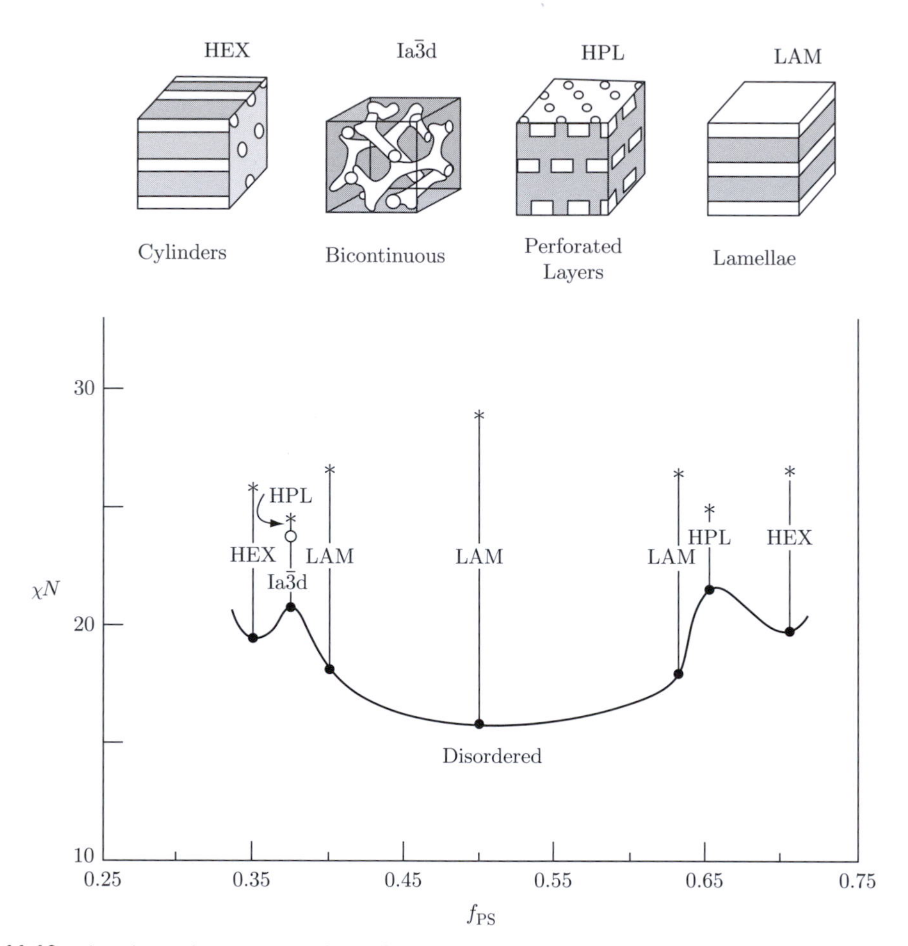

Figure 11.10 A schematic representation of block copolymer microphases. The diagram below shows experimental data for the system polystyrene–poly(2-vinylpyridine). Reprinted, with permission, from [118].

copolymer to be analogous to that of regular homopolymers. However, in situations where the two monomers are fairly different, one would expect the two blocks to become immiscible. That is indeed what is observed in experiments: below a certain critical temperature, the two blocks of a diblock copolymer phase separate but, given that they're "joined at the hip," the two blocks cannot go too far from each other. This leads to the formation of microdomains of A and B segments, or **microphase separation**.

Depending on the composition of the molecule (i.e. the value of $\bar{f}_A$), a diblock copolymer can form a variety of ordered microphases. Figure 11.10 provides a schematic representation of the microphases observed in typical diblock copolymers; these include spherical domains, cylinders, and lamellae. Figure 11.10 shows the phase diagram for a polystyrene–poly(2-vinylpyridine) diblock copolymer. For small values of $\bar{f}_A$ the system forms spheres of one block dispersed in a matrix of the other block, for slightly larger values of $\bar{f}_A$ the system prefers to form cylinders, and for symmetric molecules the material forms lamellae.

In order to understand the nature of microphase separation in block copolymers, it is useful to construct a simple phenomenological theory that is based on ideas presented earlier in this text. For simplicity, we restrict our discussion to the case of symmetric molecules. To construct such a model, we begin by enumerating the various contributions that one would expect to the free energy.

- Chain stretching: in a lamellar phase, polymer molecules are stacked on each other in a parallel fashion; the A and B blocks repel each other, leading to the formation of distinct lamellar domains. This microphase-separated state can be achieved only by stretching the molecules. The corresponding free energy can therefore be estimated as

$$\frac{F_s}{RT} = \frac{3}{8N_K a_K^2} L^2, \tag{11.52}$$

where L is the thickness (or period) of the lamellar microdomains, N_K is the number of segments, and a_K is the so-called Kuhn length of the segments. The Kuhn length represents a characteristic length beyond which groups of monomers along the molecule can be assumed to be independent of each other. For a typical polymer, the Kuhn length comprises on the order of 5–20 monomers, the precise number depending on the chemical constitution of the molecule [39, 41, 42]. This expression is an approximation to the inverse Langevin expression, valid for small extensions of the strand, which was first seen in Exercise 3.7.B. The statistical mechanical derivation of that expression was shown in Section 6.6. Exercise 6.6.D deals with the derivation of this expression.
- Interfacial free energy: the segregation of the molecules into highly organized lamellae involves the formation of an interface between the A and B blocks. This interface has a free energy of the form

$$\frac{F_\gamma}{RT} = \frac{A_{\text{chain}}\gamma_{AB}}{RT}, \tag{11.53}$$

where γ_{AB} is the interfacial tension between blocks A and B, and A_{chain} is used to denote the interfacial area per chain. This expression is covered in greater detail in the following chapter.

The total free energy of a lamellar phase is therefore given by

$$F_{\text{lamellar}} = F_s + F_\gamma. \tag{11.54}$$

The interfacial free energy can be further simplified by assuming that the volume is completely filled by chain molecules, in which case $A_{\text{chain}} = 2N_K a_K^3/L$. The interfacial tension can be related to the Flory–Huggins χ_{AB} parameter by [153, 154]

$$\gamma_{AB} = \frac{k_B T}{a_K^2}\sqrt{\frac{\chi_{AB}}{6}}. \tag{11.55}$$

Equation (11.54) can now be minimized with respect to L to arrive at an estimate of the equilibrium microdomain size (the period of the lamellae), L_0; the result is

$$L_0 = 1.03 a_K \chi_{AB}^{1/6} N_K^{2/3}. \tag{11.56}$$

Equation (11.56) indicates that the width of the lamellae scale with the molecular weight to the power 2/3. This prediction is confirmed by experimental measurements [155].

Equation (11.54) can also be used to determine the temperature at which a copolymer undergoes phase separation. To that end, we need an expression for the free energy of a disordered block copolymer melt. For simplicity, we approximate that free energy by

$$\frac{F_{\text{disordered}}}{RT} \approx \chi_{\text{AB}} \bar{f}_{\text{A}} \bar{f}_{\text{B}} N_{\text{K}}. \tag{11.57}$$

At the phase transition from a disordered to a lamellar phase, $F_{\text{disordered}} = F_{\text{lamellar}}$. By substituting the appropriate expressions for the lamellar and disordered free energies, it is found that the order–disorder transition occurs at

$$\chi_{\text{AB}} N_{\text{K}} = 10.4. \tag{11.58}$$

This simple but important result indicates that large block copolymer molecules are likely to exhibit microphase separation, while short molecules are generally miscible.

11.4 DERIVATION OF THE FLORY–HUGGINS THEORY

In order to derive the Flory–Huggins expression for the polymer-solution free energy of mixing, Eqn. (11.1) with $r_{\text{B}} = 1$, we return to the lattice model for binary mixtures in Section 9.8 and use regular solution theory. However, now we allow the polymer chains to occupy many contiguous lattice sites, as sketched in Figure 11.11. We assume that the solvent particles occupy roughly the same volume as the monomers of the chain. Thus, each solvent particle occupies one lattice site, and a polymer chain occupies $r \equiv r_{\text{A}}$ sites.

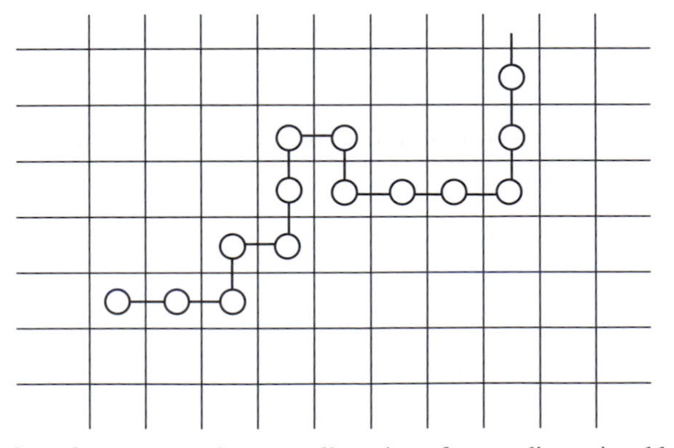

Figure 11.11 A sketch of a polymer occupying a small section of a two-dimensional lattice used in the derivation for the Flory–Huggins model. The polymer is required to occupy contiguous lattice sites. The blank sites are occupied by one solvent molecule each.

We define the following variables:

$$\begin{aligned} r_\mathrm{p} &\sim \text{number of sites occupied by a single chain,} \\ \tilde{N}_\mathrm{c} &\sim \text{number of cells on the lattice,} \\ \tilde{N}_\mathrm{s} &\sim \text{number of solvent molecules,} \\ \tilde{N}_\mathrm{p} &\sim \text{number of polymer molecules.} \end{aligned} \tag{11.59}$$

Since each cell is occupied by either a solvent molecule or a monomer,

$$\tilde{N}_\mathrm{c} = \tilde{N}_\mathrm{s} + r_\mathrm{p}\tilde{N}_\mathrm{p}. \tag{11.60}$$

The volume fraction of polymer is

$$\phi_\mathrm{p} = \frac{r_\mathrm{p}\tilde{N}_\mathrm{p}}{\tilde{N}_\mathrm{s} + r_\mathrm{p}\tilde{N}_\mathrm{p}}. \tag{11.61}$$

The lattice is at constant volume, temperature, and mole number, so the canonical partition function is

$$Q(T, V, \tilde{N}_\mathrm{s}, \tilde{N}_\mathrm{p}) = \sum_{\tilde{N}_\mathrm{ms}} \omega_\mathrm{d}(\tilde{N}_\mathrm{ms}) q_\mathrm{ms}^{\tilde{N}_\mathrm{ms}} q_\mathrm{mm}^{\tilde{N}_\mathrm{mm}} q_\mathrm{ss}^{\tilde{N}_\mathrm{ss}}. \tag{11.62}$$

The summation accounts for the interactions between monomers of the chain and solvent molecules, with the notation that $q_{ij}(T)$, $i = \mathrm{m}, \mathrm{s}$, is the partition function for the interaction of two species, either monomer or solvent, and $\tilde{N}_{ij}$ is the number of such pairs. As before, $\omega_\mathrm{d}(\tilde{N}_\mathrm{ms}; \tilde{N}_\mathrm{m}, \tilde{N}_\mathrm{s})$ is the degeneracy, or the number of conformations that lead to $\tilde{N}_\mathrm{ms}$ monomer–solvent pair interactions when there are the given numbers of monomers and solvent molecules.

We now make the same approximation as was used to derive the Margules model in Chapter 9. Namely, we assume that the monomers and solvent molecules are randomly distributed on the lattice, so that we can write

$$\begin{aligned} Q(T, V, \tilde{N}_\mathrm{s}, \tilde{N}_\mathrm{p}) &\cong \sum_{\tilde{N}_\mathrm{ms}} \omega_\mathrm{d}(\tilde{N}_\mathrm{ms}) q_\mathrm{ms}^{\langle\tilde{N}_\mathrm{ms}\rangle} q_\mathrm{mm}^{\langle\tilde{N}_\mathrm{mm}\rangle} q_\mathrm{ss}^{\langle\tilde{N}_\mathrm{ss}\rangle} \\ &= q_\mathrm{ms}^{\langle\tilde{N}_\mathrm{ms}\rangle} q_\mathrm{mm}^{\langle\tilde{N}_\mathrm{mm}\rangle} q_\mathrm{ss}^{\langle\tilde{N}_\mathrm{ss}\rangle} \sum_{\tilde{N}_\mathrm{ms}} \omega_\mathrm{d}(\tilde{N}_\mathrm{ms}) \\ &= q_\mathrm{ms}^{\langle\tilde{N}_\mathrm{ms}\rangle} q_\mathrm{mm}^{\langle\tilde{N}_\mathrm{mm}\rangle} q_\mathrm{ss}^{\langle\tilde{N}_\mathrm{ss}\rangle} \Omega, \end{aligned} \tag{11.63}$$

where Ω is the number of possible microstates and the angular brackets $\langle \ldots \rangle$ indicate an average. We approximate this average using the random-mixing approximation. $\tilde{N}_\mathrm{ss}$ is the number of solvent–solvent pairs on the lattice, and we use similar notation for the other types of pairs. We estimate these pairs assuming that the sites are randomly occupied:

$$\begin{aligned} \langle\tilde{N}_\mathrm{ms}\rangle &\cong (\text{number of cells occupied by monomer}) \\ &\quad \times (\text{number of adjacent cells}) \\ &\quad \times (\text{fraction of those cells occupied by solvent}) \\ &= r_\mathrm{p}\tilde{N}_\mathrm{p}(z_\mathrm{l} - 2)(1 - \phi_\mathrm{p}) \\ &= (z_\mathrm{l} - 2)\phi_\mathrm{p}(1 - \phi_\mathrm{p})\tilde{N}_\mathrm{c}. \end{aligned} \tag{11.64}$$

Recall that the coordination number of the lattice is assumed to be z_l. Since the chain is contiguous, there are only $(z_l - 2)$ cells adjacent to a monomer that are not necessarily occupied by another monomer. We neglect the small corrections from the chain ends. We can do something similar for the other pairs, being careful not to double count:

$$\langle \tilde{N}_{ss} \rangle \cong \tilde{N}_s z_l (1-\phi_p)\frac{1}{2} = \frac{1}{2} z_l (1-\phi_p)^2 \tilde{N}_c,$$
$$\langle \tilde{N}_{mm} \rangle \cong r_p \tilde{N}_p (z_l - 2)\phi_p \frac{1}{2} = \frac{1}{2}(z_l - 2)\phi_p^2 \tilde{N}_c. \tag{11.65}$$

To estimate the number of possible microstates, we count the number of ways in which it is possible to add the first chain, then multiply this by the number of ways to add the second chain, etc. If we employ the definition

$$\nu_i := \text{Number of ways of adding the } i\text{th chain to the lattice given that there are already } i-1 \text{ chains on the lattice,} \tag{11.66}$$

then the number of ways of arranging $\tilde{N}_p$ *distinguishable* chains on the lattice is $\prod_{i=1}^{\tilde{N}_p} \nu_i$. However, our chains are indistinguishable, so we must divide by the number of ways we can rearrange the indistinguishable chains in order to find the number of possible microstates:

$$\Omega = \frac{1}{\tilde{N}_p!} \prod_{i=1}^{\tilde{N}_p} \nu_i. \tag{11.67}$$

Now we use mean-field probability arguments to construct an expression for ν_{i+1}. We start with the first chain:

$$\begin{aligned}\nu_1 &\cong (\text{number of cells available for first monomer}) \\ &\quad \times (\text{number of cells available for second monomer}) \\ &\quad \ldots \\ &= \underbrace{\tilde{N}_c}_{\text{first monomer}} \times \underbrace{z_l}_{\text{second monomer}} \times \underbrace{(z_l - 1)}_{\text{third monomer}} \ldots \\ &= \tilde{N}_c z_l (z_l - 1)^{r_p - 2} \\ &\cong \tilde{N}_c z_l^{r_p - 1}.\end{aligned} \tag{11.68}$$

In the last line we have assumed that z_l is sufficiently large that we can make an approximation. This is done only for mathematical convenience later. We have neglected here the fact that, when we add later monomers, some cells are already occupied and therefore not available. However, we now consider later chains and account for the occupation of cells from adding previous chains. Hence,

$$\begin{aligned}\nu_{i+1} &\cong \underbrace{(\tilde{N}_c - i r_p)}_{\text{first monomer}} \underbrace{z_l \left(1 - \frac{i r_p}{\tilde{N}_c}\right)}_{\text{second monomer}} \underbrace{(z_l - 1)\left(1 - \frac{i r_p}{\tilde{N}_c}\right)}_{\text{third monomer}} \ldots \\ &\cong \tilde{N}_c z_l^{r_p - 1} \left(1 - \frac{i r_p}{\tilde{N}_c}\right)^{r_p}.\end{aligned} \tag{11.69}$$

We can now estimate the number of microstates Ω from Eqn. (11.67) and the last result:

$$\begin{aligned}\Omega &= \frac{1}{\tilde{N}_{\rm p}!}\prod_{i=1}^{\tilde{N}_{\rm p}} \nu_i.\\ &= \prod_{i=1}^{\tilde{N}_{\rm p}} \frac{\nu_i}{i}\\ &= \prod_{i=1}^{\tilde{N}_{\rm p}} \frac{z_1^{r_{\rm p}-1}\tilde{N}_{\rm c}}{i}\left(1-\frac{ir_{\rm p}}{\tilde{N}_{\rm c}}\right)^{r_{\rm p}}\\ &= (z_1^{r_{\rm p}-1}\tilde{N}_{\rm c})^{\tilde{N}_{\rm p}}\prod_{i=1}^{\tilde{N}_{\rm p}}\frac{(1-ir_{\rm p}/\tilde{N}_{\rm c})^{r_{\rm p}}}{i}.\end{aligned} \tag{11.70}$$

We used our estimate for ν_i, Eqn. (11.69), to obtain the third line. We take the logarithm of each side of this equation to obtain

$$\log\Omega = \tilde{N}_{\rm p}\log(z_1^{r_{\rm p}-1}\tilde{N}_{\rm c}) + \sum_{i=1}^{\tilde{N}_{\rm p}}\log\left[\frac{(1-ir_{\rm p}/\tilde{N}_{\rm c})^{r_{\rm p}}}{i}\right]. \tag{11.71}$$

We can approximate the summation by an integral in the thermodynamic limit of long chains to find

$$\begin{aligned}\sum_{i=1}^{\tilde{N}_{\rm p}}\log\left[\frac{(1-ir_{\rm p}/\tilde{N}_{\rm c})^{r_{\rm p}}}{i}\right] &\cong \int_1^{\tilde{N}_{\rm p}}\log\left[\frac{(1-ir_{\rm p}/\tilde{N}_{\rm c})^{r_{\rm p}}}{i}\right]di\\ &= r_{\rm p}\int_1^{\tilde{N}_{\rm p}}\log\left(1-\frac{ir_{\rm p}}{\tilde{N}_{\rm c}}\right)di - \int_1^{\tilde{N}_{\rm p}}\log(i)di\\ &= -\tilde{N}_{\rm c}\left[\frac{\phi_{\rm p}}{r_{\rm p}}\log\left(\frac{\phi_{\rm p}}{r_{\rm p}}\right) + (1-\phi_{\rm p})\log(1-\phi_{\rm p}) + \phi_{\rm p}\left(1-\frac{1}{r_{\rm p}}\right)\right].\end{aligned} \tag{11.72}$$

With our estimate for Ω above, and estimates for the average pair numbers, Eqns. (11.64) and (11.65), we can find the partition function, Eqn. (11.63):

$$\begin{aligned}F &= -k_{\rm B}T\log Q\\ &= \tilde{N}_{\rm c}\left\{z_1\left[\frac{1}{2}(1-\phi)^2\log q_{\rm ss} + \frac{1}{2}\phi_{\rm p}^2\log q_{\rm mm} + \phi_{\rm p}(1-\phi_{\rm p})\log q_{\rm ms}\right]\right.\\ &\quad\left. + k_{\rm B}T\left[\frac{\phi_{\rm p}}{r_{\rm p}}\log\left(\frac{\phi_{\rm p}}{r_{\rm p}}\right) + (1-\phi_{\rm p})\log(1-\phi_{\rm p}) + \phi_{\rm p}\left(1-\frac{1}{r_{\rm p}}\right)\right]\right\}.\end{aligned} \tag{11.73}$$

This is our estimate for the free energy of an amorphous mixture of polymer and solvent. To estimate the free energy of mixing, we need the free energies for pure solvent and pure polymer. These may be found from the above expression in the appropriate limits: $f^{\rm pure\ solvent} = F(\phi_{\rm p}\to 0)/N_{\rm c}$ and $f^{\rm pure\ polymer} = r_{\rm p}F(\phi_{\rm p}\to 1)/N_{\rm c}$. Using the definition of mixing, $\Delta F_{\rm mix} = F - N_{\rm p}f^{\rm pure\ polymer} - N_{\rm s}f^{\rm pure\ solvent}$, we obtain after some algebra

$$\frac{\Delta F_{\rm mix}}{N_{\rm c}RT} = (1-\phi_{\rm p})\log(1-\phi_{\rm p}) + \frac{\phi_{\rm p}}{r_{\rm p}}\log(\phi_{\rm p}) + \chi(T)\phi_{\rm p}(1-\phi_{\rm p}), \tag{11.74}$$

where

$$\chi(T) := \frac{z_1}{2} \log\left(\frac{q_{\mathrm{ms}}(T)^2}{q_{\mathrm{mm}}(T)q_{\mathrm{ss}}(T)}\right) \tag{11.75}$$

is called the **Flory–Huggins interaction parameter**. This derivation assumed that solvent was present instead of two polymer species. If we apply the same specification to Eqn. (11.1) and assume incompressibility, these two expressions are equivalent.

Summary

In this chapter we introduced the following.

- Mixing models for polymers, beginning with the Flory–Huggins fundamental relation for polymer liquids (Section 11.1). In Section 11.1 we used the Flory–Huggins model to predict the spinodal and binodal curves for polymer solutions. There we saw how volume fraction plays an important role in explaining the large asymmetry seen in these stability curves.

  However, the Flory–Huggins model is not sufficiently accurate to describe experimental data, so we considered also two generalizations in Section 11.2. The first adds *ad hoc* parameters to describe the rich behavior seen experimentally for nearly incompressible polymer liquids. This generalization of the Flory–Huggins model was shown to describe the miscibility curve for a mixture of copolymer poly(vinyl alcohol)$_{93}$-co-(vinyl acetate)$_7$, super-critical CO_2, and polypropylene. The Sanchez–Lacombe model generalizes the Flory–Huggins model to compressible polymer solutions, which was then used to predict the density of a mixture of polypropylene and super-critical CO_2. A modification by Lipson and coworkers involves removing the random phase approximation of the Flory–Huggins theory. The SAFT model was also introduced.
- Microphase separation in block copolymers. In Section 11.3, we showed how simple thermodynamic arguments can be used to make qualitative predictions about the formation of microphases in block copolymers.
- A statistical-mechanical derivation for the Flory–Huggins model in Section 11.4.

Exercises

11.1.A Qian *et al.* claim that the phase behavior of polystyrene in acetone can be described using their equation of state with the parameters set to $b_1 = 0$, $b_2 = 0.653$, $d_0 = -6.9933$, $d_1 = 376.2\,\mathrm{K}$, and $d_2 = 1.1$. Polymer is species A. Plot the spinodal and binodal curves for this equation of state between 276 and 430 K. Compare the curves with the experimental data for the binodal curve reported by Siow *et al.* given in Table 11.5.

11.1.B Derive an equation for the binodal points from the Flory–Huggins equation.

Table 11.5 Binodal data for polystyrene in acetone, taken from Qian *et al.* [104] (data of Siow *et al.*).

Volume fraction of polymer	Temperature (K)
0.03557	393.0
0.03874	387.2
0.04317	316.2
0.0472	323.1
0.05469	380.9
0.05521	327.4
0.05546	374.6
0.05685	334.5
0.06503	372.0
0.06726	338.8
0.2152	363.1
0.2151	343.9
0.2422	338.1
0.244	371.6
0.263	331.5
0.2704	376.4
0.2734	324.3
0.2769	381.8
0.2917	316.2
0.2953	387.8
0.3089	392.6
0.3149	310.5
0.3252	303.3
0.3265	398.9

11.1.C Using the Sanchez–Lacombe equation, find the solubility of CO_2 at a pressure of 500 MPa.

Hint: you should be able to reduce the problem to two coupled, nonlinear algebraic equations. The roots to these equations can be found iteratively by first guessing a value for the solubility.

11.1.D Prove that the expression for the chemical potential, Eqn. (11.43), is correct.

11.1.E Panayiotou and Sanchez [96] have proposed a somewhat simpler model for compressible polymer solutions:

$$G = rN\left\{P\bar{v}v^* - \frac{\epsilon^*}{\bar{v}} + RT\left[(\bar{v}-1)\log\left(\frac{\bar{v}-1}{\bar{v}}\right) - \frac{1}{r}\log\bar{v} + \frac{\phi_1}{r_1}\log\phi_1 + \frac{\phi_2}{r_2}\log\phi_2\right]\right\}. \tag{11.76}$$

The mixing rules for this model are

$$\begin{aligned}\epsilon^* &= \phi_1\epsilon_1^* + \phi_1\epsilon_2^* - \phi_1\phi_2 RT\chi_{12},\\ v^* &= \phi_1 v_1^* + \phi_2 v_2^*,\\ r &= \chi_1 r_1 + \chi_2 r_2.\end{aligned} \tag{11.77}$$

The new parameters are related to the others by the relations

$$\begin{aligned}\epsilon_i^* &= RT_i^*,\\ v_i^* &= \frac{RT_i^*}{P_i^*}.\end{aligned}$$

Find an expression for the chemical potential of species 1 as a function of T, P, and ϕ_2. Note that an implicit expression for $\bar{v}$ is needed for this model, as well.

11.1.F Verify the expression for the *PVT* equation of state from the BGY model, Eqn. (11.47).

11.1.G Find the parameters of the BGY model for poly(methyl vinyl ether). Compressibility data can be found in the paper by Tambasco and Lipson [131].

12 Thermodynamics of surfaces

The pressure inside a soap bubble is slightly larger than that outside the bubble. If those two pressures were equal, the bubble would simply collapse. The liquid film that constitutes the bubble is under tension, and is therefore able to sustain the force imbalance that arises from that pressure difference. Throughout most of this text, we have focused our attention on the properties of bulk phases. The boundaries or interfacial regions between bulk phases at equilibrium exhibit a number of interesting properties that are often different from those of the bulk. These properties are particularly important in a wide variety of technologies, including nano-fabrication, ink-jet printing, coatings, and biotechnology. This chapter discusses a few extensions of ideas presented earlier in the text to determine the behavior of interfaces.

The study of the thermodynamics of surfaces usually adopts one of two conventions. In the approach of Gibbs [17], a sharp dividing surface is introduced between two distinct phases (e.g. a liquid and a vapor). In the approach of Guggenheim [51], the interfacial region has a finite volume V^{σ}, over which the thermodynamic properties of the system change gradually. The latter approach is followed closely in this chapter because it is more physically intuitive. We note, however, that in our discussion of mixtures and surface quantities we adopt the view that the volume of the interfacial region is so small as to be negligible. In that limit, we implicitly return to the Gibbs treatment of an interface.

The thermodynamics of surfaces and interfaces have been studied for over two hundred years, and one cannot do justice to this subject in a single chapter. Readers interested in the subject are referred to the text by Hunter [67], which presents a comprehensive description of surfaces and interfaces, and their characterization.

12.1 THE INTERFACIAL TENSION OF A PLANAR INTERFACE

Consider two distinct bulk phases, α and β, separated by a planar interfacial region denoted by σ (see Figure 12.1). That interfacial region is bounded by two planes, AA′ and BB′. In any bulk phase in the absence of gravity, the pressure is uniform; that is, the force per unit area across any plane is uniform. In region σ, however, the force across any unit-area plane is not the same in all directions. If one identifies a plane of unit area *parallel* to AA′ or BB′, then that force is uniform, regardless of whether the plane is in region α, β, or σ. That force per unit area is again the pressure, and it must be constant across the interface in order for the system to be mechanically stable.

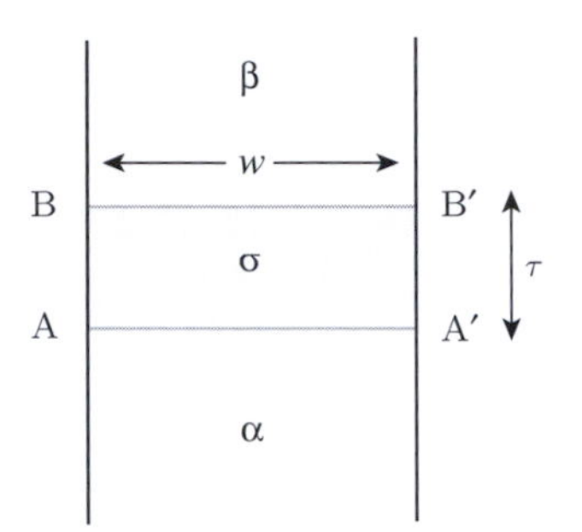

Figure 12.1 A planar interface σ between two bulk phases α and β, showing the dimensions of the surface layer σ, with volume $V^{\sigma} = wl\tau$.

If, however, a plane is chosen perpendicular to AA′ (or BB′), then the force per unit area is no longer the pressure. Consider a plane perpendicular to AA′ that extends above BB′ and AA′. Furthermore, assume that the plane has a rectangular shape with height τ (parallel to AB) and width l (perpendicular to the plane of the page), and τ extends from AA′ to BB′. As it turns out, there is an extra force on this perpendicular plane. The force across this plane is given by $P\tau l - \gamma l$, where γ is the so-called **interfacial tension** or **surface tension**. The symbol adopted in this chapter for surface tension, γ, should not be confused with the activity coefficient introduced in Chapter 9.

Because the forces are different for different planes, the work necessary to change the volume depends on how the volume changes, namely whether the change is in the direction of l, or brought about through changing the thickness. Therefore, we need to find the relationship between work and changes in both volume and thickness, or better, changes in both volume and area. This relationship can be found in the following way.

The region labeled σ in Figure 12.1 has thickness τ, and length l out of the page. If we keep l fixed, and the distance from A to A′ – we call it w – fixed, but increase the thickness by $d\tau$, then the necessary work is $dW = -Pwl\,d\tau = -P\,dV^{\sigma}$, where V^{σ} is the volume of the surface layer. On the other hand, if we change the volume by changing l, keeping w and τ fixed, then there is an additional force from the interfacial tension, and $dW = -(Pw\tau - \gamma w)dl$. Hence, the total work necessary for changing τ and l independently (with fixed w) is

$$\begin{aligned} dW &= -Pwl\,d\tau - (Pw\tau - \gamma w)dl, \quad \text{fixed } w, \\ &= -P(wl\,d\tau + w\tau\,dl) + \gamma w\,dl, \quad \text{fixed } w, \\ &= -Pd(wl\tau) + \gamma d(wl), \quad \text{fixed } w, \\ &= -P\,dV^{\sigma} + \gamma\,dA. \end{aligned} \tag{12.1}$$

The last line is our desired result. We now know the work necessary to change either the volume or the area of the surface layer at fixed temperature and mole numbers.

The differential of the Helmholtz free energy of a bulk phase is given by an expression of the form

$$dF = -S\,dT - P\,dV + \sum_i \mu_i\,dN_i. \tag{12.2}$$

For a surface phase, the work term, $P\,dV$, appearing in Eqn. (12.2) is replaced by that given in Eqn. (12.1) to give the following equation for the differential of the Helmholtz free energy of a surface phase:

$$dF^\sigma = -S^\sigma\,dT - P\,dV^\sigma + \gamma\,dA + \sum_i \mu_i\,dN_i^\sigma. \tag{12.3}$$

In fact, we now have a thermodynamic definition for the interfacial tension,

$$\gamma := \left(\frac{\partial F^\sigma}{\partial A}\right)_{T,V^\sigma,\{N_i\}}, \tag{12.4}$$

as the change in free energy with surface area at fixed temperature, volume, and mole numbers.

It is instructive to consider the molecular origins of the interfacial tension. In a bulk liquid, an individual molecule is completely surrounded by other molecules. As discussed in earlier chapters, all molecules in the liquid exert an attractive force on each other. At a liquid–vapor interface, however, this physical picture is altered. Molecules at the interface are only partially surrounded by other molecules, because of the lower density of the "vapor" side. It is therefore energetically unfavorable for a molecule to sit at the interface, and work must be done to drag a molecule from the bulk of the liquid phase to that interface; the interfacial energy can be interpreted as the energy (per unit surface area) required for such a process.

12.2 THE GIBBS FREE ENERGY OF A SURFACE PHASE AND THE GIBBS–DUHEM RELATION

In order to arrive at an expression for the Gibbs free energy of a surface phase, we note that the Helmholtz potential is extensive in the surface area:

$$F^\sigma(T, \lambda V^\sigma, \lambda A, \lambda N_i^\sigma) = \lambda F^\sigma(T, V^\sigma, A, N_i^\sigma). \tag{12.5}$$

Therefore, similar to what we have already seen for bulk systems, we can write (see the derivation of the Euler relation, Eqn. (3.20))

$$F^\sigma = -PV^\sigma + \gamma A + \sum_i \mu_i N_i^\sigma. \tag{12.6}$$

Equation (12.6) should be contrasted to the analogous expression for a bulk phase, namely $F + PV = G = \sum_i \mu_i N_i^\sigma$. For a surface phase, the Gibbs free energy differs from that of a bulk phase by an additional term consisting of the product of interfacial tension and area:

$$\begin{aligned} G^\sigma &= F^\sigma + PV^\sigma - \gamma A \\ &= \sum_i \mu_i N_i^\sigma. \end{aligned} \tag{12.7}$$

Alternatively, the above expression can be viewed as a Legendre transform of F in both volume and area (Section 3.1). With this new set of state variables, we can derive general, analogous relations. For example, the differential of the Gibbs free energy is given by

$$dG^\sigma = -S^\sigma\,dT + V^\sigma\,dP - A\,d\gamma + \sum_i \mu_i\,dN_i^\sigma. \tag{12.8}$$

Equation (12.6) can be differentiated to give

$$dF^{\sigma} + P\,dV^{\sigma} + V^{\sigma}\,dP - \gamma\,dA - A\,d\gamma = \sum_i \mu_i\,dN_i^{\sigma} + \sum_i N_i^{\sigma}\,d\mu_i. \tag{12.9}$$

Equation (12.9) can be subtracted from Eqn. (12.3) to give

$$0 = S^{\sigma}\,dT - V^{\sigma}\,dP + A\,d\gamma + \sum_i N_i^{\sigma}\,d\mu_i. \tag{12.10}$$

Equation (12.10) is the surface analog of the Gibbs–Duhem equation, Eqn. (3.22), which was presented in Section 3.1.

12.3 CURVED INTERFACES

Our discussion so far has been limited to a planar interface. The extension to curved interfaces can be presented in a relatively simple manner in terms of cylindrical coordinates. Consider the interfacial "shell" bounded by segments AA′ and BB′ shown in Figure 12.2. All distances are measured from the origin of the coordinate system O. The distance to phase α is denoted by r_α, and the distance to phase β is denoted by r_β. We now allow the pressures in regions α and β to be different: the pressure in α is P^α, and that in β is P^β. If $P^\alpha > P^\beta$, then the pressure in the interfacial region σ must vary continuously from P^α to P^β.

The pressure (or force per unit area) is uniform and isotropic in phases α and β. Now we use a force balance to relate the pressures in the bulk phases to the interfacial tension, γ. The film in Figure 12.2 is stationary, so we can use a radial force balance to relate the pressures and the interfacial tension. If the angle between the two planes is θ, then the area of the film is $r\theta l$, where $r = r_\alpha$ or r_β and l is the width of the surface coming out of the page. Hence, the force upwards from the α phase for small θ is $P^\alpha r_\alpha \theta l$. There is a similar force from the upper phase, downwards. The interfacial tension exerts a force γl at the AB surface, perpendicular to the plane. This force has a downward component, $-\gamma l \sin(\theta/2)$. There is the same force from the A′B′ plane. Hence, for small θ, our force balance is

$$0 = P^\alpha r_\alpha \theta l - P^\beta r_\beta \theta l - 2l\gamma \sin(\theta/2). \tag{12.11}$$

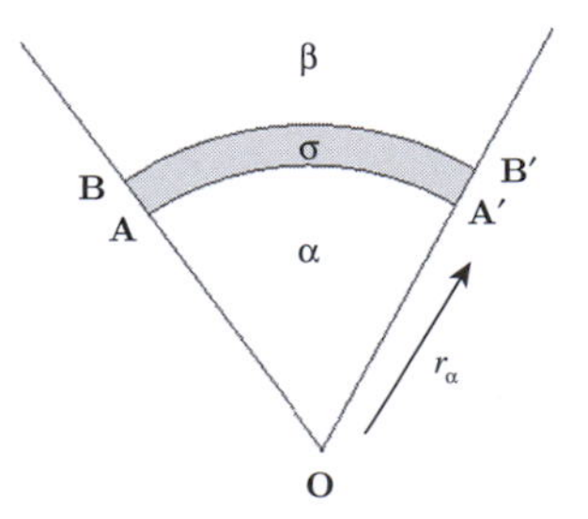

Figure 12.2 A schematic representation of an interfacial shell σ.

The thickness of the interfacial region is generally small, on the order of 100 nm or less. The difference between r_α and r_β therefore typically satisfies the inequality

$$r_\beta - r_\alpha \ll \bar{r}, \tag{12.12}$$

where

$$\bar{r} = \frac{r_\alpha + r_\beta}{2}. \tag{12.13}$$

Therefore, we replace both r_α and r_β with $\bar{r}$ in Eqn. (12.11), and take the limit $\theta \to 0$ to obtain

$$P^\alpha - P^\beta = \frac{\gamma}{\bar{r}}. \tag{12.14}$$

A similar exercise can be performed for a spherical interface, in which case Eqn. (12.14) becomes

$$P^\alpha - P^\beta = 2\frac{\gamma}{\bar{r}}. \tag{12.15}$$

The derivation outlined here can be extended to curved interfaces of arbitrary shape, with the general result

$$P^\alpha - P^\beta = \gamma\left(\frac{1}{r_1} + \frac{1}{r_2}\right), \tag{12.16}$$

where r_1 and r_2 denote the principal radii of curvature of the interface. Equation (12.16) is known as the **Young–Laplace equation**, which was first derived approximately 200 years ago.

Equation (12.15) can now be used to determine the pressure difference between the exterior of a spherical bubble, P_{ext}, and the interior of the bubble, P_{int}. Note that a bubble has two interfaces: air–film–air. If the pressure in the film itself is denoted by P', we have

$$\begin{aligned} P_{int} - P' &= 2\frac{\gamma}{r_{int}}, \\ P' - P_{ext} &= 2\frac{\gamma}{r_{ext}}, \end{aligned} \tag{12.17}$$

and therefore

$$P_{int} - P_{ext} = \gamma\left(\frac{2}{r_{int}} + \frac{2}{r_{ext}}\right). \tag{12.18}$$

By neglecting the difference between r_{int} and r_{ext}, one arrives at the following expression for the sought-after pressure difference:

$$P_{int} - P_{ext} = 4\frac{\gamma}{r}. \tag{12.19}$$

Table 12.1 gives values of the surface tension for several common liquids. Water has a relatively large surface tension. Simple alkanes have a surface tension that is typically about a third of that of pure water. Mercury, on the other hand, has a large surface tension, more than an order of magnitude larger than those of most simple fluids. This high surface tension partly explains why droplets of mercury are relatively stable under ambient conditions, and are able to roll over hard surfaces. When two small droplets of mercury come near each other, they merge into a larger droplet, thereby reducing the interfacial free energy.

Table 12.1 Surface tension with air at different temperatures for several common liquids. A more extensive compilation of data can be found in the review by Gaonkar and Neuman [46], and in an earlier paper by J. R. Dann [31].

Substance	T (°C)	γ (mN/m)
Water	10	74.2
	25	72.1
	50	67.9
	75	63.6
	100	58.9
Glycerol	25	64
Ethanol	10	23.2
	20	22.4
	25	22
	30	21.6
	50	19.9
Benzene	20	28.9
	30	27.6
n-Octane	10	22.6
	20	21.6
	25	21.1
	50	18.8
	75	16.4
Mercury	20	486.5
	30	484.5

Correlations have been developed to calculate the surface tension of liquids, yielding results that are generally within 10% of experimental values. Their use is illustrated in several exercises at the end of this chapter.

In Chapter 4, we considered a bulk liquid in equilibrium with a bulk vapor. We saw that the pressures, temperatures, and chemical potentials of the two phases were equal at equilibrium, and how we could use thermodynamic fluid models to predict the saturation pressure as a function of temperature. Because the pressures inside and outside of a droplet are different, the general equations that we arrived at in earlier chapters to describe phase equilibria must be revised, leading to interesting and important predictions. One of these predictions, for example, is that, in order for a vapor to condense into a liquid droplet, it must actually be cooled down below the bulk condensation temperature. In the remainder of this section, we consider how the "equilibrium" pressure inside a drop changes at fixed temperature, and vice versa.

EXAMPLE 12.3.1 Consider a droplet of water of diameter r suspended in air at ambient pressure and temperature. If the temperature and pressure of the droplet are in equilibrium with the moist air, are the fugacities in equilibrium? Is the effective vapor pressure of the liquid in the droplet the same as that of bulk water, at the same temperature and pressure?

Solution. In order to make a comparison between the liquid in the droplet and the liquid in the bulk, we use superscripts d and b, respectively. Furthermore, we use a prime to denote the phase in contact with the droplet and the bulk liquid. At equilibrium, the pressure above the bulk liquid or outside the droplet must be the same, i.e.

$$P^{d\prime} = P^{b\prime}. \tag{12.20}$$

Mechanical stability requires that $P^{b\prime} = P^{b}$. Equation (12.15) can therefore be written as

$$P^{d} - P^{b} = 2\frac{\gamma}{r}. \tag{12.21}$$

The pressure inside the droplet is therefore higher than in the bulk liquid. At constant temperature, the pressure dependence of fugacity is given by Eqn. (8.43)

$$\left(\frac{d\log f}{dP}\right)_T = \frac{v}{RT}, \tag{12.22}$$

where v represents the molar volume of the liquid. Assuming, for simplicity, that the liquid is incompressible, the fugacity of the liquid in the droplet and that in the bulk are related by

$$\begin{aligned}\log\left(\frac{f^{d}}{f^{b}}\right) &= \frac{1}{RT}(P^{d} - P^{b})v \\ &= \frac{2\gamma v}{rRT}.\end{aligned} \tag{12.23}$$

From this equation, we see that *the fugacity of water in the droplet and that in the bulk cannot be the same when the pressure and temperature are at equilibrium.* The surface tension is always positive; the fugacity in the droplet must always be higher, giving rise to a driving force for evaporation from the droplet. To characterize this driving force, we can consider an **effective vapor pressure**. To find the effective vapor pressure, we assume ideal-gas behavior for the vapor phase in equilibrium with the droplet. Then, the above expression simplifies to

$$\log\left(\frac{P^{d,sat}}{P^{b,sat}}\right) = \frac{2}{RT}\frac{\gamma}{r}v. \tag{12.24}$$

Equation (12.24) represents an approximate form (ideal-gas behavior was assumed) of the so-called **Kelvin equation**. These results show that, at the same external pressure and temperature, the vapor pressure (and fugacity) of a liquid in a droplet is higher than that in the bulk. This difference in vapor pressures will cause the liquid in the droplet to evaporate faster, thereby decreasing its radius and increasing the vapor-pressure difference even more. The equilibrium between small droplets and the bulk is therefore unstable, and eventually small droplets disappear. In the case of multiple droplets of different diameters in contact with the same bulk phase, the larger droplets will grow at the expense of the smaller ones.

In the particular case of water at 25 °C and 1 bar, the surface tension is $\gamma = 72$ dynes/cm. In a 1-μm droplet, the effective vapor pressure of water increases by a factor of 1.001; in a 10-nm droplet, it increases by a factor of 1.1. □

In the last example, we found that it is not possible for the temperature, pressure, and fugacity in the droplet to be in equilibrium with a vapor phase that is in equilibrium with a bulk liquid phase. If we assumed that the temperature and pressure are in equilibrium, we could find the effective equilibrium vapor pressure in the droplet at fixed temperature.

In order to arrive at a modified Clapeyron equation, Eqn. (4.45), we now consider how the equilibrium temperature changes with fixed vapor pressure, if the fugacities and pressures are in equilibrium. The chemical potentials and temperatures of the two phases are equal at equilibrium. The calculation we perform is to keep the pressure of the vapor constant, and see how the equilibrium "saturation temperature" changes with radius, at a fixed vapor pressure.

Towards this end, we begin with the Gibbs–Duhem relation for the liquid and vapor phases, Eqn. (4.44),

$$(s^{\mathrm{v}} - s^{\mathrm{l}})dT + v^{\mathrm{l}}\,dP^{\mathrm{l}} = 0, \quad \text{constant } P^{\mathrm{v}}, \tag{12.25}$$

where we have set $dP^{\mathrm{v}} = 0$ because the external pressure is constant. Furthermore, from the Young–Laplace equation we write

$$dP^{\mathrm{l}} = d\left(\frac{2\gamma}{r}\right), \quad \text{constant } P^{\mathrm{v}}. \tag{12.26}$$

Given that $s^{\mathrm{v}} - s^{\mathrm{l}} = \Delta h^{\mathrm{vap}}/T$, if we assume that the heat of vaporization and the molar volume of the liquid are independent of temperature, Eqn. (12.25) can be integrated from some finite r value to $r = \infty$ to arrive at the following expression for the dependence of the equilibrium temperature on droplet size:

$$\log\left(\frac{T}{T_0}\right) \cong -\frac{2v^{\mathrm{l}}\gamma}{\Delta h^{\mathrm{vap}} r}. \tag{12.27}$$

Here T_0 is the temperature at $r = \infty$ (and v^{l} and h^{vap} are assumed constant). Equation (12.27) is known as the **Thomson equation**. To derive it, we assumed that Δh^{vap} and v^{l} are independent of temperature.

The Thomson equation indicates that, as the size of the droplet decreases, the temperature required in order to establish *chemical* equilibrium across the interface decreases ($T < T_0$ because Δh^{vap} is always positive). For the vapor to condense into the liquid droplet, it must be cooled down to a temperature below the normal condensation temperature corresponding to a saturation pressure P^{v}. In other words, the vapor must be supercooled.

12.4 SOLID–LIQUID INTERFACES: WETTING

Liquids spreading on surfaces deserve particular attention. Our discussion so far has been largely focused on liquids, and for experiments under ambient conditions the relevant interfacial tension is that between the liquid of interest and air. There is also an interfacial tension between a solid and air, and, at the interface between a solid and a liquid, there is an interfacial tension between the solid and the liquid. In what follows, we use the subscripts L, S, and SL to denote the liquid–air, solid–air, and solid–liquid interfacial tensions, respectively. Some important phenomena, including the collapse of polymeric structures discussed earlier, the rise of water in trees, the condensation of dew droplets on plant leaves, and the effectiveness of adhesives, represent special cases in which solid–liquid interfaces play an important role.

Young's equation provides the basis for the following discussion. It relates the so-called **contact angle** to the interfacial tensions of the liquid, γ_{L}, the solid, γ_{S}, and the solid–liquid interface, γ_{SL}, according to

$$\gamma_{\mathrm{L}} \cos\theta = \gamma_{\mathrm{S}} - \gamma_{\mathrm{SL}}. \tag{12.28}$$

Figure 12.3 provides a schematic representation of a drop on a surface, showing the contact angle and the various interfaces involved in such a system. Partial wetting (i.e. $\theta < 90^{\circ}$) occurs when $\gamma_{\mathrm{S}} > \gamma_{\mathrm{SL}}$. Non-wetting (i.e. $\theta > 90^{\circ}$) occurs when $\gamma_{\mathrm{S}} < \gamma_{\mathrm{SL}}$, which corresponds to a solid-liquid interface that is not as energetically favorable as the bare solid surface. Complete wetting corresponds to the case $\theta \approx 0$. The contact angle provides a measure of the balance between a drop's tendency to minimize its own interfacial area by rounding up and its tendency to spread out and cover the surface. At a molecular level, it results from the competition between interactions of liquid molecules with themselves and with the surface material.

Table 12.2 provides specific values of the contact angle for several common liquids on representative solid surfaces.

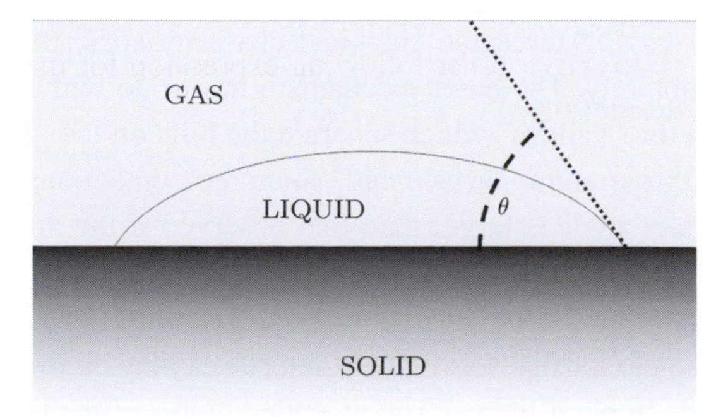

Figure 12.3 A schematic representation of a liquid droplet on a solid surface, and the definition of the contact angle θ.

Table 12.2 Advancing contact angles of several common liquids on different surfaces at 25 °C, from [31, 19].

Liquid	Solid	θ
Water	Paraffin	110°
	Polystyrene	98–112°
	PMMA	74°
	Graphite	86°
	Gold	0°
Glycerol	Paraffin	96°
	Polystyrene	71°
	PMMA	67°
n-Octane	Teflon	26–30°
Mercury	Teflon	150°
	Glass	128–148°

The values presented in Table 12.2 correspond to "advancing" contact angles. As one would expect, real surfaces are not perfect; they exhibit roughness and chemical inhomogeneities. Furthermore, the liquids whose surface tension is being measured might contain impurities (e.g. dust). All of these features lead to hysteresis in contact-angle measurements. When the drop shown in Figure 12.3 is growing in volume, one measures the so-called "advancing" contact angle. When the size of the drop is decreasing, one measures the "receding" contact angle.

Interestingly, roughness can confer water repellency on hydrophobic surfaces; if a hydrophobic surface is rough, water will exhibit an even greater contact angle than on a molecularly smooth surface having the same chemical characteristics, thereby giving the impression of greater hydrophobicity. The basic mechanism has to do with the fact that on a rough surface air is trapped in the "valleys" which separate the hills on the surface. A droplet does not come into contact with the entire surface and, since the contact angle for air is 180°, the resulting "effective" contact angle is larger than that observed if the droplet could come into complete contact with a smooth surface. The connections between roughness and hydrophobicity were originally discussed in the earlier half of the last century [23].

We mentioned earlier that solid–liquid interfaces play an important role in the condensation of water droplets in plant leaves. The so-called "lotus effect" has in fact been used to explain how the apparently extreme hydrophobicity of some plant leaves [9] leads to an intriguing self-cleaning mechanism. It turns out that the surfaces of such leaves are extremely rough,

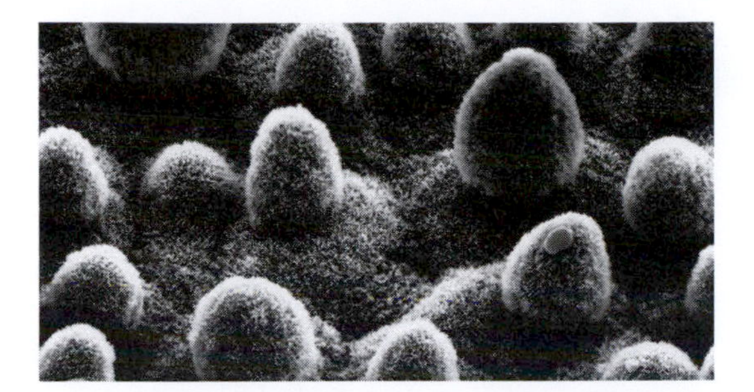

Figure 12.4 Experimental scanning electron micrographs of the leaf surface of *Nelumbo nucifera* (lotus). The field of view is approximately 400 μm × 200 μm. From [9], with kind permission from Springer Science and Business Media.

consisting of a forest of hydrophobic spikes made out of wax crystalloids. Figure 12.4, from [9], shows a micrograph of a *Nelumbo nucifera*, or lotus, leaf, where one can appreciate a rich topology on the scale of microns. That structure confers a water-repellency property on the leaves, which causes water to exhibit a large contact angle and dew droplets to adopt a spherical shape. Water droplets can then roll off the leaves' surfaces, carrying with them particles of dirt, regardless of their size, and keeping them clean. This is an area where the lessons from nature inspired artificial designs for synthetic, self-cleaning super-hydrophobic surfaces. Fürstner and co-workers [45], for example, prepared silicon surfaces consisting of arrays of hydrophobic spikes that mimic some of the characteristics of the lotus leaf, and showed that the resulting materials are indeed super-hydrophobic and self-cleaning.

12.5 CAPILLARY FORCES

Capillary forces are ubiquitous not only in science and engineering, but also in everyday life. Capillary forces are intimately related to surface thermodynamics, and in fact they provide a simple means to measure surface tension.

Figure 12.5 shows a schematic representation of a capillary tube above a liquid phase α. The surface BB′, or **meniscus**, can be regarded as a segment of the surface of a sphere with origin O. The angle between the radius of the sphere OB and the horizontal line OX is the **contact angle**, θ, introduced earlier. The radius of curvature of BB′ is given by $r/\cos\theta$.

The pressure at position A in the capillary is denoted by P^0. The pressure at the BB′ interface in the capillary is denoted by P^α in phase α (below the meniscus) and P^β in phase β (above the meniscus). The pressures of both phases, α and β, are the same at A, and are equal to P^0. If h_m denotes the height of the liquid in the capillary (i.e. the distance AB), and ρ^α and ρ^β denote the densities of phases α and β, then the pressures are just hydrostatic:

$$P^\alpha = P^0 - \rho^\alpha |\vec{g}| h_\mathrm{m},$$
$$P^\beta = P^0 - \rho^\beta |\vec{g}| h_\mathrm{m}, \qquad (12.29)$$

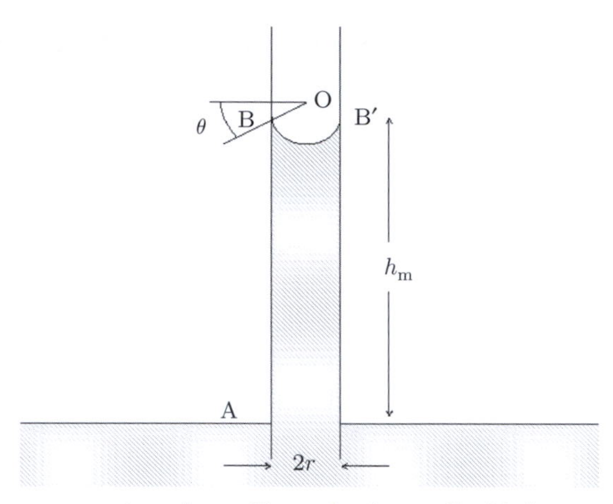

Figure 12.5 A schematic representation of a capillary tube above a liquid phase.

where $|\vec{g}|$ is the gravitational constant. Equation (12.16) applied to the system in Figure 12.5 can be written as

$$P^\beta - P^\alpha = 2\frac{\cos\theta}{r}\gamma. \tag{12.30}$$

Equations (12.29) and (12.30) can be rearranged to give

$$\gamma = \frac{1}{2}\frac{r}{\cos\theta}(\rho^\alpha - \rho^\beta)|\vec{g}|h_\mathrm{m} \tag{12.31}$$

or also

$$h_\mathrm{m} = \frac{2\gamma\cos\theta}{r(\rho^\alpha - \rho^\beta)|\vec{g}|}. \tag{12.32}$$

Equation (12.31) provides a means to calculate the surface tension from knowledge of density, contact angle, and measurement of the height of a liquid column in a capillary tube.

Capillary forces are important in a variety of applications, ranging from the study of adsorption in porous materials to pattern collapse in the fabrication of electronic devices. The following examples illustrate some of these applications.

EXAMPLE 12.5.1 Water in trees rises from the roots to the branches through small capillaries. Such capillaries, known as the xylem conduit, have diameters that range from a few microns to approximately 200 μm. How high can the water travel in such capillaries purely owing to capillary forces?

Solution. Water will rise more in smaller capillaries; to calculate an upper limit we use a capillary radius of 10 μm. The surface tension of water at 25 °C is $\gamma = 0.072$ N/m. Assuming that water wets the capillaries (i.e. $\theta = 0$), we have

$$h_\mathrm{m} = \frac{2(0.072\ \mathrm{N/m})}{(10\times10^{-6}\ \mathrm{m})(9.81\ \mathrm{m/s^2})(997\ \mathrm{kg/m^3})} = 1.45\ \mathrm{m}. \tag{12.33}$$

For an extensive discussion of water transport in trees, including some of the other factors (beyond capillary forces) that cause water to rise in plants, please see [84, 149]. A particularly important point of the calculation shown above is that, in order to rise through a capillary, water must "like" the inner surface of that capillary; that information is embodied in the contact angle, which we assumed to be zero in this case. This analysis also demonstrates that capillary forces alone cannot explain how water rises to the top of very tall trees. □

A particularly intriguing prediction of Eqn. (12.32) is that the rise of the liquid in the capillary depends only on the curvature of the meniscus through r and the contact angle θ. It is independent of the actual shape of the capillary. The capillary could consist, for example, of a narrow section, followed by a wide bulge, and then a narrow section again. The rise of the liquid in such a system would be exactly the same as in the straight capillary of Figure 12.5, provided that the meniscus at the top of the liquid column had the same dimensions. That is indeed what is observed in experiments; the liquid is of course unable to rise on its own through the wider regions, but if one finds the means to pull it up (e.g. through suction), it then remains at the height predicted by Eqn. (12.32). One could then conceive of a situation in which a very narrow pore (e.g. with a diameter on the order of 100 nm, in which case the rise in the capillary would be approximately 150 m), could be used to sustain a rather tall column of water comprised of wider pores. The only limit to the height would then be given by the tensile strength of water, which is approximately 50 atm.

EXAMPLE 12.5.2 Capillary forces can be used to "self-assemble" particles on a surface (see, for example, [19] and [86]). Consider a suspension of small particles in a liquid. As the liquid evaporates, it leaves behind a thin layer of fluid between any nearby particles (see Figure 12.6). To understand the underlying principles behind such self-assembly, one must derive an expression for the liquid-film-induced (or capillary) force between the particles.

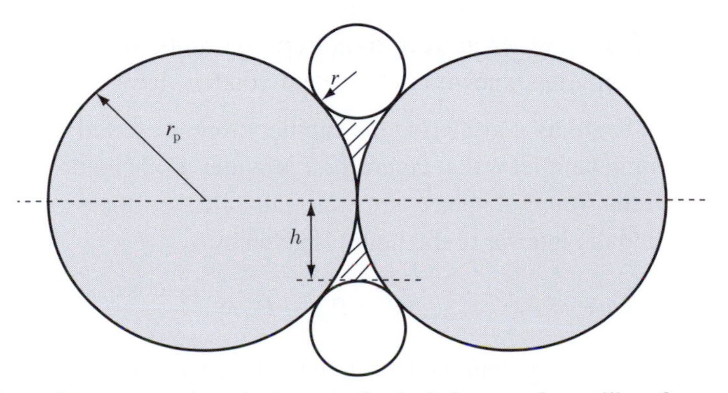

Figure 12.6 A schematic representation of spheres under the influence of a capillary force.

Solution. We assume that the particles are spherical and have uniform diameter, r_p. For simplicity, we assume that the liquid "wets" the surface of the particles, leading to a contact angle of 0. If the particles are in contact with each other, as illustrated in Figure 12.6, the radius of curvature of the liquid surface is given by r, where we have assumed that the particles are sufficiently large ($r_p \gg h \gg r$). The pressure is lower in the liquid than in the vapor by an amount $\Delta P = \gamma/r$. That pressure difference is exerted over an area $A = \pi h^2$, thereby leading to a force of attraction between the particles of magnitude $F = A\,\Delta P = \pi h^2 \gamma/r$. To relate the height of the liquid film h to the particle radius r_p we can write

$$(r_p + r)^2 = (h + r)^2 + r_p^2, \tag{12.34}$$

or, after some algebra,

$$2r_p r = h(h + 2r) \approx h^2, \tag{12.35}$$

where we have again assumed that $r_p \gg h \gg r$. We therefore have $h \approx \sqrt{2rr_p}$, and the attractive force between the particles is approximately given by

$$F \cong 2\pi \gamma r_p. \tag{12.36}$$

□

EXAMPLE 12.5.3 Electronic micro- and nano-circuits are manufactured through a lithographic process. One aspect of this process requires the formation of ultra-small polymer structures on a silicon wafer. The purpose of these polymer structures is to protect some regions of the wafer (those covered by the polymer) from subsequent etching. The speed and performance of a computer chip is largely dictated by the size of the circuit's elements; the smaller they are, the faster the device. The semiconductor industry is currently seeking to fabricate circuits having sub-50-nm dimensions. At these small length scales, the capillary forces exerted by simple solvents (e.g. water) on the "walls" of small polymeric structures can be high enough to cause these structures to fail and collapse. In this example, we examine the nature of the "collapse" problem, and we provide estimates of the maximum aspect ratio that poly(methylmethacrylate) (PMMA) polymer walls can have before the capillary forces induced by rinsing solvents (water) cause them to fail.

Solution. We begin by considering the capillary forces exerted by a liquid (water in this case) on two confining parallel walls. Figure 12.7 provides a schematic representation of the system. According to the Young–Laplace equation, Eqn. (12.16), the pressure difference between the atmosphere and the interior of the liquid is given by

$$\Delta P = P_{out} - P_{in} = \frac{2\gamma \cos\theta}{S}, \tag{12.37}$$

where S denotes the separation between the polymer walls and θ is the contact angle of water on PMMA.

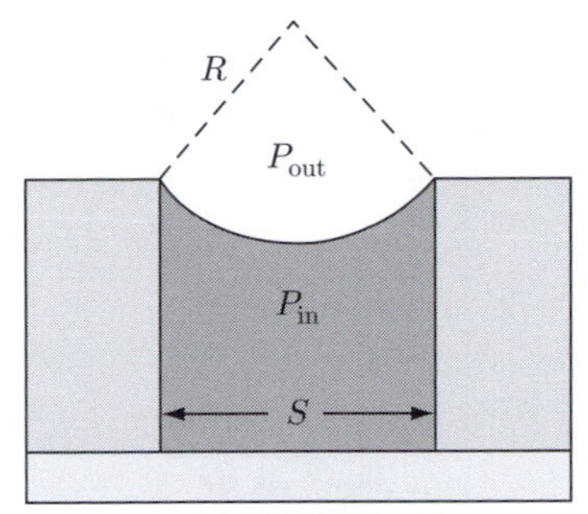

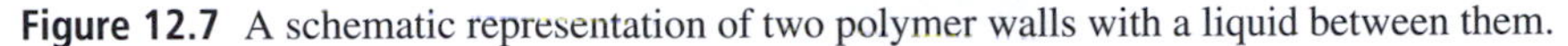

Figure 12.7 A schematic representation of two polymer walls with a liquid between them.

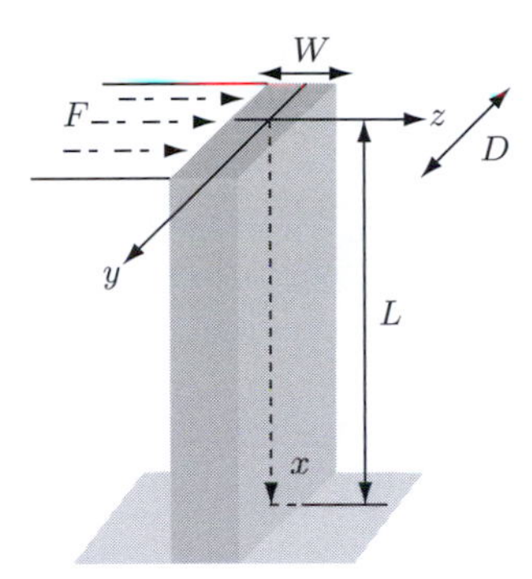

Figure 12.8 A schematic representation of a polymer wall subject to a deflecting force applied at the top.

That pressure difference gives rise to a force that pushes the walls onto each other. In order to avoid collapse, that force must be balanced by the rigidity or stiffness of the material. The following expression from elementary mechanics can be used to determine the deflection δ of a parallel plate in response to a deflecting force $\vec{F}$ applied at the top of the plate (see Figure 12.8):

$$\delta = \frac{|\vec{F}|L^3}{8EI} = \frac{3|\vec{F}|A_r^3}{2ED}, \tag{12.38}$$

where L is height of the plate, E is Young's modulus of the material (PMMA), I is given by $I = DW^3/12$, where W is the width of the polymer plate, D is the length, and A_r is the aspect ratio L/W.

For simplicity, we assume that water fills only one side of the trenches. For PMMA at 298 K, $E = 3\,\text{GPa}$, and the contact angle for water on PMMA is $\theta = 5°$ [146]. Measuring the mechanical properties of small nano-scale polymeric structures is notoriously challenging. These experiments must be conducted in a statistical manner, since minute differences between seemingly identical structures and small defects can give rise to large uncertainties. Figure 12.9 shows the percentage of polymer structures that have collapsed as a function of aspect ratio for PMMA lines of width $l = 200$ nm separated by a distance $d = 200$ nm. That figure allows us to identify a critical aspect ratio for collapse (CARC), beyond which the stiffness of the material is no longer sufficient to sustain the deflecting (capillary) forces brought about by the liquid. The results in Figure 12.9 and the electron micrographs in Figure 12.10 show that beyond a "critical" aspect ratio of 4 the polymer walls start to collapse. The sudden increase

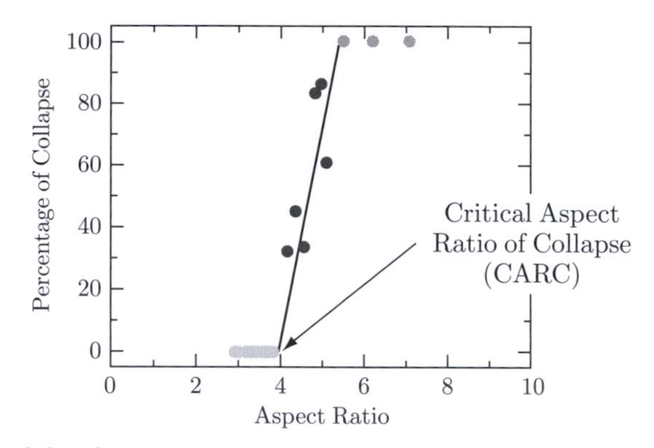

Figure 12.9 Experimental data for the critical aspect ratio for collapse (CARC) for PMMA polymer walls of line width $l = 200$ nm separated by a distance $d = 200$ nm.

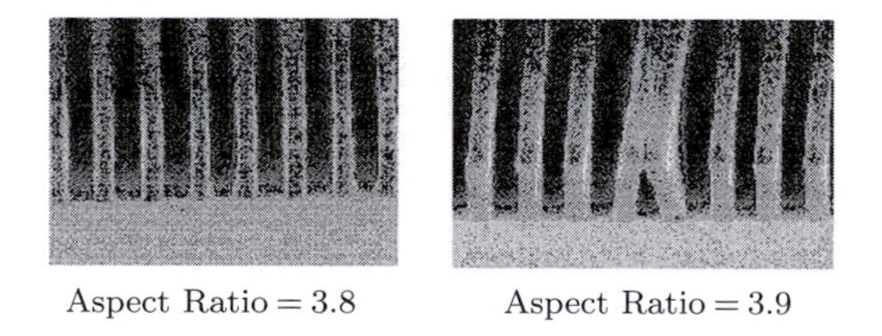

Figure 12.10 Experimental electron micrographs of PMMA polymer walls of width $l = 200$ nm separated by a distance $d = 200$ nm right below and right after the CARC.

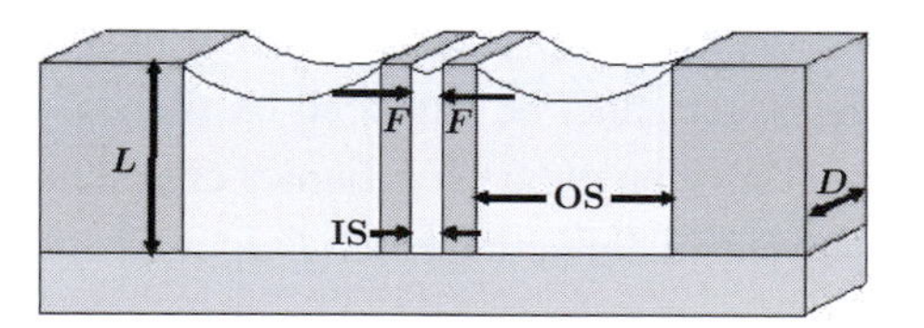

Figure 12.11 A schematic representation of a capillary-force device conceived to apply specific deflecting forces to polymer walls.

in the percentage collapse at about $A_r = 4$ is explained by the third-power dependence of the deflection on the aspect ratio in Eqn. (12.38).

Equations (12.30) and (12.38) provide the basis for the derivation of a parametric equation that permits calculation of the CARC. The derivation of that equation includes elements from mechanics that are beyond the scope of this text, and can be found in the literature [146].

The results in Figure 12.9 provide the basis for the creation of an ingenious, capillary-force-based test bed for measurement of the mechanical properties of polymeric systems in nano-scale geometries. Consider the geometries shown in Figure 12.11. Two polymer walls separated by an inner spacing IS are surrounded by two liquid pools of width OS (for outer spacing). The force exerted on the polymer walls by the liquid to their sides is given by

$$F = \Delta P \times \text{area} = 2\gamma \cos\theta \left(\frac{1}{\text{IS}} - \frac{1}{\text{OS}} \right) LD, \tag{12.39}$$

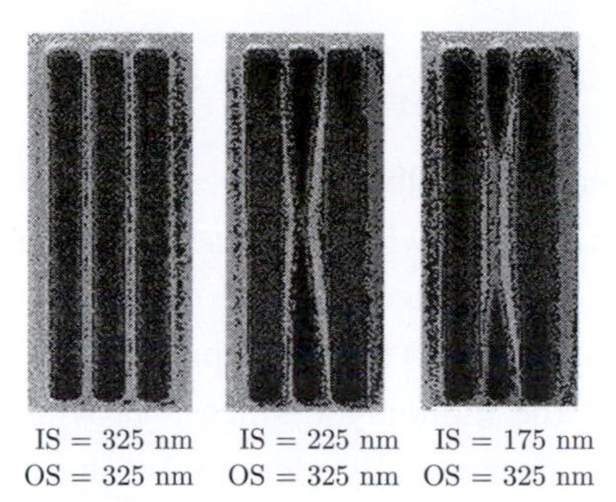

Figure 12.12 Experimental electron micrographs (top-down view) of PMMA capillary-force devices, having polymer walls of width $l = 85$ nm, separated by different inner spacings IS and constant outer spacings OS = 325 nm. The aspect ratio is 5.5. As the capillary force increases, so does the extent of collapse of the inner polymer walls.

where, as before, L and D denote the height and the length of the polymer walls. By controlling the dimensions of the capillary-force device, it is possible to tune the force that is applied to a material. When that force overcomes the stiffness of the polymer, the material collapses, as shown in Figure 12.12. In such a device, the CARC is given by substituting the force in Eqn. (12.39) into Eqn. (12.38). □

12.6 SOLID–GAS INTERFACES: ADSORPTION

In Sections 3.7.3 and 6.4 we discussed the adsorption of a gas onto a surface. The Langmuir isotherm corresponds to the case in which a monolayer of vapor-phase molecules adsorbs onto a surface. The treatment of solid–gas interfaces is slightly different from that of liquid–gas interfaces in that the area A of the solid is difficult to change. Terms of the form $\gamma\, dA$, as introduced in Eqn. (12.1), are of limited use, and common descriptions of adsorption do not invoke a surface tension.

As noted in previous chapters, solid–gas interfaces constitute an important class of systems, particularly in the fields of gas separation and catalysis. Is is therefore of interest to go beyond the simple Langmuir adsorption isotherm and consider the case in which multiple layers of gas molecules are adsorbed on the solid. The so-called "BET" isotherm, which was derived by Brunauer, Emmett, and Teller [16], corresponds to such a case. The derivation of the BET isotherm is left to the exercises (see Exercises 6.6.C and 12.6.A). The final result is

$$\frac{V}{V_{\text{mon}}} = c\frac{P/P^{\text{vap}}}{(1 - P/P^{\text{vap}})[1 - (1 - c)P/P^{\text{vap}}]}, \tag{12.40}$$

where c is a constant related to the free-energy change of adsorption, and V and V_{mon} denote the volume of gas adsorbed and the volume of gas required to form a monolayer, respectively. The BET expression is often used to determine the surface area of porous materials. To that end, it is rewritten in the form

$$\frac{x}{(1-x)V} = \frac{1}{cV_{\text{mon}}} + \frac{(c-1)x}{cV_{\text{mon}}}, \qquad x = \frac{P}{P^{\text{vap}}}. \tag{12.41}$$

This is the Lineweaver–Burke plot mentioned in Section 10.8. At sufficiently low values of x, most smooth, uniform surfaces will exhibit a linear dependence of $x/(1-x)V$ on x. The values of c and V_{mon} can then be extracted from the intercept and the slope of a plot of $x/(1-x)V$ vs. x. One can see from Eqn. (12.41) that perhaps the most important prediction of a BET analysis is V_{mon}, the volume of the monolayer, since it provides a direct measure of surface area. In practice, surface areas are generally measured using nitrogen near the boiling point, since it provides a large value of c and one can estimate area from a single data point (provided that x is sufficiently small, in the range 0.1–0.3).

It is important to note that, for porous materials, the shape of the adsorption isotherms may deviate considerably from linear behavior, particularly at large x. Adsorption isotherms may even exhibit some hysteresis. This behavior is in part related to capillary condensation; near saturation, the adsorbate begins to condense in the small pores of the material. For a discussion of the BET isotherm and its applications, readers are referred to [3].

12.7 THE TEMPERATURE DEPENDENCE OF SURFACE TENSION

In numerous applications, it is important to manipulate surface tension in such a way as to stabilize or destabilize an interface. Surface tension is particularly sensitive to temperature; when the temperature of liquid nitrogen is raised by 20 K, from 70 to 90 K, the surface tension drops by almost a factor of two, from $\gamma(70\,\text{K}) = 10.5$ mN/m to $\gamma(90\,\text{K}) = 6.2$ mN/m. It is therefore useful to derive formal expressions for the temperature dependence of γ, which can then be used to provide an additional lever for control of interface stability.

Our starting point is the differential expression for the surface tension obtained from the Gibbs–Duhem equation, Eqn. (12.10), which, for a pure component, reads

$$d\gamma = -\frac{1}{A}(S^{\sigma}\,dT - V^{\sigma}\,dP + N^{\sigma}\,d\mu). \tag{12.42}$$

For convenience, we use an asterisk to denote any specific property per unit area, i.e. $S^{\sigma,*} := S^{\sigma}/A$, except $N_i^{\sigma,*}$, for which we use the common notation, $\Gamma_i := N_i/A$. The interfacial concentration is also referred to as the surface excess. We then write

$$d\gamma = -(S^{\sigma,*}\,dT - V^{\sigma,*}\,dP + \Gamma\,d\mu). \tag{12.43}$$

At equilibrium between a liquid phase and a vapor phase, we have

$$d\mu = -s^{\text{l}}\,dT + v^{\text{l}}\,dP = -s^{\text{v}}\,dT + v^{\text{v}}\,dP, \tag{12.44}$$

where superscripts l and v refer to the liquid and vapor phases, respectively. Equations (12.43) and (12.44) can be rearranged to give

$$-\frac{d\gamma}{dT} = (S^{\sigma,*} - \Gamma s^{\text{l}}) - (V^{\sigma,*} - \Gamma v^{\text{l}})\frac{s^{\text{v}} - s^{\text{l}}}{v^{\text{v}} - v^{\text{l}}}, \tag{12.45}$$

which is analogous to the Clapeyron equation. The chemical potentials, or molar Gibbs free energies, of the liquid and vapor bulk phases are given by

$$\mu^{l} = g^{l} = u^{l} - Ts^{l} + Pv^{l},$$
$$\mu^{v} = g^{v} = u^{v} - Ts^{v} + Pv^{v}. \tag{12.46}$$

For the surface layer, we have from (12.7)

$$\Gamma\mu = G^{\sigma,*} = U^{\sigma,*} - TS^{\sigma,*} + PV^{\sigma,*} - \gamma. \tag{12.47}$$

Equation (12.45) can therefore be rearranged by eliminating entropies in favor of internal energies to give

$$\gamma - T\frac{d\gamma}{dT} = \left(U^{\sigma,*} - \Gamma u^{l}\right) - \left(V^{\sigma,*} - \Gamma v^{l}\right)\frac{u^{v} - u^{l}}{v^{v} - v^{l}}. \tag{12.48}$$

Equation (12.48) can also be written in terms of the enthalpy using $\Delta_u^{vap} = \Delta h^{vap} - P\Delta v^{vap}$:

$$\gamma - T\frac{d\gamma}{dT} = \left(H^{\sigma,*} - \Gamma h^{l}\right) - \left(V^{\sigma,*} - \Gamma v^{l}\right)\frac{h^{v} - h^{l}}{v^{v} - v^{l}}. \tag{12.49}$$

For the particular case of a vapor phase well below the critical point, $V^{\sigma,*}/\Gamma$ is comparable to v^{l}, which is much smaller than $v^{v} - v^{l}$. Equations (12.45) and (12.48) further simplify to

$$-\frac{d\gamma}{dT} = S^{\sigma,*} - \Gamma s^{l},$$
$$\gamma - T\frac{d\gamma}{dT} = U^{\sigma,*} - \Gamma u^{l}. \tag{12.50}$$

The quantity $\left(S^{\sigma,*} - \Gamma s^{l}\right)$ represents the difference between the entropy per unit area in the surface layer and the entropy that the same surface layer would have if it exhibited the properties of the bulk liquid. The surface tension decreases with temperature. Experimental data for $\partial\gamma/\partial T$ for a wide array of liquids can be found in [74]. For water under ambient conditions, the temperature dependence is approximately $\partial\gamma/\partial T \approx -138 \times 10^{-3}$ mN/m · K. For ethanol under ambient conditions, $\partial\gamma/\partial T \approx -86 \times 10^{-3}$ mN/m · K.

Several empirical correlations have been proposed to estimate the effect of temperature on surface tension. The Eötvös correlation, which was proposed more than 100 years ago [37], for example, relates γ to the liquid–vapor critical temperature and the molar volume according to

$$\gamma = k\frac{T_c - T}{v^{1/3}}, \tag{12.51}$$

where k is a constant with a value of 2.1×10^{-4} mJ/K for many liquids. Several more recent correlations are given in the exercises at the end of the chapter.

12.8 INTERFACES IN MIXTURES

As one might expect, the composition of the interfacial region in a mixture is generally different from that of the bulk. Different components exhibit different tendencies to segregate to the interface and, as a result, the surface tension can exhibit a marked dependence on composition.

The design of liquid- or vapor-phase separation processes often requires that surface tension be determined as a function of composition; we here derive formal expressions to do so. For concreteness, we derive expressions for the temperature and composition dependence in binary mixtures; their generalization to multicomponent systems is discussed in the exercises.

12.8.1 Vapor–liquid interfaces

We first consider the case of an interface between a liquid phase and a vapor phase. For a two-component system, the differential of the surface tension is given by the Gibbs–Duhem equation, Eqn. (12.10):

$$-d\gamma = S^{\sigma,*}\,dT - V^{\sigma,*}\,dP + \Gamma_1\,d\mu_1 + \Gamma_2\,d\mu_2. \tag{12.52}$$

Equation (12.52) can be simplified by assuming that the interfacial region is extremely thin, so that the term $PV^{\sigma,*}$ becomes negligible. If we assume that the volume of the interface $V^{\sigma,*}$ is small, at constant temperature that equation becomes

$$d\gamma = -\Gamma_1\,d\mu_1 - \Gamma_2\,d\mu_2,\ \text{isothermal}. \tag{12.53}$$

This equation is often referred to as the Gibbs adsorption isotherm.

Also neglecting volume, the differential of the chemical potential for components 1 and 2 in the liquid phase can be written as

$$\begin{aligned} d\mu_1 &= -\bar{s}_1\,dT + \bar{v}_1\,dP + \left(\frac{\partial \mu_1}{\partial x_2}\right)_{T,P} dx_2, \\ d\mu_2 &= -\bar{s}_2\,dT + \bar{v}_2\,dP + \left(\frac{\partial \mu_2}{\partial x_2}\right)_{T,P} dx_2. \end{aligned} \tag{12.54}$$

Here we used the fact that μ_1 can be written as a function of T, P, and x_2, plus the chain rule, to find the first line. Following our treatment for the temperature dependence of surface tension, we assume that $P\bar{v}_i$ for a condensed phase is much smaller than RT. We also assume that terms of the type $PV^{\sigma,*}$ are negligible. The two expressions above then simplify to

$$\begin{aligned} d\mu_1 &= -\bar{s}_1\,dT + \left(\frac{\partial \mu_1}{\partial x_2}\right)_{T,P} dx_2, \\ d\mu_2 &= -\bar{s}_2\,dT + \left(\frac{\partial \mu_2}{\partial x_2}\right)_{T,P} dx_2. \end{aligned} \tag{12.55}$$

The surface tension can therefore be written in terms of the chemical potentials as

$$\begin{aligned} d\gamma &= -(S^{\sigma,*} - \Gamma_1\bar{s}_1 - \Gamma_2\bar{s}_2)dT - \left[\Gamma_1\left(\frac{\partial \mu_1}{\partial x_2}\right)_{T,P} + \Gamma_2\left(\frac{\partial \mu_2}{\partial x_2}\right)_{T,P}\right] dx_2 \\ &= -(S^{\sigma,*} - \Gamma_1\bar{s}_1 - \Gamma_2\bar{s}_2)dT - \left(\Gamma_2 - \Gamma_1\frac{x_2}{1-x_2}\right)\left(\frac{\partial \mu_2}{\partial x_2}\right)_{T,P} dx_2. \end{aligned} \tag{12.56}$$

To derive Eqn. (12.56), we used the Gibbs–Duhem equation, namely

$$(1-x_2)\left(\frac{\partial \mu_1}{\partial x_2}\right)_{T,P} + x_2\left(\frac{\partial \mu_2}{\partial x_2}\right)_{T,P} = 0. \tag{12.57}$$

Equation (12.56) indicates that, at constant temperature and pressure, the surface tension depends on composition according to

$$\left(\frac{\partial \gamma}{\partial x_2}\right)_{T,P} = -\left(\Gamma_1\left(\frac{\partial \mu_1}{\partial x_2}\right)_{T,P} + \Gamma_2\left(\frac{\partial \mu_2}{\partial x_2}\right)_{T,P}\right)$$
$$= -\left(\Gamma_2 - \Gamma_1\frac{x_2}{1-x_2}\right)\left(\frac{\partial \mu_2}{\partial x_2}\right)_{T,P}. \qquad (12.58)$$

We can introduce fugacity in place of chemical potentials by using Eqn. (8.96) to arrive at

$$\left(\frac{\partial \gamma}{\partial x_2}\right)_{T,P} = -RT\left(\Gamma_2 - \Gamma_1\frac{x_2}{1-x_2}\right)\left(\frac{\partial \log f_2}{\partial x_2}\right)_{T,P}. \qquad (12.59)$$

For ideal gases, the fugacity appearing in Eqn. (12.59) can be replaced by a partial pressure P_i to yield

$$\left(\frac{\partial \gamma}{\partial x_2}\right)_{T,P} = -\frac{RT}{1-x_2}((1-x_2)\Gamma_2 - x_2\Gamma_1)\left(\frac{\partial \log P_2}{\partial x_2}\right)_{T,P} \qquad (12.60)$$
$$= \frac{RT}{x_2}((1-x_2)\Gamma_2 - x_2\Gamma_1)\left(\frac{\partial \log P_1}{\partial x_2}\right)_{T,P}. \qquad (12.61)$$

By measuring γ and f_2 or f_1 over a range of compositions, one can estimate the quantity $I = (1-x_2)\Gamma_2 - x_2\Gamma_1$. In the limit where x_2 is small, $\Gamma_2/I \to 1$, and the quantity I provides a measure of the adsorption Γ_2 of species 2 at the interface. Similarly, when $x_2 \to 1$, $-\Gamma_1/I \to 1$ and I is a measure of the depletion $-\Gamma_1$ of species 1 at the interface. As illustrated in the following examples, however, to estimate Γ_1 or Γ_2 individually, particularly for intermediate values of composition, one must utilize additional information or adopt somewhat arbitrary conventions.

EXAMPLE 12.8.1 The surface tension of liquids is often manipulated through the use of amphiphilic molecules or surfactants. Amphiphilic molecules such as sodium dodecyl sulfate (SDS), whose chemical formula is $NaSO_4(CH_2)_{11}CH_3$, exhibit a hydrophilic head-group and a hydrophobic tail that allows them to place themselves at the interface between water and air or water and a hydrocarbon. Experimental data for the surface tension of aqueous solutions of SDS are given in Table 12.6 (see the exercises). Addition of 0.5 mM SDS to water at 25 °C decreases the surface tension by approximately 3 mN/m. Estimate the surface concentration of SDS molecules at the air–water interface.

Solution. Our starting point is Eqn. (12.59). The molecular weight of SDS is 288.4 g/mol. A concentration of 0.5 mM corresponds to an extremely small mole fraction (i.e. $x_2 \to 0$). For simplicity, we may assume Lewis mixing. With these assumptions, Eqn. (12.59) simplifies to

$$\left(\frac{\partial \gamma}{\partial x_2}\right)_{T,P} \approx -RT\left(\frac{\Gamma_2}{x_2}\right). \qquad (12.62)$$

The surface concentration of SDS molecules is approximately

$$\Gamma_2 \approx -\frac{x_2}{RT}\left(\frac{\partial \gamma}{\partial x_2}\right)_{T,P} \approx \frac{1}{8.31 \times 298\,\text{J/mol}^{-1}} 3\ \text{mJ/m}^2 = 1.21 \times 10^{-6}\ \text{mol/m}^2. \tag{12.63}$$

Assuming that the entire surface is covered by SDS molecules, this result corresponds to a surface area per molecule of 1.37 nm^2. □

For a Lewis mixture in the bulk phase, Eqn. (12.58) simplifies to

$$\left(\frac{\partial \gamma}{\partial x_2}\right)_T = -RT\left(\frac{\Gamma_2}{x_2} - \frac{\Gamma_1}{x_1}\right), \quad \text{Lewis mixture, bulk.} \tag{12.64}$$

Or, for a bulk mixture whose excess molar Gibbs free energy is given by a two-suffix Margules equation, the composition dependence of the surface tension is given by

$$\left(\frac{\partial \gamma}{\partial x_2}\right)_T = -\left(\frac{\Gamma_2}{x_2} - \frac{\Gamma_1}{x_1}\right)(RT - 2A_{12}x_1x_2), \quad \text{two-suffix Margules, bulk.} \tag{12.65}$$

EXAMPLE 12.8.2 Table 12.3 gives experimental data for the surface tension of water–ethanol liquid mixtures at 25 °C as a function of composition. Use these data to estimate the concentrations of water and alcohol at the liquid–vapor interface.

Solution. The vapor phase can be assumed to be ideal. From Eqn. (12.58), we can write that the partial pressure of ethanol (2) is given by

$$\left(\frac{\partial \gamma}{\partial \ln P_2}\right)_{T,\,P} = P_2\left(\frac{\partial \gamma}{\partial P_2}\right)_{T,\,P} = -RT\left(\Gamma_2 - \frac{x_2\Gamma_1}{1 - x_2}\right), \tag{12.66}$$

where $P_i = y_iP$ denotes the partial pressure of component i. The quantity $I = x_1\Gamma_2 - x_2\Gamma_1$ is calculated in the fifth column from an estimate of the slope of a curve of γ as a function of $\log P_2$. The results reported in Table 12.3 indicate that, for small concentrations of ethanol (e.g. $x_2 = 0.1$), one molecule of ethanol occupies approximately 20 Å^2 at the interface. To arrive at an estimate of the area occupied by individual components at the interface, we will assume that the interface consists of a monolayer of alcohol and water molecules. While this is a rather severe assumption, our only purpose here is to come up with a simple measure of the composition of the surface layer. We can further assume that

$$A_1\Gamma_1 + A_2\Gamma_2 = 1, \tag{12.67}$$

where A_i represents the effective or partial area occupied by molecule i at the interface; Eqn. (12.67) basically states that the area occupied by each molecule at the interface is constant, and does not change with composition. In order to determine Γ_i, we need to assume reasonable values of A_i. If we use, as an example,

$$\begin{aligned} A_1 &= 0.04 \times 10^{10}\,\text{cm}^2/\text{mol}, \\ A_2 &= 0.12 \times 10^{10}\,\text{cm}^2/\text{mol}, \end{aligned} \tag{12.68}$$

Table 12.3 Surface tension and partial pressures of water (1)–ethanol (2) mixtures at 25 °C (from Guggenheim [51]).

x_2	γ (mN/m)	P_1 (mm Hg)	P_2 (mm Hg)	$-\partial\gamma/\partial\log P_2$ (mN/m)	I (10^{10} mol/cm^2)
0	72.1	23.7	0	0	0
0.1	36.6	21.7	17.8	19.5	7.1
0.2	29.7	20.4	26.8	14.4	4.7
0.3	27.6	19.4	31.2	12.6	3.6
0.4	26.3	18.3	34.2	11.4	2.8
0.5	25.4	17.3	36.9	10.5	2.1
0.6	24.6	15.8	40.1	9.4	1.5
0.7	23.8	13.3	43.9	8.3	1.0
0.8	23.2	10.0	48.3	7.1	0.6
0.9	22.6	5.5	53.3	5.9	0.2
1.0	22.0	0	59.0	4.7	0

Table 12.4 Surface-layer adsorption and composition of water (1)–ethanol (2) mixtures at 25 °C.

x_2	Γ_1 (mol/cm^2)	Γ_2 (mol/cm^2)	$\Gamma_2/(\Gamma_1+\Gamma_2)$
0	25.00	0.00	0.00
0.1	1.00	8.00	0.89
0.2	4.21	6.93	0.62
0.3	4.19	6.94	0.62
0.4	3.67	7.11	0.66
0.5	3.10	7.30	0.70
0.6	2.50	7.50	0.75
0.7	1.88	7.71	0.80
0.8	1.23	7.92	0.87
0.9	0.68	8.11	0.92
1.0	0.00	8.33	1.00

or also

$$A_1/\tilde{N}_A = 7\,\text{Å}^2/\text{molecule},$$
$$A_2/\tilde{N}_A = 20\,\text{Å}^2/\text{molecule}, \qquad (12.69)$$

we can calculate the sought-after interfacial concentrations from the data presented in Table 12.3. The results for Γ_1 and Γ_2 determined from Eqn. (12.67) and from the values of I reported in Table 12.3 are given in Table 12.4. The quantity $\Gamma_2/(\Gamma_1 + \Gamma_2)$, also given in the table, can be thought of as the mole fraction of ethanol in the surface layer. □

In the previous example, experimental data for the surface tension of binary mixtures of water and ethanol were measured in the laboratory. From those data we arrived at estimates of the composition of the surface layer, which, in general, turned out to be significantly different from that of the bulk liquid. It is generally desirable to develop methods for prediction of values of the surface tension of mixtures from knowledge of those corresponding to the pure components. Such methods must necessarily capture the fact that the surface layer's composition is influenced by the pure component's surface tension. Species that exhibit a small surface tension are more likely to be concentrated in the surface layer; a mixture's surface tension is therefore generally lower than that corresponding to a simple mole-fraction average of the pure-component surface tensions. Several correlations have been proposed to estimate the surface tension of multicomponent mixtures. Several of these are described in the exercises at the end of this chapter.

12.8.2 Monolayer formation on liquid surfaces

By their very nature, amphiphilic molecules exhibit a pronounced tendency to segregate towards interfaces. In the particular case of an interface between a liquid and air, that tendency will cause amphiphilic molecules to form a one-molecule-thick layer or film at the interface. The characteristics of "monolayers" at the air–water interface are important in a wide range of applications, and for that reason we include here a brief discussion of their general features.

The properties of monolayers are generally studied in a so-called Langmuir trough. Figure 12.13 provides a schematic representation of a Langmuir trough, in which insoluble amphiphilic molecules are driven to the interface, thereby forming a monolayer. The area of that interface (and the density of amphiphilic molecules at the interface) can be controlled through the lateral displacement of a barrier. In a sense, Langmuir-trough experiments constitute the two-dimensional analog of "piston" experiments introduced earlier in the text to quantify the

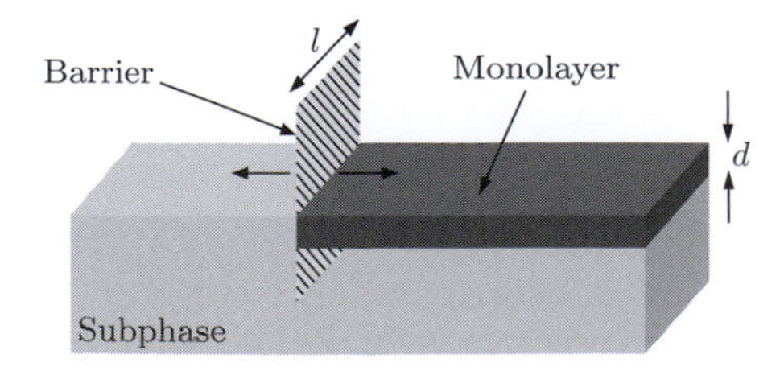

Figure 12.13 A schematic representation of a Langmuir trough.

behavior of bulk systems, including solids, liquids, and gases. The monolayer will give rise to a lateral or "film" pressure (it wants to expand as much as possible in order to decrease the free energy). That lateral pressure is defined by

$$\pi \equiv \gamma_0 - \gamma, \tag{12.70}$$

where γ_0 is the interfacial tension of the water below the monolayer (the so-called "subphase" in the parlance of Langmuir-trough measurements) and γ is the interfacial tension of the subphase when covered by amphiphilic molecules.

The film pressure π can be measured by introducing a plate in a direction perpendicular to the interface, as illustrated in Figure 12.13. The liquid will partially climb along the walls of the plate, thereby creating a force that pulls it down. That force is proportional to the interfacial tension, and it can be measured using a so-called "Wilhelmy-plate balance."

Just as we measured P–V isotherms for bulk fluids, we can measure π–A isotherms for monolayers. As we shall see below, such isotherms can exhibit a variety of behaviors, including gas-like behavior, liquid-like behavior, and coexistence between a gas and a liquid. More specifically, at low surfactant concentrations it is found that the surface tension of the system depends linearly on the concentration c, i.e.

$$\gamma = \gamma_0 - bc, \tag{12.71}$$

where b is a constant that depends on the nature of the solvent, that of the surfactant, and temperature. Equation (12.71) can be inserted into the Gibbs adsorption isotherm to yield

$$\Gamma = -\frac{c}{RT}\frac{d\gamma}{dc}. \tag{12.72}$$

The lateral pressure is given by $\pi = bc$, leading to the following expression:

$$\pi = \Gamma RT. \tag{12.73}$$

We use the symbol σ_i to denote the area per molecule; σ_i is the two-dimensional analog of the molar volume, v_i, and it is given by $\sigma_i = 1/\Gamma_i$. For the pure system considered we need consider only σ_A, the area per molecule of surfactant A. We get

$$\pi\sigma_A = RT, \tag{12.74}$$

which is reminiscent of the ideal-gas mechanical equation of state.

At higher concentrations of surfactant, one starts to observe deviations from ideal-gas behavior. In complete analogy to the behavior of gases or liquids in three dimensions, one can correct Eqn. (12.74) to incorporate the effect of the size of the molecules (or the finite compressibility of the monolayer) and the effect of interactions between surfactant molecules. A simple way of doing so is to introduce a van der Waals-like equation of state for the monolayer, of the form

$$\pi = \frac{k_B T}{\sigma_A - \sigma_{A0}} - \frac{a}{\sigma_A^2}, \tag{12.75}$$

where σ_{A0} is a substance-dependent characteristic constant that measures the area per molecule at high coverage and a is a second substance-dependent constant that characterizes the energy of interactions between molecules.

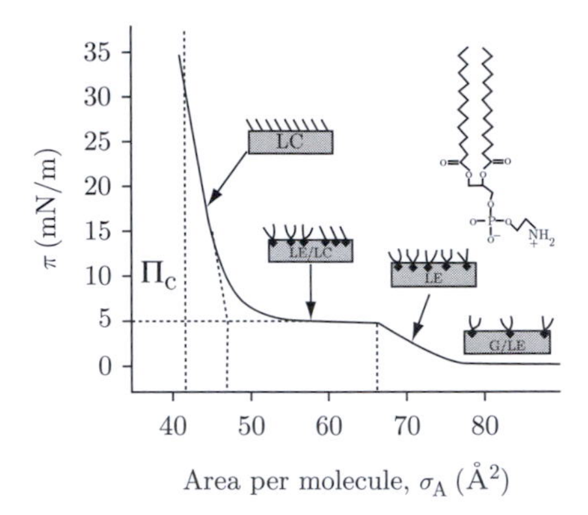

Figure 12.14 The pressure–area isotherm for 1,2-dimyristoyl-sn-glycero-phosphoethanolamine (DMPE) on a water surface at 20 °C. Reprinted from [49], with permission from Elsevier.

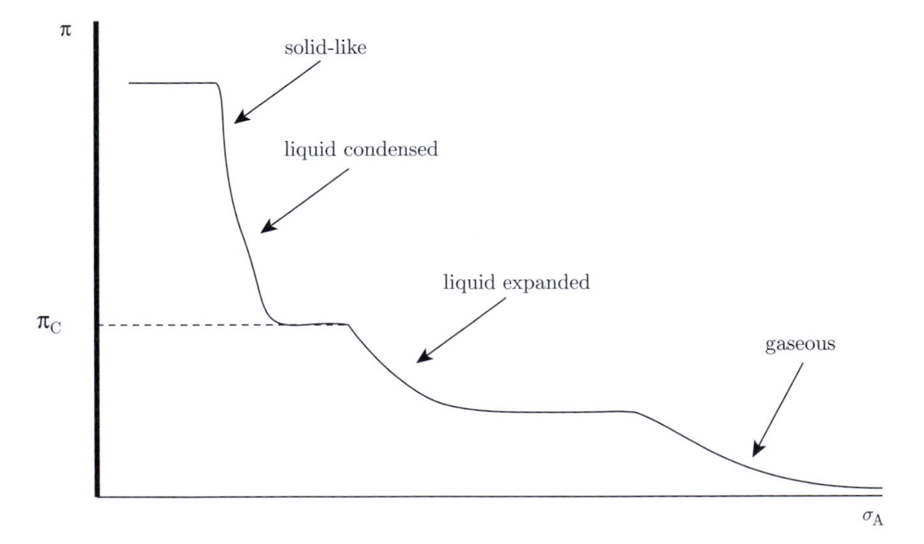

Figure 12.15 A schematic representation of a generic pressure–area isotherm for an amphiphilic-molecule monolayer at the air–water interface.

If we now go back to our discussion of phase behavior in three-dimensional fluids, it becomes apparent that the cubic functional form of Eqn. (12.75) can lead to multiple roots for a given pressure and temperature. In other words, it predicts the coexistence of high-density and low-density phases for a monolayer. Experiments confirm that prediction. Figure 12.14 shows an experimentally determined isotherm for 1,2-dimyristoyl-sn-glycero-phosphoethanolamine (DMPE) on a water surface at 20 °C. On that isotherm, one can distinguish different regimes that are reminiscent of those seen in three-dimensional fluids. Figure 12.15 provides a schematic representation of a generic, two-dimensional isotherm for an amphiphilic molecule at the air–water interface. At high surface coverage the area per molecule is small (e.g. ~40 Å^2 for DMPE), and the monolayer exhibits a solid-like behavior. As the concentration is decreased, the monolayer exhibits liquid-like behavior. At a critical film pressure π_c the film exhibits coexistence between what are known as a liquid-condensed phase and a liquid-expanded phase. For DMPE, $\pi_c = 5\,\text{mN/m}$ at $20\,^\circ\text{C}$. The area per molecule in the liquid-condensed phase is in the vicinity of $46\,\text{Å}^2$, and in the liquid-expanded phase it reaches $66\,\text{Å}^2$. At lower surface concentrations one finds coexistence between a liquid-expanded phase and a gaseous phase; and, as discussed earlier, at even lower concentrations the system exists in a gaseous phase, which is described by Eqn. (12.74).

Monolayer isotherms such as that shown in Figure 12.14 are used frequently to study the interactions of various molecules with phospholipids. Phospholipids are 1,2-di-esters of glycerol and fatty acids, and they are among the most abundant components of cell membranes. By understanding how proteins or pharmaceutical molecules interact with phospholipid monolayers, one can gain insights into the interactions of such molecules with the outer membranes of cells.

Summary

We have shown that the properties of interfaces are different from those of a bulk sample, be it a liquid or a vapor. In order to quantify such differences a new quantity was introduced, namely the so-called interfacial tension. One way to think about the interfacial tension is to view it as a "lateral pressure"; to characterize a sample that exhibits a large interface, we must consider not only the volume of the sample V, but also its interfacial area A. At constant temperature, the pressure P is the variable that controls fluctuations of the volume, and the interfacial tension γ is the variable that controls fluctuations of the area. The properties of interfaces become important when we work with thin films, droplets, or bubbles. As the thickness of a film becomes small or as the diameter of a bubble (or droplet) becomes small, the contribution of the interface to the system's thermodynamic properties can become considerable. That interface is responsible for the pressure differential that arises between the interior and the exterior of a bubble, the rise of a liquid in a capillary tube, and the instability of small liquid droplets in a vapor phase (think of mist), which, below a critical radius, simply disappear.

- The interfacial tension can be determined in a number of ways, such as from measurement of the rise of a liquid in a capillary tube or by measuring the weight of a drop of liquid as it detaches from a capillary.
- The thermodynamic analysis of a system that exhibits significant interfacial contributions is entirely analogous to that for a bulk sample, with the difference that we must now take into account the interfacial area A, in addition to the pressure P.
- The pressure inside a bubble or a droplet is different from that outside. The difference in pressure is proportional to the surface tension and is given by the Young–Laplace equation, Eqn. (12.16), namely

$$P^\alpha - P^\beta = \gamma\left(\frac{1}{r_1} + \frac{1}{r_2}\right).$$

- Another important concept that was introduced in this chapter is that of a contact angle, which determines to what extent a liquid likes a solid surface. When a liquid wets a surface the contact angle is small, i.e. $\theta \approx 0°$. When the liquid does not wet the surface, θ is large.
- When a surfactant is added to a liquid its concentration at the interface is greater than that in the bulk. The surface excess concentration can be determined from the change in surface tension with concentration.

Exercises

12.1.A Estimate the surface tension of n-hexane at 25 °C from the heat of vaporization. You will need to consider the physical picture of an interface introduced in this chapter, and estimate the energy per nearest-neighbor of an individual molecule in the liquid phase. You will also need to estimate the area occupied by an individual molecule at the surface. The interfacial tension of n-hexane at 25 °C is approximately 0.0179 N/m; how does your estimate compare with the experimental value?

12.1.B The surface tension of a liquid can be inferred from the force that a liquid exerts on a plate that is partly submerged, vertically, into the liquid. Show that, provided that the liquid "wets" the surface of the plate, a downward force of magnitude $2\gamma l$ is experienced by the plate (l is the length of the plate). This force provides the basis for the Wilhelmy-plate methods to measure surface tensions.

12.2.A Use thermodynamic stability to show that the surface tension γ is always positive.

12.3.A Consider an air bubble in water at 25 °C. What is the difference in pressure between the interior and exterior of the bubble when the diameter is 1 mm? What is that difference for a nano-bubble of diameter 40 nm?

12.3.B Consider a capillary tube of diameter 0.1 mm. A liquid is allowed to flow out of the capillary, leading to the formation of a drop at its outlet. Show that the critical radius r_c at which the drop will detach from the capillary tube is given by

$$r_c = \frac{mg}{2\pi\gamma}. \tag{12.76}$$

At room temperature, what is the size that a water droplet will have before it detaches from the capillary? Equation (12.76) provides the basis for a technique to measure the surface tension of liquids. Would there be an advantage to using smaller or larger capillary tubes? Please explain.

12.3.C Consider a bubble of air of diameter r in liquid water. Show that, assuming ideal-gas behavior, the vapor pressure inside the bubble ($P^{\text{bubble,sat}}$) is lower than in the bulk liquid by an amount

$$\log\left(\frac{P^{\text{bubble,sat}}}{P^{\text{b,sat}}}\right) = -\frac{2\gamma v}{rRT}, \tag{12.77}$$

where v is the molar volume of liquid water. Calculate the ratio of the vapor pressure inside and that outside the bubble, at 25 °C, for bubbles of diameters 10, 100, and 1000 nm. A liquid can sometimes be heated to a temperature above its boiling point. When that happens, small bubbles in which the vapor pressure is lower than that of the bulk are formed, leading to condensation of the vapor and collapse of the bubble. Equation (12.77) explains why the vapor pressure in the bubble is lower.

12.3.D Zuo and Stenby [150] proposed the following empirical correlation to estimate the surface tension of simple liquids and mixtures in terms of critical constants. A scaled surface tension is given by

$$\gamma_s = \log\left(1 + \frac{\gamma}{T_c^{1/3} P_c^{2/3}}\right), \tag{12.78}$$

where γ, T_c, and P_c are given in mN/m, K, and bar, respectively, and ω is the acentric factor of the fluid of interest. The scaled surface tension γ_s for a fluid of interest is related to that of two reference fluids, namely methane (1) and n-octane (2), according to

$$\gamma_s = \gamma_s^{(1)} + \frac{\omega - \omega^{(1)}}{\omega^{(2)} - \omega^{(1)}}(\gamma_s^{(2)} - \gamma_s^{(1)}). \tag{12.79}$$

For methane, $\gamma^{(1)}$ is given in the same units by

$$\gamma^{(1)} = 40.52(1 - T_r)^{1.287}; \tag{12.80}$$

and for n-octane, $\gamma^{(2)}$ is given by

$$\gamma^{(2)} = 52.095(1 - T_r)^{1.21548}. \tag{12.81}$$

1. Use the Zuo–Stenby correlation to estimate the surface tension of n-octane at 20 °C. The experimental value is $\gamma = 21.62$ mN/m.
2. Use the Zuo–Stenby correlation to estimate the surface tension of dodecane at 22 °C. The experimental value is $\gamma = 25.44$ mN/m.
3. Use the Zuo–Stenby correlation to estimate the surface tension of toluene at 20 °C. The experimental value is $\gamma = 28.52$ mN/m.
4. Use the Zuo–Stenby correlation to estimate the surface tension of phenol at 40 °C. The experimental value is $\gamma = 39.27$ mN/m.

Please discuss the accuracy of the correlation.

12.3.E Sastri and Rao [114] proposed the following correlation to estimate the surface tension of hydrogen-bonding liquids in terms of critical constants and the normal boiling temperature. In their approach, the surface tension is given by

$$\gamma = K P_c^x T_b^y T_c^z \left(\frac{1 - T_r}{1 - T_{b,r}}\right)^m, \tag{12.82}$$

where T_b is the boiling point, T_r is the reduced temperature, and $T_{b,r}$ is the reduced boiling point, all in K, γ, T_c, and P_c are given in mN/m, K, and bar, respectively. The constants appearing in Eqn. (12.82) are given in Table 12.5.

Table 12.5 Parameters appearing in the Sastri–Rao correlation for surface tension.

Substance	K	x	y	z	m
Alcohols	2.28	0.25	0.175	0	0.8
Acids	0.125	0.5	-1.5	1.85	11/9
All others	0.158	0.5	-1.5	1.85	11/9

1. Use the Sastri–Rao correlation to estimate the surface tension of n-octane at 20 °C. The experimental value is $\gamma = 21.62$ mN/m.
2. Use the Sastri–Rao correlation to estimate the surface tension of toluene at 20 °C. The experimental value is $\gamma = 28.52$ mN/m.
3. Use the Sastri–Rao correlation to estimate the surface tension of ethanol at 30 °C. The experimental value is $\gamma = 21.55$ mN/m.
4. Use the Sastri–Rao correlation to estimate the surface tension of acetic acid at 20 °C. The experimental value is $\gamma = 27.59$ mN/m.
5. Use the Sastri–Rao correlation to estimate the surface tension of phenol at 40 °C. The experimental value is $\gamma = 39.27$ mN/m.

Please discuss the accuracy of the correlation.

12.5.A Consider a porous material that is wetted by water. That material is placed in humid air (the level of humidity is 80%). At 25 °C and ambient pressure, what is the smallest size of the pores that can fill up with water? What happens when the level of humidity is higher (e.g. 90%)? Explain your results. This example provides the basis on which to conceive experiments for measurement of the porosity of solids. Please explain how such experiments might work, and discuss whether it would be advantageous to use a liquid or a gas with high or with low surface tension.

12.5.B A capillary of radius $r = 0.55$ mm is used to measure the surface tension of benzene. The density of benzene at 20 °C is $\rho = 0.88$ g/cm^3. The height of the benzene on the capillary is $h = 1.2$ cm. Determine the surface tension of benzene.

12.5.C What is the capillary rise of water in a tube of radius $r = 0.5$ cm at 20 °C?

12.6.A Use a kinetic argument to derive the BET isotherm for multilayer adsorption. *Hint:* let θ_i denote the surface area covered by i layers of surface molecules, where $i = 0, \ldots, n$ (n denotes the largest number of adsorbed surface layers, and 0 corresponds to the bare surface). At equilibrium, θ_0 must be constant, and therefore the rate of evaporation from the first layer must be equal to the rate of condensation onto the bare surface. If k_{-1} and k_1 denote the rates of evaporation and condensation, one can write $k_{-1}\theta_1 = k_1 P\theta_0$, where P is the pressure of the system. Similarly, at equilibrium, θ_1 must be constant, and $k_1 P\theta_0 + k_{-2}\theta_2 = k_2 P\theta_1 + k_{-1}\theta_1$. These two expressions can be combined to yield $k_{-2}\theta_2 = k_2 P\theta_1$. Similar arguments can be applied to successive layers to arrive at a recursive relation. You may use the fact that $\sum_{i=1}^{\infty} x^i = x/(1-x)$ and $\sum_{i=1}^{\infty} ix^i = x/(1-x)^2$.

12.7.A Show that the surface tension between a liquid and a vapor phase can be expressed as the difference between the Gibbs free energy of the surface layer and the Gibbs free energy which the same layer would have if it exhibited the properties of the bulk liquid, i.e.

$$\gamma = G^{\sigma,*} - \Gamma g^{l}. \tag{12.83}$$

12.8.1.A The surface tension of water at 25 °C increases at a rate of approximately $\partial\gamma/\partial C = 1.8 \times 10^{-3}$ N/m · mol upon addition of small amounts of sodium chloride. Estimate the surface tension of a 5-mM aqueous solution of sodium chloride.

12.8.1.B The surface tension of water at 25 °C decreases at a rate of approximately $\partial\gamma/\partial C = 40 \times 10^{-3}$ N/m · mol upon addition of small amounts of ethanol. Estimate the surface tension of a 5-mM aqueous solution of ethanol.

12.8.1.C Table 12.6 provides experimental data for the surface tension of aqueous solutions of SDS. Using these data, construct a plot of the surface concentration of SDS as a function of the surfactant concentration. You may assume that a molecule of SDS occupies approximately 1.4 nm^2. At an SDS concentration of approximately 10 mM, the surface tension undergoes an abrupt change. Please discuss the cause for this change. Tajima *et al.* [130] measured the excess concentration of SDS at the air–water interface by means of radio labeling. Their results are given in Table 12.7. Are your results consistent with the experimental data in that table? Please discuss your results.

12.8.1.D Hadden [53] proposed the following correlation for the surface tension γ_{m} of mixtures of simple liquids:

$$\gamma_{\mathrm{m}}^{r} = \sum_{i}^{n} x_i \gamma_i^{r}, \tag{12.84}$$

Table 12.6 The concentration dependence of the surface tension of an aqueous solution of SDS at 25 °C.

Concentration (mM)	γ (mN/m)
0.5	69.1
1.1	68.0
2	60.3
3	54.9
4	48.0
5	45.5
6	43.9
8	40.5
10	39.1
10.5	38.5
10.8	38.4
11	38.3
12	37.0
13	36.5

Table 12.7 Surface excess concentration for aqueous solutions of SDS at 25 °C.

Activity (10^{-3} M)	Γ_2 (10^{-6} mol/m^2)
0.3	1.1
0.7	1.65
1.05	2.25
1.45	2.5
2.0	2.83
2.4	3.05
3.0	3.2
4.1	3.2
4.8	3.2

Table 12.8 Surface tension of binary mixtures at 25 °C.

Mixture	x_1	γ (mN/m)
Ethanol (2)–benzene (1)	0	28.23
	0.237	27.38
	0.421	25.40
	0.564	25.00
	0.758	23.84
	1	21.90
m-Xylene (2)–benzene (1)	0.3969	28.0
	0.4655	27.92
	0.6918	28.11
	0.8428	27.85
	1	30.45
Acetone (2)–benzene (1)	0.2993	26.82
	0.5283	25.52
	0.7034	24.52
	0.8563	23.67
	1	22.72
Ethanol (2)–acetic acid (1)	0	21.90
	0.2262	24.06
	0.4241	25.75
	0.5550	26.07
	0.7414	27.10
	1	28.52

where n is the number of components, x_i is the mole fraction of component i, and r is an adjustable parameter. For $r = 1$, one recovers a linear behavior of surface tension with composition. It is found experimentally that, when linear behavior is not observed, r ranges from -1 to -3. Experimental data for the surface tension of several binary mixtures have been reported by Hammick and Andrew [54]. Table 12.8 includes some of their data for binary mixtures of ethanol–benzene, m-xylene–benzene, acetone–benzene, and acetic acid–ethanol. Please determine the values of the parameter r in Eqn. (12.84) for these mixtures.

APPENDIX A
Mathematical background

In thermodynamics we often switch between different sets of independent variables. For example, we develop the fundamental structure of the subject using internal energy as the dependent variable, whereas entropy, volume, and mole numbers are the independent variables. However, in experiments we often manipulate temperature and pressure, so we would like to use T and P as the independent variables. Therefore, it is essential that students of thermodynamics are comfortable with the manipulation of partial differentiation and the conversion between different sets of independent variables.

In this appendix we give a summary of all of the mathematical concepts necessary to perform the manipulations used in this text. As is made clear below, these manipulations can be reduced to a minimum number. When the methods described here are combined with the strategies discussed in Section 3.6, any desired simplification can be obtained with little effort, memorization, or error.

In what follows, we consider a dependent variable ψ to be a function of independent variables u, v, and w:

$$\psi = \psi(u, v, w). \tag{A.1}$$

A.1 TAYLOR'S SERIES EXPANSION

A suitably smooth function of more than one argument may be expanded in a Taylor series around the point (u, v, w) as

$$\psi(u + du, v + dv, w + dw) = \psi(u, v, w) + d\psi + \frac{1}{2} d^2\psi + \cdots, \tag{A.2}$$

where we define

$$d\psi := \left(\frac{\partial \psi}{\partial u}\right)_{v,w} du + \left(\frac{\partial \psi}{\partial v}\right)_{u,w} dv + \left(\frac{\partial \psi}{\partial w}\right)_{u,v} dw, \tag{A.3}$$

$$\begin{aligned} d^2\psi := &\left(\frac{\partial^2 \psi}{\partial u^2}\right)(du)^2 + \left(\frac{\partial^2 \psi}{\partial v^2}\right)(dv)^2 + \left(\frac{\partial^2 \psi}{\partial w^2}\right)(dw)^2 \\ &+ 2\left(\frac{\partial^2 \psi}{\partial u\, \partial v}\right) du\, dv + 2\left(\frac{\partial^2 \psi}{\partial v\, \partial w}\right) dv\, dw + 2\left(\frac{\partial^2 \psi}{\partial w\, \partial u}\right) dw\, du. \end{aligned} \tag{A.4}$$

The variables written as subscripts following the parentheses in Eqn. (A.3) are those being held constant during differentiation; thus, in the first term on the right-hand side of Eqn. (A.3), we can tell that the independent variables are u, v, and w. We have used the notation typical in thermodynamics, whereas the more usual mathematical notation is $(\partial\psi/\partial u)_{v,w} \equiv (\partial/\partial u)\psi(u, v, w)$.

The second-order differentials in Eqn. (A.4) do not have such indications, so, when we use them, we need to keep careful track of what the independent variables are, or perhaps the safest thing to do is to use the mathematical notation. If ψ is an **analytic function**, then the order of differentiation is unimportant, and we can write

$$\left(\frac{\partial^2 \psi}{\partial u\,\partial v}\right)_w = \left(\frac{\partial^2 \psi}{\partial v\,\partial u}\right)_w . \tag{A.5}$$

We will assume in this text that all thermodynamic functions are analytic. The first differential Eqn. (A.3) is important for nearly all of thermodynamics (i.e. in the integration of energy, or in order to find maxima). The second differential Eqn. (A.4) is important when considering stability criteria.

A.2 THE CHAIN RULE

Sometimes we switch from one set of independent variables, such as energy, volume, and mole number, to another set, such as temperature, pressure, and chemical potential. Then, we need to find differentials in the new set using what we know about the old set. For example, suppose that we know the relations

$$\begin{aligned} u &= u(x, y, z), \\ v &= v(x, y, z), \\ w &= w(x, y, z), \end{aligned} \tag{A.6}$$

where (u, v, w) is the original set of independent variables, and (x, y, z) is the new set. Then, we can first apply the differential definition Eqn. (A.3) to u, v, and w

$$\begin{aligned} du &= \left(\frac{\partial u}{\partial x}\right)_{y,z} dx + \left(\frac{\partial u}{\partial y}\right)_{x,z} dy + \left(\frac{\partial u}{\partial z}\right)_{x,y} dz, \\ dv &= \left(\frac{\partial v}{\partial x}\right)_{y,z} dx + \left(\frac{\partial v}{\partial y}\right)_{x,z} dy + \left(\frac{\partial v}{\partial z}\right)_{x,y} dz, \\ dw &= \left(\frac{\partial w}{\partial x}\right)_{y,z} dx + \left(\frac{\partial w}{\partial y}\right)_{x,z} dy + \left(\frac{\partial w}{\partial z}\right)_{x,y} dz. \end{aligned} \tag{A.7}$$

Then, we insert Eqn. (A.7) into the original differential expression, Eqn. (A.3), to obtain the chain rule of partial differentiation:

$$\begin{aligned} d\psi = &\left[\left(\frac{\partial \psi}{\partial u}\right)_{v,w}\left(\frac{\partial u}{\partial x}\right)_{y,z} + \left(\frac{\partial \psi}{\partial v}\right)_{u,w}\left(\frac{\partial v}{\partial x}\right)_{y,z} + \left(\frac{\partial \psi}{\partial w}\right)_{u,v}\left(\frac{\partial w}{\partial x}\right)_{y,z}\right] dx \\ &+ \left[\left(\frac{\partial \psi}{\partial u}\right)_{v,w}\left(\frac{\partial u}{\partial y}\right)_{x,z} + \left(\frac{\partial \psi}{\partial v}\right)_{u,w}\left(\frac{\partial v}{\partial y}\right)_{x,z} + \left(\frac{\partial \psi}{\partial w}\right)_{u,v}\left(\frac{\partial w}{\partial y}\right)_{x,z}\right] dy \\ &+ \left[\left(\frac{\partial \psi}{\partial u}\right)_{v,w}\left(\frac{\partial u}{\partial z}\right)_{x,y} + \left(\frac{\partial \psi}{\partial v}\right)_{u,w}\left(\frac{\partial v}{\partial z}\right)_{x,y} + \left(\frac{\partial \psi}{\partial w}\right)_{u,v}\left(\frac{\partial w}{\partial z}\right)_{x,y}\right] dz. \end{aligned} \tag{A.8}$$

If we were to insert Eqn. (A.6) into Eqn. (A.1), we would obtain ψ as a function of (x, y, z), and we could use the definition of Eqn. (A.3) to write

$$d\psi = \left(\frac{\partial \psi}{\partial x}\right)_{y,z} dx + \left(\frac{\partial \psi}{\partial y}\right)_{x,z} dy + \left(\frac{\partial \psi}{\partial z}\right)_{x,y} dz. \tag{A.9}$$

Comparison of Eqn. (A.9) with Eqn. (A.8) shows that

$$\left(\frac{\partial \psi}{\partial x}\right)_{y,z} = \left(\frac{\partial \psi}{\partial u}\right)_{v,w}\left(\frac{\partial u}{\partial x}\right)_{y,z} + \left(\frac{\partial \psi}{\partial v}\right)_{u,w}\left(\frac{\partial v}{\partial x}\right)_{y,z} + \left(\frac{\partial \psi}{\partial w}\right)_{u,v}\left(\frac{\partial w}{\partial x}\right)_{y,z}. \tag{A.10}$$

This rigorous derivation proves that we can naïvely use an algebraic trick, and still get the correct result: to obtain **the chain rule**, Eqn. (A.10), divide both sides of Eqn. (A.3) by dx, and hold y and z constant so that $dy = dz = 0$ to obtain the correct expression.

We have written Eqn. (A.8) assuming that we have three original independent variables (u, v, w) and three new independent variables (x, y, z). However, it is straightforward to generalize the result to any number of independent variables. It is also not necessary to have the same number of independent variables in the two sets; thus we could go from, say, (u, v, w) to the set (x, y) by elimination of the last line in Eqn. (A.8).

EXAMPLE A.2.1 Supposing that you had specific internal energy u as a function of specific entropy s and specific volume v, $u = u(s, v)$, and you wished to switch to temperature and pressure as the set of independent variables (T, P). Find $(\partial U/\partial T)_P$.

Solution. We use the chain rule shown in Eqn. (A.8), with the substitutions: $\psi \to u, x \to T, y \to P$, and $u \to s$:

$$\left(\frac{\partial u}{\partial T}\right)_P = \left(\frac{\partial u}{\partial s}\right)_v\left(\frac{\partial s}{\partial T}\right)_P + \left(\frac{\partial u}{\partial v}\right)_s\left(\frac{\partial v}{\partial T}\right)_P. \tag{A.11}$$

We see in the text that Eqn. (A.11) is related to measurable quantities, since, according to Eqns. (2.25) and (2.24), $T := (\partial u/\partial s)_v$ and $P := -(\partial u/\partial v)_s$. The other derivatives are related to the constant-pressure heat capacity C_P and the coefficient of thermal expansion α, so

$$\left(\frac{\partial u}{\partial T}\right)_P = C_P - Pv\alpha, \tag{A.12}$$

which follows from the definitions Eqns. (3.60) and (3.62), and from Eqn. (A.11). □

Exercise If you were given $s = s(u, v)$, how would you find $(\partial s/\partial P)_T$?

A.3 JACOBIAN TRANSFORMATIONS

Most thermodynamic manipulations can be handled in a convenient and compact manner by the use of Jacobian transformations [121]. We define the Jacobian for three independent and three dependent variables as

$$\frac{\partial(u,v,w)}{\partial(x,y,z)} := \begin{vmatrix} (\partial u/\partial x)_{y,z} & (\partial u/\partial y)_{x,z} & (\partial u/\partial z)_{x,y} \\ (\partial v/\partial x)_{y,z} & (\partial v/\partial y)_{x,z} & (\partial v/\partial z)_{x,y} \\ (\partial w/\partial x)_{y,z} & (\partial w/\partial y)_{x,z} & (\partial w/\partial z)_{x,y} \end{vmatrix}, \tag{A.13}$$

where the vertical lines $|\ldots|$ indicate taking the determinant of the matrix. Again, we have assumed that the definition for the Jacobian transformation given in Eqn. (A.13) is for three independent variables. However, it is straightforward to generalize the definition for any number of variables. From the definition of the Jacobian, it is possible to prove the following useful properties:

$$\left(\frac{\partial u}{\partial x}\right)_{y,z} = \frac{\partial(u,y,z)}{\partial(x,y,z)}, \tag{A.14}$$

$$\frac{\partial(u,v,w)}{\partial(x,y,z)} = \frac{\partial(u,v,w)}{\partial(r,s,t)}\frac{\partial(r,s,t)}{\partial(x,y,z)}, \tag{A.15}$$

$$\frac{\partial(u,v,w)}{\partial(x,y,z)} = 1 \bigg/ \frac{\partial(x,y,z)}{\partial(u,v,w)}, \tag{A.16}$$

$$\frac{\partial(u,v,w)}{\partial(x,y,z)} = -\frac{\partial(v,u,w)}{\partial(x,y,z)}. \tag{A.17}$$

Note that these properties are rather intuitive so that they are easy to remember. The last property shows that attention to the order of the variables is important for obtaining the proper sign in the answer. Systematic use of the properties (A.14)–(A.17) is made in the thermodynamic manipulations considered in Section 3.6.

Exercise Prove that relations (A.14)–(A.17) hold for a transformation in two variables from (u, v) to (x, y). Prove that property (A.14) holds for the three variable transformation, and then generalize the proof for any number of variables.

Some properties of derivatives that are particularly useful for thermodynamics are derived in the following three examples and one exercise using Jacobian transformations.

EXAMPLE A.3.1 Prove the relation

$$\left(\frac{\partial x}{\partial y}\right)_{\psi,z} = 1 \bigg/ \left(\frac{\partial y}{\partial x}\right)_{\psi,z}. \tag{A.18}$$

Solution. We start by using property (A.14):

$$\left(\frac{\partial x}{\partial y}\right)_{\psi,z} = \frac{\partial(x,\psi,z)}{\partial(y,\psi,z)} = 1 \Bigg/ \frac{\partial(y,\psi,z)}{\partial(x,\psi,z)} = 1 \Bigg/ \left(\frac{\partial y}{\partial x}\right)_{\psi,z}. \tag{A.19}$$

The second line utilizes the third Jacobian property, (A.16), and the last line utilizes the first, (A.14). □

EXAMPLE A.3.2 Prove the relation

$$\left(\frac{\partial y}{\partial x}\right)_{\psi,z} = -\left(\frac{\partial \psi}{\partial x}\right)_{y,z} \Bigg/ \left(\frac{\partial \psi}{\partial y}\right)_{x,z}. \tag{A.20}$$

Solution. We start by using property (A.14):

$$\begin{aligned}\left(\frac{\partial y}{\partial x}\right)_{\psi,z} &= \frac{\partial(y,\psi,z)}{\partial(x,\psi,z)} \\ &= \frac{\partial(y,\psi,z)}{\partial(x,y,z)}\frac{\partial(x,y,z)}{\partial(x,\psi,z)} \\ &= -\frac{\partial(\psi,y,z)}{\partial(x,y,z)}\frac{\partial(y,x,z)}{\partial(\psi,x,z)} \\ &= -\frac{\partial(\psi,y,z)}{\partial(x,y,z)} \Bigg/ \frac{\partial(\psi,x,z)}{\partial(y,x,z)} \\ &= -\left(\frac{\partial \psi}{\partial x}\right)_{y,z} \Bigg/ \left(\frac{\partial \psi}{\partial y}\right)_{x,z}.\end{aligned} \tag{A.21}$$

The second line follows from the first by virtue of the second property of the Jacobian transformation, (A.15); the third line uses the fourth property, (A.17), and in going to the fifth line we have used the third and fourth properties simultaneously. □

Exercise Prove the relation

$$\left(\frac{\partial y}{\partial x}\right)_{\psi,z} = \left(\frac{\partial y}{\partial u}\right)_{\psi,z} \Bigg/ \left(\frac{\partial x}{\partial u}\right)_{\psi,z}. \tag{A.22}$$

Exercise Using the result Eqn. (A.20) or other manipulations, prove the cyclic relation

$$\left(\frac{\partial x}{\partial y}\right)_{\psi,z}\left(\frac{\partial y}{\partial z}\right)_{\psi,x}\left(\frac{\partial z}{\partial x}\right)_{\psi,y} = -1. \tag{A.23}$$

Many thermodynamic manipulations may be accomplished either with the use of the Jacobian transformation properties Eqns. (A.14)–(A.17), or, equivalently, by use of the properties given in Eqns. (A.18), (A.20), (A.22), and (A.23).

A.4 THE FUNDAMENTAL THEOREM OF CALCULUS

The fundamental theorem of calculus states that integration and derivation are inverse operators. Namely, if one takes the integral of the derivative of a function, one obtains the function back. If this is a definite integral, the following relation results:

$$\int_a^b \left(\frac{\partial f(u,v,w)}{\partial u}\right) du = \int_a^b \left(\frac{\partial f}{\partial u}\right)_{v,w} du = f(b,v,w) - f(a,v,w). \tag{A.24}$$

This relation is used repeatedly throughout the text, and in thermodynamics in general.

A.5 LEIBNIZ'S RULE

We sometimes need to differentiate an integral where either the integrand or the limits, or both, can depend on the variable of differentiation. Leibniz's rule provides the means to evaluate such a differentiation:

$$\frac{\partial}{\partial u}\int_{A(u,w)}^{B(u,w)} \psi(u,v,w)dv = \int_{A(u,w)}^{B(u,w)} \frac{\partial \psi(u,v,w)}{\partial u} dv - \left(\frac{\partial A}{\partial u}\right)_w \psi(u,A(u,w),w) + \left(\frac{\partial B}{\partial u}\right)_w \psi(u,B(u,w),w). \tag{A.25}$$

EXAMPLE A.5.1 Find the derivative with respect to x of the integral

$$\int_0^{x^2} \sin(xy)\exp(-y^2)dy. \tag{A.26}$$

Solution. We just use Leibniz's formula, Eqn. (A.25), with the substitutions $u \to x, v \to y, w \to 0$, and $\psi \to \sin(xy)\exp(-y^2)$,

$$\frac{\partial}{\partial x}\int_0^{x^2} \sin(xy)\exp(-y^2)dy$$
$$= \int_0^{x^2} \frac{\partial}{\partial x}\left[\sin(xy)\exp\left(-y^2\right)\right]dy + 2x\sin(xy)\exp(-y^2)\big|_{y=x^2}$$
$$= \int_0^{x^2} y\cos(xy)\exp(-y^2)dy + 2x\sin(x^3)\exp(-x^4). \tag{A.27}$$

□

There is a generalization of the Leibniz formula for derivatives of volume integrals whose boundaries may be time-dependent:

$$\frac{d}{dt}\iiint_{V(t)}(\ldots)dV = \iiint_{V(t)}\frac{\partial}{\partial t}(\ldots)dV + \iint_{A(t)}(\ldots)(\vec{n}\cdot\vec{v}_A)dA, \tag{A.28}$$

where $\vec{v}_A$ is the velocity of the moving boundary at dA. This formula is used implicitly in only one location in this book, in Section 5.1, and is not essential material. The formula is sometimes also called **Reynolds' transport theorem**.

A.6 THE GAUSS DIVERGENCE THEOREM

The Gauss divergence theorem yields an equivalence between the volume integral of the divergence of a vector field $\vec{\psi}$ and a surface integral:

$$\iiint_{V(t)} \vec{\nabla}\cdot\vec{\psi}(\vec{r},t)dV = \iint_{A(t)} \vec{n}\cdot\vec{\psi}(\vec{r},t)dA, \tag{A.29}$$

where A represents the whole surface of the volume, V is the enclosed volume, and $\vec{n}$ is the unit outward normal vector of the surface element dA. This equation is also used only in Section 5.1, and is not essential.

A.7 SOLUTIONS TO CUBIC EQUATIONS

In order to predict phase changes in fluids, equations of state must predict a dependence on volume that is cubic, or higher. Hence, many equations of state are cubic, and methods for inverting cubic equations are helpful for studying stability. We first write our cubic equation for x in the form

$$x^3 + a_2x^2 + a_1x + a_0 = 0. \tag{A.30}$$

The equation may have one, two, or three real roots; we are interested only when there exist either one or three roots [1]. We can determine the number of roots from the sign of the discriminant $q^3 + r^2$, where

$$q := \frac{3a_1 - a_2^2}{9}$$
$$r := \frac{a_2\left(9a_1 - 2a_2^2\right) - 27a_0}{54}. \tag{A.31}$$

When $q^3 + r^2 > 0$ there is a single real root to Eqn. (A.30),

$$x_1 = \left(r + \sqrt{q^3 + r^2}\right)^{1/3} + \left(r - \sqrt{q^3 + r^2}\right)^{1/3} - \frac{a_2}{3}. \tag{A.32}$$

When $q^3 + r^2 < 0$, then Eqn. (A.30) has three real roots,

$$x_1 = 2\sqrt{-q}\cos\left(\frac{\theta}{3}\right) - \frac{a_2}{3},$$
$$x_2 = -2\sqrt{-q}\cos\left(\frac{\theta - \pi}{3}\right) - \frac{a_2}{3},$$
$$x_3 = -2\sqrt{-q}\cos\left(\frac{\theta + \pi}{3}\right) - \frac{a_2}{3}, \tag{A.33}$$

where

$$\theta := \cos^{-1}\left(\frac{r}{\sqrt{-q^3}}\right). \tag{A.34}$$

These equations are particularly useful for programming, for example. Table A.1 gives the coefficients necessary to find $v = v(T, P)$ for several cubic equations of state.

Table A.1 Coefficients for the cubic equations of state $v^3 + a_2v^2 + a_1v + a_0 = 0$. Use of these expressions for a_0, a_1, and a_2 yields an analytic expression for $v = v(T, P)$ in either Eqn. (A.32) or Eqn. (A.33). The Soave modification to the Redlich–Kwong equation of state makes a a function of temperature.

	Cubic equations of state		
Equation of state	a_0	a_1	a_2
Van der Waals	$-ab/P$	a/P	$-b - RT/P$
Redlich–Kwong	$-ab/(P\sqrt{T})$	$a/(P\sqrt{T}) - b^2 - RTb/P$	$-RT/P$
Peng–Robinson	$b^2(b + RT/P) - ab/P$	$a/P - b(3b + 2RT/P)$	$(b - RT/P)$
Martin's generalized	$-(\beta - b)(\gamma - b)(b + RT/P)$ $- ab/P$	$a/P + (\beta - b)(\gamma - b)$ $-(\beta+\gamma-2b)(b + RT/P)$	$\beta + \gamma - 3b$ $- RT/P$

A.8 COMBINATORICS

In statistical mechanics we often have to count large numbers of possible configurations of molecules in order to find the partition function. Many of these problems are isomorphic to the urn problems of probability. Here we summarize the problem of arranging many indistinguishable objects in distinguishable locations. We first use the binomial theorem for two such objects, and then generalize to an arbitrary number.

A.8.1 The binomial theorem

How many ways can you arrange N_w (indistinguishable) white marbles and N_b (indistinguishable) black marbles in $N_w + N_b$ (distinguishable) boxes? The answer can be found in two steps.

1. Pick the marbles blindly, one at a time, out of a bag and fill the boxes in order. The first box has $(N_w + N_b)$ possibilities. The second box has $(N_w + N_b - 1)$, since you have removed one marble from the bag to fill the first box. Keep doing this until the bag is empty. The total number of such possibilities is the product of the numbers of possibilities of each box:

$$(N_w + N_b)(N_w + N_b - 1) \ldots 1 = (N_w + N_b)!.$$

2. Since the white marbles are indistinguishable, we have over-counted these possibilities – some of the arrangements we created in the first step are redundant. Let us count these redundancies. Take a given arrangement created in the first step, and remove all the white marbles from their boxes and return them to the bag. Now do like we did before, but with just the white marbles and the empty boxes. How many different ways can we put them back?

$$N_w(N_w - 1) \ldots 1 = N_w!.$$

   Note that this result did not depend on what our chosen arrangement looks like: the same answer applies to all arrangements. We can carry out the same step for the black marbles. Hence, we have counted each unique arrangement $N_w!N_b!$ times in the first step.

Hence, we get that there are

$$\begin{pmatrix} N_w + N_b \\ N_w \end{pmatrix} := \frac{(N_w + N_b)!}{N_w!N_b!}$$

unique ways to arrange these white and black marbles, in answer to our first question.

The term on the left-hand side is called the **binomial coefficient**, since it arises in expanding powers of a sum of two terms [1, Section 3.1]

$$(a + b)^n = \sum_{i=0}^{n} \begin{pmatrix} n \\ i \end{pmatrix} a^{n-i} b^i.$$

For the example of a Langmuir adsorption isotherm, sites with a molecule adsorbed are white, and sites from which a molecule has been desorbed are black, so $N_w = N\tilde{N}_A$ and $N_w + N_b = M_s\tilde{N}_A$, where $\tilde{N}_A$ is Avogadro's number.

A.8.2 The multinomial theorem

Now generalize the binomial problem to having more than two kinds of marbles. Let's make them billiard balls with numbers painted on them. We have N_1 balls with the number 1, N_2 with number 2, etc. Then, the total number of arrangements in step 1 is $(\sum_{i=1}^{m} N_i)!$, where there are m different kinds of balls.

Again we have over-counted. Step 2 generalized leads to $N_1!N_2!\ldots N_m!$ over-counts. Hence, the total number of unique ways to arrange the balls is

$$\left(\sum_{i=1}^{r} N_i; N_1, N_2, \ldots, N_r\right) := \frac{(\sum_{i=1}^{r} N_i)!}{N_1!N_2!\ldots N_r!}.$$

This is called the **multinomial coefficient**, since [1, Section 24.1.1]

$$(x_1 + x_2 + \cdots + x_r)^n = \sum_{N_1}\sum_{N_2}\ldots\sum_{N_r}(n; N_1, N_2, \ldots, N_r)\delta\left(n, \sum_{i=1}^{r} N_i\right) x_1^{N_1} x_2^{N_2} \ldots x_m^{N_r},$$

where we used the **delta function**: $\delta(i,j) = 1$ only if $i = j$. This keeps the restriction that the sums must maintain $\sum_{i=1}^{r} N_i = n$. Or, we can write the above in shorthand form as

$$\begin{aligned}\left(\sum_{i=1}^{r} x_i\right)^n &= \sum_{\vec{N}}(n; N_1, N_2, \ldots, N_r)\delta\left(n, \sum_{i=1}^{r} N_i\right)\prod_{i=1}^{r} x_i^{N_i} \\ &= \sum_{\vec{N}} n!\delta\left(n, \sum_{i=1}^{r} N_i\right)\prod_{i=1}^{r}\frac{x_i^{N_i}}{N_i!}.\end{aligned} \tag{A.35}$$

The application of this result to the BET theory (see Exercise 3.7.E), for example, is straightforward.

Exercise Write down explicitly the probability that "heads" appears n times after N coin tosses. Find the most probable value for n.

Exercise There are two boxes with volumes V_1 and V_2 that communicate by a hole that allows molecules to pass through. There are $\tilde{N}_1$ molecules of species 1 and $\tilde{N}_2$ molecules of species 2 distributed between the two boxes. Assuming that the probability of a molecule being in a box is proportional to the volume of the box, find the probability that there are n_1 molecules of species 1 in box 1, and n_2 molecules of species 2 in box 1.

Exercise A poker hand is 5 cards drawn randomly from a deck of 52 cards. How many different hands are possible? Make your derivation of the answer clear.

Exercise Suppose that a strand of DNA contains 100 base pairs. If there are equal numbers of each kind of base pair, how many different combinations are possible?

APPENDIX B

Fluid equations of state

The text considers the ideal gas and van der Waals equations of state in some detail. In addition to these, we summarize here several more-accurate equations of state. This appendix presents but a small fraction of such equations. A more comprehensive discussion of the quality of such equations is given in [118].

Note that all fluid equations of state reduce to the general ideal gas at low densities. For small deviations from ideal behavior, the virial expansion is the most reliable. However, it is not appropriate for predicting vapor–liquid equilibria. The cubic equations of state are straightforward to use and computationally simple. However, some sacrifice must be made for accuracy. Of these, the Peng–Robinson, Soave–Redlich–Kwong, and Schmidt–Wenzel equations are usually superior. However, all cubic equations are suspect near the critical region. For an excellent review of many such equations, see [5]. If computational ease should be sacrificed for accuracy, the Benedict–Webb–Rubin and the Anderko–Pitzer equations are usually more accurate.

The parameters in the cubic *PVT* equations of state are usually determined from the critical properties of a fluid, and these equations are given. Critical values for a few substances are given in Table D.3 in Appendix D. More values can be found from the NIST web page. If the critical values of a substance are not known, they may be estimated from group methods on the basis of the chemical structure of the substance. These methods are reviewed in [102]. The additional parameters necessary for the more-accurate equations are usually of a general nature, and these are also tabulated, in Tables B.3 and B.4. However, if specific values for a substance are known, these should, of course, be used instead of the general values.

As an indication of the quality of the equations given here, Table B.1 compares the ability of a few equations to describe the vapor and liquid properties of saturated propane at 300 K.

B.1 A GENERAL IDEAL GAS

The well-known ideal-gas mechanical equation of state[1] is

$$P = \frac{RT}{v}. \tag{B.1}$$

[1] The discovery of the ideal behavior of gases is attributed to Robert Boyle, for his experiments on elasticity of air in 1662, and Joseph Gay-Lussac, for his isobaric heating of air in 1801 [122].

Table B.1 A comparison of predictions for vapor and liquid properties at saturation for several equations of state. These comparisons are made against saturated propane at 300 K, for which $v^{\text{vap}} = 2036.5$ cm^3/mol, $v^{\text{liq}} = 90.077$ cm^3/mol, and $P^{\text{sat}} = 9.9752$ bar. This table was inspired by [102, Table 4–9].

Model	v^{vap} (cm^3/mol)	Error (%)	v^{liq} (cm^3/mol)	Error (%)	P^{sat} (bar)	Error (%)
Van der Waals	1335	−34	143.4	59	14.771	48
Carnahan–Starling	1635	−20	104.5	16	12.453	25
Redlich–Kwong	3419	68	91.6	1.7	6.413	−36
Peng–Robinson	2032	−0.2	86.8	−3.7	10.003	0.3
Schmidt–Wenzel	2062	1.2	92.55	2.7	9.931	−0.4
Anderko–Pitzer	2033	−0.15	88.99	−1.2	10.25	2.8

In Section 8.2 we constructed the fundamental relation for a mixture of ideal gases:

$$f(T, v) = f_0 - s_0(T - T_0) + \int_{T_0}^{T} \left(\frac{T' - T}{T'}\right) c_v^{\text{ideal}}(T')dT' - RT \log\left(\frac{v}{v_0}\right). \tag{B.2}$$

The ideal-gas law predicts no vapor–liquid phase splitting or critical point. The simple ideal gas can be obtained by setting C_v^{ideal} to a constant CR.

B.2 THE VIRIAL EQUATION OF STATE

The virial equation of state yields an estimate for the compressibility factor as an expansion in density

$$Z = 1 + \frac{B(T)}{v} + \frac{C(T)}{v^2} + \cdots. \tag{B.3}$$

Schreiber and Pitzer [101, p. 141] recommend using a corresponding-states expression for the first virial coefficient:

$$B_r(T_r) := \frac{B(T)P_c}{RT_c z_c^*}, \tag{B.4}$$

with

$$z_c^* := 0.2905 - 0.0787\omega,$$

where B_r is found from

$$B_r(T_r) = B^0(T_r) + \omega B^1(T_r), \tag{B.5}$$

with

$$B^0(T_r) = 0.442\,259 - \frac{0.980\,970}{T_r} - \frac{0.611\,142}{T_r^2} - \frac{0.005\,156\,24}{T_r^6},$$
$$B^1(T_r) = 0.725\,650 + \frac{0.218\,714}{T_r} - \frac{1.249\,76}{T_r^2} - \frac{0.189\,187}{T_r^6}.$$

More recently, Tsonopoulos (see [138]) proposed modifications that apply to polar gases and associating fluids. For non-associating and non-polar gases, Tsonopoulos recommends

$$\frac{B(T)P_c}{RT_c} = B^0(T_r) + \omega B^1(T_r), \tag{B.6}$$

with

$$B^0(T_r) = 0.1445 - \frac{0.3300}{T_r} - \frac{0.1385}{T_r^2} - \frac{0.0121}{T_r^3} - \frac{0.000\,607}{T_r^8},$$
$$B^1(T_r) = 0.0637 + \frac{0.331}{T_r^2} - \frac{0.423}{T_r^3} - \frac{0.0008}{T_r^8}.$$

The acentric factor was first introduced by Pitzer, and is found from the behavior of the fluid at a single point on the PV diagram. Namely,

$$\omega := \log_{10}\left(\frac{P_c}{P^{\text{sat}}(T_r = 0.7)}\right) - 1. \tag{B.7}$$

Values for ω are tabulated for many substances in Appendix D.

One additional term each is necessary to describe polar and associating fluids. The coefficients, however, are not universal.

A correlation for the third term, recommended by Orbey and Vera (see [138]) is

$$C(T)\left(\frac{P_c}{RT_c}\right)^2 = C^0(T_r) + \omega C^1(T_r) \tag{B.8}$$

with

$$C^0(T_r) = 0.014\,07 + \frac{0.024\,32}{T_r^{2.8}} - \frac{0.003\,13}{T_r^{10.5}},$$
$$C^1(T_r) = -0.026\,76 + \frac{0.0177}{T_r^{2.8}} + \frac{0.040}{T_r^3} - \frac{0.003}{T_r^6} - \frac{0.002\,28}{T_r^{10.5}}.$$

These correlations are applied also for non-polar, non-associating fluids.

Experimentally determined values for the coefficients may be found in [35].

This equation of state has the fundamental relation

$$f(T,v) = f_0 - s_0(T - T_0) - RT\log(v/v_0) + \int_{T_0}^{T}\left(\frac{T' - T}{T'}\right)c_v^{\text{ideal}}(T')dT'$$
$$+ RTB(T)\left[\frac{1}{v} - \frac{1}{v_0}\right] + \frac{1}{2}RTC(T)\left[\frac{1}{v^2} - \frac{1}{v_0^2}\right] + \cdots, \tag{B.9}$$

as found from the techniques in Section 4.4.1. The residual internal energy and Helmholtz potential are

$$\frac{U^{\mathrm{R}}}{NRT} = \frac{T}{v}\frac{dB}{dT} + \frac{T}{2v^2}\frac{dC}{dT} + \cdots, \tag{B.10}$$

$$\frac{F^{\mathrm{R}}}{NRT} = \frac{B(T)}{v} + \frac{C(T)}{2v^2} + \cdots - \log\left(1 + \frac{B(T)}{v} + \frac{C(T)}{v^2} + \cdots\right). \tag{B.11}$$

The other residual properties can be found from these two. The fugacity is

$$f^{\text{pure}}(T, v) = \frac{RT}{v}\exp\left[\frac{2B(T)}{v} + \frac{3C(T)}{2v^2} + \cdots\right], \quad \text{virial expansion.} \tag{B.12}$$

There also exists a virial expansion in pressure

$$Z = 1 + B'(T)P + C'(T)P^2 + \cdots, \tag{B.13}$$

which has the fundamental relation

$$g(T,P) = g_0 - (T - T_0)s_0 + \int_{T_0}^{T}\left(\frac{T' - T}{T'}\right)c_P^{\text{ideal}}(T')dT' + RT\log\left(\frac{P}{P_0}\right) + B'(T)RTP + \frac{1}{2}C'(T)RTP^2 + \cdots. \tag{B.14}$$

B.3 THE VAN DER WAALS FLUID

Appendix B

The van der Waals fluid[2] has as its mechanical equation of state

$$P = \frac{RT}{v - b} - \frac{a}{v^2}. \tag{B.15}$$

In Section 4.4.1 we constructed the fundamental relation

$$f = f_0 - s_0(T - T_0) - RT\log\left(\frac{v - b}{v_0 - b}\right) - a\left(\frac{1}{v} - \frac{1}{v_0}\right) + \int_{T_0}^{T}\left(\frac{T' - T}{T'}\right)c_v^{\text{ideal}}(T')dT'. \tag{B.16}$$

The parameters can be found from the predicted critical state:

$$a = \frac{27R^2T_{\mathrm{c}}^2}{64P_{\mathrm{c}}}, \tag{B.17}$$

$$b = \frac{RT_{\mathrm{c}}}{8P_{\mathrm{c}}}. \tag{B.18}$$

[2] The van der Waals equation of state plays a key role in the history of thermodynamics. There has been a recent biography of Johannes Diderik van der Waals (1837–1923) [72], and a detailed history of the science surrounding the developments of thermodynamics in the Netherlands around this time [122], in which van der Waals plays a central role. Van der Waals' doctoral thesis of 1873 has been translated into English [143].

The van der Waals fluid predicts a compressibility factor at the critical point of $z_c = 3/8$, which is typically larger than values found by experiment.

We can use Eqns. (4.75) and (4.76) to find the residual properties predicted by this model:

$$\frac{F^R}{NRT} = -\frac{a}{v} - RT\log\left[1 - \frac{a(v-b)}{RTv^2}\right], \tag{B.19}$$

$$\frac{U^R}{NRT} = -\frac{a}{v}, \tag{B.20}$$

$$\frac{G^R}{NRT} = -\frac{2a}{v} - RT\log\left[1 - \frac{a(v-b)}{RTv^2}\right] + \frac{bRT}{v-b}, \tag{B.21}$$

$$\frac{H^R}{NRT} = \frac{bRT}{v-b} - \frac{2a}{v}, \tag{B.22}$$

$$\frac{S^R}{NRT} = R\log\left[1 - \frac{a(v-b)}{RTv^2}\right]. \tag{B.23}$$

The van der Waals model predicts the fugacity expression

$$f^{\mathrm{pure}}(T, v) = \frac{RT}{v-b}\exp\left[\frac{b}{v-b} - \frac{2a}{RTv}\right], \quad \text{van der Waals.} \tag{B.24}$$

Using stability analysis as shown in Section 4.2.2, this model can predict the saturation pressure, liquid density, and vapor density as functions of the sub-critical temperature. We can make an empirical fit to these numerical results of the form

$$\log\left(\frac{P^{\mathrm{sat}}[T_r]}{P_c}\right) = B(1-T_r) + C(1-T_r)^2 + D(1-T_r)^3, \tag{B.25}$$

where the van der Waals model predicts $B = -4.106\,25$, $C = -1.793\,65$, and $D = -7.971\,14$, and the result is valid for reduced temperatures between 0.6 and 1.

B.4 THE CARNAHAN–STARLING EQUATION OF STATE

Carnahan and Starling recommended the following modification to the van der Waals equation of state:

$$P = \frac{RT}{v}\left[\frac{v^3 + bv^2 + b^2v - b^3}{(v-b)^3}\right] - \frac{a}{v^2}, \tag{B.26}$$

Using the techniques of Section 4.4.1, we find the corresponding fundamental Helmholtz relation to be

$$f = f_0 - s_0(T - T_0) + RT\left\{\log\left(\frac{v_0}{v}\right) + b\left[\frac{4v-3b}{(v-b)^2} - \frac{4}{v_0}\right]\right\} - a\left(\frac{1}{v} - \frac{1}{v_0}\right) + \int_{T_0}^{T}\left(\frac{T'-T}{T'}\right)c_v^{\mathrm{ideal}}(T')dT'. \tag{B.27}$$

The parameters for the Carnahan–Starling–van der Waals equation of state can be found from

$$a \cong 0.496\,387\,7\frac{(RT_c)^2}{P_c},$$
$$b \cong 0.046\,823\,6\frac{RT_c}{P_c}. \tag{B.28}$$

These expressions are found from inverting the results of Exercise 4.2.I. The *PVT* relation yields a critical compressibility of 0.359, which is too large for most fluids.

We can use Eqns. (4.75) and (4.76) to find the residual properties predicted by this model

$$\frac{F^{R}}{NRT} = \frac{bRT(4v-3b)}{(v-b)^2} - RT\log\left[\frac{v^3+v^2b+vb^2-b^3}{(v-b)^3} - \frac{a}{RTv}\right] - \frac{a}{v}, \tag{B.29}$$
$$\frac{U^{R}}{NRT} = -\frac{a}{v}. \tag{B.30}$$

The other residual properties can be found from these two. The Carnahan–Starling model predicts the fugacity expression

$$f^{\text{pure}}(T,v) = \frac{RT}{v}\exp\left[-\frac{b\left(8v^2-9vb+3b^2\right)}{(v-b)^3} - \frac{2a}{RTv}\right]. \tag{B.31}$$

Using stability analysis as shown in Section 4.2.2, this model can predict the saturation pressure, liquid density, and vapor density as functions of the sub-critical temperature. We can make an empirical fit to these numerical results of the form

$$\log\left(\frac{P^{\text{sat}}[T_r]}{P_c}\right) = B(1-T_r) + C(1-T_r)^2 + D(1-T_r)^3, \tag{B.32}$$

where the Carnahan–Starling model predicts $B = -4.998\,03$, $C = -2.545\,47$, and $D = -13.2049$, and the result is valid for reduced temperatures between 0.635 and 1.

B.5 THE REDLICH–KWONG EQUATION OF STATE

The Redlich–Kwong equation of state [108] is given by

$$P = \frac{RT}{v-b} - \frac{a}{\sqrt{T}v(v+b)}. \tag{B.33}$$

If we use manipulations analogous to those used for the van der Waals fluid in Section 4.4.1, we find the corresponding fundamental Helmholtz relation for a Redlich–Kwong fluid to be

$$f = f_0 - s_0(T-T_0) - RT\log\left(\frac{v-b}{v_0}\right) + \frac{a}{b\sqrt{T}}\log\left(\frac{v}{v+b}\right)$$
$$+ \int_{T_0}^{T}\left(\frac{T'-T}{T'}\right)c_v^{\text{ideal}}(T')dT'. \tag{B.34}$$

The Redlich–Kwong fluid has one of the same shortcomings as the van der Waals fluid; namely, it predicts a universal value for the critical compressibility factor z_c.

Soave [127] has made a modification to this expression, which greatly improves agreement with data. He recommends allowing the parameter a to depend upon temperature according to

$$a(T_r) = a_0 \left[1 + \left(1 - \sqrt{T_r} \right) \left(m + \frac{n}{T_r} \right) \right], \tag{B.35}$$

where m and n are material parameters to be estimated from the temperature of the material at two different vapor pressures, 10 and 760 mm Hg. He describes several procedures by which to obtain m and n in his paper.

A simpler approximation is

$$a(T_r) = a_0 \left[1 + k_S \left(1 - \sqrt{T_r} \right) \right]^2, \tag{B.36}$$

where k_S depends upon the **acentric factor** ω according to

$$k_S = 0.480 + 1.574\omega - 0.176\omega^2, \tag{B.37}$$

which is a function of the material only (see Eqn. (B.7)). The remaining parameters can be found from the predicted critical state:

$$a_0 \approx 0.427\,480\,4 \frac{R^2 T_c^{5/2}}{P_c}, \tag{B.38}$$

$$b \approx 0.086\,640\,4 \frac{R T_c}{P_c}. \tag{B.39}$$

Both the Redlich–Kwong and the Soave–Redlich–Kwong model predict that the critical compressibility is approximately 0.333 (see Exercise 4.2.D).

We can use Eqns. (4.75) and (4.76) to find the residual properties predicted by this model:

$$\frac{F^R}{NRT} = \frac{a(T)}{b\sqrt{T}} \log\left(\frac{v}{v+b} \right) - RT \log\left[1 - \frac{(v-b)a(T)}{vRT^{3/2}(v+b)} \right], \tag{B.40}$$

$$\frac{U^R}{NRT} = \frac{3a(T) - 2Ta'(T)}{2b\sqrt{T}} \log\left(\frac{v}{v+b} \right). \tag{B.41}$$

The other residual properties can be found from these two. Note that $a'(T) := da/dT$.

The expression for the residual Gibbs free energy can be inserted into the definition for the fugacity, Eqn. (8.41), to obtain

$$f^{\text{pure}}(T, v) = \frac{RT}{v-b} \left[\frac{v}{v+b} \right]^{a(T)/(bRT^{3/2})} \exp\left[\frac{b}{v-b} - \frac{a(T)}{RT^{3/2}(v+b)} \right], \quad \text{Soave–Redlich–Kwong.} \tag{B.42}$$

Using stability analysis as shown in Section 4.2.2, this model can predict the saturation pressure, liquid density, and vapor density, as functions of the sub-critical temperature. We can make an empirical fit to these numerical results of the form

$$\log\left(\frac{P^{\text{sat}}[T_r]}{P_c} \right) = B(1 - T_r) + C(1 - T_r)^2 + D(1 - T_r)^3, \tag{B.43}$$

where the Soave–Redlich–Kwong model predicts

$$\begin{aligned} B &= -7.185\,14 - 5.027\,33\omega + 0.453\,954\omega^2, \\ C &= -5.047\,57 - 3.080\,97\omega + 1.363\,84\omega^2, \\ D &= -23.3129 - 35.5658\omega - 11.7846\omega^2. \end{aligned} \tag{B.44}$$

The fit to the model's prediction is valid for $0.71 \le T_r \le 1$ and $-0.3 \le \omega \le 0.3$.

B.6 THE PENG–ROBINSON EQUATION OF STATE

The Peng–Robinson PVT relation may be written

$$P = \frac{RT}{v-b} - \frac{a(T)}{v(v+b)+b(v-b)}, \tag{B.45}$$

where a has a similar temperature dependence to that used in the Soave modification,

$$a(T_r) = a_0\left[1 + k_{PR}\left(1 - \sqrt{T_r}\right)\right]^2, \tag{B.46}$$

where k_{PR} is

$$k_{PR} = 0.374\,64 + 1.542\,26\omega - 0.269\,92\omega^2. \tag{B.47}$$

We find the corresponding fundamental Helmholtz relation for the Peng–Robinson fluid to be

$$\begin{aligned} f = f_0 - s_0(T - T_0) - RT\log\left(\frac{v-b}{v_0}\right) - \frac{a(T)}{2\sqrt{2}b}\log\left[\frac{v+b(1+\sqrt{2})}{v+b(1-\sqrt{2})}\right] \\ + \int_{T_0}^{T}\left(\frac{T'-T}{T'}\right)c_v^{\text{ideal}}(T')dT' \end{aligned} \tag{B.48}$$

using the techniques in Section 4.4.1. The parameters can be found from the predicted critical state:

$$a_0 \approx 0.457\,236\,7\frac{(RT_c)^2}{P_c}, \tag{B.49}$$

$$b \approx 0.077\,796\frac{RT_c}{P_c}. \tag{B.50}$$

The critical compressibility factor is predicted by this equation of state to be

$$z_c \approx 0.307\,402, \tag{B.51}$$

which is an improvement over the Soave–Redlich–Kwong prediction, although still too large for many fluids.

We can use Eqns. (4.75) and (4.76) to find the residual properties predicted by this model:

$$\frac{F^{\mathrm{R}}}{NRT} = \frac{a(T)}{2\sqrt{2}b}\log\left[\frac{v-\left(\sqrt{2}-1\right)b}{v+\left(\sqrt{2}+1\right)b}\right] - RT\log\left[1-\frac{a(T)(v-b)}{RT(v^2+2vb-b^2)}\right], \tag{B.52}$$

$$\frac{U^{\mathrm{R}}}{NRT} = \frac{a(T)-Ta'[T]}{2\sqrt{2}b}\log\left[\frac{v-\left(\sqrt{2}-1\right)b}{v+\left(\sqrt{2}+1\right)b}\right]. \tag{B.53}$$

Note that $a'(T) := da/dT$. The other residual properties can be found from these two. In particular, the residual Gibbs free energy can be used to express the pure-state fugacity, Eqn. (8.41):

$$f^{\mathrm{pure}}(T,v) = \frac{RT}{v-b}\exp\left[\frac{b}{v-b}-\frac{va(T)}{RT(v^2+2vb-b^2)}\right]\left[\frac{v-\left(\sqrt{2}-1\right)b}{v+\left(\sqrt{2}+1\right)b}\right]^{a(T)/(2\sqrt{2}bRT)}. \tag{B.54}$$

Using stability analysis as shown in Section 4.2.2, this model can predict the saturation pressure, liquid density, and vapor density as functions of the sub-critical temperature. We can make an empirical fit to these numerical results of the form

$$\log\left(\frac{P^{\mathrm{sat}}[T_{\mathrm{r}}]}{P_{\mathrm{c}}}\right) = B(1-T_{\mathrm{r}}) + C(1-T_{\mathrm{r}})^2 + D(1-T_{\mathrm{r}})^3, \tag{B.55}$$

where the Peng–Robinson model predicts

$$\begin{aligned} B &= -5.727\,14 - 5.397\,78\omega + 0.820\,206\omega^2, \\ C &= -2.732\,59 - 1.664\,06\omega + 1.503\,36\omega^2, \\ D &= -12.3843 - 22.7863\omega - 9.012\,97\omega^2. \end{aligned} \tag{B.56}$$

The fit for this model is valid for $0.65 \le T_{\mathrm{r}} \le 1$ and $-0.3 \le \omega \le 0.3$.

B.7 MARTIN'S GENERALIZED CUBIC EQUATION OF STATE

Many cubic equations of state can be written in the general form

$$P = \frac{RT}{v-b} - \frac{a(T)}{(v+\beta-b)(v+\gamma-b)}. \tag{B.57}$$

We find the corresponding fundamental Helmholtz relation to be

$$\begin{aligned} f = {} & f_0 - s_0(T-T_0) - RT\log\left(\frac{v-b}{v_0}\right) - \frac{a(T)}{\gamma-\beta}\log\left(\frac{v+\beta-b}{v+\gamma-b}\right) \\ & + \int_{T_0}^{T}\left(\frac{T'-T}{T'}\right)c_v^{\mathrm{ideal}}(T')dT' \end{aligned} \tag{B.58}$$

by using the construction method illustrated in Section 4.4.1.

Table B.2 Parameter values for Martin's generalized cubic equation of state to recover other equations of state.

	Van der Waals	Redlich–Kwong	Peng–Robinson	Schmidt–Wenzel
β/b	1	3	$2+\sqrt{2}$	$3(\omega+1)/2 + \sqrt{[3(\omega+1)/2]^2-2}$
γ/b	1	1	$2-\sqrt{2}$	$3(\omega+1)/2 - \sqrt{[3(\omega+1)/2]^2-2}$

Martin's generalized cubic equation yields the following predictions for the critical properties of a fluid:

$$\begin{aligned} v_c &= b + \left(\beta\gamma^2\right)^{1/3} + \left(\beta^2\gamma\right)^{1/3}, \\ P_c &= \frac{a(T_c)}{\left(\beta^{2/3}+\beta^{1/3}\gamma^{1/3}+\gamma^{2/3}\right)^3}, \\ T_c &= \frac{a(T_c)\left(\beta^{1/3}+\gamma^{1/3}\right)^3}{R\left(\beta^{2/3}+\beta^{1/3}\gamma^{1/3}+\gamma^{2/3}\right)^3}. \end{aligned} \tag{B.59}$$

Hence, the critical compressibility factor is predicted to be

$$z_c = \frac{b + \left(\beta^2\gamma\right)^{1/3} + \left(\beta\gamma^2\right)^{1/3}}{\left(\beta^{1/3}+\gamma^{1/3}\right)^3}. \tag{B.60}$$

From Martin's generalized equation of state, one can recover the other cubic equations of state by proper choice for the parameters β and γ. These are shown in Table B.2. Also shown are values recommended by Schmidt and Wenzel [117]. These authors also recommend the following expression for the parameter $a(T)$:

$$\begin{aligned} a(T) &= a_0\left[1 + k_{\rm SW}(T_r)\left(1-\sqrt{T_r}\right)\right]^2, \\ k_{\rm SW}(T_r) &= \begin{cases} k_0 + (5T_r - 3k_0 - 1)^2/70, & T_r \le 1, \\ k_0 + (4-3k_0)^2/70, & T_r > 1, \end{cases} \\ k_0 &= 0.465 + 1.347\omega - 0.528\omega^2. \end{aligned} \tag{B.61}$$

B.8 THE BENEDICT–WEBB–RUBIN EQUATION OF STATE

The Benedict–Webb–Rubin equation of state has eight adjustable parameters, and is recommended for alkanes [73]:

$$P = \frac{RT}{v} + \frac{B_0RT - A_0 - C_0/T^2}{v^2} + \frac{bRT-a}{v^3} + \frac{a\alpha_{\rm BWR}}{v^6} + \frac{c}{v^3T^2}\left(1+\frac{\gamma}{v^2}\right)\exp\left(-\gamma/v^2\right). \tag{B.62}$$

We find the corresponding fundamental Helmholtz relation to be

$$f = f_0 - s_0(T - T_0) + RT\log(v/v_0) + \left(B_0RT - A_0 - C_0/T^2\right)\left(\frac{1}{v} - \frac{1}{v_0}\right)$$
$$+ \frac{bRT - a}{2}\left(\frac{1}{v^2} - \frac{1}{v_0^2}\right) + \frac{a\alpha_{\mathrm{BWR}}}{5}\left(\frac{1}{v^5} - \frac{1}{v_0^5}\right)$$
$$- \frac{c}{T^2}\left[\left(\frac{1}{\gamma} + \frac{1}{2v^2}\right)\exp(-\gamma/v^2) - \left(\frac{1}{\gamma} + \frac{1}{2v_0^2}\right)\exp(-\gamma/v_0^2)\right]$$
$$+ \int_{T_0}^{T}\left(\frac{T' - T}{T'}\right)c_v^{\mathrm{ideal}}(T')dT'. \tag{B.63}$$

Recommended values for the parameters are given in Table B.3.

We can use Eqns. (4.75) and (4.76) to find the residual properties predicted by this model

$$\frac{F^{\mathrm{R}}}{NRT} = -\frac{\left(2v^2 + \gamma\right)c}{2T^2v^2\gamma}e^{-\gamma/v^2} + \frac{c}{T^2\gamma} - \frac{a}{2v} + \frac{A_0}{v} + \frac{RTb}{2v^2} + \frac{RTB_0}{v} + \frac{a\alpha_{\mathrm{BWR}}}{5v^5}$$
$$- \frac{C_0}{T^2v} - RT\log\left[\frac{P(T, v)v}{RT}\right], \tag{B.64}$$

$$\frac{U^{\mathrm{R}}}{NRT} = \frac{a\alpha_{\mathrm{BWR}}}{5v^5} - \frac{A_0}{v} - \frac{a}{2v^2} + \frac{3c}{2T^2v^2\gamma}\left[2\left(1 - e^{-\gamma/v^2}\right)v^2 - e^{-\gamma/v^2}\gamma\right] - \frac{3C_0}{T^2v}. \tag{B.65}$$

B.9 THE ANDERKO–PITZER EQUATION OF STATE

Anderko and Pitzer [101] have recommended the following pressure-explicit equation of state that depends on the single adjustable parameter ω:

$$z = \frac{v_{\mathrm{r}} + c(T_{\mathrm{r}})}{v_{\mathrm{r}} - b} + \frac{\alpha_{\mathrm{AP}}(T_{\mathrm{r}})}{v_{\mathrm{r}}} + \frac{\beta(T_{\mathrm{r}})}{v_{\mathrm{r}}^2} + \frac{\gamma(T_{\mathrm{r}})}{v_{\mathrm{r}}^3}, \tag{B.66}$$

where

$$\begin{aligned}
b &= b_0 + b_1\omega, \\
c &= c_0 + c_1\omega + \frac{c_2 + c_3\omega}{T_{\mathrm{r}}} + \frac{c_4 + c_5\omega}{T_{\mathrm{r}}^2}, \\
\alpha_{\mathrm{AP}} &= \alpha_0 + \alpha_1\omega + \frac{\alpha_2 + \alpha_3\omega}{T_{\mathrm{r}}} + \frac{\alpha_4 + \alpha_5\omega}{T_{\mathrm{r}}^2} + \frac{\alpha_6 + \alpha_7\omega}{T_{\mathrm{r}}^6}, \\
\beta &= \beta_0 + \beta_1\omega + \frac{\beta_2 + \beta_3\omega}{T_{\mathrm{r}}} + \frac{\beta_4 + \beta_5\omega}{T_{\mathrm{r}}^2} + \frac{\beta_6 + \beta_7\omega}{T_{\mathrm{r}}^6}, \\
\gamma &= \gamma_0 + \gamma_1\omega + \frac{\gamma_2 + \gamma_3\omega}{T_{\mathrm{r}}} + \frac{\gamma_4 + \gamma_5\omega}{T_{\mathrm{r}}^2} + \frac{\gamma_6 + \gamma_7\omega}{T_{\mathrm{r}}^6}.
\end{aligned} \tag{B.67}$$

Table B.3 Values for the parameters used in the Benedict–Webb–Rubin equation of state. If temperature is in K, volume is in l/mol, and $R = 0.082\,07 \text{l} \cdot \text{atm/mol} \cdot \text{K}$, then the pressure is in atm. The values for the first nine species are taken from [55]. Their original source was [10]. The remaining values are from Felder and Rousseau [38]. Many of these values have been favorably compared with experimental results in [29]. They also studied a set of dimensionless constants for arbitrary species: $A_0' = 0.241\,809\,80$, $B_0' = 0.076\,431\,01$, $C_0' = 0.212\,117$, $a' = 0.113\,694\,78$, $b' = 0.037\,151\,71$, $c' = 0.064\,480\,01$, $\alpha_{BWR}' = 1.136\,947\,8 \times 10^{-4}$, and $\gamma' = 0.06$. The constants for species are found from these generalized constants and the critical properties: $A_0 = R^2T_c^2A_0'/P_c$, $B_0 = RT_cB_0'/P_c$, $C_0 = R^2T_c^4C_0'/P_c$, $a = R^3T_c^3a'/P_c^2$, $b = R^2T_c^2b'/P_c^2$, $c = R^3T_c^5c'/P_c^2$, $\alpha_{BWR} = R^3T_c^3\alpha_{BWR}'/P_c^3$, and $\gamma = R^2T_c^2\gamma'/P_c^2$. These values are given by [129].

	A_0	B_0	$C_0 \times 10^{-6}$	a	b	$c \times 10^{-6}$	$\alpha_{BWR} \times 10^3$	$\gamma \times 10^2$
Methane	1.855 00	0.042 600	0.022 57	0.049 400	0.003 380 04	0.002 545	0.124 359	0.600 0
Ethylene	3.339 58	0.055 683 3	0.131 140	0.259 000	0.008 600 0	0.021 120	0.178 000	0.923 000
Ethane	4.155 56	0.062 772 4	0.179 592	0.345 160	0.011 122 0	0.032 7670	0.243 389	1.180 00
Propane	6.872 25	0.097 313 0	0.508 256	0.947 700	0.022 000	0.129 000	0.607 175	2.200 00
Propylene	6.112 20	0.085 064 7	0.439 182	0.774 056	0.018 705 9	0.102 611	0.455 696	1.829 00
n-Butane	10.084 7	0.124 361	0.992 830	1.882 31	0.039 998 3	0.316 400	1.101 32	3.400 00
i-Butane	10.232 64	0.137 544	0.849 943	1.937 63	0.042 435 2	0.286 010	1.074 08	3.400 00
n-Hexane	14.437 3	0.177 813	3.319 35	7.116 71	0.109 131	1.512 76	2.810 86	6.668 49
n-Pentane	12.179 4	0.156 751	2.121 21	4.074 80	0.066 812 0	0.824 170	1.810 00	5.750 00
Benzene	6.509 97	0.069 464	3.429 97	5.570	0.076 63	1.176 42	0.700 1	2.930
CO_2	2.516 06	0.044 885	0.147 44	0.136 814	0.004 123 9	0.014 918	0.084 66	0.525 33
N_2	1.053 64	0.047 426	0.008 059	0.035 102	0.002 327 7	0.000 728 4	0.127 2	0.530 0
SO_2	2.120 54	0.026 182 7	0.793 879	0.844 395	0.014 654 2	0.113 362	0.0719 604	0.592 390

Table B.4 Constants used in the Anderko–Pitzer equation of state for normal fluids.

	0	1	2	3	4	5	6	7
b_i	0.251 896	0.048 788						
c_i	−0.080 968	−0.879 256	1.802 846	0.671581	−0.387457	−0.226976		
α_i	0.271 331	1.556 118	−2.783 816	−0.452867	−0.223685	−1.022780	−0.0051562	−0.189187
β_i	0.088 425	−0.516 256	−0.619 446	−0.319606	0.694687	1.270432	0.0068740	0.273364
γ_i	0.099 362	0.816 731	−0.259 215	−0.742537	−0.099234	−0.295499	−0.0013110	−0.071180

The constants $b_i, c_i, \alpha_i, \beta_i$, and γ_i are given in Table B.4. We find the corresponding fundamental Helmholtz relation to be

$$f = f_0 - s_0(T - T_0) + RT \left\{ \frac{1}{b} \log \left[\left(\frac{v}{v_0} \right)^c \left(\frac{v_0 - bv_c}{v - bv_c} \right)^{b+c} \right] \right.$$
$$- \alpha_{AP}(T_r) v_c \left(\frac{1}{v_0} - \frac{1}{v} \right) - \frac{\beta(T_r) v_c^2}{2} \left(\frac{1}{v_0^2} - \frac{1}{v^2} \right)$$
$$\left. - \frac{\gamma(T_r) v_c^3}{3} \left(\frac{1}{v_0^3} - \frac{1}{v^3} \right) \right\} + \int_{T_0}^{T} \left(\frac{T' - T}{T'} \right) c_v^{\text{ideal}}(T') dT'. \tag{B.68}$$

The authors claim to have achieved good fits for normal fluids for a density up to three times the critical density. They also claim that water can be fit by the equation, but with a different set of constants.

The critical point is not predicted to be exactly at the “reduced" volume (and temperature) of 1, except when the acentric factor is 0.4. Instead the critical point varies from approximately 0.98 to 1.07 as the acentric factor varies from -0.6 to $+0.6$. Therefore, some care may be necessary when using this model near the critical point.

APPENDIX C
Microscopic balances for open systems

The starting point in the derivation of all microscopic balances is a differential volume element.[1] In the simplest form, this element is a cube with edges of length Δx, Δy, and Δz (see Figure C.1).

We assume that the fluid in this element is in an equilibrium state determined by the local instantaneous values of its thermodynamic quantities (e.g. temperature and pressure). These quantities can vary with time and position. We call this the **local-equilibrium assumption**.

Suppose that, from a thermodynamics point of view, we have temperature and pressure as "independent variables." These quantities are now written as **field variables**, meaning that they vary with time and position: $T(\vec{r},t)$, $P(\vec{r},t)$. We consider a fluid that has **mass density** $\rho(\vec{r},t) = M_{\mathrm{w}}(\vec{r},t)/v(\vec{r},t)$ (mass/volume), where $M_{\mathrm{w}}(\vec{r},t)$ is the molecular weight of the fluid at position $\vec{r}$ and time t, which may vary because of composition changes, and $v(\vec{r},t)$ is the specific volume under those thermodynamic conditions.

Any quantity in this element (e.g. mass, momentum, or energy) must satisfy the following balance relation:

$$\left\{\begin{array}{c}\text{Rate of}\\ \text{accumulation}\end{array}\right\} = \left\{\begin{array}{c}\text{Flow}\\ \text{rate in}\end{array}\right\} - \left\{\begin{array}{c}\text{Flow}\\ \text{rate out}\end{array}\right\} + \left\{\begin{array}{c}\text{Rate of}\\ \text{creation}\end{array}\right\} - \left\{\begin{array}{c}\text{Rate of}\\ \text{destruction}\end{array}\right\}. \tag{C.1}$$

We first consider the mass balance in detail. It is important to recognize that thermodynamics traditionally makes extensive quantities specific by dividing by the number of moles, whereas for transport phenomena one uses mass, and for non-equilibrium thermodynamics, volume. This can make comparison more difficult.

C.1 MASS: THE CONTINUITY EQUATION

We consider each of the terms in Eqn. (C.1) as applied to the cubic control volume in Figure C.1. The total mass in the cube is density times volume, $\rho\,\Delta x\,\Delta y\,\Delta z$; hence, the left-hand side of the mass balance is

$$\left\{\begin{array}{c}\text{Rate of mass}\\ \text{accumulation}\end{array}\right\} = \frac{\partial}{\partial t}\rho\,\Delta x\,\Delta y\,\Delta z, \tag{C.2}$$

[1] Most derivations in this section are given in greater detail in [14]. This section may be skimmed by the student unfamiliar with the subject of transport phenomena.

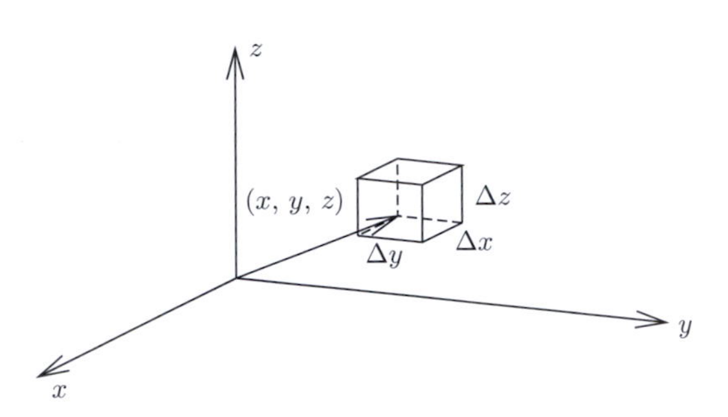

Figure C.1 A sketch of an infinitesimal volume element.

where t is time. During flow it is possible for mass to flow into or out of any of the six faces of the cubic control volume. For example, the volumetric flow rate of fluid into the cube through the side facing us in Figure C.1 is $-v_x \, \Delta y \, \Delta z|_{x+\Delta x}$, where the vertical bar | indicates where the quantity is evaluated; by considering all of the faces, we can write the first and second terms on the right-hand side of the balance equation, Eqn. (C.1), as

$$\left\{ \begin{array}{c} \text{Mass flow} \\ \text{rate in} \end{array} \right\} - \left\{ \begin{array}{c} \text{Mass flow} \\ \text{rate out} \end{array} \right\} = \rho v_x \, \Delta y \, \Delta z|_x - \rho v_x \, \Delta y \, \Delta z|_{x+\Delta x} + \rho v_y \, \Delta x \, \Delta z \big|_y - \rho v_y \, \Delta x \, \Delta z\big|_{y+\Delta y} + \rho v_z \, \Delta x \, \Delta y|_z - \rho v_z \, \Delta x \, \Delta y|_{z+\Delta z} \,. \tag{C.3}$$

Barring any nuclear reactions, to an excellent approximation we may assume that mass is neither created nor destroyed. Hence, we neglect the third (creation) and fourth (destruction) terms on the right-hand side of Eqns. (C.1). We put Eqns. (C.2) and (C.3) into the balance equation and divide by the volume of the cube, $\Delta x \, \Delta y \, \Delta z$, to obtain

$$\frac{\partial \rho}{\partial t} = -\frac{\rho v_x|_{x+\Delta x} - \rho v_x|_x}{\Delta x} - \frac{\rho v_y\big|_{y+\Delta y} - \rho v_y\big|_y}{\Delta y} - \frac{\rho v_z|_{z+\Delta z} - \rho v_z|_z}{\Delta z}. \tag{C.4}$$

If we let the volume of the cube go to zero, $\Delta x \to 0, \;\; \Delta y \to 0, \;\; \Delta z \to 0$, then we can use the definition of a derivative to write Eqn. (C.4) in differential form as

$$\frac{\partial \rho}{\partial t} = -\frac{\partial}{\partial x}\rho v_x - \frac{\partial}{\partial y}\rho v_y - \frac{\partial}{\partial z}\rho v_z = -\nabla \cdot (\rho \vec{\boldsymbol{v}}) \,. \tag{C.5}$$

Because it plays such an important role in transport phenomena, we have introduced the **vector differential operator**

$$\nabla := \vec{\boldsymbol{\delta}}_x \frac{\partial}{\partial x} + \vec{\boldsymbol{\delta}}_y \frac{\partial}{\partial y} + \vec{\boldsymbol{\delta}}_z \frac{\partial}{\partial z}, \tag{C.6}$$

where $\vec{\boldsymbol{\delta}}_x$ is the unit vector that points in the x direction. We have also introduced the fluid velocity vector $\vec{\boldsymbol{v}}$, which may be distinguished from the specific volume v which is light-faced, instead of bold-faced. We have also used the dot (or scalar) product "$\cdot$." We call Eqn. (C.5) the **microscopic mass balance**, or, more commonly, the **continuity equation**.

We may also apply the balance relation Eqn. (C.1) to the number of moles of a single species k to obtain the **continuity equation for species** k:[2]

$$\frac{\partial c_k}{\partial t} + \nabla \cdot (c_k \vec{\boldsymbol{v}}) = -\nabla \cdot \vec{\boldsymbol{J}}_k - R_k, \tag{C.7}$$

where c_k is the molar concentration (moles/volume), $\vec{\boldsymbol{J}}_k$ is the molar flux relative to $\vec{\boldsymbol{v}}$ (moles/area/volume/time), and R_k is the reaction rate (moles/volume/time), all of species k.

C.2 MOMENTUM: THE EQUATION OF MOTION

A similar balance can be performed on the momentum in the fluid.[3] However, we must now allow for the creation or destruction of momentum through forces at the face of the cube (through the **extra stress tensor** $\vec{\vec{\boldsymbol{\tau}}}$ and pressure P) and through body forces, such as gravity. Such a derivation yields the **microscopic momentum balance**, sometimes called **Cauchy's equation**:[4]

$$\frac{\partial}{\partial t}(\rho \vec{\boldsymbol{v}}) + \nabla \cdot (\rho \vec{\boldsymbol{v}} \vec{\boldsymbol{v}}) = -\nabla P + \rho \vec{\boldsymbol{g}} + \nabla \cdot \vec{\vec{\boldsymbol{\tau}}}, \tag{C.8}$$

where we have introduced the **gravity vector**, $\vec{\boldsymbol{g}}$, and the extra stress tensor, $\vec{\vec{\boldsymbol{\tau}}}$, which describes the surface forces transmitted through the fluid not from the equilibrium pressure P per unit area of contact.

If for the moment we assume that the fluid is Newtonian (e.g. water) and incompressible with viscosity η_s,

$$\vec{\vec{\boldsymbol{\tau}}} = \eta_s \left[\nabla \vec{\boldsymbol{v}} + (\nabla \vec{\boldsymbol{v}})^{\dagger} \right], \quad \text{incompressible Newtonian fluid,} \tag{C.9}$$

and has constant material properties, then we may eliminate the stress tensor in the expression to write

$$\rho \frac{\partial}{\partial t} \vec{\boldsymbol{v}} + \rho \vec{\boldsymbol{v}} \cdot \nabla \vec{\boldsymbol{v}} = -\nabla P + \eta_s \nabla \cdot \nabla \vec{\boldsymbol{v}} + \rho \vec{\boldsymbol{g}}, \tag{C.10}$$

where η_s is the (constant) **viscosity** of the fluid. Equation (C.10) is called the **Navier–Stokes equation**. The superscript dagger in Eqn. (C.9) indicates taking the transpose. We may rewrite this equation using two additional definitions that are central in transport phenomena; the **substantial derivative** is defined as

$$\frac{D}{Dt} := \frac{\partial}{\partial t} + \vec{\boldsymbol{v}} \cdot \nabla, \tag{C.11}$$

and may be thought of as finding the change in time of a quantity as one follows a fluid element along a streamline. The **Laplacian** is defined as $\nabla^2 := \nabla \cdot \nabla$, which may be used to rewrite Eqn. (C.10):

[2] See [14]: after eliminating $\vec{\boldsymbol{n}}_{\mathrm{B}}$ from Eqn. (18.1-7) using Eqn. (V) of Table 16.1-3, multiply the equation by the molecular weight of species k to arrive at this form for the continuity equation.

[3] This and subsequent subsections use differential tensor calculus, which may be reviewed in the appendix of [14].

[4] Note, however, that we use for the stress tensor the opposite sign convention to that in [14, Eqn. (3.2-10)].

$$\rho \frac{D}{Dt}\vec{v} = -\nabla P + \eta_s \nabla^2 \vec{v} + \rho \vec{g}. \tag{C.12}$$

This equation is central in momentum transport, or fluid dynamics [12, Eqn. (B); 14, Eqn. (3.2–20)].

If we take the dot product of each term in Eqn. (C.8) with the vector $\vec{v}$, it is possible to derive the **microscopic mechanical energy balance** [14, Eqn. (3.3-2)]:

$$\frac{\partial}{\partial t}\left(\frac{1}{2}\rho|\vec{v}|^2\right) + \nabla \cdot \left(\frac{1}{2}\rho|\vec{v}|^2\vec{v}\right) = -\nabla \cdot \rho\vec{v} + P(\nabla \cdot \vec{v}) + \nabla \cdot \vec{\vec{\tau}} \cdot \vec{v} - \vec{\vec{\tau}} : \nabla\vec{v} + \rho\vec{v} \cdot \vec{g}. \tag{C.13}$$

We have used the continuity equation Eqn. (C.5) to arrive at the microscopic mechanical energy balance.

Note that Eqn. (C.13) actually appears to be something like a kinetic-energy balance. Hence, we will use it below to eliminate kinetic energy from the energy balance.

C.3 ENERGY: THE MICROSCOPIC ENERGY BALANCE

Analogously, we perform a balance on the energy per unit mass ϵ in the fluid to obtain the **microscopic energy balance** [14, Eqn. (10.1-9)]:

$$\frac{\partial}{\partial t}(\rho\epsilon) + \nabla \cdot (\rho\epsilon\vec{v}) = -\nabla \cdot \vec{q} + \rho\vec{v} \cdot \vec{g} - \nabla \cdot P\vec{v} + \nabla \cdot \vec{\vec{\tau}} \cdot \vec{v}, \tag{C.14}$$

where $\vec{q}$ is the **heat-flux vector**. It describes the heat flow in the fluid Q per unit area per time with direction $\vec{q}/|\vec{q}|$. This equation plays a central role in Section 5.1 where we derive the macroscopic energy balance necessary for application in many industrial problems. It is also the starting point for deriving the internal energy balance, which is used in turn to derive the microscopic entropy balance.

If we make our **local-equilibrium approximation**, we can assume that the only energy in the fluid is the kinetic energy, and the internal energy. Hence, we write

$$\epsilon(\vec{r}, t) = u(\vec{r}, t)/M_w(\vec{r}, t) + \frac{1}{2}\vec{v}(\vec{r}, t)^2. \tag{C.15}$$

If we subtract the microscopic mechanical energy balance Eqn. (C.13) from the microscopic energy balance Eqn. (C.14), we then obtain the **microscopic internal energy balance**,

$$\underbrace{\frac{\partial}{\partial t}\left(\frac{u}{v}\right)}_{\substack{\text{internal}\\\text{energy}\\\text{accumulation}}} + \underbrace{\nabla \cdot \left(\frac{u\vec{v}}{v}\right)}_{\substack{\text{internal}\\\text{energy in by}\\\text{convection}}} = \underbrace{-P\nabla \cdot \vec{v}}_{\substack{\text{compression}\\\text{work}}} + \underbrace{(-\nabla \cdot \vec{q})}_{\substack{\text{heat}\\\text{flux}}} + \underbrace{\left(\vec{\vec{\tau}} : \nabla\vec{v}\right)}_{\substack{\text{extra}\\\text{stress}\\\text{work}}}, \tag{C.16}$$

which plays an important role in heat transport [12, Eqn. (H); 14, Eqn. (10.1–13)].

Equation (C.16) has a straightforward physical interpretation: the first term on the left-hand side represents the accumulation of internal energy in the fixed control volume; the second term on the left-hand side is the convection of internal energy into the control volume; the first term on the right-hand side is the work done in compressing the fluid; the second term on the right-hand side is the heat flow into the volume, and the last term is the work done by the surrounding fluid to deform the control volume.

We mention in passing that in transport phenomena this equation is typically used to calculate temperature fields. One does this by writing a differential for internal energy in (T, v) and using the local-equilibrium assumption to arrive at an evolution equation for temperature [14, p. 315].

C.4 ENTROPY: THE MICROSCOPIC ENTROPY BALANCE

We use the microscopic internal-energy balance to derive the microscopic entropy balance. The derivation of this equation is far less common than the above equations. Hence, we show in more detail its derivation in a simplified system.[5]

We make the following assumptions for our simplified system.

- The fluid is composed of only one component.
- The fluid is Newtonian and incompressible, so its extra stress tensor is described by Eqn. (C.9).

Under these assumptions, we can rewrite Eqn. (C.16) as

$$\rho \frac{D}{Dt}\left(\frac{u}{M_{\mathrm{w}}}\right) + P\nabla \cdot \vec{v} = -\nabla \cdot \vec{q} + \vec{\vec{\tau}} : \nabla\vec{v}, \quad \text{pure, Newtonian fluid.} \tag{C.17}$$

We have also used the continuity equation to rewrite the left-hand side of the equation. We can further simplify this equation by making use of the continuity equation (C.5) in the form

$$\nabla \cdot \vec{v} = -\frac{1}{\rho}\frac{D\rho}{Dt} = -\frac{v}{M_{\mathrm{w}}}\frac{D}{Dt}\left(\frac{M_{\mathrm{w}}}{v}\right). \tag{C.18}$$

For a pure-component system, the molecular weight is constant, and may be pulled outside the substantial derivative. Thus, Eqn. (C.17) can be written

$$\frac{1}{v}\left[\frac{Du}{Dt} + P\frac{Dv}{Dt}\right] = -\nabla \cdot \vec{q} + \vec{\vec{\tau}} : \nabla\vec{v}, \quad \text{pure, Newtonian fluid,} \tag{C.19}$$

which is the starting point for our derivation of the simplified microscopic entropy balance. The left-hand side of this equation can be simplified using the differential for internal energy, Eqn. (2.27), and the chain rule, Eqn. (A.8), to write

$$\frac{Du}{Dt} + P\frac{Dv}{Dt} = T\frac{Ds}{Dt}. \tag{C.20}$$

[5] The derivation here closely follows that of [36].

We insert Eqn. (C.20) into Eqn. (C.19), and divide by T to obtain the **simplified microscopic entropy balance**

$$\underbrace{\frac{1}{v}\frac{Ds}{Dt}}_{\substack{\text{entropy}\\ \text{accumulation}}} = + \underbrace{\left(-\nabla\cdot\frac{\vec{q}}{T}\right)}_{\substack{\text{net flux}\\ \text{of entropy}}} + \underbrace{\vec{q}\cdot\nabla\left(\frac{1}{T}\right) + \frac{1}{T}\vec{\vec{\tau}}:\nabla\vec{v}}_{\substack{\text{entropy}\\ \text{generation}}} \tag{C.21}$$

valid for a pure-component fluid when local equilibrium can be assumed. The equation has a straightforward physical interpretation: the left-hand side describes the accumulation of entropy in a fluid element convected along a streamline; the first term on the right-hand side describes the flux of entropy into the element from heat, while the second and third terms yield the creation of entropy per unit volume inside the volume element from temperature gradients and viscous dissipation, respectively.

EXAMPLE C.4.1 Calculate the entropy generation per unit volume for a Newtonian fluid of constant material properties in steady laminar flow in a pipe of constant wall temperature.

Solution. The velocity field and temperature field for the geometry of this problem are already known. The steady, laminar velocity field for a Newtonian fluid with constant viscosity and density is parabolic in the z direction of cylindrical coordinates,[6]

$$v_z = v_{\max}\left[1-\left(\frac{r}{R}\right)^2\right], \tag{C.22}$$

where R is the pipe radius, r is the radial coordinate, and $v_{\max}$ is the maximum velocity, which is known for given values of the pressure drop, pipe length and fluid viscosity. This relation is obtained by solving Eqn. (C.12) for the prescribed geometry, while assuming a steady state and neglecting end effects.

The temperature in the fluid is not uniform because of viscous dissipation. By solving Eqn. (C.16) for a fluid with constant viscosity, density, and thermal conductivity, the temperature field may be found [13, pp. 218–220]:

$$T(r) = T_{\mathrm{w}} + \frac{\eta_{\mathrm{s}} v_{\max}^2}{4k}\left[1-\left(\frac{r}{R}\right)^4\right], \tag{C.23}$$

where T_{w} is the uniform wall temperature and k is the thermal conductivity of the fluid.

From these expressions we can find the entropy generation per unit volume from viscous dissipation assuming an incompressible Newtonian fluid:

$$\frac{1}{T}\vec{\vec{\tau}}:\nabla\vec{v} = \frac{\eta_{\mathrm{s}}}{T}\left(\frac{dv_z}{dr}\right)^2 = \frac{4\eta_{\mathrm{s}} r^2 v_{\max}^2}{R^4 T(r)}. \tag{C.24}$$

Note that the entropy generation per unit volume from viscous dissipation is positive *only if the viscosity is positive*.

[6] The solution may be found, for example in [14, Eqn. (2.3-16)].

Using Fourier's law for heat conduction, the entropy generation per unit volume from temperature gradients is

$$\begin{aligned}\vec{q} \cdot \nabla\left(\frac{1}{T}\right) &= -k\,\nabla T \cdot \nabla\left(\frac{1}{T}\right) = \frac{k}{T^2}\,\nabla T \cdot \nabla T \\ &= \frac{k}{T^2}\left(\frac{dT}{dr}\right)^2 = \frac{\eta_s^2 r^6 v_{\max}^4}{kR^8 T(r)^2}.\end{aligned} \tag{C.25}$$

Note that the entropy generation per unit volume from temperature gradients is positive *only if the thermal conductivity is positive.*

To complete these expressions, it is necessary to eliminate $T(r)$ from them by using the known temperature field, Eqn. (C.23). □

A similar derivation of the entropy balance for a system with reactions and multiple components (not necessarily Newtonian) yields [33, Eqn. (III.19)] (see the exercise at the end of this appendix)

$$\frac{\partial}{\partial t}\left(\frac{s}{v}\right) + \nabla \cdot \left(\frac{s\vec{v}}{v}\right) = -\nabla \cdot \vec{J}_s + \sigma_{\mathrm{hf}} + \sigma_{\mathrm{mf}} + \sigma_{\mathrm{sw}} + \sigma_{\mathrm{r}}, \tag{C.26}$$

where we have assumed local equilibrium. The left-hand side of Eqn. (C.26) represents the accumulation of entropy in a fixed volume element plus the convection of entropy into the element; the first term on the right-hand side is the **net entropy flux**, where

$$\vec{J}_s := \frac{1}{T}\left(\vec{q} - \sum_k \mu_k \vec{J}_k\right), \tag{C.27}$$

is the **entropy flux** (entropy/area/time) relative to $\vec{v}$. This definition for the entropy flux is reminiscent of the differential for S, Eqn. (2.29),

$$\begin{aligned} dS &= \frac{1}{T}\,dU + \frac{P}{T}\,dV - \sum_k \frac{\mu_k}{T}\,dN_k \\ &= \frac{1}{T}\,(đQ + đW) + \frac{P}{T}\,dV - \sum_k \frac{\mu_k}{T}\,dN_k \\ &= \frac{1}{T}\left(đQ_{\mathrm{rev}} - \sum_k \mu_k\,dN_k\right). \end{aligned} \tag{C.28}$$

We have used energy conservation Eqn. (2.1) to obtain the second line, and assumed only reversible work to obtain the third line – which is consistent with our local-equilibrium assumption. Compare the last two equations.

The new terms on the right-hand side of Eqn. (C.26) represent the creation of entropy from, respectively, heat fluxes, molar fluxes, stress work, and reactions

$$T\sigma_{\mathrm{hf}} := -\vec{J}_s \cdot \nabla T, \tag{C.29}$$

$$T\sigma_{\mathrm{mf}} := -\sum_{k=1}^{m} \vec{J}_k \cdot \nabla \mu_k, \tag{C.30}$$

$$T\sigma_{\mathrm{sw}} := -\vec{\vec{\tau}} : \nabla \vec{v}, \tag{C.31}$$

$$T\sigma_{\mathrm{r}} := -\sum_{k=1}^{m} R_k \mu_k. \tag{C.32}$$

Equation (C.26) is the final goal of this subsection. It plays a central role in irreversible thermodynamics [33, 73, 76, Section 15.5]. A fundamental tenet of irreversible thermodynamics is that the generation terms given by Eqns. (C.29)–(C.32) should be non-negative in order to satisfy the second law of thermodynamics, Postulate IV. Hence, if we assume that heat flux is described by Fourier's law of heat conduction, then the thermal conductivity must be positive. Similarly, a Newtonian fluid must have positive viscosity in order to be consistent with the principles of irreversible thermodynamics.

Equations (C.5), (C.14), and (C.26) are used to derive the **macroscopic balance equations**, given in Section 5.1.

C.5 ENTROPY FLUX AND GENERATION IN LAMINAR FLOW

In industrial applications, an engineer does not typically know the entropy generation or entropy flux accurately. However, the following example shows how detailed information about the flow process can be used to estimate these quantities.

EXAMPLE C.5.1 Using results from transport phenomena and Appendix C, calculate the total entropy generation and entropy flux for a Newtonian fluid of constant material properties in steady laminar flow in a pipe of constant wall temperature and length L.

Solution. The entropy generation per unit volume σ_{sw} is found in Example C.4.1. We may integrate these results over the volume of the pipe to find the total entropy generation. The entropy generation from stress work is just the contribution from Newtonian viscous dissipation:

$$\begin{aligned}
\dot{\Sigma}_{\mathrm{sw}} &= \int_0^L \int_0^{2\pi} \int_0^R \frac{1}{T} \vec{\vec{\tau}} : \nabla \vec{v}\, r\, dr\, d\theta\, dz \\
&= 2\pi L \int_0^R \frac{4\eta_s r^3 v_{\mathrm{max}}^2}{R^4 T(r)}\, dr \\
&= 8\pi L k \log\left(1 + \frac{\eta_s v_{\mathrm{max}}^2}{4kT_{\mathrm{w}}}\right).
\end{aligned} \tag{C.33}$$

We used Eqn. (C.24) to obtain the second line, and Eqn. (C.23) to obtain the third. Similarly, we may find the entropy generation from heat flows:

$$\begin{aligned}
\dot{\Sigma}_{\mathrm{hf}} &= \int_0^L \int_0^{2\pi} \int_0^R \vec{q} \cdot \nabla\left(\frac{1}{T}\right) r\, dr\, d\theta\, dz \\
&= 2\pi L \int_0^R \frac{k}{T(r)^2} \left(\frac{dT}{dr}\right)^2 dr \\
&= 2\pi L \int_0^R \frac{\eta_{\mathrm{s}}^2 r^7 v_{\mathrm{max}}^4}{kR^8 T(r)^2} dr \\
&= 8\pi L k \left[\frac{\eta_{\mathrm{s}} v_{\mathrm{max}}^2}{4kT_{\mathrm{w}}} - \log\left(1 + \frac{\eta_{\mathrm{s}} v_{\mathrm{max}}^2}{4kT_{\mathrm{w}}}\right)\right].
\end{aligned} \tag{C.34}$$

Note that we used Fourier's law to obtain the second line, Eqn. (C.25) to obtain the third line, and Eqn. (C.23) to obtain the fourth.

Finally, we may obtain the entropy flux out of the walls of the pipe, since there is no molar flux at the walls:

$$\begin{aligned}
\iint_A \vec{n} \cdot \frac{k\,\nabla T\, dA}{T} &= -\int_0^L \int_0^{2\pi} \frac{q_r(R)}{T(R)} R\, d\theta\, dz \\
&= \frac{2\pi R L k}{T_{\mathrm{w}}} \left.\frac{dT}{dr}\right|_{r=R} \\
&= -\frac{2\pi L \eta_{\mathrm{s}} v_{\mathrm{max}}^2}{T_{\mathrm{w}}}.
\end{aligned} \tag{C.35}$$

To obtain the second line, we performed the integration, and used Fourier's law. We used the known temperature field Eqn. (C.23) to obtain the third.

If we combine the results in Eqns. (C.33)–(C.35), we find that

$$\iint_A \vec{n} \cdot \frac{k\,\nabla T\, dA}{T} + \dot{\Sigma}_{\mathrm{sw}} + \dot{\Sigma}_{\mathrm{hf}} = 0. \tag{C.36}$$

Hence, the macroscopic entropy balance Eqn. (5.7) states that the inlet and exit specific entropies of the laminar pipe flow with isothermal walls are the same,

$$s_{\mathrm{out}} = s_{\mathrm{in}}, \tag{C.37}$$

since mass balance requires that the mass flows are the same in and out: $\dot{m}_{\mathrm{out}} = \dot{m}_{\mathrm{in}}$.

Actually, we can derive this result more simply by integrating along a fluid streamline in the pipe. The streamlines in laminar flow are lines of constant r and θ. As the fluid element changes its z coordinate, the pressure changes but the temperature remains constant. Hence, the change in entropy for the fluid element from inlet to exit can be found as

$$\begin{aligned}
s_{\mathrm{out}} - s_{\mathrm{in}} &= \int_{P_{\mathrm{in}}}^{P_{\mathrm{out}}} \left(\frac{\partial s}{\partial P}\right)_T dP \\
&= -\int_{P_{\mathrm{in}}}^{P_{\mathrm{out}}} \left(\frac{\partial v}{\partial T}\right)_P dP \\
&= 0.
\end{aligned} \tag{C.38}$$

The second line follows from the Maxwell relation (in Section 3.3), and the third line follows because the fluid is assumed incompressible, and, hence, the specific volume v is a constant.

Therefore, we find that all of the entropy generated in the pipe is conducted out through the walls, and the fluid entropy stays constant along each streamline. □

Unlike the fluid in this example, a real fluid is compressible, of course. It is interesting to estimate the sign of $s_{\text{out}} - s_{\text{in}}$, however. Above, we found for a compressible fluid that

$$s_{\text{out}} - s_{\text{in}} = -\int_{P_{\text{in}}}^{P_{\text{out}}} v(T,P)\alpha(T,P)dP, \tag{C.39}$$

where α is the coefficient of thermal expansion defined by Eqn. (3.60). Most real fluids exhibit positive values for α. Since, $P_{\text{in}} > P_{\text{out}}$, we find that most real fluids would yield $s_{\text{out}} > s_{\text{in}}$. However, some fluids such as water near the freezing point exhibit a negative value for α; hence, these would show that entropy *decreases* along the pipe.

Exercise Derive the microscopic entropy balance Eqn. (C.26) in the following way:

1. Using equilibrium thermodynamic manipulations, show that

$$d\left(\frac{s}{v}\right) = d\left(\frac{S}{V}\right) = \frac{1}{T}\left[d\left(\frac{u}{v}\right) - \sum_i \mu_i\, dc_i\right].$$

   (*Hint:* you may need to use the Euler relation.)

2. Using the chain rule, show that this expression, within the local-equilibrium approximation, leads to

$$\frac{D}{Dt}\left(\frac{s}{v}\right) = \frac{1}{T}\left[\frac{D}{Dt}\left(\frac{u}{v}\right) - \sum_i \mu_i \frac{Dc_i}{Dt}\right].$$

3. Show that for any function ψ

$$\frac{\partial \psi}{\partial t} + \nabla \cdot (\psi \vec{v}) = \frac{D\psi}{Dt} + \psi\, \nabla \cdot \vec{v}.$$

4. Now combine the results from parts 2 and 3 (starting with $\psi = s/v$), together with the internal-energy balance, Eqn. (C.16), and the continuity equation for species k, Eqn. (C.7), to derive the microscopic entropy balance, Eqn. (C.26).

APPENDIX D

Physical properties and references

D.1 WEBSITES WITH DATA AND PROGRAMS

There are many sources available for finding physical properties, such as critical properties, heat capacities, molecular weights, etc. Besides texts, many computer packages may be accessed for the data. Below is a list of URLs that may be used to find data for many species and computer programs on the web. Most of these are free, but they may contain links to services requiring fees. Since we have no control over their content, we cannot be responsible for the accuracy or usefulness of the information supplied. URLs are not particularly stable, but they should be obtainable from a search engine, such as Google.

- The National Institute for Standards and Technology (NIST) keeps a tremendous amount of data online, and contains links to other sites. They also sell catalogs and computer software (webbook.nist.gov/chemistry).
- Data on many thermodynamic, transport, and physical properties are stored by the American Institute of Chemical Engineers at the Design Institute for Physical Property Data. A large database has free access for students (www.aiche.org/dippr).
- Free software, and data, such as molecular weights, critical pressures and critical temperatures, may be found for gases at Flexware (http://www.flexwareinc.com/).
- Professor G. R. Mansoori of the University of Illinois-Chicago maintains a website with links to many other web pages that contain data, or computer programs for thermodynamic calculation: (http://www.uic.edu/˜mansoori/Thermodynamic.Data.and.Property_html).
- The University of Texas has a Javascript program called "ThermoDex" that searches their bookshelves for thermodynamic data. The program returns only the reference for the book, however: (www.lib.utexas.edu/thermodex).
- The Center for Research in Computational Thermochemistry at the Ecole Polytechnique de Montréal has a site that contains conversion factors, heat capacities, formation properties, links, calculators, and much more (http://www.crct.polymtl.ca/fact/).

D.2 ENTROPY AND PROPERTIES OF FORMATION

Table D.1 Entropy and properties of formation for selected substances at a temperature of 298.15 K and pressure of 0.1 MPa. These data are from [25]. The reference states are the elements in their natural state at 298.15 K, and 0.1 MPa. For details, see the JANAF tables, p. 13.

Substance	$S°$ (J/(K · mol))	ΔH_f° (kJ/mol)	ΔG_f° (kJ/mol)
Al(l)	39.549	10.562	7.201
α-Al_2O_3 (cr)	50.950	−1675.692	−1582.275
δ-Al_2O_3 (cr)	50.626	−1666.487	−1572.974
γ-Al_2O_3 (cr)	52.300	−1656.882	−1563.850
κ-Al_2O_3 (cr)	53.555	−1662.303	−1569.663
Ammonia (NH_3, ideal gas)	192.774	−45.898	−16.367
Argon (Ar, gas)	154.845	0	0
CaCl (gas)	241.559	−104.600	−130.966
$CaCl_2$ (cr)	104.602	−795.797	−748.073
C (graphite)	5.740	0	0
CO_2 (g)	213.795	−393.522	−394.389
CS_2 (g)	237.977	116.943	66.816
CO (g)	197.653	−110.527	−137.163
Cl_2 (g)	223.079	0	0
$CHClF_2$ (g)	280.968	−481.578	−450.439
CH_3Cl (g)	234.367	−83.680	−60.146
CHCl (g)	234.912	334.720	319.129
$CHCl_2F$ (g)	293.260	−283.257	−252.758
Ethene (C_2H_4, g)	219.330	52.467	68.421
Ethyne (C_2H_2, g)	200.958	226.731	248.163
Formaldehyde (H_2CO, g)	218.950	−115.897	−109.921
Hydrogen (H_2, g)	130.680	0	0

Table D.1 (cont.)

Substance	S° (J/(K · mol))	ΔH_f° (kJ/mol)	ΔG_f° (kJ/mol)
HCN (g)	201.828	135.143	124.725
HF (g)	173.780	−272.546	−274.646
Hydrogen peroxide (HOOH, g)	232.991	−136.106	−105.445
Hydrogen sulfide (H_2S, g)	205.757	−20.502	−33.329
Iron oxide (FeO, g)	241.924	251.040	217.639
Iron oxide hematite (Fe_2O_3, cr)	87.4	−825.503	−743.523
Iron oxide magnetite (Fe_3O_4, cr)	145.266	−1120.894	−1017.438
Lead oxide (PbO, g)	240.039	70.291	48.622
Lead oxide red (PbO, cr)	66.316	−219.409	−189.283
Lead oxide yellow (PbO, cr)	68.701	−218.062	−188.647
Lead oxide (PbO_2, cr)	71.797	−274.470	−215.397
Lead oxide (Pb_3O_4, cr)	211.961	−718.686	−601.606
$MgAl_2O_4$ (cr)	88.692	−2299.108	−2176.621
Magnesium oxide (cr)	26.924	−601.241	−568.945
Methane (CH_4, g)	186.251	−74.873	−50.768
Nitrogen (N_2, g)	191.609	0	0
NO (g)	210.758	90.291	86.600
NO_2 (g)	240.034	33.095	51.258
NO_3 (g)	252.619	71.128	116.121
N_2O (g)	219.957	82.048	104.179
N_2O_3 (g)	308.539	82.843	139.727
N_2O_4 (g)	304.376	9.079	97.787
N_2O_5 (g)	346.548	11.297	118.013
Oxygen (O_2, g)	205.147	0	0
Ozone (O_3, g)	238.932	142.674	163.184
Silane (SiH_4, g)	204.653	34.309	56.827

Table D.1 (cont.)

Substance	$S°$ (J/(K · mol))	ΔH_f° (kJ/mol)	ΔG_f° (kJ/mol)
Silicon (Si, cr)	18.820	0	0
Silicon oxide (SiO, g)	211.579	−1000.416	−127.305
Sodium (Na, cr)	51.455	0	0
Sodium chloride (NaCl, cr)	72.115	−411.120	−384.024
Sodium cyanide (NaCN, cr)	118.467	−90.7009	−80.413
Sodium hydroxide (NaOH, cr)	64.445	−425.931	−379.741
Sulfur oxide (SO, g)	221.944	5.007	−21.026
Sulfur oxide (S_2O, g)	267.020	−56.484	−95.956
Water (H_2O, l)	69.950	−285.830	−237.141

Table D.2 Polynomial fits to reported heat capacities for selected pure substances at ideal state. If the temperature is given in kelvins, then the above values for the coefficients yield the heat capacity in $J \cdot K^{-1} \cdot mol^{-1}$, when the formula $c_p^{ideal} = A_0 + A_1 T + A_2 T^2 + A_3 T^3 + A_4 T^4 + A_5 T^5 + A_6/T + A_7/T^2$ is used. The coefficients have been fit to data taken from [25]. These expressions are not correct outside the indicated range, and care should be taken using derivatives, especially at low temperatures. Δ is the maximum error (in percent) from the measured values inside the indicated temperature range.

	$A_0 \times 10^{-1}$	$A_1 \times 10^3$	$A_2 \times 10^6$	$A_3 \times 10^9$	$A_4 \times 10^{13}$	$A_5 \times 10^{17}$	$A_6 \times 10^{-3}$	$A_7 \times 10^{-5}$	Range (K)	Δ
NH_3	1.055 2	63.584 4	−25.232 5	5.472 7	−6.349 3	3.053 3	2.891 2	−1.228 7	100–6000	0.3
CS_2	5.908 0	10.046	−6.159 9	1.835 7	−2.579 8	1.383 8	−5.645 1	2.743 7	100–6000	0.4
CO	1.708 0	22.654 4	−11.279 6	2.909 8	−3.732 2	1.881 4	2.237 5	−1.251 7	100–6000	0.5
CO_2	3.991 1	30.558	−16.516	4.500 8	−6.001 1	3.120 7	−3.969 3	2.612 3	100–6000	0.8
Cl_2	4.090 1	−0.311 86	−1.892 1	1.373 1	−2.949 2	1.964 5	−2.443 7	1.287 8	100–6000	0.2
H_2	1.783 5	9.051 1	−0.434 4	−0.488 36	1.197 7	−0.837 80	4.928 9	−7.202 6	250–6000	0.3
HCN	3.628 5	23.945	−9.158 1	1.837 1	−1.859 2	0.741 95	−2.599 7	1.655 8	100–6000	0.5
HF	2.809 7	−1.482 2	5.031 9	−2.008 3	3.205 2	−1.842 0	0.411 37	−0.300 4	100–6000	0.6
H_2O_2	0.215 31	182.97	−254.86	195.99	−765.42	1190.0	1.208 2	0.335 34	100–1500	0.1
H_2S	1.275 4	46.191	−20.693	4.896 2	−5.837 5	2.766 4	3.333 5	−1.726 2	100–6000	0.5
N_2	1.768 1	20.095	−9.152 0	2.175 3	−2.597 0	1.232 6	2.245 9	−1.297 6	100–6000	0.8
O_2	2.250 8	19.608	−10.946	3.321 5	−4.907 6	2.789 9	0.653 78	−0.175 9	100–6000	1.0
H_2O	1.915 1	25.175	−5.858 1	0.396 97	0.447 76	−0.553 72	2.693 3	−1.527 2	100–6000	0.8

Table D.3 Values for the critical temperature, critical pressure, critical density, and acentric factor for several substances. The critical temperature is in degrees Celsius (not Kelvin), the critical pressure is in atm, and the density is in grams per cubic centimeter. The critical values are taken from [34], which has a much more extensive list. The acentric factors come from [101]. An extensive list is in the appendix of [102], and more values may be found from the NIST web page.

	T_c (°C)	P_c (atm)	ρ_c (g/cm^3)	ω
Acetic acid	321.3	57.1	0.351	
Ammonia	132.4	111.3	0.235	0.250
Argon	−122.44	48.00	0.5307	−0.004
Benzene	288.94	48.34	0.302	0.212
Bromine	311	102	1.18	0.132
n-Butane	152.01	37.7	0.228	0.200
Carbon dioxide	31.04	72.85	0.468	0.223
Carbon monoxide	−140.23	34.53	0.301	0.049
Carbon tetrachloride	283.15	44.97	0.558	0.194
Chlorine	144.0	76.1	0.573	0.073
Chloroform	263.4	54	0.50	0.216
Cyclohexane[a]	280.3	40.2	0.273	0.213
Ethane	32.28	48.16	0.203	0.100
Ethanol	243.1	62.96	0.276	
Fluorine	−129.0	55	0.63	0.048
Hydrogen	−239.91	12.80	0.0310	−0.22
Hydrogen Chloride	51.40	81.5	0.42	0.12
Methanol	239.43	79.9	0.272	
Naphthalene	475.2	39.98		0.302
Oxygen	−118.38	50.14	0.419	0.021
n-Pentane	196.5	33.35	0.237	0.252
Phenol	421.1	60.5	0.41	
Propane	96.67	41.94	0.217	0.153
Toluene	318.57	40.55	0.292	0.257
Water	374.2	218.3	0.325	

[a]The source text appears to have a typographical error for this entry.

Table D.4 Values for the parameters used in the Antoine equation (4.49) for several substances. Note that these constants are for logarithms to the base 10 ($\log_{10}$), not natural logarithms, in the Antoine equation. The temperature is in degrees Celsius (not Kelvin), and the resulting pressure is in mm Hg. These values are from [34], which has a much more extensive list. Unless otherwise indicated, the values are for liquid vapor pressures. Similar expressions in reduced form can be found in Appendix B for the models given there.

	A	*B* (°C)	*C* (°C)	*T* range (°C)
NH_3	7.360 50	926.132	240.17	
Ar	6.616 51	304.227	267.32	
Br	6.877 80	1119.68	221.38	
N_2	6.494 57	255.680	266.550	
CO	6.694 22	291.743	267.99	
CO_2 (s)	9.810 66	1347.786	273.00	
Cl_2	6.937 90	861.34	246.33	
H_2	5.824 38	67.5078	275.700	
HCl	7.170 00	745.80	258.88	
HCN	7.528 2	1329.5	260.4	
I_2	7.018 1	1610.9	205.0	
O_2	6.691 44	319.013	266.697	
Acetic acid	7.387 82	1533.313	222.309	
Acetone	7.117 14	1210.595	229.664	
Acrylonitrile	7.038 55	1232.53	222.47	−20–140
Benzene	9.106 4	1885.9	244.2	−12–3
Benzene	6.905 65	1211.033	220.790	8–103
n-Butane	6.808 96	935.86	238.73	−77–19
1-Butanol	7.476 80	1362.39	178.77	15–131
Chloroform	6.493 4	929.44	196.03	−35–61
Cyclobutane	6.916 31	1054.54	241.37	−60–12
Cyclohexane	6.841 30	1201.53	222.65	20–81
Ethanol	8.321 09	1718.10	237.52	−2–100

Table D.4 (cont.)

	A	B (°C)	C (°C)	T range (°C)
Ethylene glycol	8.090 8	2088.9	203.5	50–200
Formic acid	7.5818	1699.2	260.7	37–101
Heptane	6.896 77	1264.90	216.54	–2–124
Methanol	7.897 50	1474.08	229.13	–14–65
Methanol	7.973 28	1515.14	232.85	64–110
Naphthalene	6.818 1	1585.86	184.82	125–218
Phenol	7.133 0	1516.79	174.95	107–182
Styrene	7.140 16	1574.51	224.09	32–82
Toluene	6.954 64	1344.800	219.48	6–137
Water	8.107 65	1750.286	235.0	0–60
Water	7.966 81	1668.21	228.0	60–150

D.3 PHYSICAL CONSTANTS

Table D.5 Values for selected physical constants, in various units.

Constant	Description	Value	Units
R	Ideal gas constant	0.082 057	l · atm/(mol · K)
		8.3145	J/(g-mol · K)
		8.3145	Pa · m^3/(mol · K)
		8.3145	kPa · l/(mol · K)
		0.083 145	l · bar/(mol · K)
		1.987	cal/(g-mol · K)
		1.987	BTU/(lb-mol · °R)
		0.7302	atm · ft^3/(lb-mol · °R)
		10.7316	psi · ft^3/(lb-mol · °R)
		62.3637	mmHg · l/(g-mol · K)
		62.3637	Torr · l/(g-mol · K)

Table D.5 (cont.)

Constant	Description	Value	Units
$\tilde{N}_A$	Avogadro's number	$6.022\,141\,5 \times 10^{23}$	1/g-mol
k_B	Boltzmann constant	$1.380\,65 \times 10^{-16}$	erg/K
		$1.380\,65 \times 10^{-23}$	J/K
		$1.380\,65 \times 10^{-23}$	$m^2 \cdot kg/(s^2 \cdot K)$
		$8.617\,39 \times 10^{5}$	eV/K
		0.138 066	pN · Å/K
$\|\vec{g}\|$	Acceleration due to gravity	9.8066	m/s^2
		32.174	ft/s^2
$\hbar$	Planck's reduced constant	$1.054\,571\,6 \times 10^{-34}$	J · s
e	Electron charge	1.602×10^{-19}	C
$\mathcal{F}$	Faraday's constant, $\tilde{N}_A e$	9.6492×10^{4}	C/mol
		230 60	cal/(mol · eV)
		2.8025×10^{14}	esu/mol

D.4 STEAM TABLES

Steam tables are readily available online.

REFERENCES

[1] M. Abramowitz and I. E. Stegun. *Handbook of Mathematical Functions*. Washington, D.C.: National Bureau of Standards, 1964.

[2] B. J. Alder, D. A. Young, and M. A. Mark. Studies in molecular dynamics. 10. Corrections to augmented van der Waals theory for square-well fluid. *J. Chem. Phys.*, 56:3013–3029, 1972.

[3] T. Allen. *Particle Size Measurement*. London: Chapman and Hall, 5th edn., 1997.

[4] S. Anand, J. P. Grolier, O. Kiyohara, C. J. Halpin, and G. C. Benson. Thermodynamic properties of some cycloalkane–cycloalkanol systems at 298.15 K. III. *J. Chem. Eng. Data*, 20:184–189, 1975.

[5] A. Anderko. Cubic and generalized van der Waals equations, in *Equations of State for Fluids and Fluid Mixtures*, pages 75–126. Amsterdam: Elsevier, 2000.

[6] A. J. Appleby and F. R. Foulkes. *Fuel Cell Handbook*. Morgantown, VA: U.S. Department of Energy, 5th edn., 2000. Available online at www.osti.gov/bridge/purl.cover.jsp?purl=/834188

[7] K. A. Arora, A. J. Lesser, and T. J. McCarthy. Preparation and characterization of microcellular polystyrene foams processed in supercritical carbon dioxide. *Macromolecules*, 31:4614–4620, 1998.

[8] R. D. Astumian and P. Hänggi. Brownian motors. *Phys. Today*, 55(11):33–39, 2002.

[9] W. Barthlott and C. Neinhuis. Purity of the sacred lotus, or escape from contamination in biological surfaces. *PLANTA*, 202:1–8, 1997.

[10] M. Benedict, G. B. Webb, and L. C. Rubin. An empirical equation for thermodynamic properties of light hydrocarbons and their mixtures. Constants for twelve hydrocarbons. *Chem. Eng. Progress*, 47:419–422, 1951.

[11] A. N. Beris and B. J. Edwards. *Thermodynamics of Flowing Systems with Internal Microstructure*. New York: Oxford University Press, 1994.

[12] R. B. Bird. The basic concepts in transport phenomena. *Chem. Eng. Educ.*, Spring:102–109, 1993.

[13] R. B. Bird, O. Hassager, R. C. Armstrong, and C. F. Curtiss. *Dynamics of Polymeric Liquids Vol. I: Rheology*. New York: Addison-Wesley, 2nd edn., 1987.

[14] R. B. Bird, W. E. Stewart, and E. N. Lightfoot. *Transport Phenomena*. New York: John Wiley and Sons, 1960.

[15] I. Brown and F. Smith. Liquid–vapor equilibria. The system carbon tetrachloride + acetonitrile at 45 °C. *Australian J. Chem.*, 7:269–272, 1954.

[16] S. Brunauer, P. H. Emmett, and E. Teller. Adsorption of gases in multimolecular layers. *J. Am. Chem. Soc.*, 60:309–319, 1938.

[17] H. A. Bumstead. *Josiah Willard Gibbs. The Collected Works.* New York: Longmans, Green, 1928.

[18] C. Bustamante, J. F. Marko, E. D. Siggia, and S. Smith. Entropic elasticity of λ-phage DNA. *Science*, 265:1599–1600, 1994.

[19] H. J. Butt, K. Graf, and M. Kappl. *Physics and Chemistry of Interfaces*. Weinheim: Wiley-VCH, 2003.

[20] D. Buttin, M. DuPont, M. Straumann *et al.* Development and operation of a 150 W air-feed direct methanol fuel cell stack. *J. Appl. Electrochem.*, 31:275–279, 2001.

[21] H. B. Callen. *Thermodynamics and an Introduction to Thermostatistics*. New York: Wiley, 2nd edn., 1985.

[22] N. F. Carnahan and K. E. Starling. Intermolecular repulsions and equation of state for fluids. *AIChE J.*, 18:1184–1189, 1972.

[23] A. B. D. Cassie and S. Baxter. Wettability of porous surfaces. *Trans. Faraday Soc.*, 40:546–551, 1944.

[24] D. Chandler. *Introduction to Modern Statistical Mechanics*. New York: Oxford University Press, 1987.

[25] M. W. Chase, Jr., C. A. Davies, J. R. Downey, Jr. *et al.* JANAF thermochemical tables, 3rd edn. *J. Phys. Chem. Ref. Data*, 14, Supplement No. 1, 1985.

[26] S. Jer Chen, I. G. Economou, and M. Radosz. Density-tuned polyolefin phase equilibria. 2. Multi-component solutions of alternating poly(ethylene-propylene) in subcritical and supercritical olefins. Experiment and SAFT model. *Macromolecules*, 25(19):4987–4995, 1992.

[27] N. Choudhury and B. M. Pettitt. On the mechanism of hydrophobic association of nanoscopic solutes. *J. Am. Chem. Soc.*, 127:3556–3567, 2005.

[28] R. Clausius. *Über den zweiten Hauptsatz der mechanischen Wärmetheorie.* Braunschweig: Friedrich Vieweg und Sohn, 1867.

[29] K. K. Crain. An investigation of the Benedict–Webb–Rubin equation. Masters of Engineering, University of Louisville, 1972.

[30] C. Danilowicz, Y. Kafri, R. S. Conroy *et al.* Measurement of the phase diagram of DNA unzipping in the temperature–force plane. *Phys. Rev. Lett.*, 93(7):078101, 2004.

[31] J. R. Dann. Forces involved in the adhesive process. I. Critical surface tensions of polymeric solids as determined with polar liquids. *J. Colloid Interface Sci.*, 32:302–320, 1970.

[32] M. Daune. *Molecular Biophysics: Structures in Motion*. Oxford: Oxford University Press, 2003.

[33] S. R. de Groot and P. Mazur. *Non-Equilibrium Thermodynamics*. New York: Dover Publications, 1984.

[34] J. A. Dean, editor. *Lange's Handbook of Chemistry*. New York: McGraw-Hill, 12th edn., 1978.

[35] J. H. Dymond and E. B. Smith. *The Virial Coefficients of Pure Gases and Mixtures*. Oxford: Clarendon Press, 1980.

[36] C. Eckert. The thermodynamics of irreversible processes. *Phys. Rev.*, 58:267–269, 1940.

[37] L. Eötvös. Über den Zusammenhang der Oberflächenspannung der Flüssigkeiten mit ihrem Molekular-volumen. *Ann. Phys. Chem.*, 27:448–459, 1886.

[38] R. M. Felder and R. W. Rousseau. *Elementary Principles of Chemical Processes*. New York: John Wiley & Sons, 2nd edn., 1986.

[39] J. D. Ferry. *Viscoelastic Properties of Polymers*. New York: John Wiley & Sons, 3rd edn., 1980.

[40] R. P. Feynman, R. B. Leighton, and M. Sands. *The Feynman Lectures on Physics: Volume I. Mainly Mechanics, Radiation and Heat.* Reading, MA: Addison-Wesley, 1963.

[41] P. J. Flory. *Principles of Polymer Chemistry*. Ithaca, NY: Cornell University Press, 1953.

[42] P. J. Flory. *Statistical Mechanics of Chain Molecules*. Munich: Hanser, 1988.

[43] H. S. Fogler. *Elements of Chemical Reaction Engineering*. Upper Saddle River, NJ: Prentice-Hall, 3rd edn., 1998.

[44] R. H. Fowler and E. A. Guggenheim. *Statistical Thermodynamics*. Cambridge: Cambridge University Press, 1939.

[45] R. Fürstner, W. Barthlott, C. Neinhuis, and P. Walzel. Wetting and self-cleaning properties of artificial superhydrophobic surfaces. *Langmuir*, 21:956–961, 2005.

[46] A. G. Gaonkar and R. D. Neuman. The uncertainties in absolute value of surface tension of water.*Colloids Surfaces*, 27:1–14, 1987.

[47] A. Garg, E. Gulari, and C. W. Manke. Thermodynamics of polymer melts swollen with supercritical gases. *Macromolecules*, 27:5643–5653, 1994.

[48] J. G. Gay and B. J. Berne. Modification of the overlap potential to mimic a linear site–site potential. *J. Chem. Phys.*, 74:3316–3319, 1981.

[49] K. Graf and H. Riegler. Molecular adhesion interactions between Langmuir monolayers and solid substrates. *Colloids Surfaces A: Physicochem. Eng. Aspects*, 131:215–224, 1998.

[50] J. Gross, O. Spuhl, F. Tumakaka, and G. Sadowski. Modeling copolymer systems using the perturbed-chain SAFT equation of state. *Indust. Eng. Chem. Res.*, 42(6):1266–1274, 2003.

[51] E. A. Guggenheim. *Thermodynamics*. Amsterdam: North Holland Publishing Company, 4th edn., 1959.

[52] A. F. Gutsol. The Ranque effect. *Phys. – Uspekhi*, 40(6):639–658, 1997.

[53] S. T. Hadden. *Hydrocarbon Processing Petrol. Refiner*, 45:161, 1966.

[54] D. L. Hammick and L. W. Andrew. The determination of the parachors of substances in solution. *J. Chem. Soc.*, 754–759, 1929.

[55] E. J. Henly and E. M. Rosen. *Material and Energy Balance Computations*. New York: Wiley, 1969.

[56] C. J. Henty, W. J. McManamey, and R. G. H. Prince. The quaternary liquid system benzene–furfural–iso-octane–cyclohexane. *J. Appl. Chem.*, 14:148–155, 1964.

[57] T. Heyduk and J. C. Lee. Application of fluorescence energy transfer and polarization to monitor *Escherichia coli* cAMP receptor protein and LAC promotor interaction. *Proc. Nat. Acad. Sci.*, 87:1744–1748, 1990.

[58] P. C. Hiemenz and R. Rajagopalan. *Principles of Colloid and Surface Chemistry*. New York: Marcel Dekker, 1997.

[59] T. L. Hill. *An Introduction to Statistical Thermodynamics*. New York: Dover, 1986.

[60] T. L. Hill. *Thermodynamics of Small Systems*, volumes I and II. New York: Dover, 1963 and 1964.

[61] T. L. Hill. *Statistical Mechanics. Principles and Selected Applications*. New York: Dover, 1987.

[62] R. Hilsch. The use of the expansion of gases in a centrifugal field as a cooling process. *Rev. Sci. Instrum.*, 18(2):108–113, 1947.

[63] D. M. Himmelblau. *Basic Principles and Calculations in Chemical Engineering*. Upper Saddle River, NJ: Prentice-Hall, 6th edn., 1996.

[64] J. O. Hirschfelder, C. F. Curtiss, and R. B. Bird. *Molecular Theory of Gases and Liquids*, second corrected printing. New York: John Wiley and Sons, 1954.

[65] J. Honerkamp. *Stochastische dynamische Systeme*. Berlin: VCH, 1983.

[66] X. H. Huang, R. H. Zhou, and B. J. Berne. Drying and hydrophobic collapse of paraffin plates. *J. Phys. Chem. B*, 109:3546–3552, 2005.

[67] R. J. Hunter. *Foundations of Colloid Science*. New York: Oxford University Press, 2nd edn., 2001.

[68] J. Israelachvili. *Intermolecular and Surface Forces*. London: Academic Press, 2nd edn., 1991.

[69] J. K. Johnson, J. A. Zollweg, and K. E. Gubbins. The Lennard-Jones equation of state revisited. *Mol. Phys.*, 78:591–618, 1993.

[70] S. Kapsabelis and C. A. Prestidge. Adsorption of ethyl(hydroxylethyl)cellulose onto silica particles: the role of surface chemistry and temperature. *J. Colloid Interface Sci.*, 228:297–305, 2000.

[71] A. I. Khinchin. *Mathematical Foundations of Information Theory*. New York: Dover, 1957.

[72] A. Ya. Kipnis, B. E. Yavelov, and J. S. Rowlinson. *Van der Waals and Molecular Science*. Oxford: Clarendon Press, 1996.

[73] D. Kondepudi and I. Prigogine. *Modern Thermodynamics: From Heat Engines to Dissipative Structures*. Chichester: Wiley, 1998.

[74] G. Kőrösi and E. sz. Kováts. Density and surface tension of 83 organic liquids. *J. Chem. Eng. Data*, 26:323–332, 1981.

[75] D. E. Koshland, G. Nmethy, and D. Filmer. Comparison of experimental binding data and theoretical models in proteins containing subunits. *Biochemistry*, 5(1):365–385, 1966.

[76] H. Kreuzer. *Nonequilibrium Thermodynamics and Its Statistical Foundations*. Oxford: Clarendon Press, 1981.

[77] S. M. Lambert, Y. Song, and J. M. Prausnitz. Equations of state for polymer systems, in *Equations of State for Fluids and Fluid Mixtures*. pages 523–588. Amsterdam: Elsevier, 2000.

[78] L. D. Landau and E. M. Lifshitz. *Statistical Physics*, volume 1. New York: Pergamon Press, 3rd edn., 1980.

[79] G. N. Lewis and M. Randall. *Thermodynamics and the Free Energies of Substances*. New York: McGraw Hill, 1923.

[80] G. C. Maitland, M. Rigby, E. B. Smith, and W. A. Wakeham. *Intermolecular Forces*. Oxford: Oxford Clarendon Press, 1981.

[81] J. F. Marko and E. D. Siggia. Stretching DNA. *Macromolecules*, 28:8759–8770, 1995.

[82] M. L. McGlashan and D. J. B. Potter. An apparatus for the measurement of the second virial coefficients of vapours; the second virial coefficients of some n-alkanes and of some mixtures of n-alkanes. *Proc. Roy. Soc. London A*, 267:478–500, 1962.

[83] M. L. McGlashan and C. J. Wormald. Second virial coefficients of some alk-1-enes and of a mixture of propene and hept-1-ene. *Trans. Faraday Soc.*, 60:646–652, 1964.

[84] F. C. Meinzer, M. J. Clearwater, and G. Goldstein. Water transport in trees: current perspectives, new insights, and some controversies. *Environ. Exp. Bot.*, 45:239–262, 2001.

[85] K. H. Meyer and C. Ferri. Sur l'élasticité du caoutchouc. *Helv. Chim. Acta*, 18:570–589, 1935.

[86] R. Micheletto, H. Fukuda, and M. Ohtsu. A simple method for production of a two-dimensional, ordered array of small latex particles. *Langmuir*, 11:3333–3336, 1995.

[87] M. Modell and R. C. Reid. *Thermodynamics and Its Applications*. Englewood Cliffs, NJ: Prentice-Hall, 2nd edn., 1983.

[88] M. Mohsen-Nia, H. Modarress, and G. A. Mansoori. A cubic hard-core equation of state. *Fluid Phase Equilibria*, 206(1–2):27–39, 2003.

[89] P. S. Murti and M. van Winkle. Vapor–liquid equilibria for binary systems of methanol, ethyl alcohol, 1-propanol, and 2-propanol with ethyl acetate and 1-propanol–water. *Chem. Eng. Data Ser.*, 3:72–81, 1958.

[90] H. Naghibi, A. Tamura, and J. M. Sturtevant. Significant discrepancies between van 't Hoff and calorimetric enthalpies. *Proc. Nat. Acad. Sci.*, 92:5597–5599, 1995.

[91] S. Nath, F. A. Escobedo, and J. J. de Pablo. On the simulation of vapor–liquid equilibria for alkanes. *J. Chem. Phys.*, 108:9905–9911, 1998.

[92] W. Nernst. *The New Heat Theorem*. New York: Dutton, 1926.

[93] S. L. Oswal, P. Oswal, and J. P. Dave. $V(E)$ of mixtures containing alkyl acetate, or ethyl alkanoate, or ethyl bromoalkanoate with n-hexane. *Fluid Phase Equilibria*, 98:225–234, 1994.

[94] H. C. Öttinger. *Beyond Equilibrium Thermodynamics*. Hoboken, NJ: Wiley-Interscience, 2005.

[95] A. Z. Panagiotopoulos, N. Quirke, M. Stapleton, and D. J. Tildesley. Phase equilibria by simulation in the Gibbs ensemble: alternative derivation, generalization and application to mixture and membrane equilibria. *Mol. Phys.*, 63:527–545, 1988.

[96] C. Panayiotou and I. Sanchez. Statistical thermodynamics of associated polymer solutions. *Macromolecules*, 24:6231–6237, 1991.

[97] W. M. Pardridge. CNS drug design based on principles of blood–brain barrier transport. *J. Neurochem.*, 70(5):1781–1792, 1998.

[98] W. M. Pardridge. *Brain Drug Targeting*. Cambridge: Cambridge University Press, 2001.

[99] V. A. Parsegian. *Van der Waals Forces*. Cambridge: Cambridge University Press, 2006.

[100] K. S. Pitzer and R. F. Curl. The volumetric and thermodynamic properties of fluids. III. Empirical equation for the second virial coefficient. *J. Am. Chem. Soc.*, 79:2369–2370, 1957.

[101] K. S. Pitzer. *Thermodynamics*. New York: McGraw-Hill, 3rd edn., 1995.

[102] B. E. Poling, J. M. Prausnitz, and J. P. O'Connell. *The Properties of Gases and Liquids*. New York: McGraw-Hill, 5th edn., 2001.

[103] W. H. Press, S. A. Teukolsky, W. T. Vetterling, and B. P. Flannery. *Numerical Recipes in FORTRAN: The Art of Scientific Computing*. Cambridge: Cambridge University Press, 2nd edn., 1992.

[104] C. Qian, S. J. Mumby, and B. E. Eichinger. Phase diagrams of binary polymer solutions and blends. *Macromolecules*, 24(7):1655–1661, 1991.

[105] D. Quéré. Non-sticking drops. *Rep. Prog. Phys.*, 68:2495–2532, 2005.

[106] G. Ranque. Expériences sur la détente giratoire avec productions simultanées d'un échappement d'air froid. *J. Phys. Radium*, 4:1125–1155, 1933.

[107] J. B. Rawlings and J. G. Ekerdt. *Chemical Reactor Analysis and Design Fundamentals*. Madison, WI: Nob Hill Publishing, 2002.

[108] O. Redlich and J. N. S. Kwong. On the thermodynamics of solutions. V. An equation of state. Fugacities of gaseous solutions. *Chem. Rev.*, 44:233–244, 1949.

[109] R. A. Robinson and D. A. Sinclair. The activity coefficients of the alkali chlorides and of lithium iodide in aqueous solution from vapor pressure measurements. *J. Am. Chem. Soc.*, 56:1830–1835, 1934.

[110] P. A. Rodgers. Pressure–volume–temperature relationships for polymeric liquids: a review of equations of state and their characteristic parameters for 56 polymers. *J. Appl. Polym. Sci.*, 48:1061–1080, 1993.

[111] M. Rubinstein and R. Colby. *Polymer Physics*. Oxford: Oxford University Press, 2003.

[112] D. M. Ruthven, S. Farooq, and K. S. Knaebel. *Pressure Swing Adsorption*. New York: VCH Publishers, 1994.

[113] I. C. Sanchez and R. H. Lacombe. Statistical thermodynamics of polymer solutions. *Macromolecules*, 11:1145–1156, 1978.

[114] S. R. S. Sastri and K. K. Rao. A simple method to predict surface tension of organic liquids. *Chem. Eng. J.*, 59:181–186, 1995.

[115] Y. Sato, K. Fujiwara, T. Takikawa *et al.* Solubilities and diffusion coefficients of carbon dioxide and nitrogen in polypropylene, high density polyethylene, and polystyrene under high pressures and temperatures. *Fluid Phase Equilibria*, 162:261–276, 1999.

[116] Y. Sato, M. Yurugi, K. Fujiwara, S. Takishima, and H. Masuoka. Solubilities of CO_2 and N_2 in polystyrene under high temperature and pressure. *Fluid Phase Equilibria*, 125:129–138, 1996.

[117] G. Schmidt and H. Wenzel. A modified van der Waals type equation of state. *Chem. Eng. Sci.*, 35:1503–1512, 1980.

[118] M. Schulz, A. Kandpur, and F. Bates. Phase behavior of polystyrene–poly(2-vinylpyridene) diblock copolymers. *Macromolecules*, 29(8):2857–2867, 1996.

[119] B. Schwager, L. Chudinovskikh, A. Gavriliuk, and R. Boehler. Melting curve of H_2O to 90 GPa measured in a laser-heated diamond cell. *J. Phys.: Condens. Matter*, 16:S1177–S1179, 2004.

[120] J. V. Sengers, R. F. Kayser, C. J. Peters, and H. J. White. *Equations of State for Fluids and Fluid Mixtures*. Amsterdam: Elsevier, 2000.

[121] J. Levelt Sengers. *How Fluids Unmix: Discoveries by the School of Van der Waals and Kamerlingh Onnes*, Amsterdam: Royal Netherlands Academy of Arts and Sciences, 2002.

[122] J. Levelt Sengers and A. H. M. Levelt. Diederek Korteweg, pioneer of criticality. *Phys. Today*, 55(12):47–53, 2002.

[123] A. N. Shaw. The derivation of thermodynamical relations for a simple system. *Phil. Trans. Royal Soc. London A*, 234:299–328, 1935.

[124] P. J. Sides. Scaling of differential equations: analysis of the fourth kind. *Chem. Eng. Educ.*, 36(3):232–235, 2002.

[125] D. A. Sinclair. A simple method for accurate determinations of vapor pressures of solutions. *J. Phys. Chem.*, 37:495–504, 1933.

[126] S. B. Smith, Y. Cui, and C. Bustamante. Overstretching B-DNA: the elastic response of individual double-stranded and single-stranded DNA molecules. *Science*, 271:795–799, 1996.

[127] G. Soave. Rigorous and simplified procedures for determining the pure-component parameters in the Redlich–Kwong equation of state. *Chem. Eng. Sci.*, 35:1725–1729, 1980.

[128] C. L. Stong. The amateur scientist: some delightful engines driven by the heating of rubber bands. *Scient. Am.*, 224(4):119–122, 1971.

[129] G. J. Su and D. S. Viswanath. Generalized thermodynamic properties of real gases. I – Generalized PVT behavior of real gases (generalized compressibility values at reduced temperature and pressure ranges of real gases compared to existing charts). *AIChE J.*, 11:202–204, 1965.

[130] K. Tajima, M. Muramatsu, and T. Sasaki. Radiotracer studies on adsorption of surface active substance at aqueous surfaces. I. Accurate measurement of adsorption of tritiated sodium dodecylsulfate. *Bull. Chem. Soc. Japan*, 43:1991–1998, 1970.

[131] M. Tambasco and J. E. G. Lipson. Analyzing and predicting polymer fluid and blend properties using minimal pure component data. *Macromolecules*, 38(7):2990–2998, 2005.

[132] A. S. Teja, S. I. Sandler, and N. C. Patel. A generalization of the corresponding state principle using two nonspherical reference fluids. *Chem. Eng. J.*, 21:21–28, 1981.

[133] J. W. Tester and M. Modell. *Thermodynamics and Its Applications*. Upper Saddle River, NJ: Prentice-Hall, 3rd edn., 1996.

[134] L. R. G. Treloar. *The Physics of Rubber Elasticity*. Oxford: Clarendon Press, 2nd edn., 1958.

[135] J. P. M. Trusler. The virial equation of state, in *Equations of State for Fluids and Fluid Mixtures*, pages 35–74. Amsterdam: Elsevier, 2000.

[136] Y. V. Tsekhanskaya, M. B. Iontem, and E. V. Mushkina. Solubility of naphthalene in ethylene and carbon dioxide under pressure. *Russ. J. Phys. Chem. USSR*, 38:1173–1176, 1964.

[137] C. Tsonopoulos. An Empirical Correlation of Second Viral Coefficients. *AIChE J.*, 20:263–272, 1974.

[138] C. Tsonopoulos and J. H. Dymond. Second virial coefficients of normal alkanes, linear 1-alkanols (and water), alkyl ethers, and their mixtures. *Fluid Phase Equilibria*, 133:11–34, 1997.

[139] C. Tsonopoulos, J. H. Dymond, and A. M. Szafranski. Second virial coefficients of normal alkanes, linear 1-alkanols and their binaries. *Pure Appl. Chem.*, 61:1387–1394, 1989.

[140] Udovenko, V. V. and Frid, Ts. B. Heats of vaporization of binary mixtures. II. *Zh. Fiz. Khim.*, 22:1135–1145, 1948.

[141] P. T. Underhill and P. S. Doyle. Development of bead–spring polymer models using the constant extension ensemble. *J. Rheol.*, 49(5):963–987, 2005.

[142] J. D. van der Waals. *On the Continuity of the Gaseous and Liquid States*. Translated by J. Shipley Rowlinson. Amsterdam: North-Holland, 1988.

[143] N. G. van Kampen. *Stochastic Processes in Physics and Chemistry*. Amsterdam: North-Holland, 1992.

[144] B. Widom. *Statistical Mechanics*. New York: Cambridge University Press, 2002.

[145] C. Xiao, H. Bianchi, and P. R. Tremaine. Excess molar volumes and densities of (methanol+water) at temperatures between 323 K and 573 K and pressures of 7.0 MPa and 13.5 MPa. *J. Chem. Thermodynamics*, 29(3):261–286, 1997.

[146] K. Yoshimoto, M. P. Stoykovich, H. B. Cao *et al.* A two-dimensional model of the deformation of photoresist structures using elastoplastic polymer properties. *J. Appl. Phys.*, 96:1857–1865, 2004.

[147] B. H. Zimm and J. K. Bragg. Theory of the one-dimensional phase transition in polypeptide chains. *J. Chem. Phys.*, 28:1246–1247, 1958.

[148] B. H. Zimm and J. K. Bragg. Theory of the phase transition between helix and random coil in polypeptide chains. *J. Chem. Phys.*, 31:526–531, 1959.

[149] U. Zimmermann, H. Schneider, L. H. Wegner *et al.* What are the driving forces for water lifting in the xylem conduit? *Physiologia Plantarum*, 114:327–335, 2002.

[150] Y. X. Zuo and E. H. Stenby. Corresponding-states and parachor models for the calculation of interfacial tensions. *Can. J. Chem. Eng.*, 75:1130–1137, 1997.

[151] J. H. Gibbs and E. A. Dimarzio. Statistical mechanics of helix–coil transitions in biological macromolecules. *J. Chem. Phys.*, 30: 271–282, 1959.

[152] W. G. Chapman, K. E. Gubbins, G. Jackson, and M. Radosz. SAFT: equation of state model for associating fluids. *Fluid Phase Equilibria*, 52: 31–38, 1989.

[153] A. N. Semenov. Contribution to the theory of microphase layering in block copolymer melts. *Sov. Phys. JETP*, 61: 733–742, 1985.

[154] A. N. Semenov. Phase equilibria in block copolymer–homopolymer mixtures. *Macromolecules*, 26: 2273–2281, 1993.

[155] F. S. Bates and G. H. Fredrickson. Block copolymers – designer soft materials. *Phys. Today*, 52: 32–38, 1999.

[156] In and out of cells in *Chemistry for Biologists*. Royal Society of Chemistry (2004) Cambridge, UK. http://www.rsc.org/Education/Teachers/Resources/cfb/cells.htm

[157] C. G. Hill and T. W. Root. *Introduction to Chemical Engineering Kinematics and Reactor Design*. New York: John Wiley & Sons, 2014.

[158] Landolt-Börstein. *Zahlenwerte und Funktionen*. Vol. 1, Part 3. Berlin: Springer-Verlag, 6th ed., 1951.

[159] W. A. Duncan, J. P. Sheridan and F. L. Swinton. Thermodynamic properties of binary systems containing hexafluorobenzene, Part 2 – Excess volumes of mixing and dipole movement. *Trans. Faraday Soc.*, 62: 1090–1096, 1966.

[160] J. M. Prausnitz, R. N. Lichtenthalen and E. Gomes de Azevedo. *Molecular Thermodynamics of Fluid-Phase Equilibria*. New Jersey: Prentice-Hall, 3rd ed., 1998.

[161] T. N. Olney, N. M. Cann, G. Cooper and C. E. Brison. Absolute scale determination for photoabsorption spectra and the calculation of molecular properties using dipole sum rules. *Chem. Phys.*, 223:59, 1997.

[162] S. A. Clough, Y. Beers, G. P. Klein and L. S. Rothman. Dipole moment of water from Stak measurements of H_2O, HDO and D_2O. *J. Chem. Phys.*, 59:2254, 1973.

[163] S. I. Sandler. *Chemical, biochemical and engineering thermodynamics*. New Jersey: John Wiley & Sons, 4th ed., 2006.

INDEX